The Invention of the Wireless Communication Engine

B.J.G. van der Kooij

In the Invention Series, the following books have been published:

The Invention of the Steam Engine
The Invention of the Electromotive Engine
The Invention of the Communication Engine 'Telegraph'
The Invention of the Electric Light
The Invention of the Communication Engine 'Telephone'
The Invention of the Wireless Communication Engine

This case study is part of the research work in preparation for a doctorate-dissertation to be obtained from the University of Technology, Delft, the Netherlands (www.tudelft.nl). It is one of a series of case studies about innovation under the title *The Invention Series*.

About the text: This is a scholarly case study describing the historic developments that resulted in the communication engine called the "wireless". It is based on a large number of historic and contemporary sources. As we conducted hardly any research into primary sources, we made use of the efforts of numerous others by citing them quite extensively to preserve the original character of their contributions. Where possible, we identified the individual authors of the citations. As some are not identifiable, we identified the source of the text. Facts and texts that are considered to be of a general character in the public domain (eg Wikipedia) are not cited but integrated in the text.

About the pictures: Many of the pictures used in this case study were found at websites accessed through the Internet. Where possible, they were traced to their origins, which, when found, were indicated as the source. For those that are not out of copyright, we feel that the fair use we make of the pictures to illustrate an aspect of the scholarly case is not an infringement of copyright.

Cover art is a line drawing of Marconi's Wireless Telegraph (US Patent № 586,193 from 1897). Courtesy USPTO.

Version 1.1 (August 2017)

ISBN-10: 154703937X
ISBN-13: 978-1547039371

Contents

Preface

When everything is said and done
and all our breath is gone,
The only thing that stays
Is history to guide our future ways.

My lifelong intellectual fascination with technical innovation within the context of society started in Delft, the Netherlands. In the 1970s I studied at the University of Technology, at both the electrical engineering school and the business school.[1] Having been educated as a technical student, I studied vacuum tubes, followed by transistors, and I found the change and novelty caused by the new technology of microelectronics to be mindboggling, not only from a technical point of view but because of all the opportunities it created for new products, new markets, and new organizations.

During my studies at both the school of electric engineering and the school of business administration I was lucky enough to spend some time in Japan and California,[2] where I noticed how cultures influence the context for technology-induced change and what is considered novel. In Japan, I explored the research environment. In the American Silicon Valley, meeting people like Ted Hoff—the inventor of the microcomputer—I saw the effects of the business environment. I observed the extremes: from the nuances in human interaction of the Japanese to the stimulating and raw capitalism of the United States. The microelectronic technology forecasted by my engineering thesis made the coming 'technology push' a little clearer: the personal computer was on the horizon. The implementation of

---

[1] At the present time, it is the Delft University of Technology Electrical Engineering School and the Erasmus University Rotterdam School of International Business Administration.
[2] The institutions' actual names were Afdeling Electro-techniek, Vakgroep Mikro-Electronica, and Interfaculteit Bedrijfskunde.

innovation in small and medium enterprises and the subject of my management thesis left me with a lot of questions. Could something like a Digital Delta be created in the Netherlands?

During my life's journey, innovation has been the theme. In the mid-1970s, I joined a mature electric company that manufactured electric motors, transformers, and switching equipment. Business development was one of my major responsibilities. How could we change an aging corporation by picking up new business opportunities? Japan and California were again on the agenda but now from a business point of view. I explored acquisition, cooperation, and subcontracting. Could we create business activity in personal computers? The answer was no.

I entered politics and became a member of the Dutch Parliament—a quite innovative move for an engineer—and innovation on the national level became my theme. How could we prepare a society by creating new firms and industries to meet the new challenges that were coming and that would threaten the existing industrial base? What innovation policies could be applied? In the early 1980s, my introduction of the first personal computer in Parliament caused me to be known as 'Mr. Innovation' within the small world of my fellow parliamentarians. Could we, as politicians, change Dutch society by picking up the new opportunities technology was offering? The answer was no.

The next phase on my journey brought me in touch with two extremes. A (part-time) professorship in the Management of Innovation program at the University of Technology in Eindhoven gave room for my scholarly interests. I was looking at innovation at the macro level of science. In addition, the starting of a venture company making application software for personal computers satisfied my entrepreneurial obsession. Now it was about the (nearly full-time) implementation of innovation on the microscale of a start-up company. With both my head in the scientific clouds and my feet in the organizational mud, my capabilities were stretched. At the end of the 1980s I had to choose, and entrepreneurship won for the next eighteen years. Could I start and do something innovative with personal computers myself? The answer was yes.

When I reached retirement in the 2010s and reflected on my past experiences and the changes in our world since the 1970s I wondered what had made all this happen. Technological innovation was a phenomenon that had fascinated me along my entire life journey. What is the thing we call 'innovation'? In many phases of the journey of my life I tried to formulate an answer with my first book, *Micro-computers, Innovation in Electronics* (1977, technology level), my second book, *The Management of Innovation* (1983, business level), and my third book, *Innovation, from Distress to*

Guts (1988, society level). In the 2010s I had time on my hands, so I decided to pick up where I had left off and start studying the subject of innovation again. As a guest of my alma mater, working on my dissertation, I tried to find an answer to the question 'What is the nature of innovation?'

My fascination with 'innovation' started in Delft. And seen from an intellectual point of view, *Deo volente*, it will end in Delft.

About the Invention Series

Our research into the phenomenon of innovation, focusing on technological innovation, covered quite a time span: from the late seventeenth century up to today. The case study of the steam engine marked the beginning of the series. That is not to say there was no technological innovation before that time. On the contrary, imitation, invention, and innovation have been with us for a much longer period of time and could have been investigated. However, we had to limit ourselves, as we wanted to look at those technological innovations that were the result of a 'General Purpose Technology' (GPT) of electricity—an expression that is not a part of everyone's vocabulary. As, clearly, some clarification is needed here, we will start with some definitions of the major elements of our research: innovation, product, technology, GPT, and revolution.

We define *innovation* as the creation of something new and applicable. It is a process over time that results in a new combination: a new artefact, a new service, a new structure or method. Whereas *invention* is the discovery of a new phenomenon that does not need a practical implementation, innovation brings the initial idea to the marketplace, where it can be used. We follow Alois Schumpeter's definition: 'Innovation combines factors in a new way, or…it consists in carrying out New Combinations…' (Schumpeter, 1939, p. 84). Innovation is quite different from invention. Also, for Schumpeter:

> *Although most innovations can be traced to some conquest in the realm of either theoretical or practical knowledge, there are many which cannot. Innovation is possible without anything we should identify as invention, and invention does not necessarily induce innovation, but produces of itself…no economically relevant effect at all.* (Schumpeter, 1939, p. 80)

What about invention, then? We follow here Abott Usher's interpretation, where the creative act is the new combination of the 'act of skills' and the 'act of insight': "Invention finds its distinctive feature in the constructive assimilation of pre-existing elements into new syntheses, new patterns, or new configurations of behaviour" (Usher, 1929, p. 11). Again the element of a combination is recognizable. By the way, one has to realize that these definitions arose in the early twentieth century and their meaning has shifted over time.

As a great part of our research is related to *product innovation*, we define a product as an artefact (from the Latin 'arte'—by or using art—and factum—something made) that, through its product-function, fulfils a need. As Herbert Simon stated:

> *An artefact can be thought as the meeting point—an interface in today's terms— between an 'inner' environment, the substance and organization of the artefact itself, and an 'outer environment', the surroundings in which it operates. If the inner environment is appropriate to the outer environment, or vice versa, the artefact will serve its purpose.* (Simon, 1996, p. 6)

Just imagine the product-function of timekeeping, realized by the 'inner' environment of the timepiece, that can be considered an answer to the need for timekeeping of the 'outer environment', the person who wants to know the time. Those environments, at a given moment in time, have to fit, as the example of the sundial illustrates—it is useless without the sun shining.

Those *needs* to be fulfilled by artefacts are ultimately related to human needs, from the basic need for shelter (the need for keeping warm creates the need for clothing) to derived needs (as in 'keeping the clothing closed') and aesthetical needs (as 'keeping the clothing elegantly closed'). There is a hierarchy in needs where the invention of the button certainly would fulfil a specific 'cloth-fixing need'. The concept of the product function thus can be quite abstract (as in the 'transportation' function) to quite detailed (as in the 'short-haul person and load transportation' function realized by a horse-powered cart). Basic needs are a constant, but derived needs come and go. So, over time, new product functions arise, as illustrated by the clock, which was an answer to a need when the agricultural society changed into an industrial society. The same clock—the marine chronometer—also played a vital role in navigation, used to determine longitude by means of celestial navigation.

Innovation takes place in those product functions when the artefacts change. Take the timepiece, evolving over time from those early hourglasses and sundials into the pendulum clocks, marine chronometers, and pocket watches. It was the adaptation of the 'inner' environment' of the artefact to the requirements of the 'outer' environment. For example: It is the mechanical implementation as a wristwatch that realizes a time piece the function of 'easy portable timekeeping'. Thus the product function 'time'— being a response to a universal need in our day, when the fourth dimension of time (ie in 'timing' work, travel) is dominating private and professional life—is implemented in many ways.

The realization of a certain implementation of the product-function is realized by people who know 'how to make it'. For the watch, there were people who mastered the 'fine mechanical watch technology', such as those nineteenth-century German and Swiss mechanics with their high-grade technological skills. This was soon followed by other skills when the digital timepieces were realized by people mastering microelectronic technologies. This leads us to the link between product innovation and technology.

We define *technology* as the 'knowhow' (based on knowledge) and the 'way' (based on skill) of making things. So technology—knowing how to make things—is part of the 'act of skills'. Technology is more than the 'technique'—ie a body of technical methods and procedures—from which it originates.

> *Technology is a recent human achievement that flourished conceptually in the 18th century, when technique was not more seen as skilled handwork, but has turned as the object of systematic human knowledge and a new 'Weltanschaung' (at that time purely mechanistic).* (Devezas, 2005, p. 1145)

We follow Anna Bergek and associates here:

> *The concept of technology incorporates (at least) two interrelated meanings. First, technology refers to material and immaterial objects—both hardware (e.g. products, tools and machines) and software (e.g. procedures/processes and digital protocols)—that can be used to solve real-world technical problems. Second, it refers to technical knowledge, either in general terms or in terms of knowledge embodied in the physical artefact.* (Bergek, Jacobsson, Carlsson, Lindmark, & Rickne, 2008, p. 407)

Technology can be simple. Just imagine the basic way of making coffee by pouring hot water over ground roasted coffee beans. Obviously, technology also can be quite complex. Just imagine the electronic semiconductor technologies: knowing how to make those complex integrated circuits.[3] On one hand, it is based on fine-mechanical, optical, chemical, and electronic technologies creating the objects—the production machines that work with astonishing accuracy—that developed over time as the result of impressive engineering efforts. On the other hand, it is based on the advanced knowledge of physical phenomena—the behaviour of electrons in semiconducting materials like silicon—that was acquired over decades by scientific efforts. So, as both science and engineering contribute, technology is 'knowledge' (originating from science), and 'knowhow' (originating from engineering) combined to fulfil a purpose.

---

[3] An integrated circuit (IC)—popularly known as a 'chip'—is a miniaturized electronic circuit with a specific function (eg a memory IC or a central processing unit). It is the core of modern electronic systems.

However, technology is not only 'skills'—the ability to carry out a task—and 'insight'—the understanding. As both science and engineering evolve over time, they create the infrastructure from earlier systems upon which the further developments build. The third element of technology, then, would be the 'accumulated experience' from earlier technological systems. Thus, as technologies are stacked on previous technologies, it is not hard to understand that the words 'electric' technologies and 'electronic' technologies represent, in fact, a collection of technologies: in a way, 'electricity' and 'electronics' are *meta-technologies*.[4]

A specific construct of a (meta-)technology is the concept of the *general purpose technology* (GPT). It can be defined in the following way:

> *A GPT is a single generic technology, recognizable as such over its whole lifetime, that initially has much scope for improvement and eventually becomes widely used, to have many uses, and to have many spill-over effects.* (Lipsey, Carlaw, & Bekar, 2005, p. 98)

This is a broad definition, hard to make operational or usable. Thus, in complement, we see a GPT as a cluster, or clusters, of innovations of which the fundamental new combinations, the basic innovations, have considerable impact on society. We call these basic innovations the *general purpose engines* (GPEs). Henceforth, more narrowly, we define a general purpose technology as *the collection of 'general purpose engines' appearing in a range of interrelated clusters of innovations*.[5] In other words, a GPT is a cluster, or range of clusters, of innovations around the general purpose engines. And the 'engine' is the device that transforms, such as in the transformation of heat into rotative power and from electricity into (rotative and linear) motion, light, sound. One observes that a GPT has also been defined by its spill-over effects, the GPT being 'the pervasive technologies that occasionally transform a society's entire set of economic, social and political structures' (Lipsey et al., 2005, p. 3). Lipsey also described a GPT as 'a technology that initially has much scope for improvement and eventually to be widely used, to have many uses and to have many spill-over effects' (ibid, p. 133). Thus we refined Richard Lipsey's definition, by focussing on the general purpose engines being the micro-foundations of a GPT themselves. The 'spill-over effects' then are the events that originate from the GPEs, giving the GPT their pervasive nature.[6]

---

[4] We use the word 'meta' to indicate a higher level of abstraction.
[5] This definition is more precise than the one we used in the preceding case studies as the result of new insights developed in the micro-foundations of a GPT during those studies.
[6] See: B.J.G. van der Kooij, *Lipseys Quest for the Micro-foundations of GPT—The General Purpose Engine*. Delft Repository: http://repository.tudelft.nl/islandora/object/ uuid%3A56fed0f9-8a38-487d-b93d-dd239c3e60c5?collection=research

In popular terms, a GPT is the meta-technology creating GPEs that results in techno-economic breakthroughs such as the Industrial Revolution, the Information Revolution, etc. It is the engine of economic growth but also the engine of technical, social, and political change—and it is the engine of creative destruction. The GPT is not a single-moment phenomenon; it develops over time: They often start off as something we would never call a GPT (e.g. Papin's steam engine) and develop into something that transforms an entire economy (e.g. Trevithick's high-pressure steam engine). (ibid, p. 97).

These examples of engines are our general purpose engines (GPEs).

These case studies are about observing phenomena as they occur in the real world—for example, the development of the steam engine, from which one can conclude it was a GPT according to the definition. The observation of what caused the Second Industrial Revolution shows its complexity. Is 'electricity' in its totality the GPT, or are the electro-motor and the electric dynamo, engines with a complementary power-conversion function,[7] the GPT? Does the development of the electric motor and electric light, the telegraph, telephone, and wireless illustrate the pervasive nature of the GPT-Electricity? Or can it be that the resulting development trajectories of the telegraph, telephone, and wireless, engines with a communication function responding to the basic human need to communicate, are GPT on their own? The interpretation becomes more complex, the opinions diffused, especially when one looks at the present time, for example, at the phenomenon of the Internet, part of the Information Revolution. As it is based on 'electricity', it could be considered a recent spill-over from the GPT-Electricity that started in the nineteenth century. By restricting the GPT concept to (a limited collection of) GPEs with the same product function, the concept stays within limits.

To conclude our definitions, a word about the use of the notion of *revolution*, as in 'Industrial Revolution' and 'French Revolution'. The word 'revolution' can be used to denote major social and political upheavals (eg the French Revolution) resulting in a major restructuring of society (ie regime change) or the replacement of a former ruling elite with a new one (ie government change), often with a lot of violence and casualties (ie the Madness of the Times). In that sense, a *political revolution* is an internal war—in contrast to the external wars between nations—that attempts to alter state policy, its rulers, and institutions. *Societal revolutions* are the changes in the structure of society—often originating from the oppressed or neglected classes but also as a result of the Spirit of the Times—that are related to the concept of social change that we will go and explore. The companion

---

[7] The electric motor uses electricity to create movement (rotational power). The electric dynamo uses rotational power to create electricity.

concepts of scientific change, related to *scientific revolutions,* and technical change, related to *technological revolutions,* are discontinuities outside the political and societal spheres. In a technological revolution, the ruling meta-technology is replaced, or complemented, by another meta-technology: the new general purpose technology (eg steam technologies being replaced by electric technologies). As a consequence, the technological revolution restructures the material conditions of human existence and ultimately results in *socio-economic revolutions,* just as the preceding socio-economic revolutions did originally create the context for the following technological revolutions.

These drastic changes in the societal and social structures, caused by such major technological changes, are creating a broad spectrum of technical and organizational novelty. The socio-techno-economic disruptions are based on the technical and the economic dimension of the *industrial revolution*s. Although the violence aspect on the social level is not that obvious, like the social revolution, the industrial revolutions also have 'victims'. The casualties of these socio-techno-economic revolutions—by unemployment or outdated technical knowledge and engineering practise—certainly can be identified as the victims. Schumpeter labelled the phenomenon as 'creative destruction'. New technologies created new jobs and destroyed old jobs. The lamplighter of the gaslights, the messenger boy for the telegrams, the male telegraph operator, and the female switchboard operator, they all faded away, to be replaced by totality different jobs in other technologies.

About our Research

This book is the sixth manuscript in the *Invention Series,* a series of books on inventions that created the world we live in today. In the first case study, *The Invention of the Steam Engine,* we explored a methodology to observe and investigate the complex phenomena of technological innovation as part of a general purpose technology (GPT). In that case, it was about the steam technology that fuelled the First Industrial Revolution. One could consider that case study as a trial to see if our methodology could be applied. It looked promising enough to try again. The result was a case study on electro-motive engines and a case study on electric light. Now, in this case study, to complement the preceding studies on the communication engines 'telegraph' and 'telephone', we focus on wireless communication. So, let's start to describe the basic elements of our research approach.

Our *field of interest* in the GPT-Electricity is, here in particular, the area of its application in communication over distance, aka telecommunication. To understand how this technology could fuel the next industrial revolution, we again applied the method of the case study. The case-study method offers room for context and content. The context is the real-life context: the scientific, social, economic, and political environment in which the observed phenomena occurred. The content is the description of the technical, economic, and human details of those phenomena. The reader will again recognize this content and context approach in the dualistic structure of the manuscript.

The case study is based on a specific scholarly view to observe the phenomena as they occurred in the real world. This view is based on the construct of clusters of innovations, as identified by early twentieth-century scholars active in the domain of innovation research. Among those economists was Alois Schumpeter, who related the clusters of innovations to business cycles under the influence of creative destruction:

> *Because the new combinations are not, as one would expect according to general principles of probability, evenly distributed through time...but appear, if at all, discontinuously in groups or swarms.* (Schumpeter & Opie, 1934, p. 223)

Schumpeter continues: 'the business cycle is a direct consequence of the appearance of innovations' (Ibid, pp. 227–230). For Schumpeter, it was the entrepreneur who realized the innovation and, as imitators were soon following in the entrepreneurial act, thus created the business cycles nested within the economic waves. Later it was Gerhard Mensch and Jaap van Duijn who related the basic innovation within the clusters to the long waves in the economy with respect to industrial cycles. Mensch related the cyclic economic pattern to basic innovations: 'The changing tides, the ebb and flow of the stream of basic innovations explain economic change, that is, the difference in growth and stagnation periods' (Mensch, 1979, p. 135). Duijn referred to innovation cycles (Duijn, 1983). More recently it was scholars like Utterbach and Abernathy, Suarez, Dosi, Tushman, Anderson, and O'Reilly who developed and used, as part of their view on technological revolutions and technological trajectories, the construct of the 'dominant design' being the watershed in a technology cycle (Tushman, Anderson, & O'Reilly, 1997). This dominant design is the innovation that— at a given moment in time—has become the de facto industry standard. This dominant design we considered to be the basic innovation.

Our *focus of analysis* is the cluster around the basic innovation with the preceding and derived innovations (Scheme 1). Our *units of analysis* are the contributions made by individual people resulting in inventions and innovations. Then, for our domain of analysis, we first observed contributions in the GPT-Steam (a collection of many mechanical, hydraulic, thermic, and related technologies explored in the first study), followed by our observations of the electro-motive engines in the GPT-

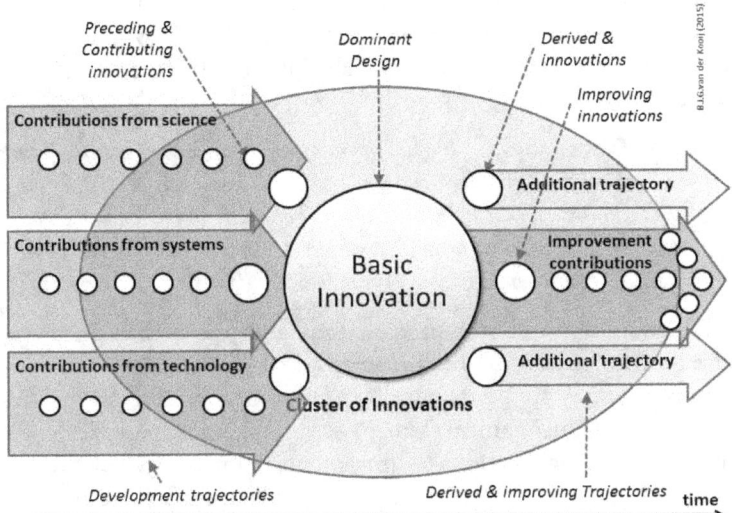

Scheme 2: The construct of the trajectories leading towards and from the basic innovation in a cluster of innovations.

patent wars (patent infringement and patent litigation) and economic booms (business creation, business and industry cycles) to identify basic innovations, this aspect is quite dominant in the study.

Considering our *unit of analysis*, in view of the previously mentioned aspect of innovation being the result of a combination, we tried to refine the cluster concept by detailing the contributing innovations into specific technological development trajectories (see Scheme 2, left):

Scientific contributions: These include the trajectory of the 'scientific contributions' concerning the basic laws of nature the curious and ingenious people in the eighteenth and nineteenth century were inquiring into. We use the definition of *science* as:

> *The intellectual and practical activity encompassing the systematic study of the structure and behaviour of the physical and natural world through observation and experiment.* (Oxford Dictionary)

This incorporates the contributions of the electro-physicists who discovered the basic principles of electromagnetism and the experimentalists who applied those principles.

Technology/Engineering contributions: Next we distinguish the technological contributions and use—in addition to our previously mentioned definition—the definition of *technology* as 'The application of scientific knowledge for practical purposes' (Oxford Dictionary) and as the knowhow (knowledge) and way (skill) of making things. Or, as Giovani Dosi puts it:

> *[We] define technology as a set of pieces of knowledge, both directly 'practical' (related to concrete problems and devices) and 'theoretical' (but practically applicable although not necessarily already applied), know-how, methods, procedures, experience of successes and failures and also, of course, physical devices and equipment.* (Dosi, 1982, p. 151)

Practical knowhow is built up over time and transferred from generation to generation by the means of the apprenticeship. Thus, this incorporates the contributions of all those instrument makers using their fine mechanical skills to create magnets, batteries, telegraph components, and telegraphic instruments, which were so essential to the creation of electrical devices. In today's terms, their activity would be called engineering.

System contributions: A third development trajectory consists of the contributions that resulted in previously developed systems. The system concept being quite general, we will be using the definition of a system as "A set of things working together as parts of a mechanism or an interconnecting network; a complex whole" (Oxford Dictionary). The keyword here is 'network', to which development so many creative minds contributed. However, these are contributions that are harder to classify. Let's, for example, consider our application area of communication (postal, optical, or electrical). Communication is always realized in a structure of several elements (parts, components) connected by a structure (network). For the classic postal system, it is the network of mail coaches, mail couriers, and the inns to change horses: the postal network. For optical communication, it is, as we have seen, the network of semaphore relay towers and the organization of telegraphists that transmitted information in the semaphore code: the semaphore network. For electric telegraphy, it is similar. The electrical components like the transmitter, the cabling, and the receiver, the code used for the transmission, and the structure of the telegraph offices created the network infrastructure for electric telegraphy: the telegraph network. People who contributed to that totality created the system contributions.

Given the genesis of the *basic innovation*, it will be followed over time by new contributions leading to other innovations (Scheme 2, right). Such as:

Improvement contributions: This includes contributions that enhance and improve upon the basic invention. The increasing knowhow of the ever-developing technology will add to the original invention step by step in in an incremental way. These improvement contributions create a technological trajectory of incremental innovations.

Derived contributions: In addition to the improvements, there will be contributions of another nature. In those cases, either to circumvent the patent protection or just by accident, the same functionality of the basic invention will be realized using a different concept, spinning off in a different trajectory. The example here is the development of the speaking telegraph (also known as the telephone) using undulatory electrical currents (ie alternating current) for the transmission, which resulted from the improvement efforts in electromagnet based telegraphy using direct electrical current. Those derived innovations will create additional trajectories when the new development is applied in other ways and other fields of application, thus showing the pervasiveness of the GPT-Electricity.

About the Context

As mentioned before, case studies are about content and context. Our specific case studies are about the *content* of technical change—they cover technological innovations—and we look at change from the perspective of the development of technological innovations themselves: the clusters of innovations. These clusters are the result of contributions of many individual persons, individuals who lived within their specific 'Spirit of the Time', often even with its specific 'Madness of the Time'. People with personal hopes and fears, drives, ambitions, and limitations, honest people and cheating people, extraverted and introverted people, people who lived in—and whose behaviour was influenced by—times of war, physical destruction, and economic stagnation, and people who lived in times of peace, creation, and progress—and people who lived in a specific society.[8]

---

[8] The word 'society' relates to human societies that are characterized by patterns of relationships (social relations) between individuals who share a distinctive culture and institutions in a specific social environment. A 'society' is different from social groups (family, tribe, or clan), the *family* being a group of people affiliated by consanguinity (by recognized birth) and by affinity (by marriage), the *tribe* being a distinct people, dependent on their land for their livelihood, who are largely self-sufficient, and not integrated into the national society, and the *clan* being a group of people united by actual or perceived kinship and descent. Clearly, the word society can have different meanings such as *national societies*, countries/states with an economic, social, industrial, and/or cultural infrastructure, and *collective societies*, like companies with an economic, social, industrial, and/or cultural infrastructure.

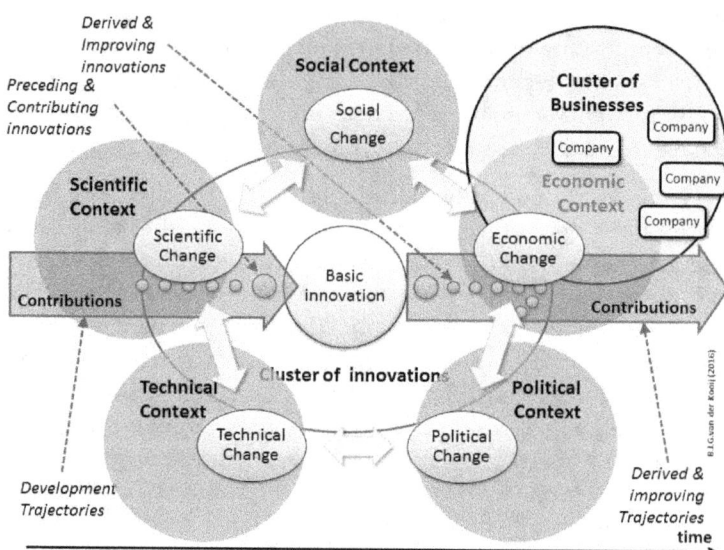

Scheme 3: The construct of change in the different contexts.

When observing the real world in all its complexity, one tries to create a mental model of that world. By definition, that mental model is simplified, limiting the complexity. The limiting is done by creating mental 'constructs'. For our contextual analysis, we will us the *construct of change*, eg the constructs called Social Change, Economic Change, Scientific Change, etc. Each of these change constructs covers a part of the total context. So— along with analysing the content, where we used the perspective of the clusters of innovations (Scheme 2)—now we analyse the *contexts* that influence the occurrence of those clusters by including the different change constructs to the cluster of innovations (Scheme 3). They are defined as follows:

Social Change: Each case takes place in the society as it existed at that moment in time. That society defined the *social context* for the individual inventor and his inventions at that given period of time—a society that itself changed constantly. Hence, we speak about the autonomous change of social structures, social behaviour, and social relations in a society as being the result of social forces. When those changes are incremental, social change is incremental and evolutionary in nature. But sometimes the changes are discontinuous and disruptive—even revolutionary. Then we talk about revolutions such as the American,

French, and Russian Revolutions. They are the drastic—even dramatic—forms of social change.

Economic Change: Part of the interaction in a society has to do with 'economic' activities like the production, distribution, trade, and consumption of goods and services. Together they create the *economic context* for innovation. Each of these activities has a dynamic of its own. Take 'trade' as an example, such as when the surplus of agrarian production was brought to the local market and traded for other surpluses, or when the local and regional trade grew into national trade and even colonial trade. These market-based local, regional, and national economies exchanged goods and services between participants by barter or a medium of exchange like currency. After the social system expanded into nations, the economies also evolved into larger structures, and each of the participating institutions developed on its own. Take the example of 'production': when the cottage industry developed into industrious mass production, creating clusters of businesses within specific areas of manufacturing.

When national economies develop, state policies—and their resulting laws—are needed to facilitate and control those economies. In creating these, the economic policies become part of the political structure, as they represent an (economic) interest. This process resulted in mercantile economic policies or—later in time—free trade policies. Economies have a dynamic of their own: they grow, and they contract, but normally, they change within limits, as the participants in the economies have an interest in maintaining a state of equilibrium. But sometimes the changes are discontinuous and disruptive—even revolutionary. Then we talk about revolutions: such as (the economic part of) the Industrial Revolution(s). It is the totality of these dynamics we consider to be economic change.

Political Change: In each society, be it small like a family, group, or organization, or be it big like a clan, tribe, or even nation, there is a power structure with centralization of authority. The power structure in a society is a means for the survival of the group. Politics is the dynamic interaction between the participants in that power structure.[9]

---

[9] The word 'politics' can have different meanings. Such as: *national politics,* the working out of forms of agreement or conflict solution between different groups of people and their specific interests; *parliamentary politics,* the formalized interaction in the structure of a democratic parliament; *party politics,* the collective views and wheelings and dealings of political parties to exercise political power; and *people's politics,* the representation of specific interests such as religious interests. We will consider the political context to be the national execution (national politics) of the representation (parliamentary politics) of specific interests (party politics and people's politics). Related is the word 'policy': a policy is a statement of

Consequently, the participants of a group create the political system together as a system of entities: the participants, their institutions, and the relations between those entities. In short, there is the *political context* for innovation of those who rule and those who are ruled, and that 'ruling' is the exercise of influence on someone's individual behaviour. This influence is based on personality, physical power, expertise, or just—historic—acceptance. That power structure is not fixed; temporary alliances between members of the group result in changes in the picking order. Hereditary power can be challenged, creating succession wars. The totality of these dynamics is considered to be political change.

Technical Change: Technology—in short, 'knowing how to make things'—is the collection of techniques, skills, methods, and processes. It is the basis of industrious human activity in the realization of goods and services. It is related to engineering, the process of designing and making tools and systems. On one hand, technology is based on the application of scientific knowledge, the understanding of the basic phenomena in nature. However, it is also based on engineering skills, acquired over time by 'doing things' and passed on by generations. Thus, technology is the combination of understanding (knowledge) and practise (knowhow). In their totality, they create the *technical context* for innovation. So, science, engineering, and technology are interrelated and undergoing changes, sometimes incremental or sometimes even disruptive. We talk about the technical part of industrial revolutions as a drastic form of technical change. When technology changes, either evolutionary of revolutionary, we call that technical change.

Scientific Change: Science is about understanding the phenomena of the natural world we live in. That understanding is—in our present world— based on the scientific method, a disciplined way to study the natural world. It results in understanding the 'nature of matter', such as the 'nature of heat', 'nature of light', etc, subjects that have evolved in the scientific disciplines of chemistry, physics, etc, each discipline having its own knowledge base. This knowledge is represented in the form of concepts, theories, models, and laws, which create the *scientific context* for innovation. When scientific knowledge changes through evolution, we call that scientific change. And when it is based on a paradigm shift,[10] we call it a scientific revolution.

---

intent and is implemented as a procedure or protocol. Examples are: public policy, foreign policy, and economic policy.

[10] A paradigm shift is a fundamental change in the basic concepts and experimental practices of a scientific discipline. In the case of a paradigm shift, the prevailing framework of shared scientific views, theories, and models is replaced by a new framework.

About our Perspective

One has to be realistic and not try to cover the developments under scrutiny in their totality and all their complexity.[11] At the time they occurred, the economic change and social change leading up to industialization received extensive scholarly attention, such as seen from a purely *economic perspective* (eg the Scot Adam Smith in the eighteenth century) or a *social perspective* (eg the German Karl Marx in the nineteenth century). Among all those manifest scholarly views, we will humbly limit ourselves to a *technological perspective*. However, in doing this, we will first focus on the *context* for technical change, a context that itself was the result of a continuous process of change.

To explore that context, we will zoom in on social change and political change, setting the stage for technical change. The role of so-called 'political' institutions is especially the focus of our attention.[12] We will observe along the timeline of history over several centuries the social transition from a feudal society characterized by inequality and dependency into a democratic society characterized by equality, individuality, and independency. This transition away from the restrictions set by church and ruling class created free, liberal societies where innovation could thrive and prosper.

Therefore, our particular perspective will be on the socio-political environment (Scheme 4) with the specific social context and political context, as well as the overall context that encapsulated the techno-economic environment for technical change and economic change. These contexts can be described as follows:

Societal context: The societal context is determined by the social world we are living in. It is our society that defines our existence, and that society is in a constant process of autonomous change. The societal context we are going to study is about (groups of) people, their (hierarchical) relations, and their collective social behavior. We look

---

[11] This general analysis draws heavily on the information as available in Wikipedia, and quite often partially edited text parts are used. As we consider this to be public knowledge, we do not quote individual pieces of text but incorporate them in our narrative. In addition, much of the details used were obtained from general sources. Finally, we quote from some general books about the Industrial Revolution.

[12] The expression 'institution' is often—confusingly—used in a double sense: the economic sense and the social sense. That is a) in the classical sense of the economical institution of the (in)formal organization, and b) the social institution in the sense of stable, recurring patterns of behaviour in a social structure. More popularly expressed, the former consists of the organized 'players of the game' (eg the organized institution of the parliament), and the latter of 'the rules of the game' (the behaviour of that parliament). Political institutions are therefore the patterns of behaviour related to dominance in a society.

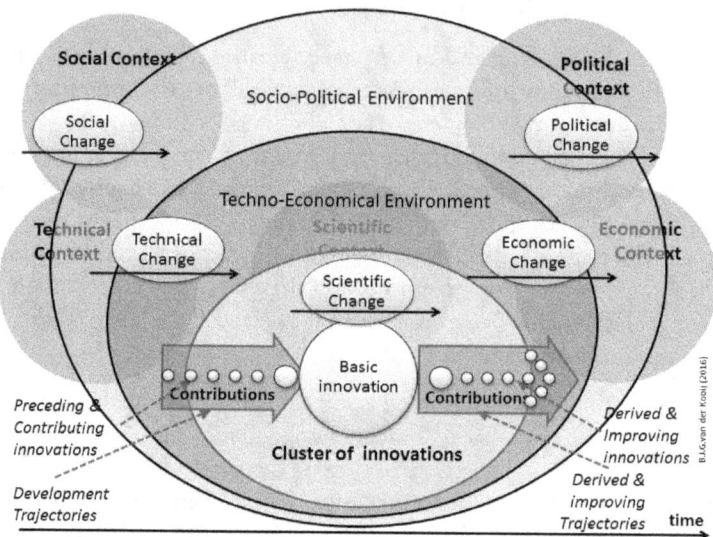

Scheme 4: The specific contexts for the cluster of innovations in its relevant environment.

The figure illustrates how the context of the techno-economic environment is shaped by the socio-political environment.

at the contextual facts themselves and the way that context changes over time. Within that domain of social change, the themes of the societal tensions and controversies in Italy were the same as those in France leading up to the French Revolution, being about the legitimacy of the *absolute monarchy*, based on the 'divine right of kings,'[13] in relation to the evolution of *parliamentarian sovereignty*, based on the natural rights of man. It was also about the relation of Church and State, the place and role of religion in society based on the 'Divinity'. In addition, it was about 'court and country': the different groups of participants in society with the accompanying tensions between the societal classes as they developed over time.

Political context: Part of the social context is the interaction between social groups, where each group has its shared, specific interests. It created the 'politics' of that time, resulting in the (national) political context.

---

[13] The doctrine of the divine right of kings—also called God's mandate—is about the royal legitimacy. It asserts that a monarch is subject to no earthly authority—only to the authority of God's representative on Earth, the pope—and that there is nothing to regulate the powers of the king, and he thus becomes an absolute power.

The political context is about 'governance'[14]: the *organizations* of governance and the *process* of governing. Firstly, it is about the role of the state in governing the society: the executive form of government.[15] Secondly, it is about the relations between the 'governors' and 'those who were governed': the class structure of society and the interaction between the classes. The basic subject is *political power*: the ability to control the behavior of others. Over the ages these political relations were based on feudal thinking.[16] However, in the period of time we are evaluating, it was the revolutionary views on the natural rights of man—as proclaimed by the Enlightenment philosophers—that were having an increasing influence on the relation between the 'rulers' and the 'ruled'.

In short, when we are looking at these social and political contexts, we look at the Constitution,[17] its form and institutions, and the holders of the related governing powers that came with it. Obviously these social and political contexts are not isolated issues. Therefore, we will—where appropriate—stray and immerse ourselves in neighbouring contexts, and we will describe some of the personalities that played an important role in it.[18]

---

[14] Governance can be defined as all processes of governing, whether undertaken by a government, market, or network, whether over a family, tribe, formal/informal organization, or territory and whether through laws, norms, power, or language (Bevir, 2012, p. 1).

[15] Government is the political system by which a country or community is administered and regulated: how it is ruled. The political system is the set of formal legal institutions (eg the parliament, political parties) that constitutes the behavior of a 'State'. Thus, the state is the political organization of society. We will use the legal institutions, like monarchy and parliament to describe the 'governmental context' (based on Encyclopedia Britannica definitions).

[16] Feudalism: The political structure in a social system in which the relations are derived from land ownership. It includes the concept of manorialism, where the landowning lords and the landworking peasants are interrelated.

[17] Constitution: A body of fundamental principles or established precedents according to which a state or other organization is acknowledged to be governed (Oxford Dictionary).

[18] Our historic observations on what happened in society are by their nature based on the observations and interpretations of others. And each of these observers has his own point of view, his perspective, and his focus—and bias. The resulting interpretations they wrote down can be biased from a personal aspect (being a liberal thinker gives a different view than being a Marxist thinker) but also from a societal aspect (the nationalistic views). By using a wide variety of sources, both from different backgrounds and from different timeframes (contemporary authors and present-day others), we hope to eliminate this bias as much as possible.

About the Interaction

In our case studies, we primarily observe the phenomena related to technical innovation, phenomena that took place within the context of their times. In a way, it is a multi-layered context (Scheme 4). For one, there is the content shaped by the *individual environment*: the technical, economic, or scientific context on the individual level for the observed cluster of innovations. In addition to that direct context, there was the more indirect *techno-economic environment* with its technical change and its economic change. And finally, there is the context of the society in which 'it all happened', the *social-political environment* with its social change and political change. Seen in this way, it is a dynamic multi-layered environment that influences the phenomena we look at. We look at each of the layers, and we look at the interaction between each of those layers.

We look at change: such as at those artefacts that started to appear and that developed over time in an evolutionary process where some artefacts survived and others disappeared from the scene due to the properties of the environment. In analogy with biology, in the hot and dry desert, even the most potent plants struggle to survive, and only the best adapted ones survive. In the fertile environment of a tropical rainforest, plants in abundance compete with each other in a battle for survival. The same goes for the appearance of new artefacts (eg innovations): some thrive and prosper; others do not survive. It is—again in Herbert Simon's concept— the properties of the 'inner environment' of the artefact that has to match the properties of the 'outer environment' if the artefact is to survive.

Therefore, trying to understand the dynamics of changes, we borrow from evolutionary biology the concept of Darwinian 'fitness for survival', which encompasses the fitness of the organism and the fitness of the environment. It is a concept that—in short—refers to the mutual relationship between organism and its relevant environment, between the properties of organisms to survive and the conditions of the environment in which the changes on a species level occur.

> *The fitness of the environment is one part of a reciprocal relationship of which the fitness of the organism is the other. This relationship is completely and perfectly reciprocal; the one fitness is not less important than the other, nor less invariably a constituent of a particular case of biological fitness.* (Henderson, 1914, p. 113)

In terms of technological innovation, this term refers to the fitness of a specific technology and its resulting artefacts in relation to the fitness of the environment in which it appears. Clearly, as we have seen before, some technologies 'make it and prosper'; other technologies proved to be 'dead ends'. They were not fit enough.[19] When the economic environment proves to be fertile—for example, in business terms—many technology-induced innovations and their artefacts will prosper.[20] That fitness of the environment may be the cause for the appearance of certain similar inventions at the same time, such as the invention of the telegraph both in England and America.[21]

As this is not the place to dwell on evolutionary biology, we focus dominantly on the fitness of the relevant socio-political environment and techno-economic environment in relation to technological innovation itself (Scheme 4). As we analysed elsewhere,[22] the development of the GPT-Electricity took place over a century, picking up in the second half of the nineteenth century. Then the GPT-Electricity, child of the first Industrial Revolution, became one of the dominant catalysts of the Second Industrial Revolution. Both industrial revolutions took place within the techno-economic environment of their time. Hence, we talk about the *technical dimension* and the *economic dimension* of these revolutions. They will make up a large part of our analysis

Also, as both the First and Second Industrial Revolution showed quite a bit of social dynamics—ie the American and French Revolutions—obviously, there is a relationship between the phenomena and its socio-political environment, part of the described relationship between content and context. Hence, we talk about the *social dimension* and the *political dimension* of these revolutions, and we include in our analysis the social and political revolutions that took place when the foundations for these industrial revolutions were created. Our analysis for patterns of change in the different contexts is quite abstract; one could say we take a helicopter view. Not so much the larger and stationary 'satellite view', nor the more detailed 'birds-eye view', this helicopter view enables us to alter between pattern and detail by zooming in or out.

---

[19] An example would be the reciprocating electromotor of the early days of the electromotive engines. See: B.J.G. van der Kooij, *The Invention of the Electro-motive Engine* (2015), pp. 72–75. Another would be the arc light, which was replaced by the incandescent lamp. See: B.J.G. van der Kooij, *The Invention of the Electric Light* (2015), pp. 55–98.

[20] Here, the example is the availability of electricity when the electric dynamo came into existence and replaced the cumbersome voltaic batteries. Then the electric light, the telegraph, and telephone started to develop in force. See: B.J.G. van der Kooij, *The Invention of the Electro-motive Engine* (2015) pp.87–125.

[21] Both the Morse telegraph (based on the relay principle) and the original Cooke & Wheatstone telegraph (based on the galvanometer principle) appeared at the same time but in different environments (ie the American and British societies). The Morse concept survived. See: B.J.G. van der Kooij, *The Invention of the Communication Engine 'Telegraph'* (2015)

[22] See: B.J.G. van der Kooij, *The Invention of the Electro-motive Engine* (2015)

Finally, a word about the use of the words *invention* and *innovation* in the case study. We described before how we intend to define them, but in the case study, we follow our sources. They use the words in the context of their time—a use that can be different from our time—for example: what would be called an invention in the early nineteenth century, could be called an innovation today. There is quite a difference between the two, and even our present-day interpretations of both words show great variance, as we found in a survey of the word *innovation* as used by innovation scholars.[23]

About this Case Study

This case study is another result of our quest to understand the Nature of Innovation. Where the other cases focused on energy—the power of steam and the power of electricity—and the application of electricity—in light and rotative power applications—in this case, it is about the forms of communication using electricity. Of the dual roles of electricity—on one side offering means for transporting power and on the other side offering means for transporting information—the latter is explored. It is an exploration in three parts covering three separated phases. The first phase was the invention of the communication engine called 'telegraph' (Morse, Cook & Wheatstone) in the early nineteenth century. The second phase was the invention of the communication engine called 'telephone' (Bell) in the second half of the nineteenth century. And the third phase was the invention of the 'wireless' communication engine (Marconi) at the end of the nineteenth century. The cases about the telegraph and telephone have been explored elsewhere.[24] Now we will analyze the context for the invention of wireless communication. Again it is about a communication engine that uses electricity as the carrier of information.

Although the birth of electricity more or less was initiated in Italy by Alessandro Volta, and Marconi lived also in Italy, much of the actual 'wireless' developments took place in the nineteenth century's Kingdom of Great Britain. That overall context will be analyzed in a following case study.[25] Here, we will pay attention to the overall context created on the Italian Peninsula from the late Roman times through the late nineteenth century[26].

---

[23] See: B.J.G. van der Kooij, 'Innovation Defined: a Survey'. Source:
http://repository.tudelft.nl/view/ir/uuid%3A6a5624c9-e64e-4426-98e9-f239f8aaba18/
[24] See: B.J.G. van der Kooij, *The Invention of the Communication Engine 'Telegraph'* (2015) and B.J.G. van der Kooij, *The Invention of the Communication Engine 'Telephone'* (2015).
[25] See: B.J.G. van der Kooij, *Context for Innovation: British (R)evolutions in Perspective* (2016).
[26] Our reader might wonder why such a long period is going to be observed. Basically, the answer would be that the foundations for our present-day technology-dominated,

The context for the discoveries: We will begin in the first section with a thorough look at the events that created the general historical context. First, we explore the Scientific Revolution, which changed how we look at the world around us, and the Enlightenment movement, with its evolving natural and liberal thinking.

Then we focus on Italy, where we describe from a particular perspective the social change and political change that occurred on the Italian Peninsula from Roman times up to World War I. For different periods, the development of Italy is analysed extensively. Specific developments of European importance are highlighted, such as the development of the Papal Authority, the Papal States, and the dominance of the Roman Catholic Church. Also, the relationship between the popes and the Holy Roman emperors has our attention. In addition, we investigate the underlying transition from a Roman imperial society to the powerful city-states. We investigate the early republics, different in character and form, with their influence in the Mediterranean Basin over time. We describe the nearly continuous domination by other European nations—Spain, France, Austria, Germany—in which the riches of the peninsula were transferred to these invaders.

At the end of our contextual analysis, we explore some of the early efforts where curious people started to try and apply the new phenomenon of electricity as they were trying to understand the 'nature of communication', just as they'd earlier tried to understand the 'nature of lightning' and the "nature of heat' before.[27]

The content of the discovery of the wireless telegraph: The second section of the case is devoted to the content of the invention of the wireless communication engine itself. We start with an analysis of the early days of wireless communication, when the Hertzian waves became the focus of attention of scholars interested in the nature of lightning. Their work became part of the wireless mania in science. Then we move on and start to focus on the contribution of Guglielmo Marconi. After describing the local Bologna context, we follow his explorations into the dawn of wireless telegraphy in Italy. Next, we examine the big decision to go to England and bring his invention there to maturity, and his subsequent pioneering years are also analysed. We describe his decision to follow the entrepreneurial

---

democratic society—with all its imitations, inventions, and innovations— were created over that period of time.

[27] See: B.J.G. van der Kooij, *The Invention of the Steam Engine* (2015); *The Invention of Electromotive Engine* (2015).

route and start a company. In a mix of technical activities and business activities, we follow Marconi in this period of time till his marriage: a hectic period in which the young man was faced with national and international opposition of quite some magnitude. We describe the creation of the Marconi monopoly and the Marconi empire, as well as his patents and the many patent conflicts he had to endure. We analyse in detail some specific aspects, such as his Act of Invention, his Act of Business, and the patent wars he became entangled in.

Marconi's technical and entrepreneurial activities cannot be separated from what happened in the early twentieth century, so we paint a picture of the cluster of businesses that emerged in Britain and America, and the fury of patenting that followed his initial patent and the wireless mania in capitalistic America. We place Marconi's invention in the perspective of the nationals' interests (eg the British state monopoly). His contribution and his priority right are discussed in detail. Finally, we try to place his contribution in the perspective of Communication Revolution, as that occurred in the nineteenth century. We show the remarkable similarities of his contribution with those of his brothers in arms Samuel Morse and Alexander Graham Bell. We end our analysis of Marconi's contribution at the time of the Great War, an event that severely damaged society, more or less halted wireless telegraphy, and also greatly influenced Marconi's later activities after he returned to Italy.

Conclusion: In our conclusions, we describe the common denominator for the total development of wireless technology seen in the context of time: the Invisible Hand of Innovation.

One might wonder why we chose to delve into Italy's history leading up to the Italian Revolution, which happened so shortly before Guglielmo Marconi appeared on stage. One reason can be found in the fact that we want to further explore the importance of the context for innovation.[28] In the earlier case studies on communication, we explored the American Revolution and the French Revolution. Now the events that could be called the Italian Revolution have our attention. That revolution created the overall context that existed for the young Marconi, but also for all those other creative and entrepreneurial people before and after him. A context that emerged in the Western world over some centuries, and was the result of the Scientific Revolution and the Enlightenment, both developments

---

[28] The broader Context for Innovation is covered in the case study about the English Revolution up to the British Industrial Revolution. See: B.J.G. van der Kooij, *Context for Innovation: British (R)evolutions in Perspective* (2016).

that could not have been in existence without the declining absolute monarchies and reclining ecclesiastical powers. A second reason can be found in that so much from Italy's history influenced the western societies: from the ideas and works of Italians like Galileo, Leonardo da Vinci, Michelangelo—to mention just a few—and the democratic structures of the city-states, to the worldly and religious influence of the Roman Catholic Church that so absolutely dominated European societies for centuries. And finally, as the many artistic inhabitants of the Italian Peninsula gave so much beauty to the Western world, we explore some cultural aspects: the beauty of Renaissance arts.

To the readers who are still sceptical about that emphasis on the context, we would like to say the following. Just imagine for a moment yourself living in the medieval times as a common serf (98% chance you were): locked into a feudal society with none of your present freedoms, depending of the whims of your landowning master, having an expected lifespan of some thirty to forty years, living a life of minimal subsistence. Then look at your life today, with the freedom to move where you want (ie the freedom of the body), the freedom to believe what you want (ie freedom of the soul) and the freedom to express yourself without fear (ie freedom of the spirit). Living a life that can be enjoyed for nearly twice the time in totally different economic, political, and social circumstances. Wouldn't you want to know what made that happen? I certainly did.

This is a sixth case study about the general purpose technology of electricity with its clusters of innovations and clusters of businesses that created the Eras of Communication and changed the world we live in. It will take you along a multitude of winding social, political, economic, technical, and scientific roads, deep into historic times and back to our present time.

B. J. G. van der Kooij

B.J.G. van der Kooij

The Invention of the
Wireless Communication Engine

Context for the Discoveries

For someone living in the pre-electric era, it would have been hard to imagine verbally communicating over longer distances the way we do today. In those times person-to-person communication was local, and the world was small. Most conversations took place around the village water flow or local water source (Figure 1), where the local women did their washing,

Figure 1: Chatting women washing clothes by a stream.

Source: Daniel Ridgway Knight (ca 1898). Wikimedia Commons.

laughing, and gossiping, and where they exchanged the latest news, which was predominantly local. In the larger villages, public proclamations were done by the local town crier—also called 'bell man'—who would walk around the village and loudly shout his message so everyone could hear it; the early form of one-to-many communication. Whatever form it had, verbal communication over some distance—what today we call telecommunication—was non-existent, as sound had a limited reach. Over distances, only the written communication would do. And that form of communication took time.

In those times person-to-person long-distance communication was with the written word. As the world was small, the messenger boys often transported the written word locally. For communication over longer distances, one could send a letter by postal messenger, who transported it on horseback in a mail pouch with other written messages (eg the American Pony Express). Or, when available, one could use the mail coach that was part of the network of royal postmasters (eg the British Royal Mail). A network for p-mail (ie postal mail) with *relays*, stations along the postal trajectory where fresh horses would be available to replace the exhausted ones (Figure 3). Whatever form the

Figure 3: Delivering the post and passengers by coach.

Source: James Edwin McConnell, http://www.scholarsresource.com/browse/work/2144689165

Figure 2: *Party Wire* by Norman Rockwell (1919).

Several subscribers connecting to the same 'party line' made any privacy hardly possible.

Source: http://www.best-norman-rockwell-art.com/norman-rockwell-leslies-cover-1919-03-22-the-party-wire.html

communication had, concerning communication over distance, it all would take time.

Nobody even dreamed that one day there would exist rather simple communication engines, such as the telegraph, that would send the message at the speed of light to faraway destinations all over the world. Neither would many people have imagined the communication engine of the telephone, which could be used to transmit gossip as instant electric speech (Figure 2). Messages and speech transmitted around the globe by a network of copper wires—spanning distances over land, crossing rivers, and bridging the continents by undersea telegraph cables—may have seemed already magic for our ancestors, but there was more to come.

Communication: from 'Wire' to 'Wireless'

By the end of the nineteenth century another medium for person-to-person communication had also developed. In addition to the cabled infrastructure used by telegraph and telephone, the wireless telegraph had emerged. Soon to be called 'radiotelegraphy', Morse codes passed through the air in a network of wireless stations, for example, to maintain shore-to-ship communications or to make wireless communication between ships possible. The importance of this new form of communication was shown when, during its maiden voyage, the British passenger ship the RMS Titanic collided with an iceberg in 1912. The heroic 'marconist' did send his Mayday messages, calling for assistance till the ship sank. Luckily, nearby ships responded and were able to save hundreds of passengers.

Wireless communication expanded rapidly. Carried by the long electric waves, transmission over long distances developed next to the submarine telegraph cables. In the United States, radio amateurs appeared in the early twentieth century. This non-commercial use of the technology by 'ham radio' amateurs—often young people who were fascinated by the new phenomenon of radio communication—was soon clogging the airwaves. By then use of the commercial-based 'marine wireless' was increasing, and 'wireless' even became business hype on the financial stock markets. Then the world became engaged in the 'Great War' of 1914–1918, which halted all non-governmental wireless communication. Military use soon dominated the ether, and the military powers tried to control the business.

Figure 4: *The Wonders of Radio* by Norman Rockwell (1922).

Source: Post Cover, May 20, 1922. Finch, Chr.: Norman Rockwell 322 Magazine Covers. P.116

After the war had ended, by the 1920s, the wireless transmission of speech—then called 'radiotelephony'—was complementing the coded information of wireless telegraphy. Many new fields of application, next to the marine applications, were developed, from single-frequency telephony between a base station and mobile units (police, taxis) to two-way radio services. As wireless messages could be listened to by anyone tuning to that frequency, a new application arose: the 'radio' systems (later called 'radio

3

broadcasting') where, in a one-way system from a central point, music and speech were transmitted to be received by the owners of the new wonder of the 'radio' (Figure 4). News now came over the air, next to the traditional printed papers of that time. True, by that time, telephone news services already supplied the spoken 'newspaper' by cabled telephone, and the 'theatrophone' brought opera to the telephone listeners at home, but that was only for the happy few. Now that the wireless 'radio' had been born people young and old, located everywhere, could listen to music and speech broadcasted over the air with just a simple device; a crystal radio, soon available to the masses. For many, it was magic that soon became part of daily life. People stayed at home to listen to radio plays like the *Familie Doorsnee* (in English: *Common Family*) in the 1950s in the Netherlands.

The Online Syndrome

Now look at our present times. People today can hardly comprehend a world without the modern communication engines, from radio and telephone to the Internet. Today the youngest generation is literally growing up with the most modern communication engine ever devised: the wireless 'smartphone'. Gone is the phone cord; the cell phone is connected wirelessly to the network, both for speech and data. Each person has his/her own connection and is constantly 'online'. This advanced device has—along with the telephone function—a range of functions that enhance its use (camera, agenda, timer, calculator, browser, etc). Other application programs—called apps—offer additional functionality (eg weather forecasting, stock market info, mobile payment with Apple Pay of Google Wallet). In less than two centuries mankind went from 'no line' to 'online'.

The overwhelming functionality of this information engine has attracted users of all ages, for example, the very young, for whom the smartphone is

used in a not-so-obvious way. There, the *pacifier* function is used to placate the crying toddler, keeping the youngster quiet for a while as he/she happily swipes and presses the screen at random (Figure 5). Thus, for many a parent, the smartphone is a tool used for getting the kids out of their hair. But in the meantime, by playing, the kid gets

Figure 5: Toddler in pram with smartphone (2016).

Source: http://mumsgrapevine.com.au/

4

acquainted on the fly with modern communication technology.

For a young person, say a teenager, in the second decade of the twenty-first century, the smartphone is a *social tool*. As he is glued to his smartphone, twittering and tweeting, using social media for hours a day, it has become an indispensable tool for social interaction. Being part of a social group is now enabled by technology. And for workers, it has become an indispensable aid as a *business tool*—even when travelling, increasing the chance of causing traffic accidents (Figure 6).

The smartphone is a communication device for all ages, including grownups, who also exchange photos, stories (blogs), and music through 'the cloud', and who use mobile internet facilities offered by 'apps'. Even the older generation has discovered internet communication, sending mail and 'apps' and chatting for hours to maintain their family network. For modern people of all ages—and not only the wealthy and

Figure 6: Social interaction replaced by mobile phone interactions.

Bottom photo taken by author in Modena (Italy), September 2016. Others photos from unknown sources.

privileged or those living in the developed countries—the smartphone is an indispensable tool. One has to be 'online' 24/7. The device is always close by, even during dinnertime, while driving to work, or when having professional meetings. It is not only used for the spoken word, also text messages, pictures and videos are exchanged and/or accessed. Tthe same time it is an agenda, photo- and video camera. And mobile financial transactions (payments, transfers) and stock market transactions are part of its many functionalities. Just to name a few.

That intensive use of the smartphone also has another, less glamorous side. One of them being the smartphone addiction where people are preoccupied by the magic of the device: the dependence syndrome. The smartphone penetrates social life (Figure 6) from early morning to bed time, and have adverse effects on social relationships, face-to-face conversations being replaced by cybernetic ones. And 'cyber-crime' is not something from the SF movies; hacking and fishing are the pests of modern-day communication. Even worse, individual privacy is lost when the device also transmits information on all our activities and whereabouts to (unknown) parties online: from personal data to financial transactions, physical locations, and search behaviour. Big Brother seems to be advancing rapidly.[29]

This short introduction of the progress of modern telecommunication in the early twenty-first century, which one can observe in his environment on a daily basis, originated in the late eighteenth century when optical communication systems (aka the semaphore) were being developed. Then the foundations were laid to communicate over distance with the help of technical means. From there on technology has started to play a dominant role in communication. It was the start of the Age of Communication that was fuelled by (the general purpose technology of) electricity.

The Age of Communication

The Age of Communication facilitated by technical devices—called communication engines—has arrived without people realizing it. It all started in the early nineteenth century when 'electricity' came about with Alessandro Volta discovering the electric battery. From then on it took decades to transform a scientific curiosity into applications usable in daily life: the electric light lengthened the days; people were transported with the electric tram; in hot climates, they were kept cool by the electric fan and cool drinks from the electric refrigerator; in cold climates, the electric heater

---

[29] Big Brother is a fictional character and symbol in George Orwell's novel *Nineteen Eighty-Four*. In the society that Orwell describes, every citizen is under constant surveillance by the authorities.

kept the cold away. That discovery of the 'voltaic' battery was the dawn of the enormous impact of the introduction of electricity in society. But that was only the beginning.

Soon the application of electricity in communication had a massive impact. This became clear in the first half of the nineteenth century, when 'telegraphy' came about and conquered the communication field in a couple of decades, starting the *First Phase* of the Communication Revolution. And it became even clearer in the late-nineteenth century when 'telephony' revolutionized private and professional communications. That was the *Second Phase* of the Communication Revolution. Then came the replacement of the cabled infrastructure when Marconi invented 'wireless' telegraphy. This *Third Phase* of *the Communication Revolution started* in the early twentieth century. In less than a century the world had changed totally. And it would continue to do so as electricity-facilitated communication continued maturing up into the late twentieth and early twenty-first centuries, creating the communication-dominated world we are living in today.

From those early days when electricity was discovered to the electricity dependent societies of today, it took some time before this all came to happen. It took human curiosity leading to scientific discoveries, and a lot of ingenuity and engineering effort, to develop reliable communication systems. But, next to the tinkering and thinking, there was also the entrepreneurial side, where a different breed of people successfully brought the technical fruits to the marketplace, where latent human needs developed into massive markets for telecommunication services in all their heterogeneity.

One could wonder how this all came about. Surely, the many individual contributions—from the scientists to the engineers—making their discoveries and creating the technical artefacts, developing the individual technologies, and creating the technical innovations, were a factor of importance. But there was more, as these people lived in the societies of their time, societies with their own dynamics of war and peace, societies that created the context for their lives.

So, it's worth investigating among that long winding historic road of the Communication Revolution some of the basic societal developments that created the foundations of our present-day communications-dominated societies. And—adding to that—we will investigate the context in which those developments took place. For that, we have to go back in time and delve into history and look at what changed over time. Let's start by looking at the overall introductory picture of the relevant social, political, technical and scientific changes.

Change: Social, Political, Economic, Technical, and Scientific

We used the word 'Communication Revolution' to indicate the massive social changes that resulted from the development of the 'communication engines': the telegraph, the telephone, and the wireless. Their origin—as we will see further on in detail—lies in the early nineteenth century. By then many scholars had been experimenting with the 'electric fish', different species of fish that produced with their electric organs shocks when touched, from the electric eel to the electric ray. They could produce electric shocks up to 220 volts, strong enough to fell a human adult. Then the *general purpose technology* (GPT) of electricity came into existence when the secrets of the 'electric fire' were slowly unravelled by different kinds of experimental scientists (Figure 7, left).

The early experimenting of curious and ingenious people called the electro-physicists resulted in an early form of electricity: among which was static electricity. But that changed around the turn of the nineteenth century, when Alessandro Volta developed the electro-chemical cell. His discovery of the 'voltaic' battery created a scientific frenzy—also known as

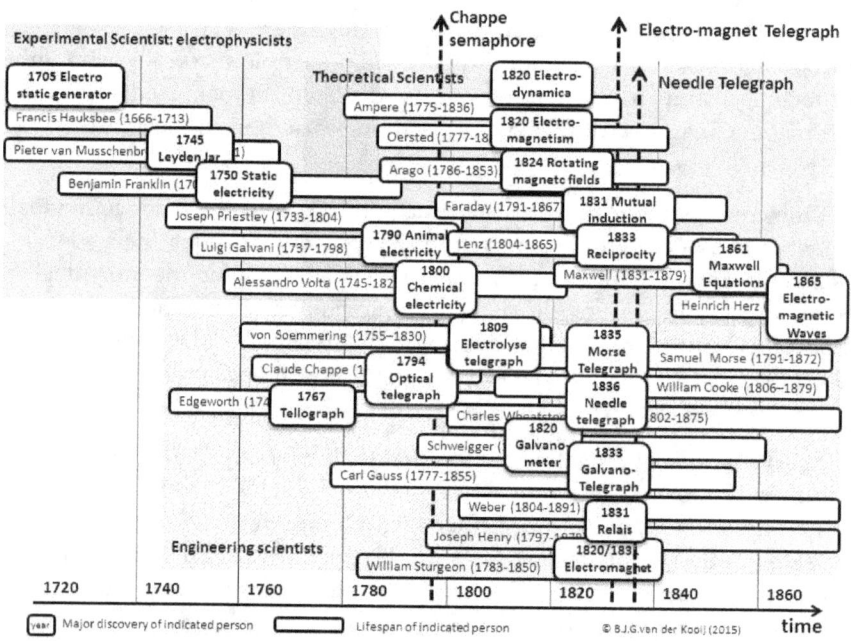

Figure 7: Science discovers electricity and telegraphy.

Source: Figure created by author.

the Battery Mania—that resulted in the first novel applications: the electromagnet, the 'direct current' electric motor, and the early spark lights. It was a trajectory of constant discovery as the properties of electricity that had been explored by the experimental scientists (eg Hauksbee, Volta) were applied by the engineering scientists (eg von Soemmering, Sturgeon) and explained by the theoretical scientists (eg Ampere, Faraday).[30]

Along with the electric motor, their work gave birth to the invention of the distant writing 'communication engine' that was called 'telegraph'. It was the parallel development in Great Britain and in America that resulted in distant writing with lightning speed in the mid-1830s: Cooke & Wheatstone's telegraph and Samuel Morse's telegraph. Soon the further development of the telegraph focussed on two development trajectories: the technical trajectory (with technical improvements such as communication speed, distance, and readably text), and the economic trajectory (lower cost, higher capacity). The later trajectory was dominated by the push to find a solution for the 'economics' of the telegraph system when the need for more capacity arose and the high cost of the long copper

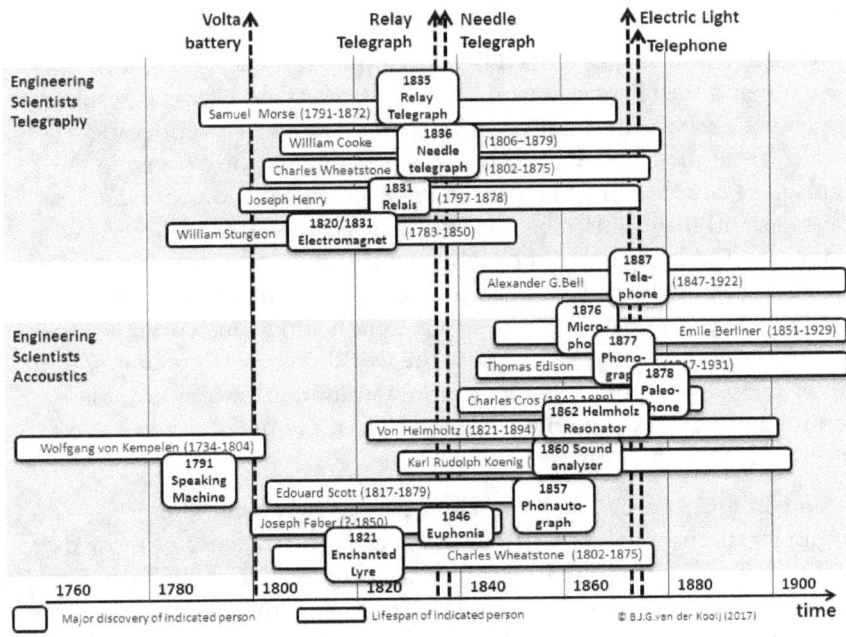

Figure 8: Science discovers telephony.

Source: Figure created by author.

---

[30] See for more details: B.J.G. van der Kooij, *The Invention of the Communication Engine 'Telegraph'* (2015).

telegraph lines was hindering further applications of the new communication medium. This led to the search for the 'harmonic telegraphy', where one telegraph line could transmit more telegraph signals at the same time.[31]

In a totally different field of physics—the field that came to be known as 'acoustics'—the experimenting of many curious engineering scientists into the producing (eg Helmholtz) and the recording of sound (eg Edison), had created artefacts such as the 'Helmholtz resonator' and the 'phonograph' (lit. 'sound writer'). In addition, the work of the more theoretical scientists had resulted in a better insight in the Nature of Sound and had created theories of sound (eg by John Strutt). Then it was the fusion of the built-up knowhow of 'telegraphy' and the built-up knowledge in the field of 'acoustics' (Figure 8) that created something unheard of: 'electric speech'. It was Alexander Graham Bell—the teacher of deaf fascinated by telegraphic communication over distance—who, after much experimenting with the 'harmonic' telegraph, transmitted audible sounds over a considerable distance. The new phenomenon of 'distant speaking' complemented 'distant writing'. Communication over distance had found a new dimension.

This, in a nutshell, described the development trajectoryies of the telegraph and the telephone from the early days of electricity. It resulted in a worldwide *cabled* communication infrastructure that—over decades— evolved from the early point-to-point telegraph lines, first covering the highly populated regions, then expanding to regional and continental coverage, and finally attaining transatlantic transmission, connecting the continents. Cables spanned the globe, transmitting messages that crossed the continents. Cables offered the solution for distant communication, but they were also the problem, both in a technical and in an economic sense. By the end of the nineteenth century the world was eagerly waiting for an alternative solution for the cabled communication infrastructure. The Communication Revolution might have passed the first two phases; now it was time for the third phase.

One has to realize that all those efforts that resulted in the Communication Revolution had taken place in the societies of those days: both the Old World of the European continent as well as the New World of North America. In the European societies the remnants of earlier times still existed, the former monarchical societies—the kingdoms with their feudal origins—that were followed by the eighteenth and nineteenth century empires (the British Empire, the French Empire, etc.) with their

---

[31] See for more details: B.J.G. van der Kooij, *The Invention of the Communication Engine 'Telephone'* (2015).

mercantilism and protectionism,[32] and later imperialism and colonialism. These societies had seen mass disruptions caused by social revolutions, from the American Revolution and French Revolution to the European Revolutions of 1848.

These societies all experienced conflicts with the 'madness of the times': war, revolt, and turmoil at sea and on land; conflicts arising locally in regional wars—often related to the succession of the ruling elite (eg the many 'Succession Wars')—sometimes even on a broader scale, such as wars between nations fighting for territories. And the sea was there, as always, for the picking by the strongest, from the Vikings raiding the Northern European coasts to Barbary pirates operating on the Mediterranean Sea, the Dutch raiding the Spanish fleet, and up to the empires colonizing the East.

It is these societies, which created the context for the discoveries, inventions, and innovations to come, that we are going to investigate as we try to unravel a complex and continuously changing, dynamic context with many interrelations. Realizing that we can only hope to decipher some of those relationships into broader patterns. We will look at different types of changes: scientific change and technical change leading up to the Scientific Revolution, and social change, political change and economic change leading to the Societal Revolution.

So, let's start and go back in time, to the times when the ancient Greek philosophers like Socrates, Plato and Aristotle were contemplating about the world they were living in.

---

[32] Mercantilist policies were aimed at creating overseas colonies, excluding them from trading with other nations, monopolizing markets with staple ports, and forbidding trade to be carried in foreign ships. Thus, governments protected their merchants—and kept others out—through trade barriers, regulations, and subsidies to domestic industries in order to maximize exports from and minimize imports to the realm. This created a massive redistribution of wealth from the colonies to the motherland. The goal of mercantilism was to run trade surpluses so that gold and silver would pour into London.

Scientific Change and Technical Change

For centuries scholarly thinking was dominated by the Aristotelian paradigm of looking at the artificial: the 'world of things'.[33] Philosophers looked at the world's change and growth,[34] and their thinking was dominated by Aristotelian physics and mechanics. Later in time the Roman Claudius Ptolemy would create a model of the heavenly bodies that could explain their motions as they circled in the skies around the Earth: the geocentric model of the universe.[35] Already, in earlier days, the thinkers of the priest class thought about the issues of 'time and distance'. They recognized the pattern of the moon faces, noticed the seasons, followed the star movements in the sky, and they created the lunar calendars with the lunar month and week. They divided the daytime into twelve hours with a sundial, and the twelve night hours with a water clock. Their concept of time lasted until our present times in the form of the 24-hour day.

They even created the Antikythera Mechanism; an ancient celestial calculator from ca 200 B.C. that mechanically displayed the date, positions of the sun and moon, lunar phases, a 19-year calendar and a 223-month eclipse prediction dial. It was believed to be an ancient astronomy calculator that shows the four-year cycle of the early Greek competitions that inspired today's Olympic Games. Inscriptions on the device list names

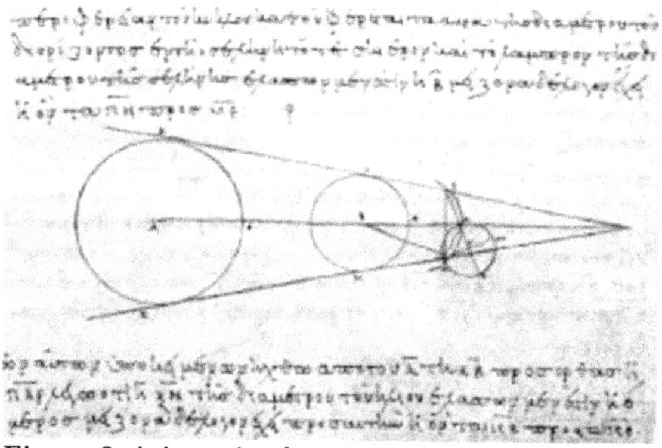

Figure 9: Aristarchus's calculations.

Figure shows calculations on the relative sizes of, from left, the Sun, Earth and Moon, from a tenth-century Greek copy.

Source: Wikipedia Commons.

---

[33] Aristotle argued that there are four earthly elements: earth, air, water, fire (combinations of the 'contrarieties' hot, cold, wet, and dry), in addition to "aether" of the heavens. The fundamental principles of Aristotelian physics were matter, form, and privation.

[34] Aristotelian change is the result of four causes. (1) the material cause, which explains the physical component of the entity; (2) the formal cause, which explains the form or shape to which a thing corresponds; (3) the efficient cause, which is what we generally mean by 'cause', the original source of the energy that allows for the change; and (4) a final cause, which is the purpose it fulfils.

[35] As described in his work *Almagest*; a mathematical and astronomical treatise of the apparent motions of the stars and planets.

linked to the Olympiad cycle of games.[36]

Those ancient Greek thinkers like Socrates, Plato, and Aristotle, looking at the world around them and at the skies above them, created their rational views of the natural world. Next to the early perception of 'time', there also was this other subject of their interest, and that was 'distance'. They deliberated how far away the sun and moon were in the sky. Their thoughts[37] created the foundations of mathematical astronomy (Figure 9).

Jumping ahead in time we see that the dominance of the Ancient Greek worldview changed over time, other views emerged, and by the end of the sixteenth century natural philosophy was no longer purely identified with the Aristotelian system. As the result of revolutionary views of people like *Nicolaus Copernicus* (1473–1543), *Galilei Galileo* (1564–1642), and *Isaac Newton* (1642–1727), in the fifteenth through seventeenth centuries another way of looking at the world around us developed. Later this was to be called the period of the Scientific Revolution, a period of fundamental importance that would start with the 'Copernican Revolution'.

It was Nicolaus Copernicus (1473–1543), a third-order Dominican, who created a totally different view of the universe. Educated in both Krakow (Poland) and Bologna (Italy), he studied—among many other fields like medicine—astrology. In 1514 he wrote about his ideas in a first outline called *Commentariolus* (Little Commentary). Ultimately, his astronomical work and intense observational activity on the movements of the planets and the sun resulted in his particular view of the universe: the heliocentric view with the sun at its centre. This 'Copernican Revolution' was the paradigm shift from the Ptolemaic model of the heavens, which described the cosmos as having Earth stationary at the centre of the universe, to the Heliocentric model with the sun at the centre of the solar system.

Being a cleric himself, Copernicus was well aware that this view was conflicting with Christian dogmas (among which was the dogma that the Earth was created in seven days[38]), so he delayed the publication of his views in *De Revolutionibus Orbium Coelestium* (On the Revolutions of the Heavenly Spheres) till just before he died in 1543. His work on the

---

[36] Source: Bradford, A.: 'Antikythera Mechanism: Ancient Celestial Calculator'. *LiveScience* May 18, 2016. http://www.livescience.com/54782-antikythera-mechanism.html. Antikythera Mechanism Research Project, source: http://www.antikythera-mechanism.gr/

[37] As described in the work of Aristarchus of Samos: 'On the Sizes and Distances (of the Sun and Moon)'.

[38] The Story of Creation (Book of Genesis 1:1; 3:24): "In the beginning, God created the universe." "God fashioned two great lights—the larger light to shine during the day and the smaller light to shine during the night—as well as stars. God placed them in space to shine on the earth, to differentiate between day and night, and to distinguish light from darkness. And God saw how good it was."

Figure 10: *The Alchemist, in Search of the Philosopher's Stone.*

Source: Wikimedia Commons. Artist: Joseph Wright of Derby, http://www.derbymuseums. org/joseph-wright-gallery/

mechanism of the universe is considered to be the start of the Scientific Revolution.

The early phase of that Scientific Revolution—also called the *Scientific Renaissance*—can be seen as a transition phase. This was not only due to the new views about the universe; it also marks the transition of the ancient alchemist (Figure 10)—still focussing on antiquity's belief in the four elements (and, often in secrecy, trying to create gold from lead)—into the 'scientific' investigator and 'engineering' experimenter. From the old 'metallurgists', with their techniques of trial and error mixed with theological beliefs, would develop the systematic experimenters who would analyse their empirical observations free of religious dogmas and prejudices. It was the transition into the time of experimental investigation, trying to create insight in the 'matter of things'.

Many of those scholars—from early philosopher to alchemists searching for the philosopher's stone[39]—had been contracted by the monarchs and aristocratic elite for practical purposes related to mining, medical, and military services, and the production of chemicals, medicines, metals, and gemstones. Their work became the origin of the 'mechanical philosophy' (aka 'corpuscular philosophy'), the belief that all natural phenomena can be explained in terms of mechanical laws. These explanations would become the 'laws of nature', where mathematically formulated regularities govern all bodies.

---

[39] Philosopher's stone: A legendary alchemical substance capable of turning base metals such as mercury into gold or silver. It is also able to extend one's life and is called the elixir of life, useful for rejuvenation and for achieving immortality; for many centuries, it was the most sought-after goal in alchemy.

Thus, over time the views of the natural philosophers replaced the Aristotelian notions of form, matter, and privation. The mechanical philosophers with their early experiments in the Nature of Motion created the foundations. Their experiments resulted in a range of theories, models, and laws on the Nature of Matter. They explained a lot, but still there were underlying questions about time and distance, as motion means covering a certain distance in a certain time. Let's zoom in along our perspective and have a more detailed look at some of these developments.

Nature of Motion: Mechanisms

The precursor of many of the explorations into the Nature of Matter was wonderment at the phenomenon of 'motion'. From the movement of the physical body to the movement of (celestial) objects. As illustrated by the life and works of *Leonardo da Vinci* (1452–1519) and *Galileo Galilei* (1564–1642), early experimenters in the mechanical arts of that time who lived and experimenting in Florence and Pisa. These early Italian philosophers were curious about the Nature of Motion. To satisfy that curiosity, they created their mechanical constructions and mechanical experiments.

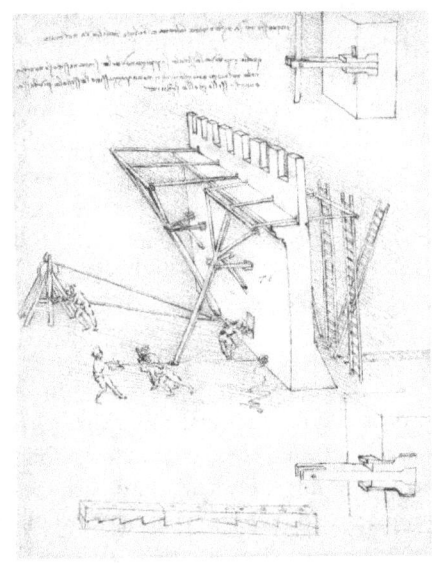

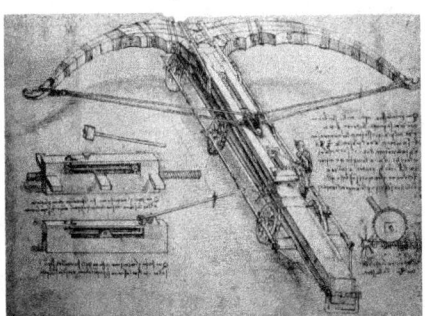

Figure 11: Leonardo da Vinci's siege defence (top), giant crossbow and air screw (bottom) and giant crossbow.

Source: Cianchi, M.: *Leonardo da Vinci's Machines.*

15

Leonardo da Vinci, for example, was fascinated by machines (Figure 11). Obviously, in a time of frequent warfare, these were 'machines of war', like his ideas for the rapid-fire crossbow (aka mitrallieur), the catapult, the giant crossbow, canon towers, and machine guns. In addition, he designed hydraulic machines, excavating machines, siege machines, siege defence systems and 'machines for flight' (such as his airscrew). Many of the designs were just ideas, maybe to impress his ruling clients, as a suitable technology to realize them at that time would not have been available. Others were quite practical though, as they were used in the local wars where the fortified cities were besieged. That resulted in his siege machines and his siege defences (Figure 11, top) constructed in wood.

Leonardo was also interested in the mechanism of the human body. He studied the mechanical functions of the skeleton and the muscular forces that are applied to it in a manner that prefigured the modern science of biomechanics. He drew the heart and vascular system, the sex organs, and other internal organs, making one of the first scientific drawings of a foetus in utero. And he made extensive studies of the skeleton and its musculature (Figure 12).

To earn his bread, Leonardo offered his services as military engineer to Ludovico Sforza, duke of Milan. For the duke, he worked as military engineer, travelled, and painted (1482–1499). Next, after the French king Louis XII captured Milan in 1500, he went to Venice, where he was employed as a military architect and engineer, devising methods to defend

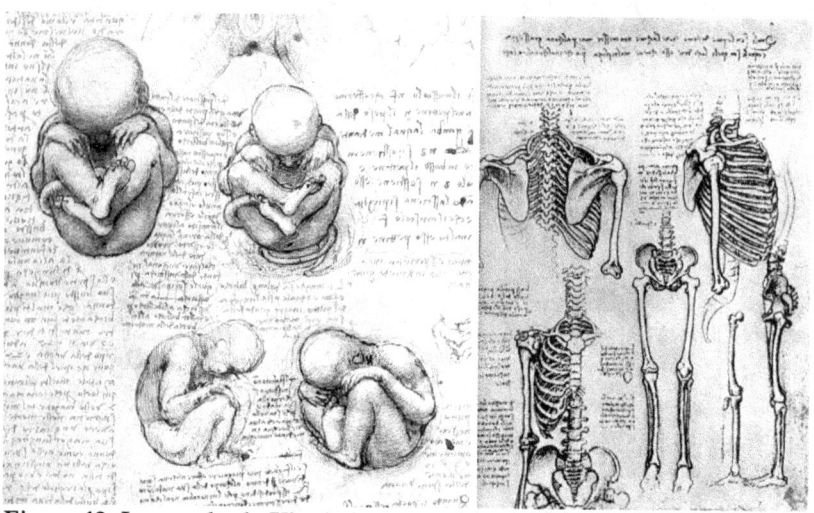

Figure 12: Leonardo da Vinci (fetus in utero, 1501, left) and studies of the skeleton (right).

Source: http://www.drawingsofleonardo.org

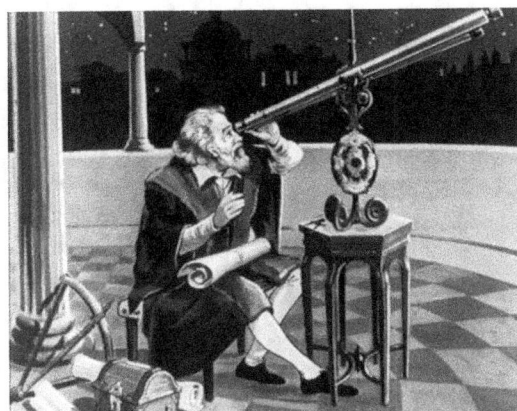

Figure 13: Galileo and his telescope (1609).

Source: Wikimedia Commons

the city from naval attack. Later in life he worked for the pope in Rome and the king of France, Francis I. There, he died in 1519.

Also, nearly a century later, the Italian *Galilei Galileo*, observing the world at hand, was interested in motion, distance, and time. He began his own experimental studies of motion while serving as a young mathematics professor at Pisa. He continued his experiments during nearly two decades of teaching at the University of Padua, near Venice. There, he measured the swinging of pendulums until he could describe their periods by a mathematical law (1602), he rolled bronze balls down inclined planes a thousand ways to derive the rate of acceleration in free fall (1604), and he explored the parabolic path of projectiles (1607/1608). Again these were all experiments executed in times of warfare and related to warfare.

> *Galileo studied the nature and laws of motion throughout his scientific career, eventually formulating in the Discourses and Mathematical Demonstrations (Leyden, 1638) an entirely new dynamics, based on mathematical reasoning and innovative experiments. In this ground-breaking work, the Aristotelian concept of motion is replaced by a vision founded on new principles: the acceleration of natural motion and its proportionality to time from rest, the parabolic trajectory of projectiles, the infinite force of impact. For Galileo, the radical reform of Aristotelian dynamics opened the way to the definitive affirmation of the Copernican system...*[40]

An important side effect of the experiments was the development of tools to execute those investigations: the instruments like Galileo's telescope, designed in 1609 (Figure 13).[41] With this telescope, Galileo

---

[40] Text originating from the exhibition in the Florence Galileo Museum (visited October 2015). I am indebted to Professor Emeritus Thomas B. Settle, who was more than willing to share his insights with me and to educate me in the history of science, especially Galileo's role in the Tuscan context of that time. (See also: http://www.imss.fi.it/~tsettle/index.html.
[41] The discovery of the telescope/microscope is attributed to Dutch spectacle makers. The original Dutch telescopes were composed of a convex and a concave lens—telescopes that are constructed this way do not invert the image. Lippershey's original design had only 3x

Figure 14: Galileo displaying his telescope to the doge of Venice (1609).

Source: Wikimedia Commons. Artist: H.J. Detouche

studied the cosmos, discovering the satellites of Jupiter, the spots on the sun, the phases of Venus, and the craters on the moon. In other words, the cosmos was opened up because the new instrument of the telescope made it possible to bridge distance visually. It made him not only famous (Figure 14), but it was also his study of the cosmos that brought him in conflict with the dogmas of the Roman Catholic Church.

> ... *The affirmation of the experimental method in the 17ᵗʰ century and the development of new instruments stimulated significant progress in the investigation of natural processes, helping to discover the laws that governed them and to unveil invisible phenomena. The barometer was used to reveal the effects of atmospheric pressure and to measure variations in it caused by changes in the weather. The graduated thermometer was used to measure temperature objectively and even more precisely. The microscope and the telescope enormously enhanced the powers of eyesight, revealing hitherto unknown phenomena of the micro cosmos and the macro cosmos. Lastly, combinations of lenses, prisms and mirror led to the progress in the science of optics.* (ibid)

His work was picked up by others. One of Galileo's students was *Benedetto Castelli* (1578–1643), a Benedictine monk. In 1612 Castelli helped see Galileo's *Discorso intorno alle Cose che Stanno in su l'Acqua* (Discourse on Floating Bodies)—with reports of his telescopic observations—through the press and published the reply to the polemics against it. Castelli was also

---

magnification. Telescopes seem to have been made in the Netherlands in considerable numbers soon after the date of their invention, and they rapidly found their way all over Europe. Galileo improved on the concept by creating telescopes that magnified up to 33x.

active in the initial stages of Galileo's sunspot research in 1612, coming up with the method of projecting the sun's image through the telescope. Later, as an advisor to the pope, he wrote the book on hydraulics *Della Misura dell'Acque Correnti* (Mensuration of Running Water)[42].

Nature of Motion: Celestial Mechanics

With aid of these new instruments, an important focus of the natural philosophers was the universe (aka cosmos); both in its material as well as its spiritual form. As a result of their explorations, they created their views and theories on the Nature of Motion. Take the example of the 'clockwork universe theory', where the universe was seen as a mechanical clock wound up by a divine power and ticking along, as a perfect machine, with its gears governed by the laws of physics. This was a theory that—in some way or the other—was connected to religion, as illustrated by Voltaire declaring: "I cannot imagine how the clockwork of the universe can exist without a clockmaker."[43]

Originating from the medieval texts on astronomy—such as Johan de Sacrobosco's *On the Sphere of the World* (1230), which spoke of the universe as the 'machina mundi'—other mechanistic views of the cosmos were developed. One example is the previously mentioned view of Nicolas Copernicus (1473–1543)—in his *De Revolutionibus Orbium Coelestium* (On the Revolutions of the Celestial Spheres, 1543), a heliocentric view of the universe (Figure 15). Copernicus reasoned that all of the motions that he observed in the sky could be accounted for with a sun-centered system and without the necessity of complicated epicycle motions.[44]

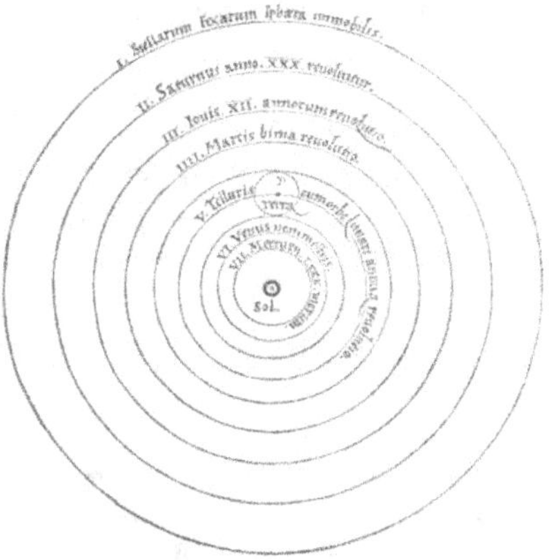

Figure 15: Copernicus's heliocentric model of the universe.

Source: Wikimedia Commons

---

[42] At that time there were no standardized units for volume, time, or speed. Galileo, for example, used to measure time in heartbeats when he experimented with the pendulum.
[43] Quoted in *The Book of Nature and the Bible* (1988) by the Watchtower Bible and Tract Society of New York Inc.
[44] The epicyclic motion was a geometric model in the Ptolemaic system of astronomy used to explain the variation in speed and direction of the Sun, Moon and the five planets known at that time.

Next, Renee Descartes (1596–1650)—in his *Principia Philosophiae* (1644)—gave a mechanistic account of the universe and described the laws of physics. Also, Isaac Newton (1643–1727)—in his *Philosophiae Naturalis Principia Mathematica* (Mathematical Principles of Natural Philosophy, 1687)—applied the mechanistic perspective with his 'laws of motion'. Together with other contributions (such as Leibnitz's *Discourse of Metaphysics* of 1686, concerning physical substance, motion, and resistance of bodies), they created the idea that the universe is governed by precise mathematical laws. This would result in the views on 'celestial mechanics' of the mathematical philosophers.

Remarkably, many of the advocates of the mechanical philosophy and the clockwork metaphor had an ecclesiastical background, such as the contributions from the French Catholic clerics: the friar Marin Mersenne (1588–1648), the priest Pierre Gassendi (1592–1655), and the Jesuit-educated René Descartes (1596–1650).

> *Mersenne contends for a clockwork analogy of the universe convinced that the mechanical philosophy can serve as a defense of theistic belief. Gassendi, famous for reviving and Christianizing Epicurean atomism, compares the wisdom evident in creation with the intentionally designed clock. As for Descartes, while not advocating a cogged machine per se, he does describe the world as a machine and is fond of describing animals as clocks and humans as clocks with souls.*
> (Snobelen, 2012, p. 152)

By the 1660s to the 1680s, mechanistic approaches on the Nature of Motion were found virtually throughout Europe and dominated intellectual discourse held at the contemporary centres of learning. By and large, the natural philosophy and mechanical philosophy also flourished outside the universities, first in salons and private academies, later in the learned societies such as the English *Royal Society of London* (1661) and the *French Académie Royale* (1665).

In Italy, the natural philosophy had developed in a similar way, also creating a host of academies. Among them were the early philosophical and cultural academies of Florence patronized by the de Medici family, such as the *Accademia Fiorentina* (1541) and, later in time, the private *Accademia dei Lincei* (1603) in Rome created by Federico Cesi, an aristocrat from Umbria, with the objective of exploring natural sciences. Galileo was inducted into the exclusive academy on April 25, 1611, and became its intellectual centre.

> [In Italy] *Outside of universities and schools, there were also other places where natural philosophy was cultivated, particularly in academies and learned societies. Even in literary societies such as the Accademia Fiorentina (1541), patronized by Duke Cosimo de' Medici and his descendants, scientific matters (e.g., alchemy*

or spontaneous generation) were occasionally debated, often in the context of commentaries on the poems of Dante and Petrarch. The Accademia dei Lincei, founded in 1603, on the other hand, was exclusively interested in the sciences: as their statute dictated, the Lincei had no interest in any controversy which was not scientific or mathematical, and they avoided involvement in political matters.[45]

Nature of Void: Vacuum Mechanics

It was the work of Galileo's student Evangelista Torricelli (1608–1647) that became one of the cornerstones of mechanical philosophy: the study of the Nature of Void. It was part of the *atomistic view* of the word, where the atomist philosophers theorized that nature consists of two fundamental principles: *atom* and *void*. In that view, the philosophical atoms came in an infinite variety of shapes and sizes, each indestructible, immutable, and surrounded by a void where they collided with the others or hooked together to form a cluster. Galileo Galilei was an advocate of atomism in his *Discourse on Floating Bodies* (1612).

One of the biggest and most controversial debates in those days was about the existence of vacuums. In the 1630s to the 1640s that empty space of the void was experimentally explored—later called barometric experiments—by many natural philosophers: French philosophers like

Blaise Pascal (1623–1662), English philosophers like *Robert Boyle* (1627–1691), and in Germany by philosophers like *Otto von Guericke* (1602–1686). The latter published *Experimenta Nova Magdeburgica de Vacuo Spatio* (1672), in which he explored the nature of space and the possibility of void: his experiments with the Magdeburg Hemispheres.

In Italy, the Jesuit-educated Torricelli, acting as Galileo's amanuensis during the end of his life, experimented with columns of water and mercury (Figure 16). In 1643 he created a tube approximately one meter

Figure 16: Torricelli and his mercury barometer (1643).

Source: www.chemistryworld.com

---

[45] Eva Del Soldato, *Natural Philosophy in the Renaissance* (2016).
Source: http://plato.stanford.edu/entries/natphil-ren/#DefRenNat

long, filled it with mercury, sealed it at the top, and set it vertically into a basin of mercury. The column of mercury fell to about 76 cm, leaving a Torricellian vacuum above. He had created the first mercury barometer.

Nature of Heat: Chemical Mechanics

The natural philosophers also were interested in exploring the Nature of Heat, trying to answer the question: why does fire burn? Their explorations resulted in the early theories of heat. Among the many theories that arose was the *phlogiston theory*, which tried to explain burning processes. Fire was thought of as a substance—the phlogiston[46]—and burning was seen as a process of decomposition which applied only to compounds. The theory was developed by the German thinkers Johann Becher (1635–1682), publishing his ideas in *Physica Subterranea* (1667), and his student Georg Ernst Stahl (1659–1734) who published the theory in 1723 in *Fundamenta Chymiae*.

The phlogiston theory remained the dominant theory until the 1780s, when Antoine-Laurent Lavoisier—the father of modern chemistry—

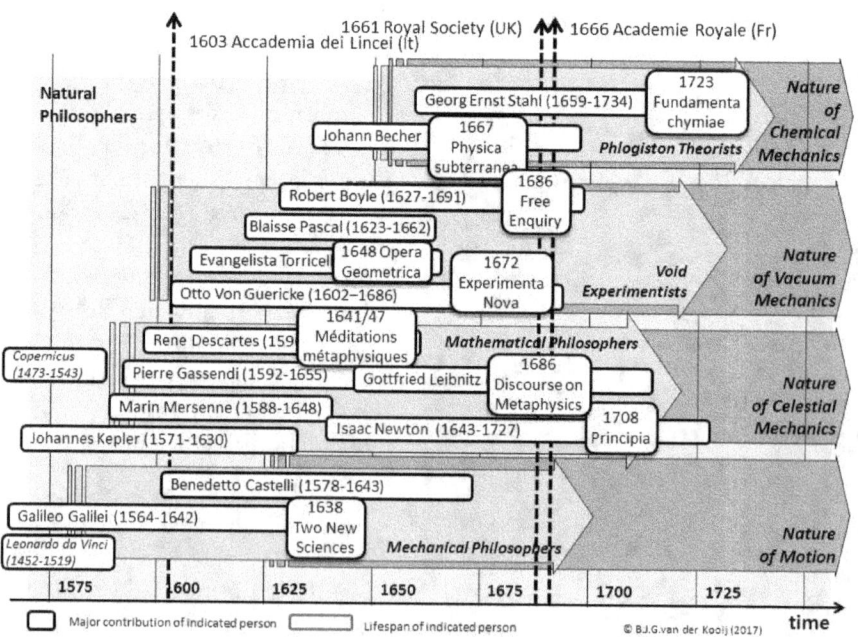

Figure 17: The Scientific Revolution: Overview of some of the natural philosophers in the 16–17th centuries.

Figure created by author.

---

[46] When metallic iron becomes red rust, it loses its phlogiston.

showed that combustion requires a gas that has mass (oxygen) and could be measured by means of weighing closed vessels. His view was part of the (first) Chemical Revolution (1770–1790).

Another theory was the *caloric theory*, which arose from the phlogiston theory but was contradicted by Lavoisier's insights. The caloric theory was based on the idea that a hypothetical, weightless caloric fluid flows from hot to cold bodies, creating latent heat that can be absorbed or released by a substance. Like the phlogiston theory, the caloric theory did not survive in later times. Thus, leaving the medieval 'science' of alchemy, chemistry gradually separated from physics—understood as the 'science of bodies subject to movement'—and positioned itself as the 'science of bodies associated and dissociated'. This was the beginning of the quest into the nature of chemical mechanics.

The preceding overview of the development of the work of natural philosophers shows—painted in rough brushstrokes—a picture of the early thinking in terms of 'mechanics': the mechanisms behind the phenomena the philosophers observed (Figure 17). Their work created the foundations for more explicit 'scientific' thinking in the seventeenth/early eighteenth century. Then the Aristotelean perspective was replaced by the scientific perspective, and the ancient methods of thinking were replaced by the scientific method, which became the foundation of the *arts of mechanics*; the behaviour of physical bodies when subjected to forces or displacements.

Natural Philosophers in Conflict with Religion

It is clear that many thinkers and experimenters of the seventeenth century studying the Nature of Motion wished to understand the secrets of nature. Previously, that knowledge had, in fact, been central to magic and alchemy, a field which soon had an ambiguous relationship to natural philosophy when the natural philosophers tried to understand nature. Sometimes natural philosophers were (financially) supported by the rulers of their time, as the fruits of their works could be advantageous to their worldly affairs (eg warfare). As we have seen, around 1500, Leonardo da Vinci was employed as military engineer by Ludovico Sforza, the ruler of Milan, for whom he designed his military machines. And in the late 1600s, the polymath Gottfried Leibnitz was under the patronage of the dukes of Brunswick-Lüneburg from the House of Hannover.

Not too surprisingly, the explanations given by natural philosophers did not always concur with the contemporary religious beliefs. In addition, the mechanical philosophy may have explained the mechanical world, but some scholars searched for the unexplainable, looking for the 'spirit of nature'. What was causing those bodies to move? What was causing the natural phenomena like lightning and thunder? And there, they touched on the spiritual domain that was claimed by religion, such as the heathen religions.

Henceforth, as a result of their mechanistic views of the world, the natural philosophers often became, just as Galileo had experienced with his heliocentric ideas, in conflict with religion, in particular with the Christian dogmas.[47] Also, the new discoveries of the Scientific Revolution threatened the supremacy of Church doctrine considerably. Most devastating was the philosophical approach, which led to conclusions that God either did not exist or at least did not play as much of a role in daily life as the clerical class of the Roman Catholic Church claimed.[48]

Thus, over time the views of the philosophers collided with the doctrines of the Roman Catholic Church. And although the Church tried to check and curb their views (eg by means of the horrendous Inquisition and the Index of Forbidden Books), nevertheless, their work progressed over time independent from religious beliefs:

> *Galileo denied the reality of the physical elements of the Aristotelian world and the theory of their natural movements, and replaced them with corporeal matter, whose properties and motions could be described in mathematical terms. Furthermore, by relying on new instruments such as the telescope, Galileo was also able to make new observations which revealed the imperfections of the supralunar world. Galileo and the Copernican theory met with the resistance of the Church, but also of the universities, whose professoriate was not eager to renounce one of the central pillars of its teaching.*

Although many of the emerging doctrines of natural philosophy contrasted with the doctrines and teachings of the Church, in the reality of those days, scholars found a number of possible solutions to this problem. Some scholars appealed to the radical difference between the realms of faith and philosophy. Other scholars tried instead to genuinely

---

[47] Such as the dogma that the Earth is the centre of the universe, and the dogma of the creation of the world (eg the seven-days concept of the creation of the world).

[48] The power of the Roman Catholic Church, exercised by the local priest, was encapsulated in the Sacraments. It had been indoctrinated into the people that they could only get to heaven via the Church. The Church covered life from birth (Sacrament of Baptism) to death (Sacraments of the Death). Confirmation, Communion, and Eucharist brought the infant into the Church, marriages were subjected to the Church (Sacrament of Matrimony), people had to confess their sins (Penance and Reconciliation), and when they were sick, they were anointed with oil (Anointing of the Sick). When they wanted to divorce, they needed the Declaration of Nullity. And when they did not conform its rules, they were excommunicated by the Church and would not go to Heaven after death. This relationship between people and Church was essentially based on money and labour: tithes (a tenth of the annual income), work on church land for free (for a specified number of days per week), and indulgences (money paid to be pardoned for sins).

reconcile philosophy and faith, particularly during the periods of doctrinal conflict and religious warfare that followed the Protestant Reformation. But over time science and religion would split as the Holy Roman Church lost its grip on society.[49] It is not too surprising then that especially Britain (ie both England and Scotland), after having severed ties with the Roman Catholic Church in the mid-sixteenth century, would become the nucleus for the Enlightenment movement.

As a result of all that collective curiosity in Europe after the sixteenth century, the more dramatic effects of this specific change in scientific thinking—later called the paradigm shift of the Scientific Revolution (c. 1550s–1800s)—were noticeable. The rapid accumulation of knowledge, free from religious overtones, saw science start to split into separate disciplines as the age of the great polymaths ended.[50] Developments in astronomy, biology/anatomy, and chemistry transformed the scholarly views of society and nature. Other philosophers—who we will meet later on—rebelled against the restrictions of Christianity and used science and metaphysics to question and probe the universe.

Scientific Revolution as a Catalyst

As explained before, the departure from the ancient Aristotelian physics would become known as the Scientific Revolution.[51] The new philosophy of using an *inductive* approach to nature—to abandon assumption and to attempt to simply observe with an open mind—was in strict contrast with the earlier Aristotelian approach of *deduction*, by which analysis of known facts produced further understanding. Deduction was then used to explain the nature and complexity of all matter in terms of simpler substances, using the classical elements of earth, water, air, and fire. Now the Scientific Revolution resulted in a new view of nature in which science over time became an autonomous discipline, distinct from philosophy and

---

[49] In the twentieth century, Pope John Paul II regretted the treatment which Galileo received, recognizing the value of science, in a speech to the Pontifical Academy of Sciences in 1992, observing '… There exist two realms of knowledge, one which has its source in Revelation and one which reason can discover by its own power. To the latter belong especially the experimental sciences and philosophy. The distinction between the two realms of knowledge ought not to be understood as opposition. …' (Cowell, A. 'After 350 Years, Vatican Says Galileo Was Right: It Moves'. New York Times Oct. 1, 1982. Source: http://www.nytimes.com/1992/10/31/world/after-350-years-vatican-says-galileo-was-right-it-moves.html).

[50] A polymath is a person whose expertise spans a significant number of different subject areas; such a person is known to draw on complex bodies of knowledge to solve specific problems. The term is often used to describe great thinkers of the Renaissance and the Enlightenment who excelled at several fields in science and the arts.

[51] Aristotelian physics relates to the general principles of change as developed by the philosopher Aristotle.

technology. It also gave rise to something different: the art of creation in which the subordinate imitation of the Middle Ages was complemented by the striving for change and improvement. It would become the time of 'imitatio and emulatio': from the time of learning by imitation (the 'imitatio') now came the times with favourable conditions for improvement (the 'emulatio').

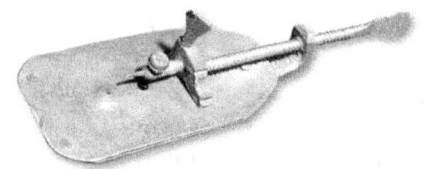

Figure 18: Microsope by Antonie van Leeuwenhoek (c 1668).

Source: Wikimedia Commons

> *Among the most conspicuous of the revolutions which opinions on this subject* [the progress of physical science] *have undergone, is the transition from an implicit trust in the internal powers of man's mind to a professed dependence upon external observation; and from an unbounded reverence for the wisdom of the past, to a fervid expectation of change and improvement.* (Whewell, 1858, p. 318)

Natural philosophers had started studying the physical universe by looking at the Nature of Matter (ie the different natures of heat, light, and lightning). It was the study of nature on a grand scale: the macro-cosmos of astronomy and cosmology as observed through a *telescope*. But it also was the study of nature on a small scale: the micro-cosmos as observed through a *microscope*.

The microscope was the result of the experimenting of lens grinders, people who worked with a piece of glass to create a lens. Among the many artisans who created lenses all over Europe, it was the Dutch *Antoni von Leeuwenhoek* (1632–1723), cloth tradesman by origin and self-taught man in science, who started experimenting with making simple microscopes (Figure 18). Van Leeuwenhoek's vindication resulted in his appointment as a Fellow of the Royal Society of London. After his appointment to the society, he wrote approximately 560 letters to the society and other scientific institutions over a period of 50 years, detailing the subjects he had investigated.

The erudite efforts of all those scholars involved in the mechanistic natural philosophy—like those English 'gentlemen of science' who met, discussed, and studied the physical world in the Royal Society—had created quite different views of world, for example, in the fields of what came to be known as physics and chemistry. The period resulted in a fundamental transformation in scientific ideas across mathematics, physics, astronomy,

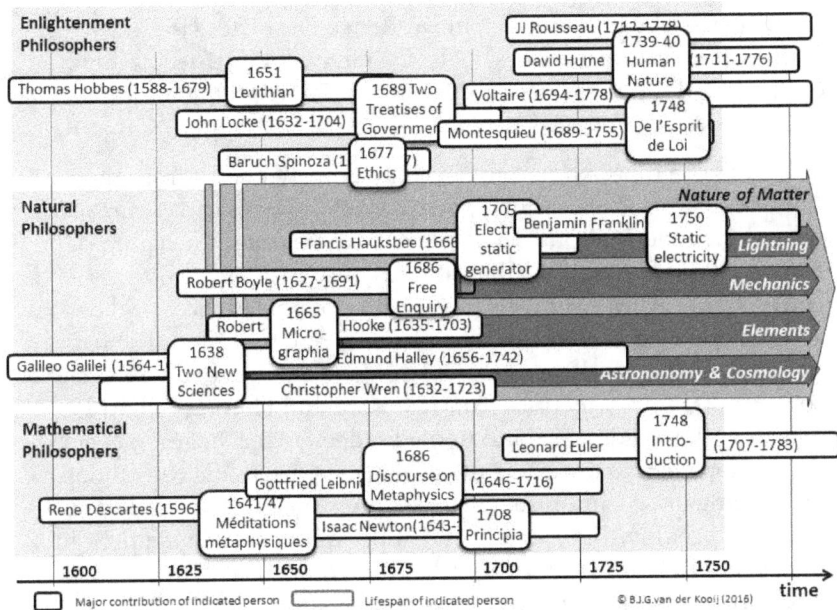

Figure 19: Overview of some of the natural philosopher in relation to the mathematical and Enlightenment philosophers in the 17–18th centuries.

Figure created by author.

and biology in institutions supporting scientific investigation and in the more widely held picture of the universe (Figure 19, middle). Also many Italian scientists—living and working in cities like Milan, Rome, Venice, Bologna and Florence—contributed to the development of medicine, physics and natural sciences. Among them the contributions of the physician and biologist *Luigi Galvani* (1737–1798), and the physicist and chemist *Alessandro Volta* (1745–1827). The polemic on animal electricity versus chemical electricity carried on between Alessandro Volta and Luigi Galvani—as we will later on in more detail—became part of a discussion which impassioned *l'Europe savante* during the eighteenth century and even on into the nineteenth century. (Moravia & Breidenbach, 1969)

Without question, the Scientific Revolution was a fundamental phenomenon that influenced mankind dramatically. It gave rise to understanding the natural world and created insight in its functioning. That understanding of the Nature of Matter was largely supported by the mathematization—use of the abstract language of mathematics—of the theories that resulted from mechanical views. This work was done by a

27

different group of philosophers: the *mathematical philosophers*—by that time called scientists—such as the Frenchman Renee Descartes (1596–1650), the Englishman Isaac Newton 1643–1727), the German Gottfried Leibnitz (1646–1716), and the Swiss Leonard Euler (1707–1783) (Figure 19, bottom).

But there were more important developments that came from the Scientific Revolution. Soon thinkers started reflecting on the *Nature of Society* and created their views about the nature of existence and the nature of government. They became the philosophers of the *Enlightenment* movement,[52] an intellectual movement which dominated the world of ideas in Europe in the eighteenth century—with contributions from Frenchmen like Condorcet, Voltaire, Rousseau, Diderot, and Montesquieu; Englishmen like Locke and Hume[53]; and Germans like Kant and Herder (Figure 19, top). Their work would contribute to the fundamental changes in society: how it functioned and how it was ruled. Thus, the Scientific Revolution became accompanied by the Enlightenment movement, which soon questioned the century-old foundations of the rulers: the 'divine right of kings' in its feudal context. It would take some time, but ultimately, they were replaced by the 'right of people' as democratic societies developed.

Social Change, Political Change, and Economic Change

Looking back in time one could observe that for the average person, the world was quite small some five hundred years ago (ie the 1500s). One mostly lived his/her whole life as member of a family, group, or clan in a small community, where existence was based on agricultural production or, later in time, on one's skills as an artisan. Climate and season dictated the pace of life and sometimes disrupted life with famines and pandemic diseases like the Black Death. One was governed by the remnants of the feudal system, with its 'privilege and inequality', where the few ruled the many. Religion was the dominant force over the short average lifespan, up to some forty years.

Technology—knowing how to make things—was transferred over the generations by imitation, as apprentices copied the master's skills. Artisan

---

[52] The Enlightenment included a range of ideas centred on reason as the primary source of authority and legitimacy, and came to advance ideals such as liberty, progress, tolerance, fraternity, constitutional government, and separation of church and state. In France, the central doctrines of *les Lumières* were individual liberty and religious tolerance in opposition to an absolute monarchy and the fixed dogmas of the Roman Catholic Church. The ideas of the Enlightenment undermined the authority of the monarchy and the Church and paved the way for the political revolutions of the eighteenth and nineteenth centuries.

[53] The English and Scottish Enlightenment movements are covered in the next case study: *Context for Innovation: British (R)evolutions in Perspective.*

activity was encapsulated by guilds, which protected the professions and set the norms and values. Trade was local: agricultural surplus was brought by horse-powered cart to local markets. Life was physically harsh; climate dominated the pattern of life, and the modern comforts of housing and appliances had not yet been developed. Life was also short for men and women: the modern forms of medicine were not available, and modern day's food abundance did not exist. Also, life was not that safe at all, as no 'police' structure had been developed yet.[54] The world was also small in terms of communication; news from outside the community was scarce, and only important news was distributed by the town crier. The majority of people were bound and restricted physically, spiritually, and mentally.

Live is quite different for people living in the Western hemisphere in present times. Take the people living in the geographic areas covered by the United States or the European Union. Living in a democratic system, they can participate in their governing. Secularization has created empty churches; the Roman Catholic Church has been criticized in many ways. People were liberated as they obtained 'freedoms', such as their freedom to speak their mind and share their ideas—also their political ideas—with others. Their freedom to travel—for pleasure or work—is unlimited. To support one's existence, one can live and work where one choses. The modern means of transportation to move—from car and train to boat and airplane—are widely available to the masses. Markets are international, connected by massive logistics systems, making produce from regions far away available throughout the year. As a result of years of education, the occupational choices are broad. One is free to pursue one's own happiness over an average lifespan of some eighty years. Life is comfortable, supported by good housing and made comfortable by modern technologies. Life is—relatively—safe; military and police forces are institutionalized, as the state has taken the monopoly on violence. Life is also much longer, and even prolonged by advanced medicine. News information is not any longer locally based, and modern broadcasting systems cover the world. Modern people, young and old, are privately and professionally always 'online', facilitated by (personal) computers, tablets, and smartphones. For the average person, the freedom of body, spirit, and soul creates a quite large world.

This illustrates how, over that period of a couple of centuries, life changed massively for the average person. These differences between 'then' and 'now' did not occur accidently. They were the result of many

---

[54] Police: body of government employees trained in methods of law enforcement and crime prevention and detection, and authorized to maintain the peace, safety, and order of the community.

developments. On the one hand, there were the *evolutionary developments* in the societal institutions such as the political power structures. On the other hand, there were the *revolutionary developments* in societies, such as the American, French, and Russian Revolution evolutions accompanied by the Industrial and Information Revolutions. These revolutions were the mechanisms of social change, political change, and economic change that fuelled a process of constant change.

Mostly change was a gradual process covering generations, in which people relatively easily adapted to their changing social, political, and economic environment. But sometimes it was about disruptive change, disruptive for existing societies and economies. Some change was initiated by natural causes and climate change creating famine and disease, shaking societies and delimitating population. However, many of those social changes were also initiated by technological changes resulting from the inventive activity of curious minds. New methods of production, tools, and machines made mass-produced products available outside the circles of the happy few—a situation created by technology-driven novelty. But this also led to *creative destruction*, where new technologies created new industrious activity by eliminating their precursors, leaving behind the economic casualties as collateral damage—such as the profession of the candle seller, which disappeared when gaslight came available, or the profession of the (gas fuelled) streetlamp lighter, which disappeared when electric lights started illuminating the streets. Many changes were catastrophic on the individual level, but for society as a whole, it meant progress.

The motor of change was mankind itself. The curiosity, creativity, and entrepreneurship of the thinkers and the tinkerers fuelled a process of constant improvement over the centuries: the previously mentioned 'emulatio'. Enabled by the evolutionary changes in political and social structures, the resulting freedom to undertake experiment, create, and initiate blossomed in the times of the Scientific Revolution. Facilitated by the evolutionary changes in transportation, energy, and communication infrastructures, their knowledge and knowhow was disseminated over larger areas of the world.

Clearly, living in the social, political, and economic context of some five hundred years ago certainly was different form the context which dominates life in the present time. It took the Scientific Revolution to change that context. As we have seen, science—in all its forms a knowledge-based activity—became the organized form of curiosity, creating answers to the phenomena of the real world. Technology—in its broad definition as 'knowing how to make things'—became fuelled by scientific discoveries that were enhanced with mechanisms resulting from human creativity (aka engineering). That real world was the domain of interest of the natural philosophers, who observed natural phenomena. Starting with phenomena like the movement of the sun and moon (originally the exclusive domain of the priest caste) and the phenomena of the seasons and weather, with the frightening thunder and lightning (originally the domain of the medicine men) but soon growing into a quest into a broad range of fields within the Nature of Matter': such as the 'nature of heat' studying the powers of steam and the 'nature of lightning' studying the powers of electricity.

In addition to the works of these natural philosophers, there were the *mathematical philosophers*. Trying to understand the mechanisms of natural phenomena, the natural philosophers were faced with two problems: firstly, the problem of methodology, how to study the phenomena properly and orderly in a way that others could reproduce the findings; secondly, how to describe the phenomena in a universal way. What was needed was a way of notation for the phenomena that gave insight and made predicting future behaviour possible. That search became the beginning of the universal language of mathematics and the rise of the mathematical philosophers (Figure 20, bottom).

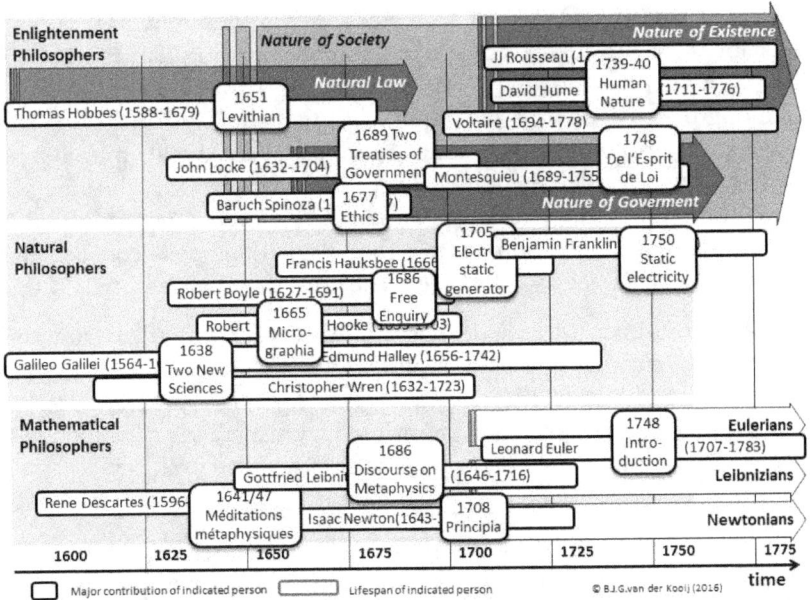

Figure 20: Overview of some of the Enlightenment philosophers in relation to the natural and mathematical philosophers in the 17–18th centuries.

Figure created by author.

However there was more than just the Scientific Revolution caused by the natural philosophers that changed the previously mentioned context. In addition, over time, when societies developed and the powers of autocracy and theocracy became challenged, there emerged another kind of philosopher alongside the natural philosophers.[55, 56] Studying the Nature of Society a range of different thinkers we would like to call the 'social philosophers' started observing the phenomena in the human world around them. Soon they were entering the domains of life originally claimed by the ruling class and the priest class. They would become known as the *Enlightenment philosophers* (Figure 20, top).

This is not the place to delve deeply into the history of these philosophers. But to illustrate the influence of the Enlightenment thinkers upon the development of society, we will highlight some early philosophers who marked milestones in the Enlightenment movement.[57]

Nature of Government: Questioning the State

Many of the early social philosophers looked at the society they were living in, whose people were harassed by the pitiless cycles of epidemics, famines, risky life, and early death. These were often troubled times, where war and turmoil dominated, insecurity and injustice were the fate of the masses, and privilege and inequality created vastly different living circumstances for the aristocracy and the common people. Times where the extravagance of court and king contrasted sharply with the poverty and pests among the common people trapped in the feudal system. In England, displaced squatters infested or produced stinking slums, while in France, rural unemployed wandered about the countryside in begging, thieving bands. In these times, the emerging nations of France and Britain often were in a state of military conflict, wars that resulted either as consequence of succession disputes or as the consequence of territorial disputes.

Also on the British Isles, the Enlightenment movement arose from the thinking of the scientists partaking in the Scientific Revolution. Already, in the early seventeenth century, the English philosopher *Thomas Hobbes* (1588–1679), in his younger years confronted with scholarly thinking, mathematics, and physics, developed his initial views on political

---

[55] An *autocracy* is a system of government in which supreme power is concentrated in the hands of one person, whose decisions are subject to neither external legal restraints nor regularized mechanisms of popular control.

[56] A *theocracy* is a form of government in which a deity is the source from which all authority derives.

[57] The life and work of the early natural philosopher Galileo Galilei has been covered in the early case study on the telephone. Isaac Newton will be covered in the case study of the British Industrial Revolutions.

philosophy. As a tutor, he accompanied his aristocratic master on a 'grand tour' through Europe (1610) and was confronted with emerging scientific thinking. Later he visited Paris (1631) and Florence (1637), which awakened his philosophical interests. His was the time of the First Civil War (1642–1651) in England, followed by the English Revolution with the execution of King Charles II in 1651.[58] Due to the subsequent rise of parliamentary powers, many royalists fled to Paris. Among them was Hobbes, who became acquainted there with the views of people like Descartes.

In 1651 Hobbes, originally interested in the physical doctrine of motion and physical momentum but shifting his focus to societal phenomena, published his book *Leviathan*—subtitled *The Matter, Form and Power of a Commonwealth: Ecclesiastical and Civil*. It was a reflection on his view on civil government that arose from the socio-political crises he had observed. In this book, he compared the state to a monster (Leviathan) composed of men, created under pressure of human needs, and dissolved by civil strife due to human passions. The secularist spirit—ie separating state from religion—of his book severed his link with the exiled royalists, who might well have killed him. In addition, his book greatly angered both the royalist Anglicans and the French Catholics. Threatened, Hobbes appealed to the revolutionary English government for protection and fled back to London in the winter of 1651. Hobbes's *Leviathan* was about the foundation of states and legitimate governments and would lay the foundations for the later emerging 'social contract theory'.[59] It was the beginning of the basic discussion of the natural rights of people under a sovereign authority (both by monarchy and papacy).

In France the tensions in society in the seventeenth century were observed by *Charles-Louis de Secondat*, Baron de La Brède et de Montesquieu (1689–1755, aka Montesquieu). Montesquieu's early life occurred at a period in time of significant governmental change. England had declared itself a constitutional monarchy in the wake of its Glorious Revolution (1688). In France, the long-reigning Louis XIV—the Sun King—died in 1715 and was succeeded by the five-year-old Louis XV. Being a member of the old aristocracy himself, Montesquieu left the practise of law to devote himself to study and writing. It resulted in works covering the spectrum of natural law and natural rights. In 1721 he published the epistolary novel *Persian Letters*, a range of letters recounting the experiences of two fictitious Persian noblemen, Usbek and Rica, who were traveling through France.

---

[58] See: B.J.G. van der Kooij: *Context for Innovation: British (R)evolutions in Perspective*. (2016) pp. 145–163.
[59] Within a social contract individuals have consented to surrender some of their freedoms and submit to the authority of the ruler or magistrate (or to the decision of a majority), in exchange for protection of their remaining rights.

Montesquieu used the travellers' account to comment on a variety of subjects: from governmental institutions to salon caricatures. The book became largely admired as the true interest of the work was for its factitious 'oriental' impression of French society, along with political and religious satire and critique. Next, studying the rise and fall of the Roman Empire, he published *Considérations sur les Causes de la Grandeur des Romains et de leur Décadence* (Considerations on the Causes of the Grandeur and Decadence of the Romans, 1734). He observed:

> *The greatness of the state led to the greatness of private fortunes. But, as opulence consists in manners and not in riches, the wealth of the Romans, which admitted no bounds, produced an unlimited luxury and profusion in life.* (Baron de Montesquieu, 1965, p. 202).

Quite some time later, after 21 years researching and writing, he published in his *De l'Esprit du Lois* (The Spirit of the Laws, 1748) his ideas about separation of political powers (aka trias politica).[60] In this political treatise, Montesquieu pleaded in favour of a constitutional system of government and the separation of governing powers, the ending of slavery, the preservation of civil liberties and the laws, and the idea that political institutions ought to reflect the social and geographical aspects of each community. Overnight, *The Spirit of the Laws* became the political bible of learned men and would-be statesmen everywhere in Europe and beyond. To avoid censorship, it was published anonymously. Despite critics in France and his work being banned by the Catholic Church, it was popular in the rest of Europe and influenced America's founders.

In Italy, it was Cesare Beccaria, Marquis of Gualdrasco and Villareggio (1738–1794), born in Milan and educated by the Jesuits, who was influenced by the views of Montesquieu, Voltaire, and Denis Diderot. He wrote the treatise *Dei delitti e delle pene* (On Crimes and Punishments, 1764), in which he condemned torture and the death penalty. It became highly popular among French and British philosophers, was translated in several languages, and influenced thinking on criminal justice.

These are a few of the many examples of scholarly thinking that was a reaction to social circumstances of that time, where the former feudal structure still executed a power structure based on absolute monarchy and the rule of aristocracy and theocracy A rule that could be harsh and brutal, sometimes even cruel and deadly.

---

[60] The trias politica model of government is based on a division into three branches: a legislature, an executive, and a judiciary. Separation of powers, therefore, refers to the division of responsibilities into distinct branches to limit any one branch from exercising the core functions of another. The intent is to prevent the concentration of power and provide for checks and balances.

Nature of Being: Questioning Existence

In France in the seventeenth century, there were several philosophers of the Nature of Existence such as *René Descartes* (1596–1650), educated at the French Jesuit Collège Royal. He studied military engineering after he had enlisted as a soldier in the Dutch Army in Breda, Holland.[61] He became acquainted with the philosopher Isaac Beeckman (1588–1637), who convinced him to devote his studies to a mathematical approach of nature.

> *He introduced Cartesian geometry, which incorporates algebra; through his laws of refraction, he developed an empirical understanding of rainbows; and he proposed a naturalistic account of the formation of the solar system, although he felt he had to suppress much of that due to Galileo's fate at the hands of the Inquisition. His concern wasn't misplaced—Pope Alexander VII later added Descartes' works to the Index of Prohibited Books.[62]*

He sought to uncover the meaning of the natural world with a rational approach, through science and mathematics. He abandoned the publication of his *The World* (written in 1629–1633), as he considered its heliocentric approach too dangerous after he heard of Galileo's conviction by the Inquisition. Instead, he published his two treatises *Meditationes de Prima Philosophia* (Meditations on First Philosophy, 1641) and *Principia Philosophiæ* (Principles of Philosophy, 1644). His interest into the Nature of Matter, especially medicine, brought him to write *La description du Corps Humaine* (The Description of the Human Body) in 1648, which again brought him to reflections on the universe.

> *Although Descartes nominally subscribed to the Biblical story of creation, in his natural philosophy he presented the hypothesis that the universe began as a chaotic soup of particles in motion and that everything else was subsequently formed as a result of patterns that developed within this moving matter. Thus, he conceived that many suns formed, around which planets coalesced. On these planets, mountains and seas formed, as did metals, magnets, and atmospheric phenomena such as clouds and rain. The planets themselves are carried around the sun in their orbits by a fluid medium that rotates like a whirlpool or vortex.[63]*

Descartes's approach of combining mathematics and logic with philosophy to explain the physical world turned metaphysical when confronted with questions of theology; it led him to a contemplation of the Nature of Existence. Descartes shifted the authoritative guarantor of truth

---

61 Descartes spent about 20 years of his life in the Dutch Republic. But he also stayed for periods of time in the service of the Duke of Bavaria.
62 Source: 'René Descartes Biography'. http://www.biography.com/people/ren-descartes-37613#becoming-the-father-of-modern-philosophy.
63 Source: 'René Descartes'. https://plato.stanford.edu/entries/descartes/

from God to humanity; the human being equipped with autonomous reason ("I think; therefore I am"). He questioned the nature of the soul and envisioned a purely physical and mechanical universe, postulating that animals and the body were automatons, with only the soul elevating humanity. Not too surprisingly, this brought him conflict with the prevailing doctrines of the Church, and in 1663 his works were placed on the Index of Prohibited Books.

> ..., he offered a new vision of the natural world that continues to shape our thought today: a world of matter possessing a few fundamental properties and interacting according to a few universal laws. This natural world included an immaterial mind that, in human beings, was directly related to the brain; in this way, Descartes formulated the modern version of the mind–body problem (Ibid.)

So, over time, his thinking moved from mathematics, through natural philosophy, to metaphysics. As a consequence, during the course of his life, he was a mathematician first, a natural scientist or 'natural philosopher' second, and a metaphysician third. Descartes died of pneumonia in 1650 in Sweden at the age of 54 while tutoring the queen, Christina. Being a Catholic in Protestant Sweden, he was interred in a graveyard mainly used for unbaptized infants. His legacy was both the Cartesian coordinate system—a form of mathematics—as well as Cartesianism, the metaphysical link between social philosophers and natural philosophers.

Nature of Religion: Birth of Pantheism

So, it was in the Age of Science that metaphysics—the study beyond the phenomena of the natural world, the physics—saw a revival in which many scholars took part. Along with the before mentioned Descartes, it was philosophers like Gottfried Wilhelm Leibnitz and Imanuel Kant who created their views. Among them, we also find the Dutch lens grinder and philosopher of Sephard/Portuguese origin: *Baruch Spinoza* (1632–1677), who developed—certainly for his time—highly controversial ideas regarding the authenticity of the Hebrew Bible and the nature of the divine. In short, he developed the view that all of reality is identical with divinity.

Spinoza's ancestors were of Sephardic Jewish descent and were a part of the community of Portuguese Jews that had settled in the city of Amsterdam in the wake of the Portuguese Inquisition (1536–1821), which had resulted in many victims (est. 40,000), forced conversions, and expulsions from the Iberian Peninsula. The Netherlands in the mid-seventeenth century was at the zenith of its economic, military, and political golden age. As a result it saw a flourishing commercial activity that encouraged a culture relatively tolerant of the play of new ideas, to a considerable degree sheltered from the censorious hand of ecclesiastical authority. And as a consequence of those relative liberal attitudes, it was place for many to flee to.

The 1650s started turbulent times for the Dutch Republic, a confederation of the Seven Provinces. Dutch politics of this era were quite complicated. The government of the provinces traditionally consisted of of independent 'states' (eg the States of Holland); a body of representatives of the nobility and the cities of the province. Representatives of those states, assembling in the States General, created the national government. Next, as a heritage of the medieval Burgundian institutions, there was the office of the 'stadhouder' (stadholder), the former representative of the king. However, there was no king ruling anymore, as, after the Dutch Revolt (1568–1648) and the subsequent *Plakkaat van Verlatinghe* (Act of Abjuration, 1581) had the position of the king abolished. The leader of the revolt, the aristocrat William the Silent (1533–1584) from the House of Orange, was elected stadholder in the most provinces. After his assassination in 1584 in his Delft residence, several other members of the House of Orange-Nassau, were elected as stadholder until 1650. Then came the *First Stadtholderless Period* (1650–1672), with the internal conflicts between the factions of the 'Staatsgezinden' (Republicans, or the ruling regents of the urban merchant class) and the 'Orangist' (aristocratic followers of the Prince of Orange). This was a conflict structure of the ruling class, of people versus hereditary aristocracy, that was found in many other places (eg the Italian city-states).

Internationally, there was also quite some turbulence. On the one hand, this was caused by the English as they went through their English Revolution, the English Republican Civil Wars (1642–1651), and their subsequent problems with the Dutch Republic in the different *Anglo-Dutch Wars* (1652–1674), during which the Dutch admiral Michel de Ruyter defeated the Royal Navy in the *Raid on the Med*way (1667). In the process hurting England's pride and reputation considerably. It was a time of strong commercial competition between the Dutch and English. In addition, there was turbulence caused by the French, who considered the Dutch not only to be trading rivals, but also seditious Republicans and Protestant heretics. The result was the *Franco-Dutch War* (1672–1674). But there was more than commercial rivalry and religious separatism; it was also about government. The French had something to settle with the Dutch Republic after their claims to the succession of the Spanish Netherlands (nowadays Belgium), had been blocked by the *War of Devolution* (1667–1668). Then the cooperation between the Dutch Republic and Spain—their former enemies during the Eighty Years' War's (1568–1648) fight for independence—had proven disastrous for the French expansion to the north. Not too surprisingly, the French and English had found each other in the secret *Treaty of Dover* (1670), which was about the return of England to

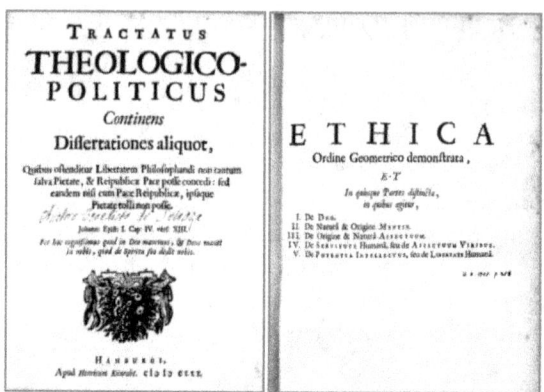

Figure 21: Title pages of Spinoza's *Tractatus Theologico-Politicus* (1670) and *Ethics* (1677)

Source: Wikimedia Commons.

Catholicism and defeating those stubborn Dutch.

It was in this context that Spinoza came of age. In 1653, at the age of 20, Baruch Spinoza began studying Latin with Francis van den Enden, a notorious free thinker, former Jesuit, and radical democrat who likely introduced Spinoza to scholastic and modern philosophy. Many of his friends belonged to dissident Christian groups which met regularly as discussion groups and which typically rejected the authority of established churches as well as traditional dogmas. Soon Spinoza became branded as a heretic, and his clashes with authorities became more pronounced.

> *His extremely naturalistic views on God, the world, the human being and knowledge serve to ground a moral philosophy centered on the control of the passions leading to virtue and happiness. They also lay the foundations for a strongly democratic political thought and a deep critique of the pretensions of Scripture and sectarian religion. … It is hard to imagine a more passionate and reasoned defense of freedom and toleration than that offered by Spinoza.[64]*

Spinoza called into question the tenets of both Judaism and Christianity: he believed in God but denied that the Bible was divinely inspired and rejected the concept of miracles and the religious supernatural. As a result of these conflicting views, at the age of 23, he was expelled from the Jewish community in Amsterdam. After also being expelled by the municipal authorities from Amsterdam, Spinoza lived in other Dutch cities, Rijnsburg and Voorburg. In 1663 he published his first philosophical work on the views of Descartes: the Cartesian philosophy. By 1665 he already had a nearly completed draft of his later publication, named *Ethics*. But times were not ready yet, and Spinoza was dogged by charges of atheism by the religious authorities. Then, in 1670, Spinoza finished his *Korte Verhandeling van God, de mensch, en deszelfs Welstand* (Short Treatise on God, Man, and His Well-Being), which was published anonymously as *Tractatus Theologico-Politicus* (Theologico-Political Treatise, 1677) (Figure 21). In the treatise,

---

[64] Source: Baruch Spinoza. https://plato.stanford.edu/entries/spinoza/

Spinoza put forth his most systematic critique of Judaism and all organized religion in general.

> *Spinoza presented a new account of the relation between State and religion, according to which religious authorities were to subservient to the rulers of the state and freedom of thought and speech were seen as essential to the well-being of the state.* (Della Rocca, 2008, p. 31)

Soon his Tractatus was condemned; in 1673 by the Synod of the Reformed Church and formally banned in 1674. In the meantime Spinoza kept working on his ethical vision:

Figure 22: Detail of the lynching of the Dutch Gebroeders De Witt (1672) in The Hague, Holland.

Source: Rijksmuseum, artist Romeyn de Hooghe

> *God is no longer the transcendent creator of the universe who rules it via providence, but Nature itself, understood as an infinite, necessary, and fully deterministic system of which humans are a part. Humans find happiness only through a rational understanding of this system and their place within it.* (Dutton, nd)

By that time of the 1670s, the Netherlands were in political turmoil, being attacked from England, France, and the German prince-bishops of Munster and Cologne. In the 'Rampjaar' (Disaster Year) 1672, when the English, French, and German forces invaded the Netherlands, the country was 'reddeloos' (beyond rescue), the government 'radeloos' (desperate), and the people 'redeloos' (irrational). In addition, there were the internal conflicts between Republic and Orangist factions, resulting in a popular insurrection that led to the gruesome murder of the brothers De Witt (Johan being the 'raadspensionaris', the grand pensionary, and his brother Cornelius a military man) (Figure 22).

Understandably, by this time Spinoza had become very cautious, aware of the political and religious opposition. However, earning a living as a lens grinder, he still was continuing his philosophical work till his death in 1676. Spinoza's views evidently appeared at the wrong place at the wrong time. Spinoza's magnum opus, *Ethics demonstrated in Geometrical Order*—aka *Ethics* (Figure 21)—which he started sometime in 1664, was published posthumously in 1677, only to be put on the Catholic Church's Index of Forbidden Books.[65, 66] In the *Opera Postuma*, his works were posthumously published by his friends.

The views expressed in *Ethics* exposes Spinoza's 'naturalism': everything that exists is part of nature, and everything in nature follows the same basic laws. In this perspective, human beings are part of nature, and hence, they can be explained and understood in the same way as everything else in nature. In his monist view, God is the natural world: God is nature, and nature is God. Based on this concept, he presented his view on human freedom and virtue and the path to happiness.

The initial distribution of his political philosophy might have been influenced by the troubled times he lived in, the later spreading of his views were due to the then more liberal environment in the Netherlands. And the impact of his work was a catalyst for the Enlightenment, as Spinoza's philosophy provided to late-eighteenth-century European philosophers an alternative to materialism, atheism, and deism.[67]

Enlightenment Movement as a Catalyst

These are just a few of the early Enlightenment philosophers, who arose from the natural philosophers—sometimes even being the same person—and complemented their views. These philosophers were interested in the Nature of Government and the Nature of Being, practising the 'science of freedom', covering key issues in the development of democratic societies, such as the *economic freedom* from the limitations of movement and labour that the state—the absolute powers of monarchy and the aristocracy, combined with the guilds—enforced. Over time economic freedom would become a human right:

---

[65] The *Index Librorum Prohibitorum* (English: *List of Prohibited Books*) was a list of publications deemed heretical, anti-clerical, or lascivious and therefore banned by the Catholic Church. It existed from the ninth century up to 1966.
[66] The Library of the Vatican possesses, anno 2017, a handwritten copy of Spinoza's *Ethics*.
[67] Materialism considers the substance of nature in the material world, physical matter is all there is. Atheism is the denial of the existence of gods, sometimes associated with worldviews as materialism or naturalism. Deism is the belief in a God who made the world but who never interrupts its operations with non-natural events.

Economic freedom is the fundamental right of every human to control his or her own labor and property. In an economically free society, individuals are free to work, produce, consume, and invest in any way they please, with that freedom both protected by the state and unconstrained by the state. In economically free societies, governments allow labor, capital and goods to move freely, and refrain from coercion or constraint of liberty beyond the extent necessary to protect and maintain liberty itself.[68]

In addition, they advocated *civil freedom*—freedom from the state's abuse and the right to democratically participate in government. The right to life led to the freedom from life-threatening state behaviour (eg torture, death penalty) and state corruption. The right to a fair trial protected the people from legal abuse. The freedom of speech, assembly, and association gave rise to democracies, states in which the population at large participated in their government. Their work also fuelled *religious freedom*; the freedom to change religion or belief, to exercise it, and the right to have no religion. It freed society from the tyranny and constraints by which institutional Christian religion had encapsulated it. And finally, the Enlightenment philosophers created *intellectual freedom*: the freedom to hold, receive and disseminate ideas without restriction. It was the freedom of the spirit, where 'innovation', originally a word to indicate social abuse, became a word of praise for those creative thinkers and tinkerers.

Concluding

So far we have painted in rough brushstrokes a picture of the changing society under the influence of specific developments now labelled the Scientific Revolution and the Enlightenment movement.

Techno-economic change: The Scientific Revolution resulted in the analytical approach to methodically observe and interpret the natural world around us. From it emerged the science of physics, ultimately leading to the development of all those new technologies and subsequent economic effects.

Socio-political change: From the Scientific Revolution also emerged the 'science of freedoms' practiced by the Enlightenment philosophers. Their work resulted in changing societal structures and new power structures in those societies.

It was the period of the Scientific Revolution and the subsequent Enlightenment (roughly 1550–1800) that would create the foundations of the Era of Industrialization to come. It ended the powers of the absolute monarchies of state and Church, but also of the guilds. From that

---

[68] Source: 2017 Index of Economic Freedom. http://www.heritage.org/index/about

perspective, it ended the 'Times of Imitatio and Emulatio'. It gave people their civil and intellectual freedoms that would facilitate the coming 'Times of Invention'.

Having explored from a bird's-eye perspective some of the historical relationships between social, technical, political, and scientific change, now it is time to look at what happened in the 'real world' that created the context for the technical developments that would change society by setting the social 'scene for change'. So, it is time to explore the developments that resulted in the third phase of the Communication Revolution, especially the events that took place on the Italian Peninsula.

Italy's Early Times

In Southern Europe at the beginning of our era, the Romans ruled their mighty Roman Empire, stretching over large parts of Europe up to the British Isles and lands bordering the Mediterranean. The Mediterranean, the great trade route of the then known world, was theirs, and the countries bordering upon it had become provinces of the Roman Empire. Although the realm was conquered by sword, fhe first two centuries (27–180) had been the time of the *Pax Romana*, a period of relative peace throughout the Mediterranean under Roman rule. Over time the mighty empire lost control over their vast territory as the result of a series of governmental crises, civil wars, corruption of its bureaucracy, and pandemic diseases such as the smallpox pandemic of the Antonine Plague (165–180) and the Plague of Cyprian (250). The empire collapsed after the death of Emperor Theodosius I (347–395), and by early fifth century had become divided into the (Latin) *Western Roman Empire*—ruled from Rome by his son Honorius— and the (Greek) *Eastern Roman Empire*—ruled from Constantinople, later called the Byzantine Empire, by his son Arcadius (Figure 23).

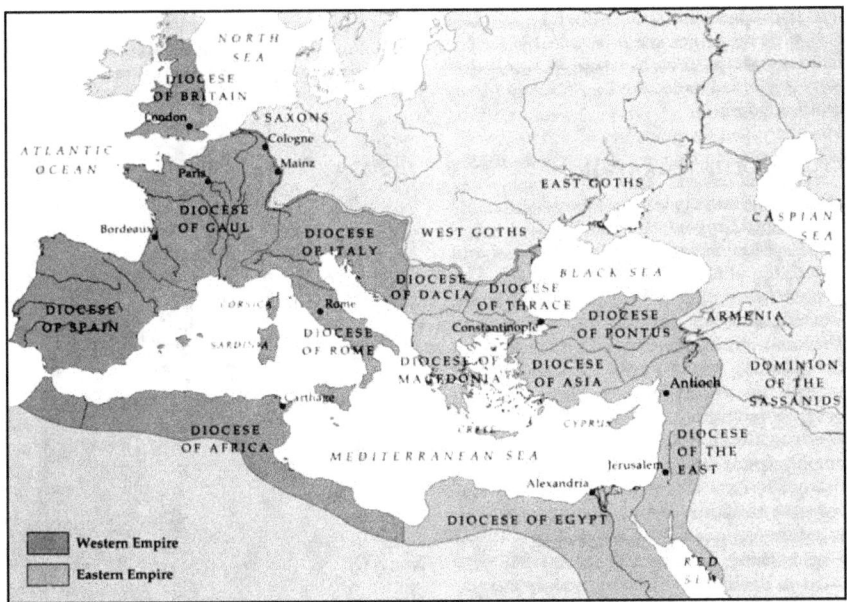

Figure 23: The Western Roman Empire and the Eastern Roman Empire (c. 400).

The map shows the administrative division of the Roman Empire in administrative regions called 'diocese', grouping several provinces.

Source: www.usu.edu

It was upon Rome and the *Western Roman Empire* that the full force of the subsequent barbarian onslaught fell. The *Eastern Roman Empire* (aka Byzantine Empire with its capital Constantinople) would, during the following centuries, develop in completely different ways than the Western Empire. During most of its existence, the Eastern Empire was the most powerful economic, cultural, and military force in Europe. The borders evolved significantly over its times, as it went through several cycles of decline and recovery, until it fell to the Ottoman Turks in 1453.

Centuries of Barbarian Invasions: From Visigoths to Normans

After the fifth century, the Western Roman Empire began to collapse. Next to civil conflicts and economic downturn, this was partly due to the migrating 'barbarians'[69]: peoples such as the Goths, Vandals, Angles, Saxons, Lombards, Suebi, Frisii, Jutes, and Franks, who moved to other parts of Europe and in turn were later pushed westwards by the Huns, Avars, Slavs, and Bulgars. All in all, a massive 'barbarian' immigration had taken place during the *Migration Period* in the first half of the millennium. Among them were the migrations of two branches of the Goths: the Visigoths and the Ostrogoths (Figure 24). Both were to play important roles in the fall of the Roman Empire.

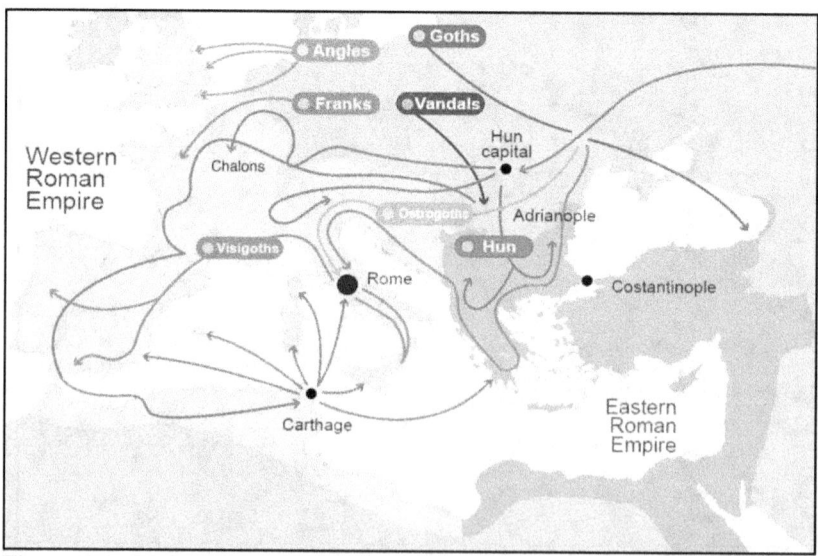

Figure 24: Migration Period: Invasions of the Roman Empire (100–500).

Source: http://www.tesorodialarico.com/

---

[69] The term 'barbaric' is based on the Greek interpretation: those who do not speak the Greek language. In Roman times it referred to the non-Roman people living in the Roman Empire or near its borders.

Visigoths invasion: The East-Germanic people of the *Visigoths* were allowed around 375—after they became foederati in return for subsidies[70]—to settle in the Roman territory of Thracia at the mouth of the Danube in the Black Sea. However, mistreatment by provincial Roman governors soon led to widespread discontent among the Visigoths, and by 376 open rebellion had broken out. The Visigoths plundered the neighbouring Roman towns, growing in power and wealth as they went and settled in Illyria along the eastern coast of the Adriatic Sea. This was not for long, as in 401 they started a new migration, moving to Italy and ultimately to Gallia, on the way sacking Rome in 410. It was the first time in almost 800 years that Rome had fallen to a foreign enemy. Then the Visigoths continued climbing up the Italian peninsula. In 476 they marched on to Ravenna, where they disposed of the—sixteen-year-old—Emperor Romulus Augustulus. This ended the rule of the Roman Empire.

Continuing up north and west, they finally settled down in Southern Gallia (present day Provence up to Toulouse), where they later founded the Kingdom of the Visigoths, the first barbaric reign on the territory of the Roman Empire. The Roman way of life was going to be changed.

> *Throughout the Empire, societal change had occurred to begin to integrate the Gothic peoples into Roman populations, especially in Italy. In Gaul and Spain, widespread looting and conflict with the Rhine-crossers was occurring, perhaps not damaging in itself in the long term, but the sudden decrease in security experienced by the inhabitants of these provinces arguably proved detrimental for the Roman lifestyle there. ... [In Italy] The resettlement of their soldiers and families in amongst the existing Roman population triggered in certain areas a change in culture. The period following 476 ... experienced a pronounced economic regression and a disappearance in specialisation and quality of tradable goods, two hallmarks of Roman economic prosperity.* (Iball, n.d. , pp. 5, 14)

Ostrogoths invasion: Originally, the *Ostrogoths* were settled in present-day Romania, which became the Roman province of Dacia, during 106–274. The Romans undertook a massive and organized colonization of Dacia. New mines were opened, and ore extraction intensified, while agriculture, stock breeding, and commerce flourished in the province. Dacia began to supply grain not only to the military personnel stationed in the province but also to the rest of the Balkan area. It became an

---

[70] To protect their borders of their immense imperium, the Romans had a foreign policy of treaties with the peoples of the neighbouring regions of their provinces. In return for certain subsidies (eg grain), these 'foedera' were required, on demand, to supply troops of 'foederati' for major campaigns to the Roman army in which they usually served as auxiliary forces and were employed to protect the frontiers from external enemies.

urban province of the Roman Empire. However, during the late fourth century, the invasion of Huns from the East forced them to move on. Many of them migrated into Roman territory in the Balkans, while others remained north of the river Danube under Hunnic rule. After the revolt against the Huns and the Hunnic Empire collapsing in 454, large numbers of Ostrogoths settled in the Roman province of Pannonia (present-day Hungary/Austria/Croatia) as 'foederati' to the Byzantines.

That relationship did not last, though, as a dispute about the subsidies brought the Ostrogoth army to the gates of Constantinople. There, Byzantine Emperor Leo I (401–474) agreed to pay an annual subsidy of gold. Next, the Ostrogoths, in 488, moved into Italy, laying siege to Ravenna in 492. By the turn of the sixth century the Ostrogoths had created their *Ostrogothic Kingdom* over Italy and large parts of the Balkans. Their rule ended with the *Gothic Wars* (535–554) when the Byzantine Empire, ruling from Constantinople, launched in 535 a re-conquest of Italy. It was a range of wars that took decades and devastated Italy's political and economic structures.[71]

Figure 25: The Kingdom of the Lombards (700).

Source: Encyclopedia Britannica Inc.

Lombards Invasion: Just as the Gothic wars wound down, the German tribe of the Lombards, originating from Hungary, invaded the peninsula from the north and faced little if no resistance from the Byzantine border forces. The Lombards managed to annex Northern Italy, quickly creating the *Kingdom of the Lombards*. Byzantine control was left in the Ravenna area up to Rome. However, the Lombard conquest

---

[71] By the fifth century, lacking the wealth needed to pay and train a professional army, the Western Roman military strength was almost entirely reliant upon foederati units. Hence, in the Migration Period the Roman rulers saw their former armies of the foederati become their enemies.

did not stop short in the north, as next the *Exarchate of Ravenna* was conquered. Soon Central Italy, with the *Duchy of Spoleto* and the *Duchy of Beneveto*, and South Italy, with the Duchies of Naples and Calabria, were also overrun (Figure 25).

The Lombards assimilated with the Roman people and gradually adopted Roman titles, names, and traditions. The Gothic Wars and the Lombard conquest had destroyed the landed nobility, and for the next two centuries, the Lombards controlled much of the peninsula until the mid-eighth century. Initially the Lombard were Arians at odds with the Papacy, both religiously and politically. By the end of the seventh century their conversion to Catholicism was all but complete.

Frankish Invasion: Then came the Germanic tribes of the *Franks*, originating from the region of Austrasia (Figure 26). By the third century Frankish Belgea lived in *Germanie Inferrior* (present-day Belgium), and the Salians lived in *Germania Superior* (present-day Northern France). And they expanded their territory over time in a range of conquests (Figure 26), first adding, in the early sixth century, Swabia, Neustria, and Aquitaine, followed later by the conquest of Burgundy, Swiss, and Gascony and, in the eight century, Saxony and Bavaria. When the Franks invaded the regions occupied by the Roman Empire they did so in a manner

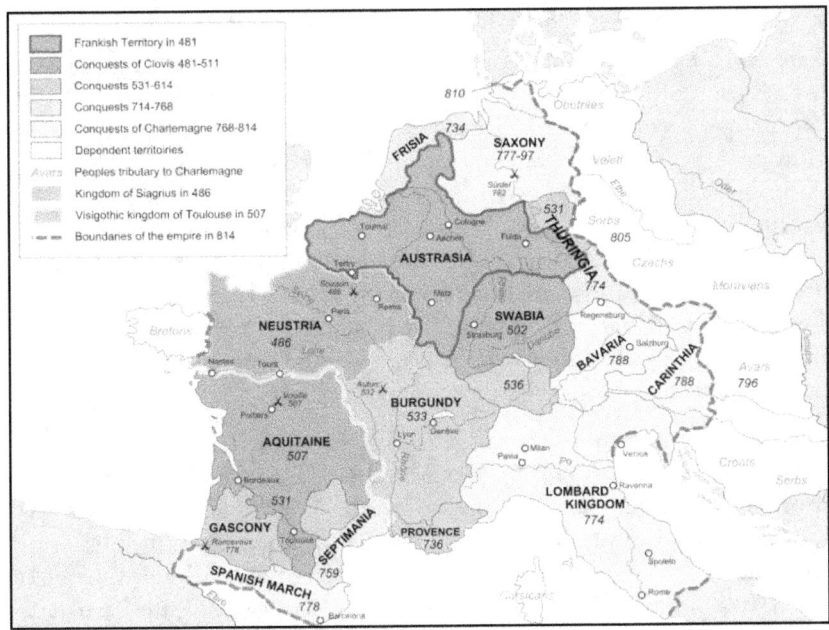

Figure 26: Frankish Empire (481–814).

Source: Wikimedia Commons.

different from that of the other Teutons. They did not cut themselves off from Germany. They did not wander far into the empire, making conquests now here, now there. They simply crossed the border and, taking possession of a small portion, settled there. The Salian-Franks did become foederati, the tribes who were bound by treaty as allies to Rome, supplying contingents of fighting men in times of trouble. However, they—under their ruler Clovis—turned against the Roman commanders in the province of Belgica Secunda. Ultimately, by the early fifth century the Merovingians, with Clovis and his sons, established Frankish hegemony over most of Gaul, from the Rhine to the Pyrenees. Their *Merovingian* rule ended in 754, when the Carolingian monarchs took over.

It was the Carolingian ruler *Charlemagne* (742–814) who invaded Northern Italy in 773. There, the Kingdom of the Lombards collapsed; the region became divided between the Franks and the pope, creating the Papal States. From then the Franks ruled Northern Italy, and the pope ruled his Papal States for centuries to come. Members of the Carolingian dynasty continued to rule Northern and Central Italy until the deposition of Charles the Fat in 887. In 961 King Otto I of Germany, already married to Adelaide, widow of a previous king of Italy, invaded the kingdom and had himself crowned in Pavia on December 25. He continued on to Rome, where he had himself crowned emperor on February 7, in the year 962.

Norman Invasions: The invasion of the Normans was one of the last 'barbaric invasions' on the southern part of the Italian Peninsula (Figure 27). The Norman-Vikings, the traders and raiders coming from the Northern European Scandinavian region, spread over sea to large parts of Europe. After invading coastal areas of Bretagne in the 800s, they managed to invade England, Wales, and Ireland (after the Battle at Hastings, 1066). Already, from the early eleventh century, in the Mediterranean, they arrived in the South Italian region as mercenaries in the service of Lombard and Byzantine factions. They liked what they saw and were communicating news swiftly back home about opportunities in the Mediterranean.

For their masters, they fought several wars in the 1022–1046 period: Aversa (1030), Capua (1038), and Apulia (1042). Soon they started to create fiefdoms and states of their own: the County of Melfi (1046–1059) and the County of Apulia (1049–1098). Their last attack was on the Duchy of Naples, in the meantime fighting the Muslim influence on the island of Sicily. With the conquest of Naples in 1077, they created their rule in Southern Italy. All in all, the original Norman mercenaries

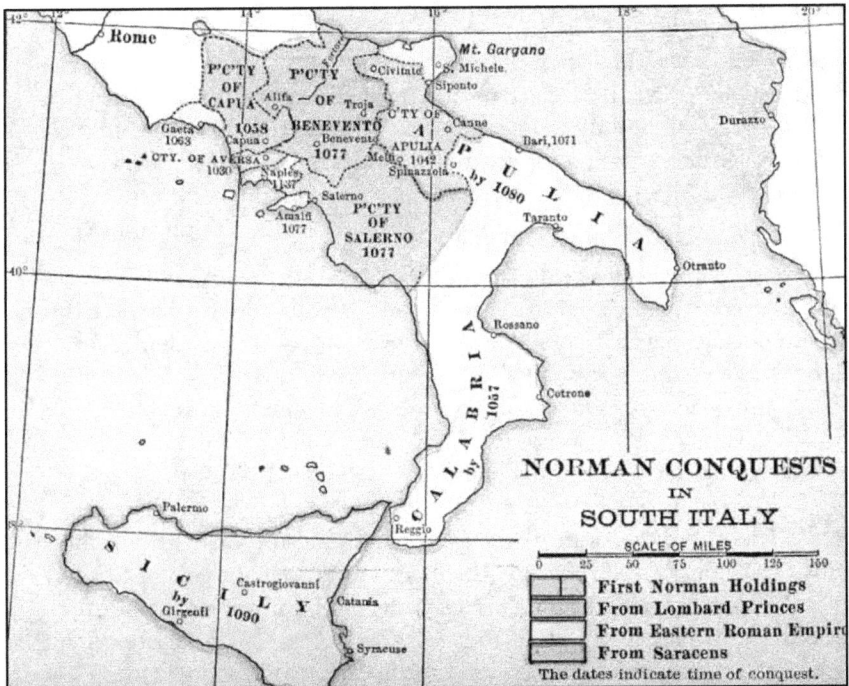

Figure 27: The Norman conquest of Southern Italy (1030–1137).

Source: http://www.roangelo.net/valente/conquest.html

united and elevated their status to de facto independence within fifty years of their arrival, the result of a range of random battles.

The invasions, from the Visigoths to the Normans, contributed to the demise of the Western Roman Empire and its successors. They all left their traces in the Italian society, not only in the generations to come—of which many had blue eyes and blond hair—but also in the geopolitical situation at the end of the eleventh century. By then the cooperation between the Papacy and the Frankish Empire had resulted in the Holy Roman Empire. Northern Italy saw the creation of the Kingdom of Lombardy, the Marquisate of Verona, and the Venetian Republic. Central Italy had the Marquisate of Tuscany, the Papal States, Romagna/Pentapolis, and the Duchy of Spoleto. Southern Italy was under the influence of the Byzantine Empire and would become dominated by the Normans.

New Rulers for the Peninsula (950s–1050s)

Seen from a satellite's perspective, by the turn of the millennium, two foreign secular powers were active on the Italian Peninsula: the Frankish rulers in the north and the Norman rulers in the south. In addition, locally, the Italian families, originating from roman descent, ruled their own territories. And there were the ecclesiastical powers in the spheres of influence of the Roman pope (north) and the Byzantine pope (south).

In parallel to all these invasions in the first millennium, which dismembered over the centuries the former Roman Empire, there had been other important developments on the European continent, changes that related to the disintegration of the Roman Empire that were felt on the Italian peninsula, where imperial interests and papal interests coincided and conflicted. Obviously, disunion in the Roman Empire contributed to disunion in the Roman Catholic Church. The separation of the former Roman Empire into the Western-Roman Empire and Eastern-Roman Empire (aka the Byzantium Empire) would lead also to a schism in the Roman Catholic Church. On the one hand, there was the See of Rome,[72] the ecclesiastical jurisdiction of the Catholic Church in Rome, and the independent sovereign entity of the Church of Rome. On the other hand, there was the equivalent See of Constantinople and the Church of Constantinople in the Byzantium Empire. Conflicts between the two churches, similar to those between the two empires, would ultimately lead to the East-West Schism of 1054.

In the Western-Roman Empire, the conversion into Christianity was orchestrated from Rome, where the pope, as the supreme authority, not only ruled the Church, but also the Papal States. Much of this conversion took place within the realm of the Carolingian Empire, brought to power by Charles the Great (748–814, aka Charlemagne). Charlemagne had united large parts of Europe under his rule, but after the Carolingian Civil War (840–843), ended by the Treaty of Verdun in 843, his realm was divided. It became three regions ruled under his sons: West Francia (that which developed into today's southern and western France[73]), Middle Francia (Lotharingia, centred on Lorraine, Burgundy, and Northern Italy's Lombardy), and East Francia (the Kingdom of Germany) (Figure 28).

---

[72] See the authority, jurisdiction, and governmental functions associated with the papacy.
[73] This area plus England and parts of Ireland and Wales would later be incorporated by the Angevins of the House of Plantagenet in the Angevin Empire (although it did not have an emperor).

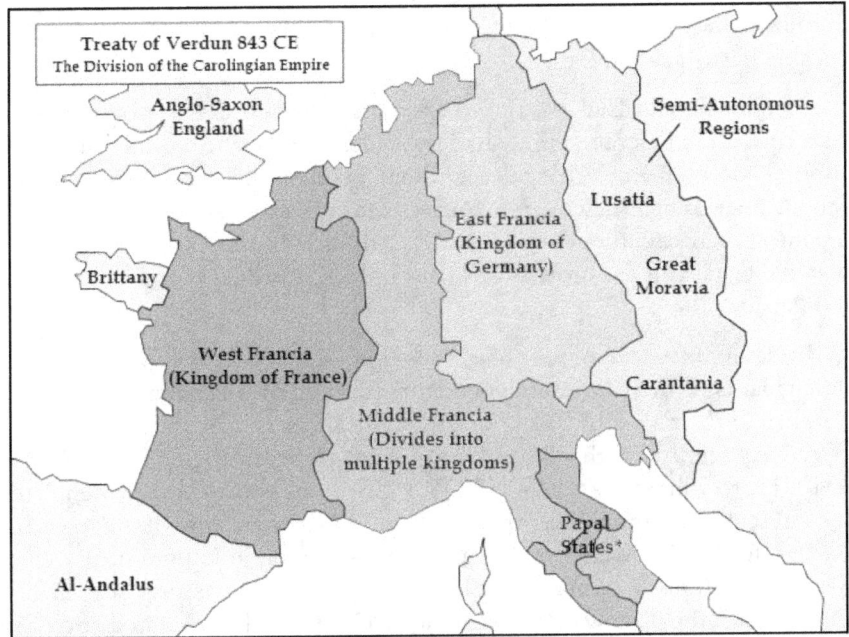

Figure 28: The division of the Carolingian Empire by the Treaty of Verdun (843).

Source: Wikimedia Commons.

The fortunes of the three countries carved out of the Empire of Charlemagne were widely different. France slowly, but surely, became welded into a nation, but Germany remained merely a conglomeration of independent states. For while France struggled towards unity, Germany chased after the phantom of world dominion, claiming with the title of emperor the right to rule over Italy. This claim brought great evil to Italy, it brought scarcely less evil to Germany. It produced endless wars and strife with the Church, it was a constant hindrance to the real progress of Germany, and for nine hundred years it prevented Italy from becoming a united nation. (Marshall, 1920)[74]

It was the East-Francia region that would become the kernel of the later Holy Roman Empire by way of the Kingdom of Germany enlarged with some additional territories from the Middle Frankish Realm. And it was the Holy Roman Emperors who cast an eye on the riches of the Italian Peninsula, riches to be found in its earthly wealth gathered over time, but

---

[74] Chapter 'The Holy Roman Empire—Saxon Emperors'. Source:
http://www.mainlesson.com/display.php?author=marshall&book=sketches&story=saxon

also in political rewards due to it religious affluence. All they needed was the alliance with the Bishops of Rome, rulers of the Church of Rome who became known as the Papacy.

As elsewhere in Europe, Italian aristocracy had spread its power base. As a rule, an aristocratic family had the oldest son be the worldly ruler, another son be the celestial ruler, and one be the military ruler. As religion was such a dominant element in societies, they had entered into the organizational structure of the Roman Catholic Church, not only at the regional level with the crown-cardinals, but also at the top of the Church: the Papacy.

In the *Saeculum Obscurum period* (904–964), it was the Theophylacti family, descendants of the former independent city-state of Tusculum that nominated their sons to be popes. As political rulers of Rome, they effectively controlled the election of the new pope: from Sergius III (ruling 904–911) to John XII (ruling 955–964). The same was the case when, later, the Crescentii clan came to power in Rome, also controlling the Papacy. Next, after 1012, the counts of Tusculum—a region in Latium near Rome—again had a dominant influence on the Papacy, as they supplied three popes: the *Tusculan Papacy* (ruling 1012–1048). They created and perfected the political formula of the noble-papacy, wherein the pope was arranged to be elected only from the ranks of the Roman nobles. Obviously, the Holy See and its Papacy was a pawn not only in the power struggles between the regional aristocratic families, but also in the larger geopolitical scene.

> Take, for example, the Tusculan Pope Benedict VIII (reigning from 1012 to1024), whose powerful protector was the King of the Germans, Henry II. When his predecessor, Sergius IV (1009–1012)—a puppet ruler for Rome's ruler Crescentius III—died in 1012, Benedict's candidacy was opposed by the anti-pope, Gregory VI, installed by the Cressentii. The anti-pope looked for support from the German King Henry II. However, that effort failed, and Benedict became pope. In return for his protection, he crowned Henry II to Holy Roman Emperor in Rome in 1014. Consequently, the Cresentii rulers were forced out of Rome.

Obviously, religious affairs and worldly affairs became mixed, and power struggles around the Papacy dominated the political scene. Then the Gregorian Reforms, initiated by Pope Gregory VII (1073–1085), challenged the authority of the monarchs in the papal selection, resulting in the *Investiture Controversy*, a major power dispute between Roman Church and Holy Roman Emperor. The *Concordat of Worms* (1122) finally affirmed the primacy of papal authority and the new canon law governing the election of the pope by the College of Cardinals. The Church had gained an independency on its own (with the help of God, surely).

Feudal Society

To understand how this loose complex of societies could function, one has to understand the mechanism of the feudal society behind it[75]. Simply put, the social structure of the feudal system was about wealth and power. On the one hand, there was the natural wealth that was transferred from the bottom of the feudal pyramid (Figure 29)—the peasantry and artisans—to the top: the aristocracy, royalty, and clergy up to king and pope. On the other hand, there was the power structure that went down from the monarch through the knight as warrior, to the peasantry paying tribute and fighting (part-time, if needed) as soldiers.

It was this feudal system in which the Church surprisingly played an important role as absolute monarchs. Over time the bishops, being princes of the religious empire, had secured many privileges and had become, to a great extent, feudal lords themselves over great districts of the imperial

Figure 29: The feudal pyramid.

The picture shows the layers of civil classes (left) and ecclesiastical classes (right) in medieval society. On top the layer of the Monarchy and the pope of the Church, in the middle layer the Nobility and the Clergy, and in the bottom layer the Artisans and Peasantry.

Source: ilustrared.cl/v3/uploads/galerias/00193/01072/7773.jpg

---

[75] This topic is extensively covered in the next case study: *Context for Innovation: British (r)evolutions in Perspective.(2017)*

territory. Many bishops and abbots were usually part of the ruling nobility; since the eldest son would inherit the aristocratic title, other siblings often also found careers in the church.

So, it started with small societal units in which a ruler dominated (chiefdoms), that combined to ever greater units when neighbouring societies were absorbed and a ruler (aka king) was elected. The control of these greater units of economic and military power was for the king a question of primary importance in an interdependent power structure. The king needed vassals (aka lords) who supported him. It was essential for a ruler to appoint (or sell the office to)—by means of the power of 'investiture'[76]—someone who would remain loyal. So the monarchy was based on vassalage: a social structure in which a local ruler willingly recognized his subordinate status and pledged his loyalty to his lord in return for the military protection and societal stability the lord provided.

Within this feudal social structure, there were the religious authorities with a similar structure intertwined within the secular power structure. It was the Roman Catholic Church that had developed over time into the dominant faith. It became organized in a similar structure of archbishops (ruling a religious province), bishops (ruling a diocese), and priests and rectors (ruling a parish). On the top of the religious pyramid was the pope, representing God on earth. The Church painted God as the greatest lord of all, with every person on earth as vassals owing Him honour and service and loyalty. A given church official, therefore, could be the servant of the pope at the same time he also was the vassal of a king. State and Church were intertwined. Over time various religious institutions, such as monasteries and convents, had become important, quite rich and powerful. Since a substantial amount of wealth and land was usually associated with the office of a bishop or abbot, the sale of Church offices (a practice known as 'the rule of simony'[77]) was an important source of income for leaders among the nobility.

The Roman Catholic Church behaved like an absolute monarchy. Through the sacraments, it exercised complete control over the people,

---

[76] *Investiture:* The term is used to describe the installation of individuals in institutions that usually have been extant from feudal times. For example, the installation of heads of state and various other state functions with ceremonial roles when they are *invested with office.* Usually, the investiture involves ceremonial transfer of the symbols of the particular office (such as the crown).

[77] *Simony:* The act of selling church offices and roles. It originates from the New Testament— the Acts of the Apostles 8:9-24— where Simon Magus offers two disciples of Jesus, Peter and John, payment in exchange for their empowering him to impart the power of the Holy Spirit to anyone on whom he would place his hands. The term also extends to other forms of trafficking for money in 'spiritual things'.

from birth to death, in the process collecting tithes, free labour, and indulgences. And becoming richer and richer in the process.

Not Holy, Not Roman, Not an Empire

By the eleventh century, the Kingdom of Germany was a conglomerate and assemblage of a number of once separate and independent 'gentes' (peoples) and 'regna' (kingdoms). The Carolingian dynasty had ruled until 911, after which the kingship became elective. The initial electors were the rulers of the 'stem duchies': the prince-electors who generally chose one of their own to be become their king.

Figure 30: Holy Roman Empire (ca 1000).

Source: Wikimedia Commons.

Under the different rulers, the Kingdom of Germany expanded, for example, during the military campaigns of Otto I (912–973). It was the Italian Campaigns that brought Italy into the sphere of influence of the Ottonian Dynasty. Again Italy was invaded. In accordance with Lombard tradition, Otto was crowned with the Iron Crown of the Lombards on October 10, 951. Like Charlemagne before him, Otto was now concurrent King of Germany and King of Italy. Later his successors of the Salian Dynasty would add Bohemia (after 1004) and Burgundy (after 1032). The union of the crowns of the parts of the former Frankish Empire with the remnants of the former 'Empire of the Romans' became known as the Holy Roman Empire (Figure 30).

That Holy Roman Empire was to be a loose, multi-ethnic complex of territories with local rulers covering a large part of central Europe. The empire evolved into a decentralized, limited *elective monarchy* composed of hundreds of societal sub-units called principalities, duchies, counties, free imperial cities, and other domains. As a consequence, the power of the emperor was limited, and while the various princes, lords, bishops, and cities of the empire were vassals who owed the emperor their allegiance, they also possessed an extent of privileges that gave them de facto independence within their territories. The Holy Roman Emperor was, in that organizational structure, the 'primus inter pares' (first among equals). In its origin, it was not holy, not Roman, and it was not an empire.[78]

Holy Roman Empire (1000–1500)

The preceding picture shows feudal Europe from feudal England to feudal Russia around the year 1000. It's a picture with fundamental differences, though, especially when it concerned Central Europe, where this loose federation called the *Holy Roman Empire* was located. In contrast to the Western European society that had evolved from Roman times with a *centralized form* of government (ie the Roman Empire), this middle European society had evolved from *localized governance*. Here, it was the powerful local, quite independent rulers of large territories that elected their emperor (thus, they were known as the prince-electors[79]).

---

[78] Quote: « Ce corps qui s'appelait et qui s'appelle encore le saint empire romain n'était en aucune manière ni saint, ni romain, ni empire. » (*Voltaire: Essai sur l'histoire générale et sur les mœurs et l'esprit des nations*, Chapter 70 (1756)).

[79] Later in time a document called the 'Golden Bull' (1356), described seven prince-electors: the prince-archbishops of Main, Trier, and Cologne, the king of Bohemia, Count Palatine of the Rhine, the duke of Saxony, and the margrave of Luxemburg. In the seventeenth century the duke of Bavaria and the duke of Brunswick Luneburg were added. Their domains were to become the later federal states like Austria, Bohemia, and Prussia. As we will see, it was

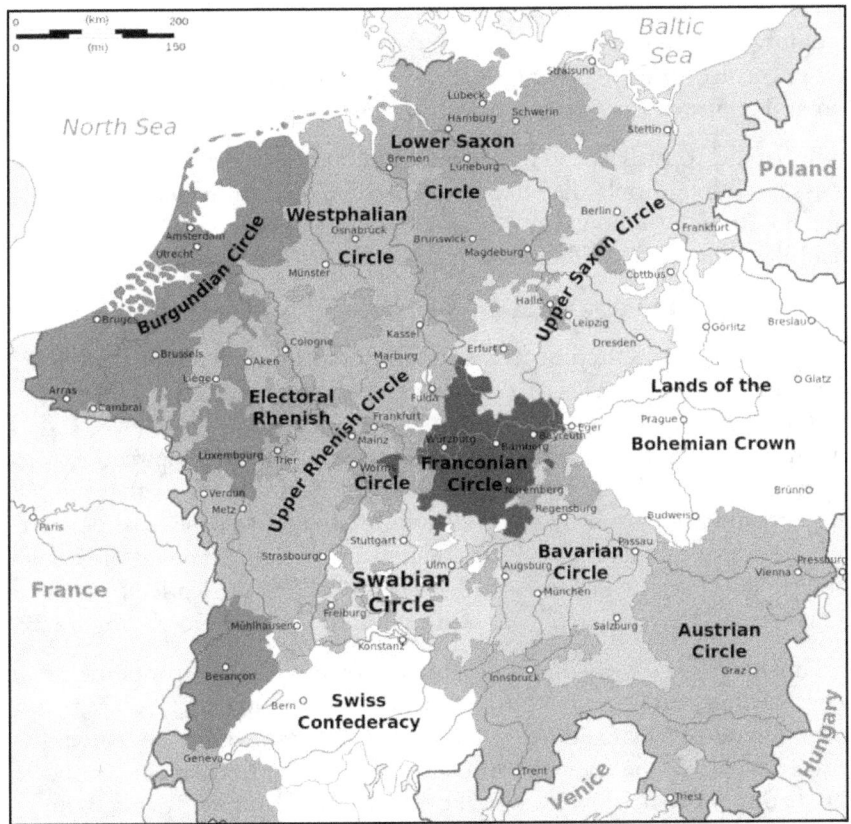

Figure 31 : Imperial circles of the Holy Roman Empire (c. 1512).

Source: Wikimedia Commons.

Looking ahead in time at the development of the Holy Roman Empire, it became an important geopolitical force in Central Europe. Its development intertwined with the emerging societies in the Italian Peninsula. The Holy Roman Empire had enough problems of its own. Dukes often went into feuds against each other that, more often than not, escalated into local wars. In the fifteenth century, which was a time of crisis for the Roman Catholic Church, with different popes claiming the Papacy, the Holy Roman Empire was also in a governmental crisis. In the *Reichsreform* (Imperial Reform) after the *Riechstag* at Worms (1495), king and dukes agreed on four bills, restructuring the disintegrating empire. This

---

this governing structure of the Holy Roman Empire that would play an important role in the geopolitical development of Italy.

resulted in the evolution of the kreis; the 'imperial circles' as administrative groupings for defence and taxation (Figure 31). The empire also received its new title as ruler of the Heiliges Römisches Reich Deutscher Nation ('Holy Roman Empire of the German Nation').

In parallel to the Central European development from feudal medieval times, something similar had happened in the dominant religion of that time: the Roman Catholic Church and its worldly possession of the so-called Papal States (aka Church States) located on the Italian Peninsula.

The Birth of the Papal States

Originally, the Church of Rome did not hold property. That all changed when, under the patronage of Emperor Constantine the Great (272–337) and the rule of the last Roman Emperor Theodosius I (347–395), Christianity became the dominant religion of the Roman Empire: the *Triumph of the Church*.[80] It was Constantine who declared in 321 that the church could hold property. Soon the church became a private landowner as the result of the numerous land donations of wealthy families of Roman nobility. This became known as the *Patrimonium Petri*, the landed possession and revenues that belonged to the Church of St. Peter at Rome (Figure 32).

Thus, over time the Church became the owner of many properties, from Sicily, Sardinia, and Corsica to the Neapolitan, Tuscan, and Tivoli regions. It was Theodosius who added to that the religious power, as he issued between 389–392 the 'Theodisian Decrees' that effectively made orthodox Christianity the official state church of the Roman Empire. It ended the practises of the Roman religion, closing Roman temples and disbanding the Vestal Virgins. It was the start of the domination of the Roman Catholic Church.

As a result of the *Gothic Wars* (535–554) between the Byzantine Empire and the Lombards, the Duchy of Rome was created. It became part of the Byzantine Empire till 751, when the Lombards conquered the Exarchate of Ravenna. The Byzantine rule over the Papacy—aka the *Byzantine Papacy* (537–752)—meant that the pope was appointed by the Byzantine emperor, who was located in Constantinople. Then came the time of the *Frankish Papacy* (756–857). It started when the Byzantine rulers, after the rise of the Kingdom of the Lombards, were unable to govern the Rome-Ravenna corridor (Figure 32). Thus, the pope himself, although being the prelate, also became the ruling authority in the area around Rome, the Duchy of Rome.

---

[80] The Edict of Milan (313) had established tolerance for Christianity. Then emperor Theodosius made Christianity legal within the Roman Empire. And by the *Edict of Thessalonica* (380)—De Fide Catholica—Christianity became the sole authorized religion.

In 751 the *Exarchate of Ravenna* (the area governed by the Byzantine Exarch) fell to the Lombards, and this ended the Byzantine rule in Northern Italy. Next, the Lombards also demanded the submission of Rome. In response to this threat, Pope Stephen II (715–757) made an unusual journey north of the Alps to visit the Frankish king, Pepin III, to seek his help against the invading Lombards. After anointing Pepin in a coronation rite in 754, in return, Pepin and his Frankish army forced the last Lombard king to surrender his conquests in 756, and Pepin officially conferred upon the pope the territories belonging to Ravenna. These *Donations of Pepin* in 756 made the pope a temporal ruler and created the Papal States (Figure 32). It was the next step of the domination of the Roman Catholic Church.

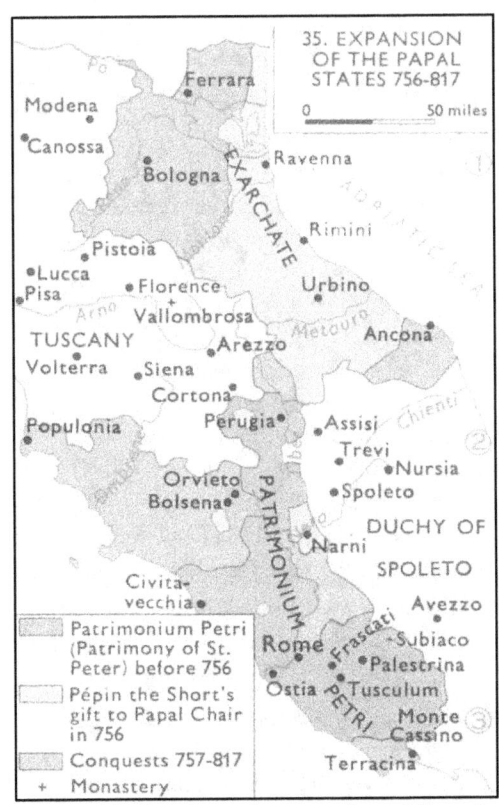

Figure 32: Origin and expansion of Papal States (756–817).

Source:
  http://pages.uoregon.edu/mapplace/EU/EU19%20-%20Italy/Italy-REC-JN.html

Then the Lombards invaded the papal territories again. After Pepin's son Charlemagne became ruler of Northern Italy in 781, he pledged his loyalty to the pope and codified the regions over which the pope would be temporal sovereign. The Duchy of Rome was the core, but the territory was expanded to include Ravenna, the Duchy of the Pentapolis, parts of the Duchy of Benevento, Tuscany, Corsica, Lombardy. and a number of Italian cities. In 787 he further enlarged the states of the Church by numerous further donations. In 800 Charlemagne again travelled to Rome to support Pope Leo III in the papal conflicts. As a reward for his loyalty, he was crowned at St. Peter's Basilica in Rome by Pope Leo III and acclaimed emperor.

This public alliance between the pope and the ruler of a confederation of Germanic tribes—the Frankish Empire—was a reflection of the reality of political power in the west. The coronation launched the concept of the new Holy Roman Empire, which would play an important role throughout the Middle Ages.

Clash within the Church: East-West Schism

As observed before, disunion in the Roman Empire contributed to disunion in the Church and estranged the Greek Christians from the Latin Christians. It caused ecclesiastical differences and theological disputes between the Greek East and Latin West. Similar to the territories of the Eastern Roman Empire and the Western Roman Empire (Figure 23), the two halves of the Church were naturally divided along similar lines. Over time they developed different rites and different approaches to religious doctrines. Ultimately, this growing estrangement and gradual separation led to a schism in the Church in 1054 (Figure 33).

The primary causes of the split were disputes over conflicting claims of ecclesiastical jurisdiction. That was more than a discussion about icons: the so-called *Iconoclastic Controversy*,[81] or a discussion about rites and doctrines. The controversy was influenced when the Western Roman Empire saw the invasions of the Normans on Byzantine territory in Southern Italy (Figure 27), heralding the declining Byzantine power in that region. By that time, while the Eastern Roman Empire considered itself to be the 'inheritor of the Roman Empire', Constantinople had become the largest (with a population of 400,000) and wealthiest city in Europe between the ninth and eleventh century. During this period, the Byzantine Empire employed a strong civil service staffed by competent aristocrats that oversaw the collection of taxes, domestic administration, and foreign policy. The Greek-speaking Byzantines had combined the imperial tradition of the Roman world with an intense Christianity. Religious controversies sometimes spilled over into politics for the simple reason that the (Byzantine) emperor represented the highest power of Church and State. Clearly, the controversy was as much about doctrinal disputes as it was political machinations.

> *The schism in the Church was a symptom of a deeper malaise in the body of Christendom. The balance of power was shifting from East to West. The westerners were full of new ideas and new vitality. But the Byzantines could not bear to think of them as equal partners: nor could they go back on their own past by renouncing their claim to be the only true heirs of the Cristian Roman Empire.* (D. Baker, 2009, p. 4)

---

[81] Iconoclastic Controversy: a dispute over the use of religious images (icons) in the Byzantine Empire in the eighth and ninth centuries.

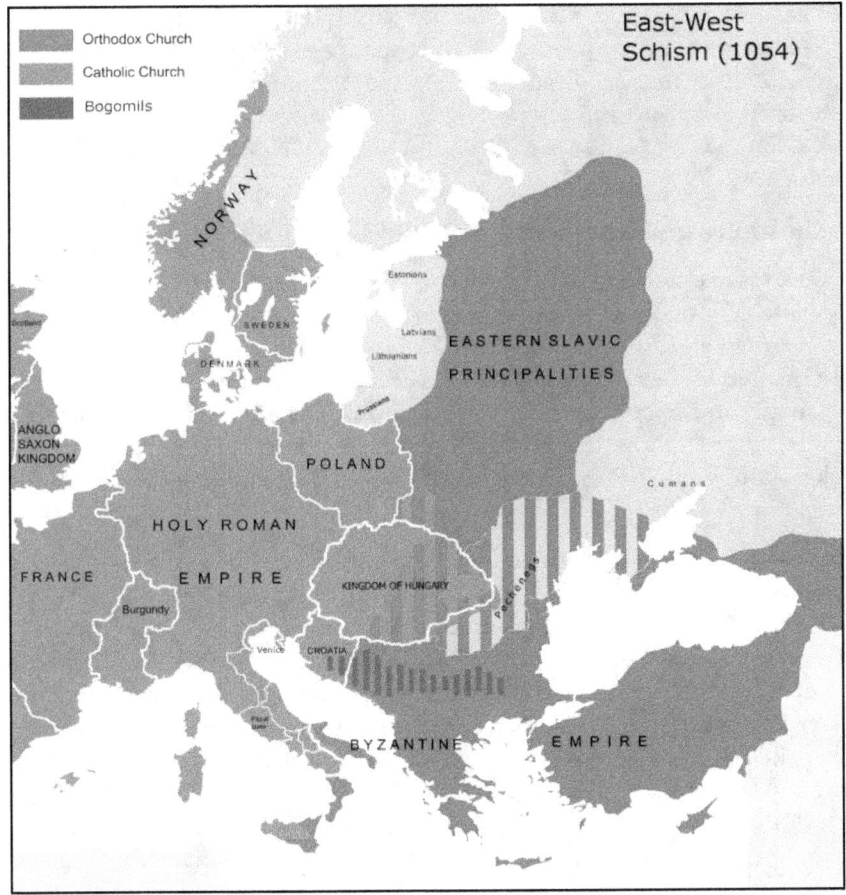

Figure 33: East-West Schism of Roman Church (1054).

Vertical shading near Croatia indicates Bogomil area.

Source: Wikimedia Commons.

The *East-West Schism* manifested itself in 1054 when the western Pope Leo IX excommunicated the patriarch of Constantinople, the leader of the Eastern Church. The patriarch condemned the pope in return. This resulted in the division of the spheres of religious influences: the Catholic Church focussed on the Western European area, and the Orthodox Church on the Eastern-European area (Figure 33). And in between lay the Balkan area with the dualist religious-political sect of the Bogomils.[82] The schism has never been healed.

---

[82] Bogomilism was a Cristian neo-Gnostic religio-political sect founded in the tenth century in present-day Macedonia on the Balkan Peninsula.

The eleventh century saw the decline of the Byzantine Empire. In the east, the empire was confronted by the expansionism of the Seljuq Turks, who made their first explorations across the Byzantine frontier into Armenia in 1065 and 1067. It was their ruler Saladin (1137–1193) who would, in the wars against the Christian crusaders, capture the Holy City of Jerusalem.

Clashes between Roman Church and Holy Roman Empire

Over time the (elected) rulers of the Holy Roman Empire and the (appointed[83]) popes became involved in a complex relationship with its own power struggles. Originally, the Holy Roman Emperors considered it their right granted by God to name Church officials within their territories (such as bishops) and to confirm the papal election. This imperial supremacy had resulted, after all those Italian popes, in many popes with a Germanic background: such as Pope Clement II (1046–1047), Pope Leo IX (1048–1054), Pope Victor II (1055–1057), and Pope Stephen IX (1057–1058).

However, in the eleventh and twelfth centuries, a series of popes challenged the authority of European monarchies. At issue was who, the pope or the monarchs, had the authority to appoint (invest) local Church officials such as bishops of cities and abbots of monasteries—and...get the money related to the simony. But there was more: who was the ultimate power, the (elected) Holy Roman Emperor appointing the pope, or the (independent) pope crowning the emperor? And then, who was the one to claim the taxes to be collected from the clergy with their large landholdings? In addition, one should not forget that the Papacy was an Italian affair among the large Italian families, and they liked to keep it that way.

In 1059 a Church council in Rome declared with the papal bull *In Nomine Domini* that leaders of the nobility would have no part in the selection of popes. In addition, they created the *College of Cardinals* as a body of electors made up entirely of Church officials—that is to say, ruling Italian houses who supplied the religious rulers. Once Rome regained control of the election of the pope, it was ready to attack the practice of investiture and simony on a broad front. In 1075, Pope Gregory VII composed the papal bull *Dictatus Papae*. It declared that the Roman Church had been founded by God alone, that the papal power was the sole universal power, that he alone could depose or reinstate bishops, and that it was permitted to him to depose emperors. Also, the tithes, taxes, and simonies were to be his. The conflict between Roman Catholic Church and the Holy Roman Emperors was born.

---

[83] Before 1059 there was no formal process of papal selection. Instead, the method of papal appointment in which secular rulers appointed (or confirmed) the pope, was applied.

In 1074 this directly created the conflict between Holy Roman Emperor Henry IV (1050–1106) and the Italian Pope Gregory VII (1015–1085); it became the struggle for supremacy after, in the process of appointing bishops, Henry withdrew his imperial support of Gregory as pope. In 1076 Gregory responded by excommunicating Henry and deposing him as German king, releasing all Christians from their oath of allegiance.

That shifted the balance of power, when—as there was little German loyalty to a ruler who claimed the world as his dominion—the already rebellious German aristocrats now openly rebelled under this religious excuse. Henry had to back down, and in 1077 he travelled to Canossa in Northern Italy to meet the pope and apologize in person in what has become known as the *Walk to Canossa*. As penance for his sins, he dramatically wore a hair shirt and was forced to humiliate himself on his knees, waiting for three days and three nights before the entrance gate of the castle while a blizzard raged in January 1077 (Figure 34).

This 'Investiture Controversy' continued for several decades, as each succeeding pope tried to diminish imperial power by stirring up revolt in Germany. These revolts were gradually successful, till the controversy—after some 50 years of struggle—was settled in the *Concordat of Worms* (1122). There, the king was recognised as having the right to invest bishops with secular authority ('by the lance') in the territories they governed, but not with sacred authority ('by ring and staff'). The result was that bishops owed allegiance in worldly matters both to the pope and to the king, and in religious matters to the pope. The pope emerged as a figure above and out of the direct control of the Holy Roman Emperor. The outcome seemed mostly a victory for the pope and his claim that he was God's chief representative in the world. Nevertheless, the emperor did retain considerable power over the Church. This agreement did function for some centuries to come. It was not to last, however, as, in England, it was Henry VIII (1491–1547) who opposed it successfully.[84]

Figure 34: Henry's Walk to Canossa (1077).

Source: http://wikivisually.com/

---

[84] See: B.J.G. van der Kooij: *The Context of Innovation: British (R)evolutions in Perspective.* (2016) pp. 80-86.

Renaissance Papacy: Growth of Papal States

The pope, being the Bishop of Rome, did not always stay in Rome. Especially after Rome's decline, he had moved to Ravenna and back. Political papal instability in thirteenth-century Italy forced the papal court to even move to several different locations outside the Italian peninsula. From 1309 to 1377 several popes resided in Avignon: the *Avignon Papacy*.

Next, the *Papal Schism* (1378–1417) resulted in rival claims to the papal throne; three men simultaneously claimed to be the true pope. The conflicts quickly escalated from a church problem to a diplomatic crisis that divided Europe. Secular leaders had to choose which claimant they would recognize: Avignon or Rome. For nearly forty years the Roman Catholic Church had two papal curies and two sets of cardinals, each electing a new pope for Rome or Avignon when death created a vacancy. Each pope lobbied for support among kings and princes, who played them off against each other, changing allegiance according to political advantage. Driven by politics rather than any theological disagreement, the schism was ended by the *Council of Constance* (1414–1418).

During the *Renaissance Papacy* (1417–1534) the Papal States began to resemble a modern nation-state, and the Papacy took an increasingly active role in European wars and diplomacy. The Renaissance Papacy was perceived to be an Italian institution simply by virtue of the fact that the control of the Papal States gave a pontiff and his family colossal power in the Italian Peninsula itself, both in terms of direct political influence and in terms of familial aggrandizement. Thus, the competition was great between the leading Italian families in the central region (Figure 35) to have one of their clerical members being elected pope. However, they were not

Figure 35: The extent of the Papal States (c. 1400).

Source: http://www.kingscollege.net/
gbrodie/Italy%201400%20showing%20papal%20s
tates.jpg Artist: Christos Nussli. (Map adapted)

always successful in their efforts, as can be concluded from the rise to power of the Spanish Borgia Pope.[85]

The Borgia-family—a family that originated from Valentia, Spain, and had already furnished Alonse de Borja as Pope Calixtus III (1455–1458)— saw the immensely rich Cardinal *Rodrigo Borgia* (1431–1503) elected as Pope Alexander VI in 1492. This cardinal obtained the Papacy through considerable bribing of his fellow cardinals in the election process.[86] With political manoeuvring, he managed to get the crucial support of Cardinal Asconia Sforza, brother of Lodovico il Moro of the influential Milanese Sforza family. Already, as a cardinal, he had fathered several children, which he used to establish and safeguard his power base.

Among his children was his son Cesare Borgia (1475–1507), made cardinal by his father, who created himself a reputation for savage megalomania by murdering his adversaries and who created a private fiefdom within the Papal States in Romagna and Marche. Another son, Giovanni Borgia (1474–1497), was to follow a military career as gonfalonier and captain general of the Church,[87] but that was cut short by his murder in 1497. A third son, Gioffre Borgia (1481–1516), was married at the age of twelve to Sancha, daughter of Alfonso II of Naples.

Rodrigo's daughter Lucrezia Borgia (1480–1519) was married out three times to powerful families. Firstly, she was married to the cruel Giovanni Sforza—also member of the powerful House of Sforza—but that marriage was annulled in 1497 on claims of his (supposed) impotence. The second marriage was with Alfonso d'Aragon, ruler of the bordering Kingdom of Naples, who was murdered in 1500. Then a third marriage was arranged, now with Alfonso d'Este, Duke of Ferrara. That marriage lasted till her death in 1519.

Till that time, obtaining the Papacy was the result of the power struggles between the local feudal families of the regions in Central Italy. Each of them had seen their second/third son into the clergy, such as the Orsini family, the Colonna family, the Savelli family, and the Caetani family,

---

[85] This part is included to give a glimpse of the complex societal situation in which different parts of society fought each other by all available means. The sources for the given facts are numerous, among them: Meyer, G.J., *The Borgia, the Hidden History* (2013, Bantam Books, New York); Chamberlin, E.R., *The Fall of the House of Borgia* (1974, Dial Press) and Bond, J., *In the Pillory: The Tale of the Borgia Pope* (1929).

[86] All but five of the 23 cardinals took enormous bribes. Cardianl della Rovere received 200,000 ducats from the French monarchy and 100,000 ducats from Genova. Rumours spread that Rodrigo gave Ascanio Sforza four mules loaded with gold, his palace, and the position of vice chancellor.

[87] The gonfalonier was a military and political office of the Papal States. The captain general was the commander-in-chief of the papal armed forces.

supplying a stream of popes and cardinals over the ages. In the game of power, they used the tools of marriage, military force, and money to dominate their neighbours. But there was more than political machination to try and obtain dominance in the Church hierarchy.

The election of the sixty-one-year-old Rodrigo Borgia in 1492 has to be seen in the context of that time, with its own norms and values. As marriage was a political instrument for aristocracy to create alliances, it had a different connotation. Celibacy was required for the clergy, but having a mistress was common. By then, poverty and violence was part of urban life, and military force was by lance and sword.

> *... there is no doubting that Alexander VI was a lusty and sexually adventurous pope. He openly acknowledged fathering a bevy of children by his mistress, Vannozza dei Cattanei, and later enjoyed the legendary affections of Giulia Farnese, renowned as one of the most beautiful women of her day. But here again, Alexander was merely following the norms of the Renaissance papacy, and it is telling that Pius II had no shame about penning a wild, sexual comedy called Chrysis. Popes and cardinals were almost expected to have mistresses. Julius II, for example, was the father of numerous children, and never bothered to hide the fact, while Cardinal Jean de Jouffroy was notorious for being a devotee of brothels. Homosexual affairs were no less common, and in that he seems to have limited himself to only one gender, Alexander VI almost seems straight-laced. Sixtus IV was, for instance, reputed to have given the cardinals special permission to commit sodomy during the summer, perhaps to allow him to do so without fear of criticism, while Paul II was rumoured to have died while being sodomised by a page-boy.[88]*

During his reign Alexander VI was faced with many crises, many of them being instigated by a competing cardinal who lost the 1492 election when Pope Innocent VIII died. This was Cardinal *Giuliano della Rovere* (1503–1513), who was considered Innocent's logical successor. However, he was outmanoeuvred in the Papal Conclave of 1492 by Cardinal Rodrigo Borgia, whom he accused of foul play. As a result of his accusations, Giuliano felt it safer to absent himself from Rome. So, in 1493 he left Rome, sought support in Florence and Milan, and went to Paris to the French King Charles VIII. This visit was not by accident, as France and Spain were both claiming the Kingdom of Naples.[89] Della Rovere convinced Charles to realize his old claim on the Kingdom of Naples by

---

[88] Source: Lee, Alexander. 'Were the Borgias really so bad?' http://www.historytoday.com/alexander-lee/were-borgias-really-so-bad. (Accessed December 2016)

[89] This period saw the beginning of the Great Italian Wars, a conflict between the Papal States, the city-states, the Republic of Venice, and the major Western European states like France, Spain, the Holy Roman Empire, and England.

military force. So, Charles raised a powerful army,[90] and after free passage through the Duchy of Milan, he sacked Lucca and entered Florence without bloodshed.[91] He arrived at the gates of Rome with an army of 25.000 soldiers (among which were 8,000 Swiss mercenaries) armed with powerful canons (Figure 79).[92] Deserted by his allies, the Sforzas from Milan, Alexander VI seemed defeated.

Facing a crisis without real military power, Alexander VI reverted to his papal power. As a result, the French king did not dispose of Alexander but accepted the pope's offer of crowning him as both King of France and King of Naples, thus gaining prestige in France and conquering Naples without bloodshed. The king had chosen the side of the pope, and Alexander VI had won an unthinkable victory by mere political manoeuvring.

But the French king did not enjoy the quite poor and desolate Kingdom of Naples for long. Being frightened by the show of French military power, an anti-French coalition was created in 1495. It was the *Holy League*, a first example of uniting the Holy Roman Empire, the Kingdom of Spain, the Republic of Venice, the Duchy of Milan, the Republic of Florence, the Papal States, and other smaller duchies against the French. When the French army marched up north, this created panic in Rome. However, tricked by Cesare with mock-up wooden cannons, they left the city unharmed and continued up north into the Po delta. There, the forces of the Holy League managed to confront the French in the open field—which made their real cannons useless—at the *Battle of Fornovo* (1495). The battle was an indecisive one, but the French lost their valuable booty. Thanks to their supremacy in military power, the French managed to travel further north, back to France.

At about the same time as Charles was battling at Fornovo, the Spanish Ferdinand II was being welcomed back into the city of Naples, and with Spanish support, he was soon able to regain control of his kingdom. For the Borgia pope in Rome, the threat was over, but the Italians were shocked by the realization that, for all their virtues, talents, wealth, past glory, and experience, they had been unable to withstand the ruthless men—and their military power—from the north.

---

[90] As a result of the Gunpowder Revolution, the French army was superior. In two centuries, gunpowder altered the battlefield beyond recognition, as new troop types, tactics, and organization hierarchies were introduced.

[91] Florence had to billet the French troops and had to pay the enormous sum of 400,000 ducats to pay for the cost of maintaining the army.

[92] The French had improved upon the new weapons called 'cannons': they created the artillery that could bombard the defences of cities and castles.

Pope Alexander VI, relieved of the pressure from the French army, was still faced with the cardinals (ie from the prominent Italian families) that opposed him. Already from 1493 on, he nominated (often for a simony) many new cardinals (among them many members of the Borgia family, like Cesare Borgia) to diminish the Italian influence. But the Italian families stayed influential. Then, in 1497, disaster struck the Borgia family when Alexander's son Giovanni Borgia was found murdered in the Tiber. Who had killed him was a mystery. Alexander was devastated, but in town, his death went not unnoticed.

> *When the gruesome discovery became known in the city the excitement of the populace passed all bounds. Everywhere the masses were surging through the streets, business was suspended, every store closed its doors. The crowds scarcely attempted to conceal their joy at this misfortune to the hated house. Spanish soldiers and gentlemen were seen passing up and down in the streets with drawn swords, crying and cursing. The body of the murdered youth was brought to the Castle in a barge and was then laid out in his uniform of a "captain of the church" and allowed to lie in state for several hours, if such a phrase can be used on such an occasion. The funeral took place at night—the uncoffined corpse, ghastly pale in the glare of the torches that led the way, was carried to Santa Maria del Popolo.* (Bond, 1926)

Figure 36: Savonarola preaching in Florence.

Source: Wikimedia Commons. Artist Nikolay Lomtev.

At the same time the Dominican friar *Girolamo Savonarola*—predicting the future under Borgia rule and condemning the immorality of Florence under Medici rule—preached revolution in Florence against the Papacy in Bologna (Figure 36).

Savonarola's prophesies seemed on the verge of fulfilment when the army of Charles VIII reached the gates of Florence. While Savonarola intervened with the French king, the Florentines expelled the ruling Medici and, at the friar's urging, established a popular republic. In 1495, when Florence refused to join Pope Alexander VI's Holy League against the French, the Vatican summoned Savonarola to Rome. He disobeyed and further defied the pope by preaching under a ban, highlighting his campaign for reform with processions, bonfires of the vanities, and pious theatricals. In retaliation, the pope excommunicated him in May 1497 and threatened to place Florence under an interdict. Cesare Borgia was sent to capture him from Bologna, and—after thorough inquisition—Savonarola was burned at the stake in Florence (Figure 57). Cesare retired as Cardinal and became a military ruler.

It was also the time the conflict between Spain and France intensified during the *Second Italian War of Louis XII* (1499–1503). Pope Alexandre, hearing of Louis's wish to be divorced, used his papal authority to annul that marriage. Sending Cesare on a mission to France with the Bull of Annulment, Cesare succeeded in obtaining in return French military support, and a wife and Duchy for himself. By 1499 Louis's army came to Italy, took Milan, and marched up to the region of Romagna. There, Cesare, empowered by the support of the French, began to attack the turbulent cities one by one in his capacity as nominated gonfaloniere of the church. Soon the estates were seized, and their wealth brought to Rome. But the expulsion of the French from Milan and the return of Lodovico Sforza interrupted Cesare's conquests, and he returned to Rome early in 1500. That was the year that masses of pilgrims visited Rome in the 'Holy Year', bringing richness to the pope's coffers. A papal bull offered Christians who did not make the pilgrimage jubilee indulgences for money. The revenues were added to the Church's tax collections for the war against the Turks. The money enabled the pope to finance the papal army.

In the same time, after a failed attempt by Della Rovere to poison Alexandre with 'cantarella' (arsenic poison), the Borgia revenge with the Purge of the Cardinals started. The cardinals were accused of conspiracy, their lands and wealth were confiscated, and in 1500 some 13 new cardinals were ordained (followed by nine new cardinals on May 31, 1503). Cardinal Ordini was cast into the dungeon and died, the papal army under Cesare's command went in January 1500 into the campagna, crossed the Apennines, and went up north to Forli. As a result, the two great houses of Orsini and Colonna were subjugated. In a second campaign, Cesare also commanded the French troops and laid siege to Naples, which was captured in 1501, causing the collapse of the Spanish rule there. By the year 1501 Cesare was master of all the usurped papal territory and was made Duke of Romagne (Figure 37).

Figure 37: The extent of the Papal States (c. 1700).

Source: http://www.kingscollege.net/gbrodie/
Italy%201700%20showing%20papal%20states.jpg
Artist: Christos Nussli (Map adapted)

The humiliation of the Roman aristocracy was complete; for the first time in the history of the Papacy the pope was, in the fullest sense, ruler of his states. Next, to clean up the last resistance, Cesare began his third campaign into the region of Urbino on June 12, 1502, with Leonardo da Vinci as his chief engineer, and besieging Giulio Orsini's fortress at Ceri on March 14 using Leonardo's war machines.

God seemed to smile upon the Borgias' fate, but that was to change. Although Alexandre survived the first attempt of poisoning organized by Cardinal Della Rovere, that was not the case when he and Cesare drank poisoned wine at a dinner in 1503. Their attempt to poison Cardinal Adrian failed, and at the same dinner, the two Borgias drank the poisoned wine themselves. On August 18, 1503, Alexander died (after a prolonged agony), and Cesare barely survived. The next morning, according to custom, the dead body of Alexandre was carried into St. Peter's.

> The ambassador of Venice writes home to this effect: "Today the pope, according to custom, was carried to St. Peter's and shown to the people, but it was the most loathsome and the most monstrous and frightful corpse that had ever been seen. It had neither the form nor the aspect of a human being." Another ambassador writes that the body was horribly swollen and entirely black, "many think that he died of poison." Another eyewitness declares that the body looked "more black than the devil." (Bond, 1926)

Now that the pope was dead, a new pope had to be elected. As a result of the Papal Conclave of October 1503, in which Cardinal Della Rovere in turn used massive bribery, he was elected Pope Julius II (1503–1513). He became known as 'the Warrior Pope' for his use of bloodshed to increase the territory and property of the Papacy. Julius II tricked Cesare Borgia into supporting him by offering him money and continued papal backing for Borgia policies in Romagna, promises which he disregarded upon election. Soon Cesare's authority in Romagna crumbled after Julius raised a powerful army, which he led himself into Romagna to confiscate Cesare's lands, and with which he regained control of Bologna, Perugia, and many smaller towns in 1506. Cesare was arrested and packed off to prison in Spain. He managed to escape, but he died there in a skirmish in 1507. The Borgia dynasty was no longer a factor in papal politics.

Clearly, the militant politics of those times ruled the Renaissance Papacy. But there was also another side, as this was the time that popes became supporters of art. However, the papal patronage of arts and literature, such as the commission of Michelangelo to paint the Sistine Chapel ceiling, Raphael to paint the Stanze di Raffaello in the Vatican, and Bramante to begin the new St. Peter's Basilica, could not cover up the papal abuses.

> *However brutal and savage the popes were after their return to Rome* [from Avignon], *and however many lives their ambition claimed, they knew well that the careful manipulation of artistic patronage could gloss over their sins and claim violence, murder, and conspiracy as laudable necessities for the greater glory of the Church. It was a paten lie, of course, but, in a sense, that was what Papal Renaissance was all about.* (A. Lee, 2015, p. 370)

To finance the construction of the new St. Peter's Basilica, the practise of 'indulgences' was intensified. Any devoted Catholic parishioner could pay money to the church in exchange for the forgiveness of sins. The de Medici Pope Leo X (1513–1521), short for cash due to his warfare and spending, sold 'cardinal hats', sold memberships of 'Knights of St. Peters', borrowed from bankers, curials, princes, and Jews, and cashed in on indulgences, papal taxes, jubilees. The normal income of the Papacy rose to some 580,000 ducats by 1517, but that was still not enough to cover his expenses.

Next, in the sixteenth century, the *Reformation Papacy* (1517–1585) was faced with the Protestant Reformation and the consequent Counter-Reformation within the church. The attacks from the protestant reformers (such as Martin Luther) on the practises of the Church (eg the system of indulgences), led to a religious crisis in which Northern Europe came under the influence of Protestantism. The Church responded to the Protestant Reformation with a Church reform that started with Pope Paul III (1534–1549). Although by the end of the sixteenth century the prestige of the Papacy was diminished by the spread of the Reformation, and the southern growth of the Papal States was limited by the conflicts with the Spanish power on the Italian Peninsula in the seventeenth century, the Papal States enlarged steadily (Figure 37). At their greatest extent, in the eighteenth century, the Papal States included most of Central Italy: Latium, Umbria, Marche and the Legations of Ravenna, Ferrara, and Bologna, extending north into Romagne. It was, as we will see later on, the French Revolution that proved as disastrous for the temporal territories of the Papacy as it was for the Roman Church in general.

Roman Catholic Church (1050s–1250s)

The Italian Peninsula had been confronted over the centuries by the rise of the Holy Roman Empire and by the growth of the Papal States. In addition, the influence of the Roman Catholic Church over Western Europe had steadily grown. As we have seen, the papal authority had developed into both a religious and secular power. Not without power struggles of its own, the 'papal supremacy' had grown above the realm of the Church itself and had often clashed with the monarchs of the day. Monarchs who, in turn, were involved in a continuous power play for their ruling positions. Moreover, both sides used each other to reach their objectives: power over a greater realm, whether it be the power of the Roman Catholic Church or the power of the Holy Roman Emperor. Thus, from 1048 to 1257, the Papacy experienced increasing conflicts with the leaders and churches of the Holy Roman Empire and the Byzantine Empire. Longstanding divisions between East and West also came to a head in the previously mentioned East-West Schism of 1054, and continued with the Crusades, which started in 1095.

Religious Wars: The Crusades

The crusades were a series of religious military campaigns—armed pilgrimages—initiated by the Roman Catholic Church in several forms during the late eleventh and fourteenth century (Figure 38). One reason for this was that the Crusades were part of the power play by the pope to expand the Papacy's sphere of influence in Europe beyond the Papal States. The pope mingled with the affairs of the temporal rulers.

> *The Popes offered privileges, both spiritual and temporal, to all who took the Cross. Because of the intense enthusiasm for the crusades and also because of the weakness of most of the monarchs in Western Europe during the first half of the twelfth century, the Church, and especially the Pope, were allowed through these privileges to encroach upon the sphere of the temporal authorities. All crusaders were given the protection of the ecclesiastical courts; thus when a vassal took the cross he might escape to a considerable extent from the jurisdiction of his feudal lord. Moreover, his family and property were taken under the protection of the Church and in this way many cases were taken from the feudal court. … Other instances might be cited to show the manner in which the Popes added to their*

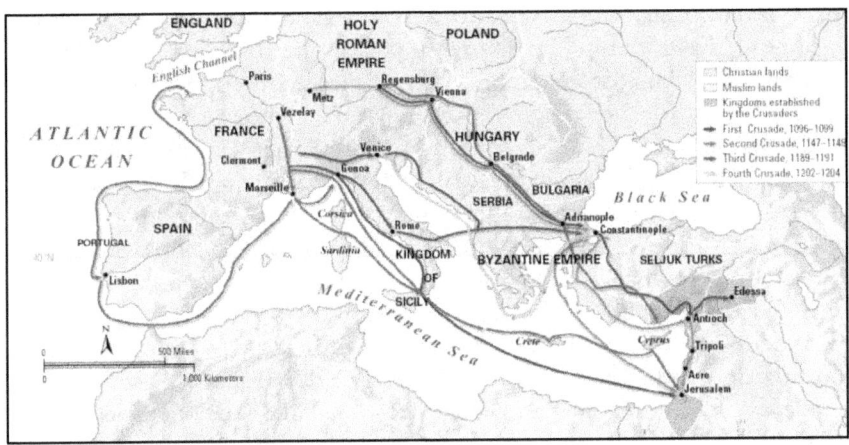

Figure 38: Map of the first four Crusades.

Source: www.thinglink.com

temporal power and control over those who were not members of the clergy, so that after a century of crusading activity the Pope's power had been enormously enhanced. (D. C. Munro, 1916, pp. 350, 351)

Also, there was the wish to bring the Greek Church back under the Roman curia. And the Crusades were an outlet of the religious piety that preceded them. People were gripped by a religious passion. The crusader, as soldier of the Church, would receive 'remission of sins'; absolution and indulgence, and dying during the war as martyrs allowed for automatic entry into heaven.[93] For the Church, the crusade proved a profitable business.

The crusades also brought to the Church and to the Popes an enormous increase in wealth. Crusaders gave freely to the Church before starting for the East; they also mortgaged or sold their property to ecclesiastical foundations under conditions very advantageous to the latter. The Orders of the Temple and Hospital received great endowments and became very wealthy. Men who had taken the Cross and were unable to go, purchased exemption from their vow. Taxes for the crusades were frequently collected and handled by the Church. It is not possible to give any estimate of the total amount which the Church received through the crusades, but it was enormous. Consequently the Popes became much more powerful, especially through their control over the appointment of the officials who profited from this wealth. (D. C. Munro, 1916, p. 352)

---

[93] There seems to be a striking parallel with today's religious wars conducted by Muslims: the religious duty called 'jihad' referring to spreading the Islamic religion by the mujahideen. These martyrs are also promised Heaven.

The *First Crusade* (1096–1099) arose after a call to arms in a 1095 sermon to a large clerical assembly by Pope Urban II in the southern French town of Clermont. Calling for liberation of the Eastern churches, he urged military support for the Byzantine Empire and its emperor, *Alexius I Comnenus*, who needed reinforcements for his conflict with westward migrating Seljuk Turks in Anatolia, Asia Minor. To that goal, he added the objective of winning back the Holy Land and the most hallowed site in the Christian universe: Jerusalem, at that time ruled by the Muslims. So, the months and years that followed saw disjointed waves of departure. In total, some 60,000–100,000 Latin crusaders answered Urban's call, among them five princes with some 7,000–10,000 knights. For these potentates, their followers, and perhaps even the poorer classes, the process of joining the crusade involved a dramatic and often emotional and religious-laden ceremony.

Arriving in Constantinople, they offered their services to emperor Alexius. To the sophisticated Byzantines, it looked like a great barbaric invasion, now the West invading the East. Continuing their journey, the crusaders left Constantinople for the Holy Land across Asia Minor. After a range of battles and pillages, they arrived at the gates of Jerusalem on June 7, 1099. By July 15 they took the city, creating a bloodbath among the Muslim inhabitants fleeing to the Temple Mount. Then the crusaders prayed as pilgrims at the Holy Sepulchre, as had been their objective from the start.

Soon most of the surviving crusaders—some 20,000—returned to their homes in Europe, where they were welcomed as heroes. However, the material success of the first crusade was small. The Byzantine Emperor Alexius had recovered small parts of Asia Minor. The crusaders had created the four *Crusader States* in the Eastern Mediterranean: the Country of Edessa, the Principality of Antioch, the County of Tripoli, and the Kingdom of Jerusalem (Figure 39), each with its own aristocratic ruler.

Then came the *Second Crusade* (1147–1149), which started after the County of Edessa was recaptured by the Turks. But now the armies of two kings were marching separately across Europe: one French army under Louis VII and a German army of 20,000 soldiers under Conrad III of Germany. From Constantinople, the two armies again separately progressed to Antioch. Continuing their journey after many losses during their voyage, they reached Jerusalem in April 1148. The nobility of Jerusalem welcomed the arrival of troops from Europe, and at the Council of Acre, it was decided to attack Damascus. The attack failed, however, and the disintegrated army went back to Jerusalem and, from there, to their home countries. Thus, the Second Crusade became a failure for the crusaders and

Figure 39: Crusader States (ca 1100).

Source: Wikimedia Commons.

a great victory for the Muslims. It would ultimately have a key influence on the fall of Jerusalem. That happened in 1187, when the Muslims of Saladin's army took the city. There was little bloodshed, and ransoms were paid, allowing the crusaders to leave the city.

As a reaction to the fall of Jerusalem, Pope Gregory VIII called for a new crusade, the *Third Crusade* (1189–1192), which started with separate contingents led by Richard the Lionheart (king of England), Philip Augustus (king of France), and Frederick Barbarossa (emperor of the Holy Roman Empire). The large German army of some 100,000 men reached Anatolia, but during a river crossing their leader, Emperor Barbarossa, fell from his horse on the rocks and drowned (Figure 40).

Much of the army returned to Germany; only some 5,000 crusaders continued. By that time the English army of some 10,000 men reached—after a delay to fight Muslims in Lisbon, Portugal—the region of Acre. That was also the case with the French army, which consisted of 650 knights, 1,300 horses, and 1,300 squires. Philip had hired a Genoese fleet for the transport to the Holy Land by way of Sicily. By June 1194 the first confrontation with the Turks began: the *Siege of Acre*.

After negotiations the city surrendered, and the first prisoners of war were exchanged. But when further negotiations failed, Richard ordered the massacre of 2,700 Muslims prisoners. The Kingdom of Jerusalem was re-established in Acre and became the *Kingdom of Acre*. Although more successful than the previous crusade—the crusaders managed to maintain considerable states in Cyprus and on the Syrian coast—the Third Crusade ultimately failed to reconquer the Holy City of Jerusalem.

Figure 40: Death of Barbarossa after drowning (1190).

Source: Wikimedia Commons. Artist: Gustav Dore

Thus, the *Fourth Crusade* (1202–1204) was planned to reconquer Muslim-controlled Jerusalem (Figure 41). Already called for by Pope Innocent II in 1198, the idea of a new crusade was largely ignored by the European monarchs. They were too busy fighting with each other: the Germans were struggling against papal power, and England and France were still engaged in warfare against each other. Then the tide turned.

> *The Fourth Crusade was actually conceived in 1199 at a jousting tournament held by Thibaut, Count of Champagne, at Ecry-sur-Aisne in northern France. There, in a sudden wave of mass emotion, the assembled knights and barons fell to their knees weeping for the captive Holy Land. ... Rather than wear out their army by a long land march through hostile territory, the leaders decided to reach Egypt by sea. A delegation of six trusted knights went to Venice, the leading seafaring city of Western Europe, to arrange for passage. ... In Venice, Villehardouin and his fellow envoys hammered out an agreement with Doge Dandolo and his council. Venice would provide transport ships, crews and a year's provisions for 4,500 knights with their mounts, 9,000 squires and sergeants (feudal men-at-arms of less than knightly rank), and 20,000 ordinary footmen, for a total of 33,500 men and 4,500 horses. The price for this armada would be 84,000 marks of silver. And the old doge made Venice not a mere supply contractor, but a full partner in the crusade. In return for a half-share of*

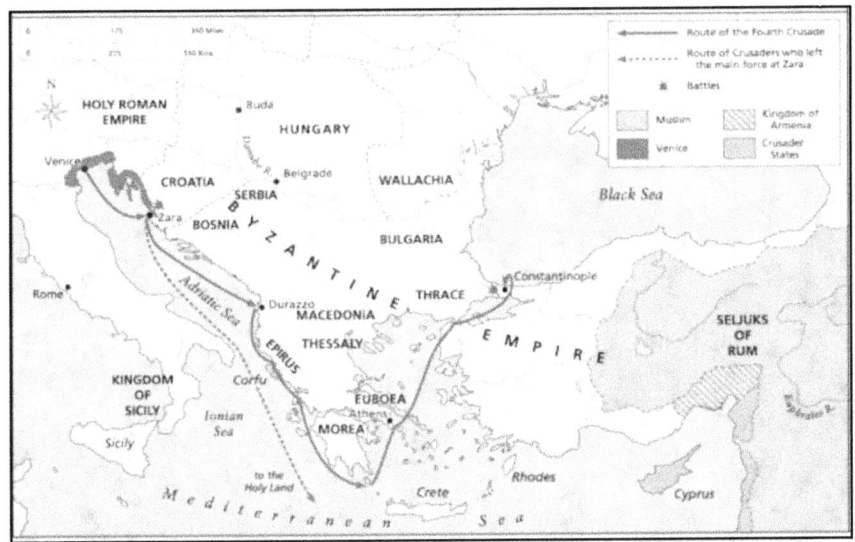

Figure 41: Route of the Fourth Crusade (1202–1204).

Source: www.haikudeck.com

*all conquests, Venice would provide an escort force of 50 fully manned war
galleys. The great fleet was to sail in the summer of the next year, 1202.*[94]

Thus, during the late spring of 1202 the crusaders began to gather at
Venice. However, the crusade turned out only some 12,000 men and could
not raise the original price agreed upon. Although the Venetians had
suspended their regular commerce to build and equip an immense fleet of
the promised 50 war galleys and 450 transport ships, they made a new offer.
The Venetians would suspend the unpaid balance of the transport charge in
return for a small favour: the crusaders' assistance in conquering the
rebellious city of Zara (later to become Zadar, Yugoslavia), a Hungarian-
owned port on the Dalmatian Coast of the Adriatic. So, the crusaders
attacked and sacked Zara, stayed the winter, and planned to continue their
voyage to Jerusalem.

However, the rulers of the Byzantine Empire had problems of their
own. In January 1203, en route to Jerusalem, the majority of the crusader
leadership entered into an agreement with the Byzantine prince *Alexios
Angelos* to divert to Constantinople and restore his deposed father as
emperor. The intention of the crusaders was then to continue to the Holy
Land with the promised Byzantine financial and military assistance. On

---

[94] Source: 'Fourth Crusade'. www. Historynet.com

Figure 42: Siege of Constantinople by the crusaders (1204).
Source: Wikimedia Commons. Artist: Palma Le Jeune.

June 23, 1203, the main crusader fleet reached Constantinople. Smaller contingents continued to Acre. By August 1203, Alexios was crowned co-emperor with crusader support. But not for long, as a popular uprising between pro-crusader Latins and anti-crusader Greeks caused his downfall and murder on February 8, 1204.

The crusaders, now quite angry at not having received the promised payments, decided to act. On April 12, 1204, they conquered the city from the seaside (Figure 42). The crusaders looted, terrorized, and vandalized Constantinople for three days, during which many ancient and medieval Roman and Greek works were either stolen or destroyed.

> One of the great libraries of the ancient world, the *Imperial Library of Constantinople*, was significantly destroyed by the illiterate Frank crusaders. Many papyrus scrolls were stripped from their (rich) covers and burnt.

Others, such as the Archimedes Palimpsest[95]—containing seven treatises of work by Archimedes of Syracuse, the mathematician of antiquity, in a tenth-century copy—travelled to Jerusalem and survived into our times. Much of the later knowledge of the Greek classics came from the earlier Byzantine copies originating from this library.

The civilian population of Constantinople was subject to the crusaders' ruthless lust for spoils and glory; thousands of them were killed in cold blood. Despite their oaths and the threat of excommunication, the crusaders systematically violated the city's holy sanctuaries, destroying or stealing all they could lay hands on; nothing was spared, not even the tombs of the emperors inside the St. Apostles Church.[96]

So, the Fourth Crusade went wrong and resulted in an attack on the Byzantine Empire instead of Islam with the sacking of Constantinople, the largest and most sophisticated city in Christendom. The sack weakened the once-mighty Byzantine Empire and instead established a *Latin Empire* in the East ruled by crusader participants.[97] The Latin emperor controlled one-fourth of the Byzantine territory, the city state of Venice three-eighths (including three-eighths of the city of Constantinople), and the remainder was divided among the other crusade leaders. Most of the crusaders involved never arrived at the kingdom of Acre, but went home with the spoils of war. It was a turning point in the decline of the Byzantine Empire; as with the events of 1204, the schism between the churches in the East and West was not just completed but also solidified.

The *Fifth Crusade* (1213–1221) was a new attempt by Western Europeans to reacquire Jerusalem and the rest of the Holy Land by way of Egypt. The first contingents of the Fifth Crusade, led by King Andrew of Hungary, reached Acre in the fall of 1217. Andrew accomplished little, however, before departing in January 1218. A large fleet of Frisian, German, and Italian crusaders arrived in April and joined the remnants of Andrew's force. In May the combined army set out for Egypt under the leadership of John of Brienne (the titular king of Jerusalem from 1210). They arrived in Egypt, did some fighting and negotiating, but that was all. Then the flooding Nile sealed their fate, and they were forced to surrender. Always

---

[95] Source: The Archimedes Palimpsest is a medieval parchment manuscript, now consisting of 174 parchment folios. http://archimedespalimpsest.org/about/

[96] Rather than wantonly destroying all around like their comrades, the Venetians stole religious relics and works of art, which they would later take to Venice to adorn their own churches, such as the famous bronze horses from the Hippodrome that were sent back to adorn the façade of St Mark's Basilica in Venice, where they still are today.

[97] The Latin Empire ruled by the Latin emperor was located around the Sea of Marmara. Isles like Crete and Rhodes were under Venetian control.

Figure 43: Knights Templar in battle.

on the verge of success, the Fifth Crusade failed largely because of divided leadership and the frequently unwise military decisions.

Other crusades followed in the coming years, such as the Bosnian Crusades, but none were focussed on the Holy Land. The crusades gave rise to the military orders of religious fighters—such as the Knights Hospitallers and the Knights Templars—though.

> Take, as an example, the development of the religious order of the military *Knight Templars*,[98] the monk elite fighting force of their day, highly trained, well-equipped and highly motivated (Figure 43). Originally, the order started as an institution in which the knights protected pilgrims on their journey to visit holy places after the First Crusade (1095–1099), certainly a service that was needed at the time, as robbery threatened travellers everywhere on their long voyages. Though initially an order of poor monks, the official papal sanction made the Knights Templar a charity across Europe. Traveling pilgrims could draw on funds deposited in a Templar branch in their home country when passing branches underway.

> By 1150 the order's original mission of guarding pilgrims had changed into a mission of guarding their valuables through an innovative way of issuing letters of credit, an early precursor of

---

[98] Another order was the Knights Hospitallers.

modern banking. Further resources came in when members joined the order, as they had to take oaths of poverty and therefore often donated large amounts of their original cash, properties, land, and business dealings to the order. In 1139 Pope Innocent II issued the papal bull *Omne Datum Optimum*. It stated that the Knights Templar could pass freely through any border, owed no taxes, and were subject to no one's authority except that of the pope. Soon the Templars became involved in business, as knights who went on a crusade put their business interests under the power of attorney of the Templars. The order also gave the pope his military power.

During the 180 years of the Crusades, the Templar wealth grew into a huge fortune. From donations, they owned over 9,000 manors and castles across Europe (Figure 44), all of which were tax-free. Each property was farmed and produced revenues that were used to support the largest banking system in Europe. From conquests, they owned tracts of land in the Middle East, built churches and castles, bought farms and vineyards, were involved in manufacturing and import/export, had their own fleet of ships, and, for a time, even 'owned' the entire island of Cyprus. The Templars were a 'state within a state', were institutionally wealthy, paid no taxes, and had a large standing army which by papal decree could move freely across all European borders.

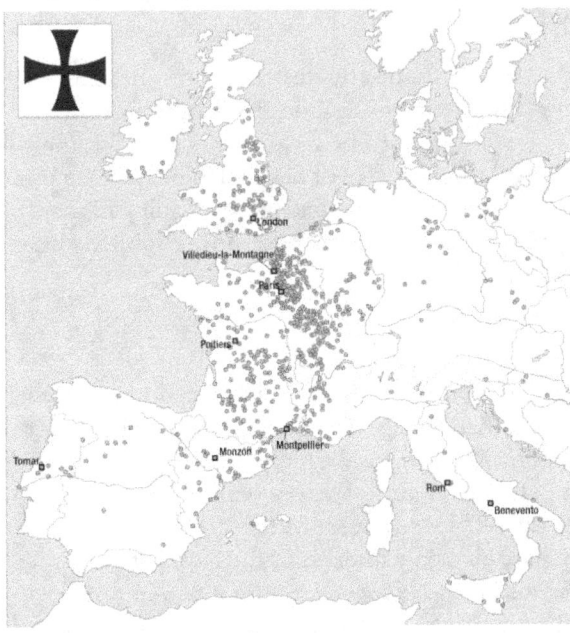

Figure 44: Posessions of Knights Templar in Europe (c. 1300).

Source: Wikimedia Commons. Author: Marco Zanoli based on Grosser Historischer Weltatlas. Bayrischer Schulbuch-Verlag (Hg.). Bd. 2, Mittelalter. München 1970, S. 82.

The Crusades had far-reaching impact on European countries, politically, religiously, scientifically, and economically. Not only did they introduce new forces in the political realm of the Eastern Mediterranean, they consolidated the papal leadership of the Latin Church, reinforcing the link between Western Christendom, feudalism, and militarism which manifested itself in the habituating of the clergy to violence.

Also, the growth of the system of indulgences, among others used to finance the Crusades, was a catalyst for the Protestant Reformation. In addition, the access to (copies of) classical Greek and Roman texts brought back from the East allowed Europeans to rediscover the pre-Christian philosophies of Aristotle, Socrates, and Plato. And quite importantly, the Crusades also opened the 'Europe Overseas' (aka Outremer) to commerce and trade. The Italian city-states of Genoa and Venice especially flourished, creating profitable trading colonies in the eastern Mediterranean. The architecture, Eastern cultures, and advances in science and medicine that the crusaders were exposed to had an influence when they came back. But it was also the start of the Muslim invasions of Europe later on.[99]

> *The crusades, which had been directed against the Muslims for the rescue and protection of the Christian and their holy places, in the end did more damage to the Christendom than to the Islam. The Muslims ultimo triumphed with the conquest of Asia Minor, of the Balkans, and of Constantinople itself in 1435* ... (D. Baker, 2009, p. 7)

Due to the failure of the early Crusades by mistrust and conflicting interests, much of the control of the later Crusades had come from the popes, for they considered the Crusades to be God's work and that they— the popes– were his agents. Although the Crusades had increased the power and wealth of the Roman Catholic Church considerably, it still had major problems with the worldly powers of that time. And those problems were related to the developments of the Italian city-states.

The City-states (1100s–1500s)

Up to the Late Middle Ages (fourteenth and fifteenth century), Italy had been a geographical region that had seen a range of different political forces at work in a complex geopolitical framework. On one hand, there were the worldly powers such as the Holy Roman Emperors with their actions. On the other hand, there were the religious powers of the Roman Catholic Church mixed up with the papal worldly interest in the Papal States. It created both *imperialistic domination* and *religious domination*, two important components of the overall geopolitical dynamics of that time. In addition, on the Italian Peninsula, independent civil developments had taken place

---

[99] In 1453 the Ottomans ended the Byzantine Empire with the conquest of Constantinople. From then on the Ottoman Empire would grow into Europe, including Greece and the Balkan Peninsula up to Hungary.

that had introduced a power of their own in the society: the city-states with their *republican domination* (Figure 45).

By the eleventh century, many cities—including Venice, Milan, Florence, Genoa, Pisa, Siena, Lucca, Cremona—had become large trading metropolises, able to obtain independence from their formal sovereigns. Most were old habitations. Genoa originated from Etruscan times and had

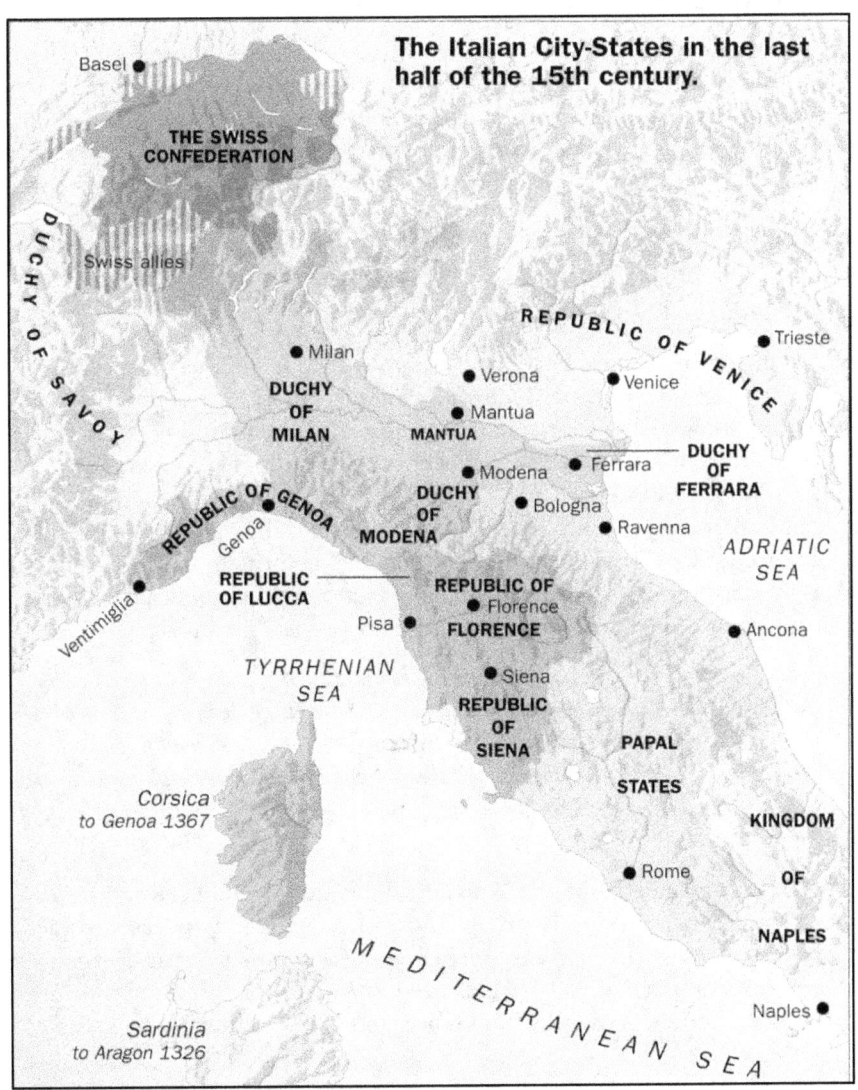

Figure 45: Italian city-states (15th Century).

Source: University of Oregon. http://pages.uoregon.edu/mapplace/EU/EU19%20-
   %20Italy/Maps/EU19_128.jpg

become a Roman port. Venice was founded in the fifth century by people fleeing from Attila the Hun. They settled on a group of islands on the northeastern edge of the Italian peninsula. Florence, the 'city of flowers', located in the Etruscan hill country of north-central Italy, was started when former legionnaires were allotted former Etruscan lands in the rich farming valley of the Arno. It prospered because of the early wool industry.

Republican Rule in the Oligarchic City-state

The societal development in the early history of Italy as we know it today, shows a deviation from the feudal patterns in the rest of Europe, in which the absolute monarchy of king and Church, and aristocracy and clergy, ruled within the remnants of the feudal system. In Italy, however, during the eleventh century an elaborate pattern of communal government began to evolve under the leadership of a burgher class (aka 'comuni') grown wealthy in trade, banking, and such industries as wool textiles. Many cities, especially Milan, Genoa, Venice, Florence, and Pisa, became powerful and independent city-states (Figure 45).

Resisting the efforts of both the old nobles and the emperors to control them, these 'comuni' promoted the end of feudalism in Northern Italy, replacing it with a deeply rooted identification with their city as opposed to the larger region or country. Public business was conducted by and for citizens—the 'popolo'—who knew each other, and civic issues were a matter of widespread and intense personal concern. The darker side of this intense community life was conflict: conflict between ruling families, ruling factions, and even between rivalling cities. It resulted in a politically fragmented, but quite democratic, northern part of the peninsula.[100]

> *Most notably during the later Middle Ages and the Renaissance, at precisely the time when the great western monarchies were consolidating, the political system of central and northern Italy was characterized by the city-states' great fragmentation and spirit. … The cities transformed themselves precociously into city-states with corresponding territorial dimensions and political functions.* (Chittolini, 1989, p. 659)

The early communities in the Northern Italian delta had known, after Roman's demise, a continuous struggle for independence. Especially in the medieval times, they tried to gain independence from their common enemy, the German 'Swabian' rulers—such as the House of Hohenstaufen—of the Holy Roman Empire. Around 1100 Genoa and Venice emerged as

---

[100] Sources include: 'The Rise of the Italian City States' (www.arcaini.com), 'City-states in Italy' (http://www.wga.hu/). Signoria (Encyclopedia Britannica). Gilbert, W., *Renaissance and Reformation* (1998), Chapter 3: The Italian States of the Renaissance. (http://vlib.iue.it/carrie/texts/carrie_books/gilbert/)

independent maritime republics. By the same time Lucca and Pisa had also rejected bishopric and imperial overlord-ship in favour of establishing their own communes, and promoted protection by regional nobility. In the Po delta, similar developments took place. The idea of the commune, a consortium of guilds, spread beyond Lombardy and Tuscany during the late eleventh and twelfth centuries. These communes were evolving gradually into formal city-states, each with its own republican institution of government.

Sharing the same problems against the Holy Roman Empire rulers, many of the northern city-states sought cooperation. It had resulted in the 1160s in the creation of the medieval alliance of the *Lombard League* (Italian: 'Lega Lombarda'), which included most of the republican communes in the Po delta of Northern Italy including Milan, Piacenza, Cremona, Mantua, Crema, Bergamo, Brescia, Bologna, Padua, Treviso, Vicenza, Verona, Lodi, Reggio Emilia, and Parma. These opponent cities, supporting the pope, who had his own worldly agenda, and defending the liberties of the urban communities against the emperors, were called the 'Guelfi' (aka the Church Party). More to the south, the central Italian Peninsula up to the fifteenth century was also a patchwork of independent city-states, republican communes that were the survivors of the towns that had existed during the Roman Empire, like Florence, Pisa, Siena, Lucca, Ancona, and many others. Some had become large trading metropolises, able to obtain independence from their formal sovereigns. These cities, known as the 'Ghibellini' (aka the Imperial Party), were supporting the imperial aspiration of the Holy Roman Emperors.

The division between Guelfi and Ghibillini developed its own dynamic in the politics of medieval Italy, and it persisted long after the direct confrontation between emperor and pope had ceased. Smaller cities tended to be Ghibelline if the larger city nearby was Guelph, as can be observed in the fractious relationship between the Guelph Republic of Florence and the Ghibelline Republic of Siena. They faced off at the *Battle of Montaperti* (1260), the bloodiest battle fought in medieval Italy, with more than 10,000 fatalities. Pisa maintained a staunch Ghibelline stance against her fiercest rivals, the Guelph Republic of Genoa and Florence.

However, over time the republican institution of government was replaced by the rule of individuals: the 'signori'. Originating from the military with the power to safeguard the city-states, they gradually extended their authority until it was permanent and made hereditary in their families. The Lombard League passed out of existence after the death of Holy Roman Emperor Frederick (Barbarossa) in 1250. But its legacy was the independent oligarchic city-state. Once the machinery of the individual rule came in place, in commune after commune, from the late thirteenth century, the local oligarchs accepted a powerful leader as their signori and subsequently allowed the post to remain with a family. In this way, the near-democracy of the early Italian communes reverted gradually to princely rule. During the fifteenth century the mainland was dominated by five great powers: Venice, Milan, Florence, the Papal States, and Naples (Figure 45).

This description of the republican development of the city-states gives the general overview. It is time to explore the historic development of some of these city-states over the centuries in more detail.

Popolo, Podesto, and Signori

At the end of the fifteenth century we find the northern and central part of the Italian Peninsula to be a patchwork of city-states. The fertile Po plain area, the mountainous Mid-peninsula, and the empty and poor South were a melange of political and cultural elements, developed over time from Mediterranean influences (Phoenician, Greek, Byzantine), invasions (Goths and Norman) and German imperial conquests (Frank, Carolingians, Ottonians, and Hohenstaufens). In general, the rule of the city-states was a heterogeneous mixture of 'popolo'—non-noble people ruling communities—and the 'podesta',[101] the rule of imperial magistrates.[102]

The people of the most city-states of the era were composed of four social classes: the nobles, the merchants, the tradesmen, and the unskilled workers. Milan had, around 1400, over 100,000 inhabitants. Genoa, Bologna, and Verona each had over 50,000 inhabitants. Cities based on trade and commerce, such as the merchant republics of Venice and Genoa. By that time most cities—except Venice, Florence, and Lucca—were ruled by signori: Milan by the Visconti family and later by members of the House of Sforza, Ferrara by the House of Este, Verona by the Della Scalas, and Padua by the House of Carraras, just to name a few of the many local families. However, other governing bodies also existed, such as in Florence and Venice.

---

[101] In November 1158 Frederick appointed in several major cities imperial podestàs 'as if having imperial power in that place'.

[102] It was the 'popolo'—the heterogeneous middle ranks of urban society and families of artisans organized in guilds, merchants, teachers, and doctors—who dominated the local political scene in many city-states. And to govern the city, they assigned the 'capitano del popolo' as municipal officeholder. However, soon this function became a breeding ground for despotism and hereditary lordship. By gaining control of the election process for choosing the title-holder, many influential families (including aristocrats that the establishment of this office had contributed to keeping out of power) gained control over their cities and towns of residence, thus assuring their long-lasting influence and progressively transforming the commune into a signoria.

In the thirteenth century Europe in general had been experiencing an economic boom. Also, the city-states of Italy expanded greatly during this period and grew in power to become *de facto* fully independent of the Holy Roman Empire. During this period the modern commercial infrastructure (including the land and sea trading routes) developed: a system with joint stock companies, an international banking system, a systematized foreign exchange market, insurance, and government debt. Milan was dominant among the Lombard communes and was soon to make its bid for hegemony in Northern Italy. In Tuscany, Florence became the centre of the financial industry, and the gold florin became the main currency of international trade. It had increased in size, wealth, and power to become the dominant force in Tuscany and the chief bulwark against the ambitions of the Visconti, the powerful Milanese family. In Rome, the Papacy vacillated under Angevin influence, while the kings of Naples carried on a vain war to retake the island of Sicily from the Aragonese. Venice came to dominate the rich trade with the Byzantine Empire, especially after the Fourth Crusade (1204). Genoa, which eclipsed Pisa in the latter part of the century, expanded its trade in the western Mediterranean and in the Provence.

These were the times of the emerging powerful city-states in Northern and Central Italy amidst the greater geopolitical forces as executed by the Holy Roman Emperors and the popes of the Holy Roman Church. Most of the cities had poor defence systems, especially when one considers the development of military technology elsewhere in Europe. It made them vulnerable to foreign invasion again. One of those cities was Milan.

The Duchy of Milan

The Milanese Commune, emerging from Longobardian and Carolingian domination, had by 1042 seen the rise of the 'cives' (merchants, artisans, judges, notaries). These contested the 'podesta' that were appointed by the Holy Roman Emperor and represented his imperial power. By the end of the eleventh century the urban middle class organized its first governing council ('Consulatus Civium', 1097), laying the foundation for the creation of an independent commune. But that independence did not come easy, as in 1154 the German ruler Frederick Barbarossa crossed the Alps in his *First Italian Campaign* (1154–1155) and went to Rome, where he was crowned by Pope Adrian IV as Holy Roman Emperor. In June 1158, Frederick set out on a second campaign, seizing Crema in 1159 and next besieging Milan. It took three years, until in 1162, for the city to capitulate to an emperor incensed beyond mercy.

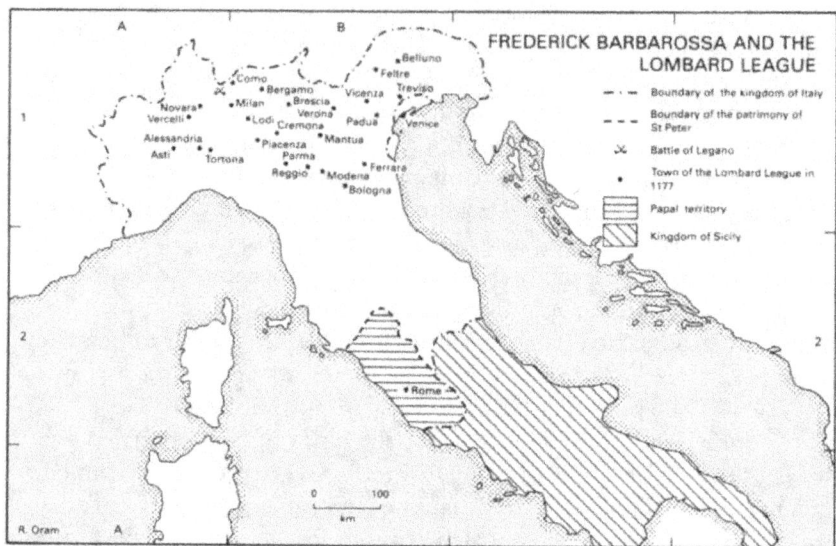

Figure 46: The cities of the Lombard League (1176–1250).
Source: http://www.worldhistory.biz/uploads/posts/2015-07/375q-54.jpg

The city was first pillaged, and then given over to the hands of the Lombards, who—such was the diligence of hatred—are said to have done more in six days than hired workmen would have done in as many months. The walls and forts were torn down, the ditches filled up, and the once splendid city reduced to a frightful scene of ruin and desolation. Then, at a splendid banquet at Pavia, in the Easter festival, the triumphant emperor replaced the crown upon his head. ...

For five years Milan lay in ruins, a home for owls and bats, a scene of desolation to make all observers weep; and then arrived its season of retribution. Frederick's downfall came from the hand of God, not of man. A frightful plague broke out in the ranks of the German army, then in Rome, carrying off nobles and men alike in such numbers that it looked as if the whole host might be laid in the grave. Thousands died, and the emperor was obliged to retire to Pavia with but a feeble remnant of his numerous army, nearly the whole of it having been swept away. In the following spring he was forced to leave Italy like a fugitive, secretly and in disguise, and came so nearly falling into the hands of his foes, that he only escaped by one of his companions placing himself in his bed, to be seized in his stead, while he fled under cover of the night.[103]

---

[103] Source: 'Frederick Barberaossa and Milan'.
http://www.mainlesson.com/display.php?author=morris&book=german&story=milan

This gave the city some relief, and Milan was rapidly rebuilt. United against a common enemy, together with other city-states—such as Cremona, Verona, Padua Bergamo, Genoa—the Milanese created the *Lombard League*, aka the 'Societas Lombardie' (1176–1250) (Figure 46). And when the Holy Roman Emperor Frederick Barbarossa returned in 1174 to Italy with a new army on a new campaign, they defeated him in the *Battle of Legano* in 1176. As Barbarossa also had other things to attend to in Germany, a truce of six years followed. It resulted in the *Peace of Konstanz* (1183), giving the cities of the Lombard League a form of independence, the regalia or privileges, such as the important freedom to elect their own councils and the freedom to enact their own legislation. Which they did.

By 1198 the plebeian merchants and artisans had created the 'Credenza di Sant'Ambrogio'; the wealthier burghers, merchants, and men of liberal profession had created the 'Credenza della Motta'; the lower nobles, the 'Credenza di Valvassorri'; and the higher nobles and archbishop, the 'Credenza dei Gagliardi'. Together, they were the four 'credenza' constituting the town council of *Credenza*. This institute fuelled, despite the controversy between the powerful nobles and the growing class of burghers, the renewed city with industrialists, traders, and artisans.

Milan was commendably full of various religious orders dedicated to different activities: Hospitallers and Sisters of Charity, the lords and intermediaries in the wool industry (the Humiliati), mediators of international politics and great

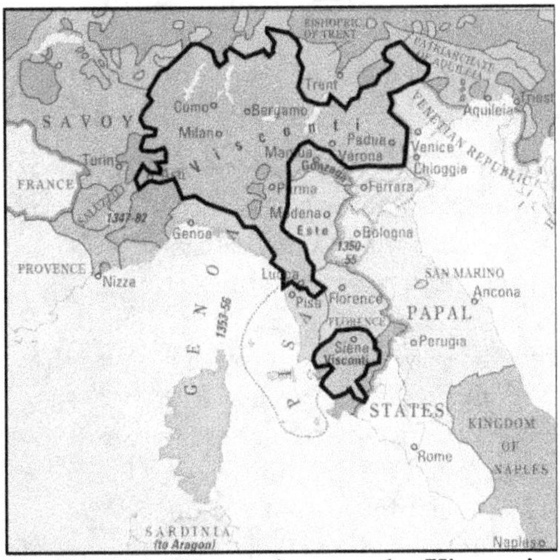

Figure 47: Milan territory under Visconti rule (c. 1350–1400).

Source:
http://pages.uoregon.edu/mapplace/EU/EU19%20-%20Italy/Maps/EU19_02.jpg. (Map adapted)

architectural experts (the Cistercians), contemplative and active friars. In particular the city was a centre for the production and export of metalwork and weapons …, but on the whole works and jobs were extremely diversified. … (Lopez, 2012, pp. 13-15)

By the end of the thirteenth century, however, the growing decadence of the commune had resulted in decades of civil war and in the rise of the *Visconti dynasty* of rulers when Ottone Visconti (1208–1295), appointed Archbishop by Pope Urban IV, emerged victorious and established a dynasty. In 1311

Matteo Visconti (1250–1322) consolidated his family's power over the city, and by 1395 Gian Galeazzo Visconti (1351–1402) had become the first duke of Milan. Between those two rulers, the Viscontis rose to power and expanded Milan's territory (Figure 47). Again it was a violent time of political intrigues and epidemics, and Milan was the major enemy of the Papal States, the Savoy, Venice, and Florence.

One of those Visconti rulers was the despot Barnabò Visconti (1323–1385). From 1360 to 1378 he was in constant conflict with three popes. It started in 1360, when he was declared heretic by Innocent VI at Avignon and condemned by Emperor Charles IV. Next, his refusal to return Bologna gained him excommunication by Pope Urban V. And lastly, after warring against the Papal Mantua, Reggio Modena, and Ferrara he was excommunicated by Pope Gregory XI. To finance al this warfare, he taxed his vassals two or three times a year, a half or a third of their wealth, driving the Milanese population into poverty. Then, in 1385 he was ousted in a coup d'état by his nephew Gian Galeazzo Visconti. After being poisoned Bernabò left 36 children alive when he died and 18 women pregnant by him.

Under later Visconti rulers—now called dukes after Gian Galeazzo Visconti (1347–1402) bought that title from the Holy Roman Emperor Wenceslaus—Milan expanded. Soon other cities of Northern Italy came under Milan's influence—Verona, Vicenza, Padua—and Milan was controlling the entire valley of the river Po. Gian Galeazzo Visconti pursued his dream of unifying Northern Italy into one kingdom and attacking Bologna and Florence, but he died in 1402, victim of the plague. The Visconti dynasty died with Filippo Maria Visconti in 1447.

By that time the people wanted peace and created the short-lived 'Aurea Republica Ambrosiana' (1447–1450). Then Captain General Francesco Sforza (1401–1466) rose to power when, after years of famine, riots raged in the streets of Milan and the city's senate decided to entrust to him the duchy. Sforza entered the city as duke on February 26, 1450. That was the start of the rule of the *House of Sforza*. During Sforza's reign, Florence was under the command of Cosimo de' Medici, and the two rulers became close friends. This friendship eventually manifested in, first, the *Peace of Lodi* (1454), which resulted in the Italian League (1450–1494), a multi-polar defensive alliance of Italian states that succeeded in stabilizing almost all of Italy for its duration. The league provided enough stability to allow the peninsular economy to recover from the population loss and economic depression caused by the Black Death (1350s) and its aftermath, leading to an economic expansion that endured until the first part of the seventeenth century. It was, for example, Ludovico Sforza (1452–1508), also called Ludovico il Moro and married to Beatrice d'Este, who took Leonard da Vinci as a military engineer in his service in 1482. For the Sforzas, Leonardo designed mechanical constructions like dredging machines, canal bridges, and sluices, along with his 'machines of war' (Figure 11).

The city too, as a whole, showed signs of great fervour: Ludovico stimulated the building industry not only with great projects of Santa Maria del Grazie or of the ducal complex at Vigevano, but also with civic constructions and ... agricultural structures and buildings from the pre-Alps to the plain crossed by the waters of the Naviglio Grande ...

However, for the ordinary people bread continued to be up to 90 per cent of the daily nourishment, the rest being wine and little more. Hence the fundamental importance of cereals (millet and rye, wheat was already a luxury) not only in everyday life but also in war. Whoever did not have enough supplies of cereals and wood to keep warm in the bad years was destined to die. ...

So, in Beatrice's days, the Court of Milan was the leader in the art of gracious living, elegance, jewellery, furnishing, decorating and amusements, and the city appeared in the avant-garde for techniques. In other words, a more favorable ambience Leonardo da Vinci could not have found: the Leonardo who would magnify Ludovico's fame for centuries to come. ... the city became a centre of attraction for artists, mathematicians, engineers and humanists. It was a place where it was possible to discuss alchemy, esoteric philosophies and practise a certain syncretism. It was surrounded by a territory where water ran a network of

Table 1: Rulers of Duchy of Milan (1395–1796).

Nation	Ruler	House	Period
-	Different local rulers	Visconti	1395–1447
-	Different local rulers	Sforza	1450–1499
France	Louis XII	Orleans	1499–1512
-	Massimiliano Sforza	Sforza	1513–1515
France	Francis I	Valois-Angouleme	1515–1521
-	Francesco II Sforza	Sforza	1522–1535
Spain	Philip II	Habsburg-Spain	1540–1598
Spain	Philip III	Habsburg-Spain	1598–1621
Spain	Philip III	Habsburg-Spain	1621–1665
Spain	Charles II	Habsburg-Spain	1665–1700
Spain	Philip IV	Bourbon-Spain	1700–1706
Austria	Charles VI	Habsburg	1707–1740
Austria	Maria-Theresia	Habsburg	1740–1780
Austria	Joseph II	Habsburg-Lorraine	1780–1790
Austria	Leopold II	Habsburg-Lorraine	1790–1792
Austria	Francis II	Habsburg-Lorraine	1792–1796

Source: Wikipedia

canals, embankments and locks, where the best forges of Europe where to be found and where the most beautiful brocades in the Western world were made. (Lopez, 2012, pp. 34-35)

After 1499, with a short interruption by Massimillian Sforza from 1513 to 1515, the Duchy of Milan was to be ruled by foreign powers (Table 1), as we will see later on in detail.

The Republic of Venice

The city of Venice originated as a collection of lagoon communities banded together for mutual defence from the Lombards, Huns, and other invading peoples as the power of the Western Roman Empire dwindled in Northern Italy. Its geographical location and the evolving powers over time placed it in a complex position. The Byzantines claimed it, the Lombards surrounded it, and the Franks threatened it. As a result, in the political community, there were the pro-Byzantine faction, the pro-Lombard faction, the pro-Frankish faction, and the Republican faction. They were engaged in complex politics, but they agreed on one thing: the independence of their city from external rulers.

The Republic of Venice was founded in 697, when the Byzantine Empire's province of Venetia elected their own leader, called Ursus, who was confirmed by Constantinople as a 'doge of Venice' (equivalent to duke). Its early riches had come from the salt trade in the Comacchio Rialto lagoons. Sometime later Charlemagne's efforts to subdue the city to his rule failed. Next, with the elimination of pirates along the Dalmatian Coast, the city became a flourishing trade centre between Western Europe and the rest of the world (especially the Byzantine Empire and Asia) with a naval power protecting sea routes from Islamic piracy.

> Constantinople, by that period of time, had become the end of the *Silk Road*, a network of trade routes formally established during the Han Dynasty of China that linked the regions of the ancient world in commerce. While many different kinds of merchandise travelled along the Silk Road, the name comes from the popularity of Chinese silk with the West, especially with Rome. The Silk Road routes stretched from China through India, Asia Minor, up throughout Mesopotamia, to Egypt, the African continent, Greece, Rome, and Britain. The Chinese had very purposefully kept the origin of silk a secret and, once that secret was out, carefully guarded their silkworms and their process of harvesting the silk.[104]

---

[104] Source: Mark, J.J.: Silk Road. http://www.ancient.eu/Silk_Road/

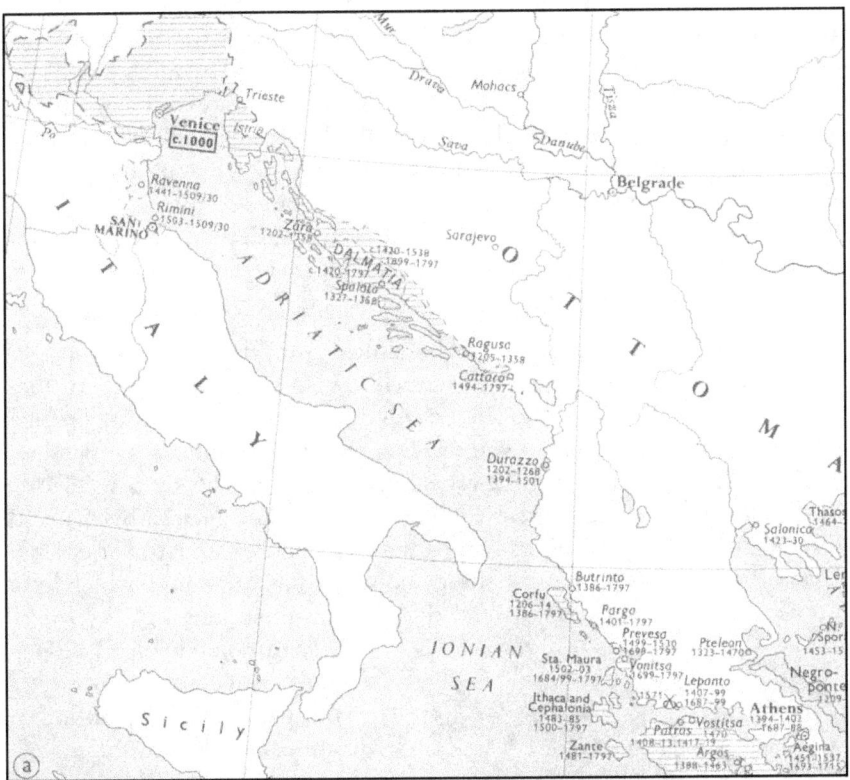

Figure 48: Seaward expansion of the Venetian Republic (1000).
Source: University of Oregon (http://pages.uoregon.edu/mapplace/EU/EU19%20-
%20Italy/Maps/EU19_91.jpg).

Over the centuries Venice had developed a formidable maritime
presence in the Adriatic Sea, trading with the Byzantine Empire and
colonizing the coasts, fighting the Croatian and Albanian pirates, creating
their *Domini dal Mar* ('Domains of the Sea'). Their support of the Byzantine
Emperor in the wars against the Normans had earned them the exemption
from taxes for its merchants in the whole Byzantine Empire, the so-called
Golden Bull of 1082. This became a considerable factor in the city-state's
later accumulation of wealth and power serving as middlemen for the
lucrative spice and silk trade. As a result, Venice became extremely wealthy
through its control of trade between Europe and the Levant. In the
eleventh century it had even began to expand into the Adriatic Sea and
Aegean Sea (Figure 48).

The chief activity of Venice was trade, and this brought commercial relationships. Next to mixed relations, such as those with the Byzantine emperor at Constantinople, they sometimes also became hostile as with their commercial rivals in Italy, chiefly Genoa and Pisa. Nevertheless, they managed to receive trading privileges from the Byzantine and German emperors, and even from the Mohammedans. In 1167 Venice, facing Frankish rule, participated in the previously mentioned Lombard League, opposing the powers of the Holy Roman Empire. A decade later the *Treaty of Venice* in 1177 established a six-year truce. This gave Venice the chance to focus on furthering its maritime endeavours.

Venice was involved in the Crusades almost from the very beginning. By 1204—as we have seen before—it even had transported the fighters of the Fourth Crusade (1202–1204) to the Levant. The crusaders, short of finance, could not pay for the ships but were willing to pay with military services. Thus, in 1202 they recaptured the rebellious city of Zadar (aka Zora, Zadra) on the Dalmatian Coast for the Venetians. Zadar fell on November 24, and the Venetians and the crusaders sacked the city (Figure 49). After spending the winter in Zadar the Fourth Crusade continued its campaign, which led to the Siege of Constantinople. After the subsequent sacking of Constantinople in April 1204, Venice gained control over major parts of the Byzantine Empire, and thus of the trade along the Silk Road that arrived in Constantinople

Figure 49: Siege of Zadar during the Fourth Crusade (1202).

Source; Wikimedia Commons. Artist Andrea Vicentino.

It was Marco Polo, born in a merchant family in Venice in 1254, who travelled the Silk Road. His father and uncle often traded on the west coast of the Mediterranean Sea. On one fortuitous occasion, they went to China and met with Kublai Khan, an emperor of the Yuan Dynasty. In 1269 they returned to Venice with a letter Kublai Khan had written to Pope Clement IV. In 1271, with a letter in reply from the new Pope Gregory X, and with valuable gifts, the Polos set out eastwards from Venice on their second trip to China. After 3.5 years they reached the Kublai in his palace. Becoming a highly valued official at the court, it was not before 1292 that Kublai Khan agreed to let Marco Polo, his father, and uncle return home. In 1295 they finally reached Venice by sea via the Black Sea and Constantinople after travelling a distance of 24,000 km. The information about China and some Asian states they brought back aroused great interest among the Venetians.[105]

In the early years of the republic, the doge ruled Venice in an autocratic fashion, the 'Commune Venciarum'. Until the eleventh century the doge's power increased, but then it weakened until he became only the presiding officer in a state governed by an oligarchy. Then his powers were limited by the 'promissione ducale', a pledge he had to take when elected limiting his powers and succession.. As a result, by 1172 governmental powers were held by the Maggior Consiglio, or Great Council, composed of 480 members taken from patrician families. One member of the great council was elected 'doge', or duke, the chief executive, who usually held the title until his death. The creation of the Great Council was followed by establishing the Minor Council (1175), composed of six advisers to the doge, and the Quarantia (1179) as a supreme tribunal. In 1223 these institutions were combined into the Signoria, which consisted of the doge, the Minor Council, and the three leaders of the Quarantia. The political power was held by the 'popolo', but that power did not last.

The victory of the merchant nobility was sealed by the 'Closing'—the Serrata—of the Great Council (established in 1172) in 1297. The essence of this act was to restrict eligibility for the Great Council to the members of about two hundred of the great merchant families. This restriction of political rights caused some discontent, which led, in 1310, to a serious conspiracy against the government. On July 19, 1315, a book of Italian nobilities was established. Only those listed in the book and above 18 years of age were eligible for the position in the Major Council. By that act, the new merchant nobility—some 2000 to 3000 families—had obtained its power base in the Great Council.

In the second half of the thirteenth century the great conflict began with Genoa, which was becoming a dangerous rival for the eastern trade. In 1294 Venice declared war, and in the following century there was constant conflict between the two cities. It would last till 1381, when peace was finally made and Venice was enabled to regain her former prosperity. The Genoese, however, because of internal troubles and French domination, which began in 1396, found themselves at a disadvantage.

---

[105] Source: https://www.travelchinaguide.com/silk-road/history/traveler-marco-polo.htm

Figure 50: The formidable warship of the Venetians: the Galeass.
Source: http://www.cogandgalley.com/2015/08/a-floating-fortress-galleass.html

At the end of the thirteenth century Venice was the most prosperous city in all of Europe. At the peak of its power and wealth, it had 36,000 sailors operating 3,300 merchant ships (many convertible to men-of-war), dominating Mediterranean commerce. Ships built at the Arsenal Nuovo; a complex of shipyards and armories employing some 16.000 people. It was one of the earliest industrial enterprises where galleys were built in mass production, owned by the Republic.

The Venetian Arsenal was not only able to function as a major shipyard, but was also responsible for these routine maintenance stops that most Venetian galleys required. This required financing, for which the Venetian government spent almost 10% of its revenues. This naval power resulted in the domination of Mediterranean commerce. Especially the wind and manpowered 'galleass' were floating fortresses (Figure 50).

Venice's leading families vied with each other to build the grandest palaces and support the work of the greatest and most talented artists. In a less public arena, they fought each other for political power.

Take the example of *Marino Faliero* (1274–1355), who was elected—during his absence as ambassador to the pope in Avignon—as doge on September 11, 1354. He belonged to one of the oldest and most illustrious Venetian families and had served the republic with distinction in various capacities (governor, ambassador, and soldier). In 1346 he had led the Venetian land forces (again) at the siege of rebellious Zara, where he'd been attacked by the Hungarians under King Louis the Great and had totally defeated them; this victory led to the surrender of the city. His election took place in a period with internal power struggles restricting the powers of the doge, a declining Levant trade due to raids on Venetian shipping, a disastrous war with the Genovese, and the recurrence of the Black Death. The merchants, excluded from government, were discontent with the governing caste, creating a tense political atmosphere.

Not content with the ruling nobility, after his election soon several conspiracies developed, among them the conspiracy to murder ruling members of the patrician class. The plot was that rumours of an imminent attack by a Genoese war fleet were to be spread for the night of April 15, 1355. In the ensuing alarm and confusion, a band of armed men, led by a member of the doge's family and ostensibly acting to protect him, were to kill as many nobles as they could and then proclaim Faliero prince of Venice.

However, the plot was discovered early, and the Council of Ten acted swiftly.[106] All those thought to be implicated were arrested, and ten of the conspirators were promptly hanged in a row from the windows of the doge's palace. Also, the doge, being suspected of involvement, was arrested and

Figure 51: The execution of the Venetian Doge Marino Faliero (1355).

Source: Wikimedia Commons. Artist Eugene Delacroix (1827).

---

[106] The formal task of the Council of Ten was to maintain the security of the republic and preserve the government from overthrow or corruption.

questioned. At the subsequent trial he confessed everything and was condemned and executed on April 17, 1355, on the landing-place of the stone staircase where he had taken his coronation oath and received the ducal bonnet (Figure 51). After the execution of the doge, other trials, executions, confiscations, and sentences of banishment followed in rapid succession.[107]

By the end of the fourteenth century, Venice had acquired mainland possessions—with diplomacy and money—in Italy, annexing Mestre and Serravalle (1337), Treviso and Bassano del Grappa (1339), Oderzo (1380), and Ceneda (1389). In the early fifteenth century Venice acquired—not without quite some fighting—additional territory in its economic hinterland, the Terraferma, not only to protect the Alpine trading routes as a buffer against belligerent neighbours, but also as a supplier of mainland wheat on which the city depended.

The growing intervention of the Venetians on the mainland was unwelcome to the other Italians, who had long thought of Venice as a foreign state and now feared her power. In the latter part of the fifteenth

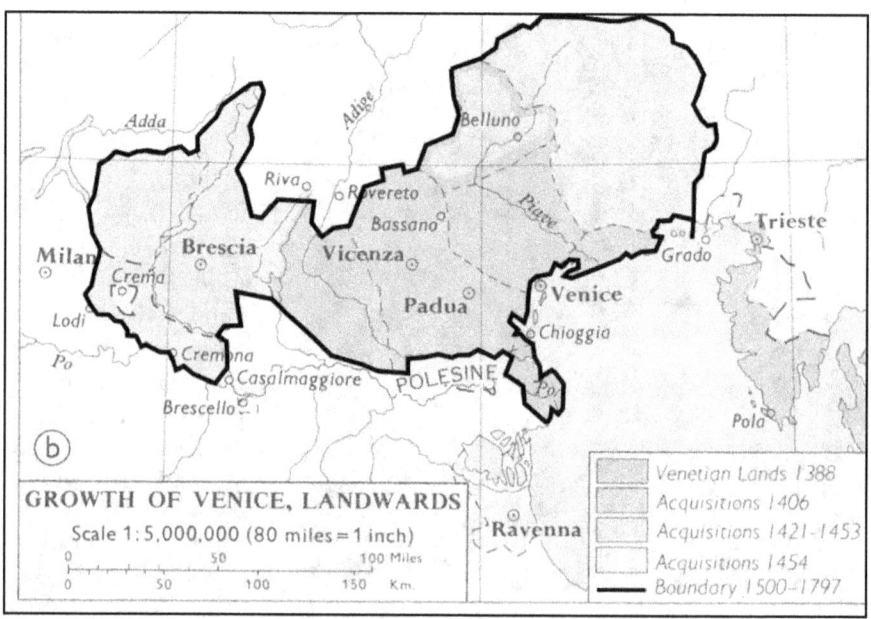

Figure 52: Venetian landward expansion (1400s)

Source: University of Oregon (http://pages.uoregon.edu/mapplace/EU/EU19%20-%20Italy/Maps/EU19_91.jpg). (Map adapted)

---

[107] Source: 'Venetian Inquisition: Marino Faliero (1279–1355)'. http://www.istrianet.org/istria/history/1000–1799AD/venetian_inquisition/1355_faliero.htm

century Venice came to be regarded as the great threat to the peace and freedom of Italy, as had been true earlier of Milan. The aggressive policies of both Venice and the Visconti made it inevitable that Venice and Milan would become enemies. During the last years of Visconti rule, Venice allied with Florence against them, but with the coming of Francesco Sforza, Florence abandoned the Venetian alliance and aided the new ruling house of Milan. By 1454, when the Treaty of Lodi ended the struggles between Milan and Venice, Venice controlled a large part of the Po delta up to Milan (Figure 52).

At the close of the fifteenth century, Venice was at the zenith of its power and splendour (Figure 53). With 180,000 inhabitants, Venice was the second largest city in Europe after Paris and probably the richest in the world. Its gold coin, the ducat, minted since 1284, was valued everywhere for its sound, reliable worth. It was the greatest trading centre in Europe. Venice became rich on trade, and the guilds in Venice produced superior silks, brocades, goldsmith jewellery and articles, armour, and glass in the form of beads and eyeglasses.

One of the early Venetian industries was glassmaking. Early glassmakers came to Venice after the Sack of Constantinople in 1204, followed by a second wave of glassmakers after the Ottomans took Constantinople in 1453. Due to the fire hazard, glassmaking moved some 1.5 km from Venice, to the island Murano, where they produced crystalline glass, enamelled glass, multi-coloured glass, milk glass, and imitation gemstones (for their famous glass chandeliers). By the end of the sixteenth century 3,000 of Murano Island's 7,000 inhabitants were involved in some way in the glassmaking

Figure 53: Procession in front of St. Mark's Square, Venice (1496).
Source: Wikimedia Commons. Artist Gentile Bellini.

industry. Murano's glassmakers were soon the island's most prominent citizens. They were forbidden, however, from leave the republic, as their technology would then become known to others.

In the eighteenth century, Venice's powers would decline after the Ottoman Turks won the Turkish-Venetian War (1714–1718). By the year 1792 the once-great Venetian merchant fleet had declined to a mere 309 merchantmen. By 1796/97 the city fell to the overpowering French armies under Napoleon.

The Republic of Florence

One of those city-states that rose to great prominence was Florence, located in Tuscany and originally founded as a settlement for former soldiers who were allotted land by Julius Caesar in the rich farming valley of the river Arno. By the beginning of the eleventh century Florence, with some five thousand people under feudal rule, had become an important trading centre for the fertile valleys of Tuscany. Many of the invading Lombards had married local rich men and women and mixed their families together. A few rich families—part Lombard and part Roman—ruled most of northern Italy as small independent countries. Supposedly, these countries were part of the Holy Roman Empire, but in reality, they did pretty much whatever they wanted to.

The biggest of these countries was Canossa, and in 1000 *Boniface III* (985–1052) was the Count of Canossa. Count Boniface ruled not just Canossa itself, but also many other cities of northern Italy, including Florence and Pisa.[108] After his death in 1052 his daughter Mathilda succeeded him. Tuscany was now under the rule of a countess who controlled vast territories (Figure 54) as a vassal of the Holy Roman Empire under Emperor Henry IV. She became involved in the conflict between the pope and emperor, chose the side of the pope, and it was in her castle that the pope received Henry V after his 'Walk to Canossa' in 1077 (Figure 34).

Quite occupied in those tumultuous days with military conflicts between emperor and pope about the 'Investiture Controversy', she had left much to the people of Florence and Pisa to govern themselves. So, it is not too surprising that after her death in 1115, the Florentine people rebelled against reinstating imperial rule: the roots for the commune had been created. A Florence that was already growing.

---

[108] Such as Brescia, Canossa, Ferrara, Florence, Lucca, Mantua, Modena, Pisa, Pistoia, Parma, Reggio, and Verona. Much of his territory was between the North and the Papal States, so he was involved in both Papal and Holy Empire intrigues. He was assassinated in 1052.

The production of textiles gave the Florentine economy a solid industrious base that few other Italian cities enjoyed. More than any other activity, it generated the extraordinary growth of the city's wealth. (J. H. Munro, 2011, p. 48).

Its geographic position gave Florence the potential for industrial growth. … the region had the resources to get an industry up and going in a number of towns— the possibility of transhumance of sheep and the availability of alum and vegetable dyes. In addition, Florence, more than any other Tuscan town, had the advantages of the Arno—water, power, and transport. Tuscany had enough of the basic materials, both wool and dyestuff, to give rise to a healthy local industry. (Goldthwaite, 2009, pp. 265-267)

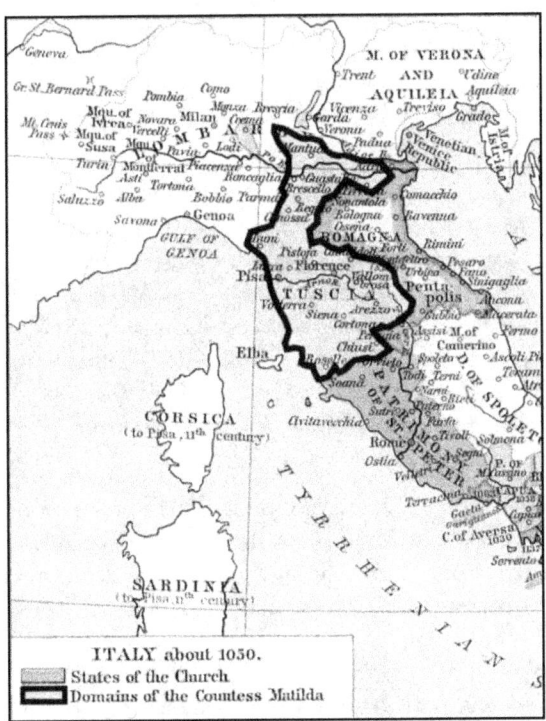

Florence's population boomed from 50,000 inhabitants in 1200 to 120,000 in 1300. Originally production was limited primarily to wool cloth, but in the late fourteenth century it would expand into to silks. The Florentine wool industry had its origins in the exploitation of resources of the region, but in the thirteenth century English wool was imported. Not surprisingly, the *Guild of the Wool Merchants* ('Il Arte della Lana') founded in the late twelfth century, was one of the most powerful guilds in Florence.[109]

Figure 54: The Domain governed by Mathilda (1050).

Source: Wikimedia Commons. (Map adapted)

---

[109] The guild structure consisted out of seven large guilds—the Arti Maggiori—and fourteen minor ones. The Arti Maggiori were: *Arte di Calimala* (Guild of Workers in Cloth), *Arte della Seta* (Guild of Silk Weavers and Merchants), *Arte della Lana* (Guild of Wool Manufacturers and Merchants), *Arte dei Giudici e Notai* (Guild of Judges, Lawyers, and Notaries), *Arte del*

The members of the greater guilds, such as the 'lanaiuoli' of the wool guild, were commonly known as the 'popolo grasso', or fat people, as distinguished from the 'popolo minuto', or little people, who belonged to the lesser guilds. Among the greater guilds were those of the wool manufacturers, the wool finishers, the silk merchants, and…the bankers. It was the popolo grasso who dominated the government of Florence from this time on, though the popolo minuto were allowed some authority. Thus arose a communal government—aka 'republic'—that was ruled by a council known as the Signoria. The Signoria chose the 'gonfaloniere' (titular ruler of the city), who was elected every two months.

Reviving from the tenth century imperial rule and governed from 1115 by an autonomous medieval commune, next the city was plunged into internal strife by the thirteenth century struggle between the Ghibellines, supporters of the Holy Roman Emperor, and the pro-papal Guelphs. It was the usual clash between the aristocratic nobles and the rich merchants and craftsmen of the city guilds.

> In the fourteenth century, Florence had already extended its control over various centers in Tuscany, from Pistoia to Arezzo, from Prato to San Gimignano and Volterra, using forms that stood midway between a true dominium and a simple protectorate. Over about thirty years during and just after the period of Milanese expansion under Gian Galeazzo Visconti, however, Florence occupied various other territories, notably the Pisano, Cortona, and some places in the Appenines; it expanded its territory to about 12,000 square kilometres. More important, it launched an administrative and fiscal reorganization designed to secure much more effective control over it. (Chittolini, 1989, p. 697)

Political conflict did not, however, prevent the city's from becoming one of the most powerful and prosperous in Europe, assisted by her own strong gold currency. The 'fiorino d'oro' of the Republic of Florence, or florin, was introduced in 1252, the first European gold coin struck in sufficient quantities to play a significant commercial role since the seventh century. By the early fourteenth century two families in the city, the Bardi and the Peruzzi, had grown immensely wealthy by offering financial services. Many Florentine banks had branches across Europe, facilitating trade and lending to royalty.

After periods of aristocratic dominance—for example by the Gherardini family, one of the founding families of the Republic of Florence—political power shifted from the aristocracy to the mercantile elite and members of

---

Cambio (Guild of Bankers and Money Lenders), *Arte dei Medici e Speziali* (Guild of Physicians and Pharmacists), *Arte dei Vaiai e Pellicciai* (Guild of Skinners and Furriers). In the sixteenth century these were complemented by the *Arte delgi Orafi* (Guild of the Goldsmiths).

organized guilds. The growing prosperity of the Florentine class of merchants and bankers eventually enabled them to take over the government of the city in 1282/1283 from the originally ruling class of the Magnates that was weakened by its own internal disunity. This shift was sealed by the anti-Magnate *Ordinances of Justice* in 1293. In these statutory laws, noble people were excluded from holding office. From then the commune of Florence was ruled by the Signori, which consisted of nine members—the Priori—chosen from the guilds of the city.

Of a population estimated at 80,000 before the Black Death of 1348,[110] about 25,000 are estimated to have been engaged in the city's wool industry. It would be a founding stone for the later development of Florence and its famous, and rival, families, such as the Albizzi family, the Alberti family, and the Medici family, who were at the centre of Florentine industrial oligarchy from 1382 on. Much of their wealth derived from the textile trade. They traded the cloth produced by the wool guild *Arte della Lana*. The Arte della Lana controlled all the processes from the raw baled wool through the final cloth, woven at numerous looms scattered in domiciles throughout the city.

Figure 55: Black Death in Florence (1348).

Wikimedia Commons. Artist L. Sabatelli.

---

[110] The Black Death or Black Plague was one of the most devastating pandemics in human history, resulting in the deaths of an estimated 75 to 200 million people (40–45% of Europa's population) and peaking in Europe in the years 1346–1353.

By 1300, Florence was producing some of the choicest fabrics in Europe, including fine imitations of Chinese brocades utilizing cloth of gold and silver. Florentine textiles were especially renown for the vibrancy and delicacy of their dyes. When the rest of Europe had difficulty producing red, Florentine guildsmen used a lichen called oricello to obtain the most striking crimson on the continent. … Florence also made early use of various machines. In this respect it was following the lead of the great religious corporations, which pioneered the mechanization of the textile industry. … Although the Florentine machinery was often powered only by hand – or at best, the waters of the Arno – this primitive industrial equipment nevertheless represented a major advance … By 1338, Florence's annual textile production was estimated at 100,000 pieces of cloth worth perhaps $10 million in modern currency. At that time, Florence boasted approximately 300 textile "factories" employing as many as 30,000 workers. Most of these workers never saw the inside of anything resembling a modern factory, however. They worked at home under a "cottage industry" system, where members of the major guilds supplied the raw materials, paid the worker by the piece, and finally marketed the finished product. (Brown, 2003 Chapter Nine)

In about 1338 Florence had reached a peak of prosperity. Its wealth, based largely on woollen cloth, was immense. It is also true that Florentine bankers were among the greatest of all, involved in far-flung transactions. However, at the beginning of the period, in the 1340s, two catastrophic events helped to shatter the well-being of the city. First, the weakened financial structure finally collapsed when the Bardi and Peruzzi went bankrupt, dragging many other firms down with them. Secondly, in 1348 came the Black Death, which may have killed as many as 50,000 Florentines (Figure 55). These disasters contributed considerably to the internal strains in Florence during the following years.

In 1346 the epidemic of the plague spread over Europe, reaching Italy in 1348. Florence was also hit, and the Italian writer Giovanni Boccaccio lived through the plague as it ravaged the city. He described it:

One citizen avoided another, hardly any neighbour troubled about others, relatives never or hardly ever visited each other. Moreover, such terror was struck into the hearts of men and women by this calamity, that brother abandoned brother, and the uncle his nephew, and the sister her brother, and very often the wife her husband. What is even worse and nearly incredible is that fathers and mothers refused to see and tend their children, as if they had not been theirs. …

Many ended their lives in the streets both at night and during the day; and many others who died in their houses were only known to be dead because the neighbours smelled their decaying bodies. Dead bodies filled every corner. Most of them were treated in the same manner by the survivors, who were more concerned to get rid of their rotting bodies than moved by charity towards the dead. With the aid of porters, if they could get them, they carried the bodies out of the houses and laid them at the door; where every morning quantities of the dead might be seen.[111]

From 1300–1454 the adjacent regions also came under Florence's control (Figure 56): Arezzo was overcome by a Florentine army in 1384. Pisa, a great prize, was taken in 1406. Livorno, of immense value as a seaport, was purchased in 1421. It saw the rise of new informal but powerful rulers who stood up to the papal authority and the expansion of the Papal States in the *War of the Eight Saints* (1375–1378).

Internally, Florence faced the *Revolt of the Ciompi* (1378–1382), when the large body of unrepresented labourers—aka the ciompi—revolted. It was the result of a power struggle between Florence's ruling elites, the established artisan guilds of Florence, and sotto posti (or un-guilded), which included the ciompi: mainly a group of low-wage textile workers employed in Florence's thriving wool industry. The new regime did not last long and was gone by 1382, when the elite families intervened. Yet the rebellion left a permanent scar in the minds of the Florentine elites (both the new and the old nobility) and created their everlasting fear and hatred toward the ciompi. This eventually gave rise to the Medici family.

Cosimo de' Medici (Cosimo the Elder, 1389–1464) was the first Medici family member to essentially control the city from behind the scenes. Although the city was technically a

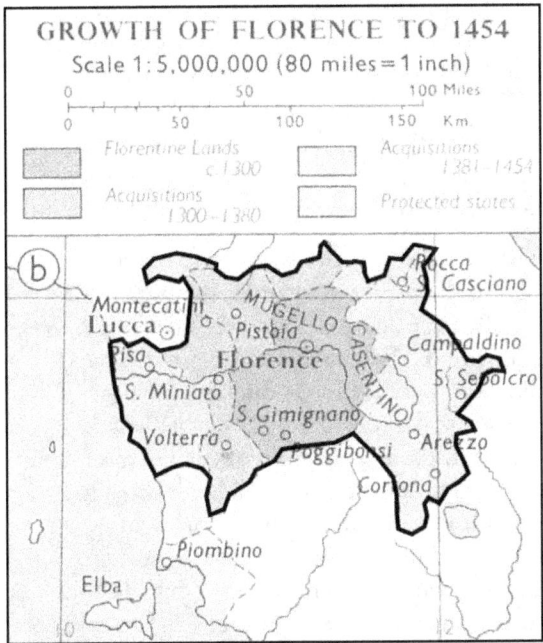

Figure 56: The growth of Florence up to 1454.

Source: University of Oregon
(http://pages.uoregon.edu/mapplace/EU/EU19%20-%20Italy/Maps/EU19_91.jpg). (Map adapted)

---

[111] Source: Boccaccio, Giovanni, *The Decameron Vol. I.* Source: 'The Black Death, 1348,' EyeWitness to History, www.eyewitnesstohistory.com (2001).

democracy of sorts, his power came from a vast patronage network along with his alliance to the new immigrants, the new citizenry of the 'gente nuova'. The fact that the Medici were bankers to the pope also contributed to their rise. Florentine leadership in the arts was well established by the time of Cosimo's rise to power in 1434. His patronage brought much work to the city's painters, sculptors, and architects. But he also greatly encouraged another strand of the Renaissance in which Florence played a major role: the scholarship of humanism.

By the 1430s Cosimo had obtained political power in Florence by participating in the Signoria. Then the anti-Medici faction—headed by the Albizzi family—managed to have him exiled to Venice. However, within a year, the flight of capital from Florence was so great that the decree of exile had to be lifted. Cosimo returned a year later, in 1434, to greatly influence the government of Florence again. In 1464 he was succeeded by his son *Piero di Cosimo de' Medici* (Piero I, the 'Gouty', 1416–1469), who was shortly thereafter succeeded by Cosimo's grandson, *Lorenzo de' Medici* (Lorenzo the Elder, called the 'Magnificent', 1449–1492) in 1469. Lorenzo, like his grandfather, was a great patron of the arts and surrounded himself with a web of advisors, friends, and clients. It was Lorenzo the Magnificent who served as patron to Leonardo da Vinci (1452–1519) for seven years. Michelangelo Buonarroti (1475–1564) also produced work for a number of Medicis. Lorenzo, more capable of leading and ruling a city, neglected the family banking business, leading to its ultimate ruin.

Over time their monetary wealth permitted the Medicis to exert influence through their banks, not only in Florence and Italy, but all over Europe. The Medici Bank, originally founded by *Giovanni di Bicci de' Medici* (father of Cosimo the Elder), was one of the most prosperous and most respected institutions in Europe. They had branches opened in Bruges, London, Pisa, Avignon, Milan, and Lyon. At one point, the Medicis managed most of the great fortunes in the European world, from members of royalty to merchants. However, their unwisely loaning large sums to secular rulers,[112] a group notorious for their delinquencies, brought problems. The London branch finished its liquidation in 1478, and the Bruges branch was liquidated in 1478 with staggering losses. By 1494 the Milan branch of the Medici bank also ceased to exist. After Charles VIII of France's 1494 invasion of Italy all the branches were declared dissolved. The invasion also ended the first period of Medici rule with the restoration of a republican government.

---

[112] For example, loaning £10,500 to the English monarch Edward IV in the 1460s which he failed to pay back.

During the rule of the Medicis, Florence was shortly a republic (1494–1498) that went through quite some religious and political turmoil. In the late fifteenth century it was the priest *Girolamo Savonarola* (1452–1498) who preached against the corruption and immorality of the clergy. Soon his aggressive and passionate style granted him a large following, making him a political factor. After Charles VIII (King of France) went on a military campaign and passed Italy and entered Florence, the Florentines drove the Medici from the city in 1494, and Savonarola was commissioned to organize the republic. Savonarola's ulterior goal, however, was to transform Florence into a 'City of God', the New Jerusalem. Florentines stopped wearing garish colours, and many women took oaths to become nuns. Although Savonarola inspired much of the political reforms of this period, he made enemies. The Borgia Pope Alexander VI, needing Florence's support, excommunicated him in 1497, and popular opinion turned against his religious endeavours. On May 25, 1498, he was burned as a heretic in the Piazza della Signoria (Figure 57). This act ended the Florentine Republic.

After Lorenzo de Medici (1449–1492) died, his son *Piero II, the Unfortunate* (1471–1503) was responsible for the expulsion of the Medicis. Due to his failure to negotiate with the invading French Charles VIII, accommodating him in all his demands, the Medicis were exiled again. The

Figure 57: Execution of Savonarola (1498).
Source: Wikimedia Commons. Artist Stefano Ussi.

following civic uproar resulted in the first period of the Medicis' exile (1494–1512), which coincided with the rise of the Borgia family to the Papacy in Rome.

Florence had during that time a diplomat called *Nicollo Machiavelli* (1469–1527). Soon after the execution of Savonarola, Machiavelli was appointed to an office of the second chancery, a medieval writing office that put him in charge of the production of official Florentine government documents. Shortly thereafter, he was also made the secretary of the council *Dieci di Libertà e Pace* (Office of the War and the Interior). In the first decade of the sixteenth century he carried out several diplomatic missions: most notably to the Papacy in Rome, to the French court of Louis XII, and to the Spanish court. Hence, from 1502 to 1503, he witnessed the brutal reality of the state-building methods of Cesare Borgia (1475–1507) and his father, Pope Alexander VI, who were then engaged in the process of trying to bring a large part of Central Italy under their control. Being exiled into forced idleness, he described his political views in the treatise called *The Prince*,[113] referring to the 'innovator' of the 'new prince' when discussing the conquest of 'princedoms'.

> *Machiavelli writes—(Chapter VI)—that reforming an existing order is one of the most dangerous and difficult things a prince can do. Part of the reason is that people are naturally resistant to change and reform. Those who benefited from the old order will resist change very fiercely. By contrast, those who can benefit from the new order will be less fierce in their support, because the new order is unfamiliar and they are not certain it will live up to its promises. Moreover, it is impossible for the prince to satisfy everybody's expectations. Inevitably, he will disappoint some of his followers. Therefore, a prince must have the means to force his supporters to keep supporting him even when they start having second thoughts, otherwise he will lose his power.*[114]

In *The Prince*, Machiavelli wrote:

> *And it ought to be remembered that there is nothing more difficult to take in hand, more perilous to conduct, or more uncertain in its success, than to take the lead in the introduction of a new order of things. Because the innovator has for enemies all those who have done well under the old conditions, and lukewarm defenders in those who may do well under the new. This coolness arises partly from fear of the opponents, who have the laws on their side, and partly from the*

---

[113] Machiavelli did not write *The Prince* to become famous but instead to achieve a position in the new Italian government formed by the Medici family, now under Lorenzo di Piero de' Medici (1492–1519). However, he failed to return to favour and died in 1527, and the book was not published until five years after his death. *The Prince* was written in times of political upheaval as a political handbook for rulers and has been used this way for many centuries.
[114] Source: http://politicalethics.org/2016/03/22/the-prince-by-machiavelli/

*incredulity of men, who do not readily believe in new things until they have had a long experience of them. Thus it happens that whenever those who are hostile have the opportunity to attack they do it like partisans, whilst the others defend lukewarmly, in such wise that the prince is endangered along with them. It is necessary, therefore, if we desire to discuss this matter thoroughly, to inquire whether these **innovators** can rely on themselves or have to depend on others: that is to say, whether, to consummate their enterprise, have they to use prayers or can they use force?[115]* (Machiavelli, 1908, pp. 24-25) (bold by author)

In 1512 the Florentine Republic was overthrown by a Spanish army that Pope Julius II had enlisted into his Holy League. The Medici family returned to rule Florence, and Machiavelli, suspected of conspiracy, was imprisoned, tortured, and exiled in 1513. He had time on his hands to write, but his famous book was not to be published until some years after his death in 1527.

During their rule, the Medicis also exercised considerable political power outside Florence. They produced three popes of the Catholic Church—*Pope Leo X* (1513–1521), *Pope Clement VII* (1523–1534)—rulers of both Rome

Table 2: Rulers of Duchy of Florence (1382–1801).

Nation	Ruler	House	Period
-	Different local rulers	Albizzi	1382–1434
-	Different local rulers	Medici	1434–1494
-	Republic of Florence	na	1494–1512
-	Different local rulers	Medici	1512–1532
-	Dukes of Florence	Medici	1532–1569
-	Grand Dukes of Florence	Medici	1569–1737
Austrian	Francis I[1], HRE	Habsburg Loraine	1737–1765
Austrian	Leopold I, HRE	Habsburg Loraine	1765–1790
Austrian	Ferdinand III	Habsburg Loraine	1709–1801

1) Husband of Maria Theresa, monarch of Austria. HRE: Holy Roman Empire.

Source: Wikipedia

---

[115] Original text in Italian (emphasis by author): « E debbasi considerare come non è cosa più difficile a trattare, né più dubia a riuscire, né più pericolosa a maneggiare, che farsi capo ad introdurre nuovi ordini. Perché lo introduttore ha per nimici tutti quelli che delli ordini vecchi fanno bene, et ha tepidi defensori tutti quelli che delli ordini nuovi farebbono bene. La quale tepidezza nasce, parte per paura delli avversarii, che hanno le leggi dal canto loro, parte dalla incredulità delli uomini; li quali non credano in verità le cose nuove, se non ne veggono nata una ferma esperienza. Donde nasce che qualunque volta quelli che sono nimici hanno occasione di assaltare, lo fanno partigianamente, e quelli altri defendano tepidamente; in modo che insieme con loro si periclita. È necessario per tanto, volendo discorrere bene questa parte, esaminare se questi **innovatori** stiano per loro medesimi, o se dependano da altri; ciò è, se per condurre l'opera loro bisogna che preghino, ovvero possono forzare. Nel primo caso capitano sempre male, e non conducano cosa alcuna; ma, quando dependono da loro proprii e possano forzare, allora è che rare volte periclitano. »

and Florence—, and *Pope Leo XI* (1605) and two regent queens of France—
Catherine de' Medici (1547–1559) and *Marie de' Medici* (1600–1610), and the
family became hereditary dukes of Florence in 1531. In 1569 the duchy was
elevated to a grand duchy after territorial expansion that gave Florence
access to the sea through the port of Pisa (Figure 56). They ruled the Grand
Duchy of Tuscany from its inception until 1737. The eighteenth century
saw the fall of the dynasty, when the Holy Roman Emperors from the
Austrian House of Habsburg-Loraine took over (Table 2).

The Republic of Genoa

The barbaric invasions had also hit Genoa, and the Lombards and
Franks had ruled it. Like other cities, by 1100 Genova had emerged as an
independent city-state. Trade, shipbuilding, and banking helped support
one of the largest and most powerful navies in the Mediterranean. As the
commerce of the city increased, so did the territory of the republic. Along
the peninsula's coast, they created footholds, like the protected harbour of
Portofino. Soon the Republic of Genoa extended over modern Liguria and
Piedmont, Sardinia, Corsica, and Nice and had practically complete control
of the Tyrrhenian Sea. Genoese merchants pressed south, to the island of
Sicily, and into Muslim North Africa, where Genoese established trading
colonies, pursuing the gold that travelled up through the Sahara.

During the First Crusade (1059–1099), the Genoese fleet transported
and provided naval support to the crusaders. As a result, many settlements
in the Middle East were given to Genoa as well as favourable commercial
treaties. Over the course of the eleventh and particularly the twelfth
centuries, Genoa became the dominant naval force in the Western
Mediterranean. By the end of the thirteenth century, Genoese colonies were
established in the Middle East, in the Aegean, in Sicily, and in Northern
Africa (Figure 58).

Genoa, and to a lesser extent Pisa, was the commercial and cultural rival
of Venice. Over time the balance of power between the two maritime
republics shifted, the result of a range of Venetian-Genoese Wars for
dominance in the Mediterranean Sea (1256–1381) that culminated in the
Battle at Chioggia (1380). Venice was left severely debilitated but was
gradually able to rebuild its public finances and to take advantage of the
weaknesses of its mainland rivals and redress its losses. Genoa had less
success in dealing with the debts accumulated during these wars and fell
into deepening financial incapacity over the following decades.

Political power in Genoa was held by rich merchant families; first was
the Grimaldi family, followed by the Pallavicino, Doria, and Spinola
families, and finally the Balbi and Durazzo families. These families had

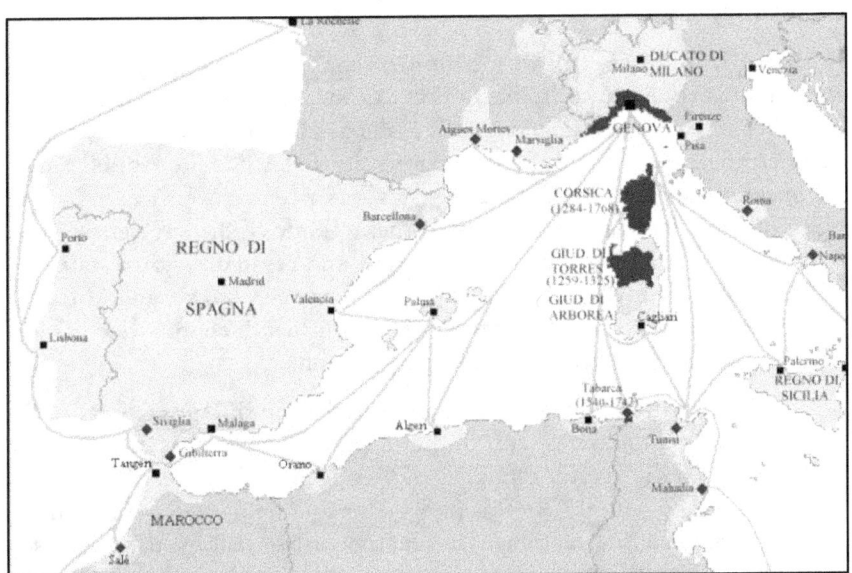

Figure 58: The extent of Genoese influence in the Western Mediterranean (c. 1400).

Source: Wikimedia Commons, Codex Parisinus latinus (1395).

amassed tremendous fortunes rooted in commerce and civic service rather than agriculture and military service. So prosperous had the city become that it was known to its citizens by the simple nickname 'La Superba', a phrase which literally means 'The Superb' but also 'The Proud'. That prosperity did not last, as by 1347 the Black Death also reached Genoa. Following the subsequent economic and population collapse, Genoa adopted the Venetian model of government and was presided over by a doge.

After a period of French domination from 1394–1409, Genoa came under rule by the Visconti of Milan (1436–1458). Genoa lost Sardinia to the Spanish House of Aragon, Corsica to internal revolt—and later, in 1768, to France—and its Middle Eastern, Eastern European, and Asia Minor colonies to the Turkish Ottoman Empire. During the 1450s and 1460s, the republic became a pawn in the struggle between France and Aragon for power and influence in Italy. After a gloomy fifteenth century marked by plagues and foreign domination, the city regained self-government in 1528 through the efforts of admiral Andrea Doria. Then Genoa became a satellite of the Spanish Empire. Thereafter, Genoa underwent something of a revival as a junior associate of the Spanish Empire, with Genoese bankers, in particular, financing many of the Spanish crown's foreign endeavours.

Other City States

Along with the previously mentioned republics, there were other city-states that rose to power in this period of time, such as the maritime republics of Pisa, Amalfi, and Ancona, all based on trading in the Mediterranean Basin, and the land-based city-states like the republics of Sienna and Lucca. Each of the city-states was a centre of power in its own right, ruled by its own form of government. Some of the city-states dominated others by sheer size and activity and created regional states, such as the previously mentioned city-states of Florence, Venice, and Milan, which each had over 100,000 people in the thirteenth century due to urbanization and the growth of their populations.

Take the Republic of Pisa—located around the river port of Pisa at the coast at the mouth of the river Arno—that was caught between the greater republics of Genoa and Florence. On sea, they competed with the increasingly powerful Genoese in controlling the Tyrrhenian Sea in a range of military confrontations: the Genoese-Pisa Wars. Pisa's role as a mercantile naval republic ended in 1258 with their defeat in the *Battle of Meloria*. On land, they faced the expansionism of Florence and Lucca, which longed for sea access. By the early fifteenth century they were controlled by Florence, after mercenary forces took the city in 1406.

Most of these city-states had one important activity in common: the merchant activities—and related banking activities—in connection to trading over the Mediterranean. Originating in trading woollen textiles, later in time these were complemented by ceramics, glassware, lace, silk, and other riches from the East that came over the Silk Route to the West. They also had something else in common; they hardly had any armies of their own, but used their wealth to create temporary mercenary armies. By doing so, though, they became easy pickings for the large nations around them.

The South of Italy

From early times the coasts of the southern part of the Italian Peninsula—aka Mezzogiorno—had seen many Greek invasions and colonization, hence its name *Magna Graecia* (Greater Greece). Later it become populated by Byzantine Christian Greeks, and after the time of the Norman invasions, it came under the nominal rule of the Byzantine Empire supported by Norman mercenary forces.

Due to its location—heart of the Mediterranean Sea—it was an interesting region to many parties: the Muslims, the French, and the Spaniards. As a result, the islands of Sicily, Sardinia, and Corsica, as well as the land later known as the Kingdom of Naples, were subject to those foreign powers at different moments in time, powers that had to cope with the local lords who maintained their feudal rule over their domains. In other words, that part of the Mediterranean Sea had a complicated geopolitical history.

In addition, its geological characteristics made the situation even worse, with active volcanoes dotting the southern peninsula. Along with the regular eruptions of the volcanoes Vesuvius (near the Gulf of Naples), Stromboli (volcanic island off the north coast of Sicily), and Etna (volcano on Sicily), all situated on the Campanian Volcanic Arc, the peninsula saw massive geological movement over the ages up to the present day (Figure 59).

Figure 59: Italy's Volcanoes.

Source: http://geology.com/volcanoes

The island of Sicily—its volcano Mount Etna being one of the most active volcanoes in the world, in an almost constant state of activity—over those ages had become a melting pot of people of Norman, Latin, Greek, and Arab origin. This process started with the Norman Count of Sicily, Roger I, descendent of early Norman invaders, conquering it from Muslim rule. In 1130 his son Roger II became, after establishing his centralised government, ruler of the Kingdom of Sicily. In 1136 Sicily was invaded by armies from the Holy Roman Empire, supporting a rebellious population, and in 1138 the pope invaded the kingdom with a large army.

By succession, Sicily came in 1194 under Germanic rule of the House of Hohenstaufen, starting with Henry VI, Holy Roman Emperor at that time. In 1266 it was Charles I, duke of Anjou, who conquered Sicily and was crowned king of Sicily by Pope Clement IV. However, in 1282 the locals, weary of taxation, rebelled on the island of Sicily, massacring thousands of Frenchmen in a couple of weeks. It was the start of the *War of the Sicilian*

Vespers (1282–1302), a conflict between locals, the kings of France, and the kings of Aragon. In the *Peace of Caltabellota* (1302), spheres of influences were decided upon. Charles II of the House of Anjou got the peninsular territories (aka the Kingdom of Naples), and the island territories (aka the Kingdom of Trinacria) came under Aragonese rule.

> *Kingdom of Sicily:* The island of Sicily came under the Aragonese rule as an independent kingdom. After the Spanish houses of Castille and Aragon had merged, Sicily was ruled by monarchs from the Crown of Aragon (1442–1458) and the House of Tastamere (1458–1501).

> *Kingdom of Napels:* Next, the part of the peninsula confusingly also known as the Kingdom of Sicily was the subject of both French Angevin and Spanish Angevin claims. Subsequently, the kingdom saw a range of different rulers. As a result of intrigue and murder, Charles III of Spain became king of Naples after the *War for Naples* (1379–1381). By 1442 the Kingdom of Naples was again under Aragonese rule when Alfonse V conquered the island. And after Charles VIII of France invaded Italy in 1494, resulting in the Italian Wars (1494–1559), the kingdom came under French rule.

After the *Battle of Garigliano* (1503), in 1504 the *Kingdom of the Two Sicilies* reunited the Kingdom of Sicily and the Kingdom of Naples, as Ferdinand II of Aragon united them under Spanish rule. The kingdom continued to be a focus of dispute between France and Spain for the next several decades, but French efforts to gain control of it became feebler as the decades went on, and Spanish control was never genuinely endangered. Thus Spain ruled the south of Italy for the next two centuries, till the *War of the Spanish Succession* (1701–1704).

Italian Renaissance (1300s–1500s)

During the years in which the merchant oligarchy governed Florence, and in the early period of Medici rule, the increasingly frequent contacts with examples of Greek and Roman antiquity gave rise to a new spirit.[116] Florence became the centre in which Renaissance humanism was forged.[117] Literary culture, the sciences, arts, and human activities came to the forefront, and it was a golden period in European intellect and culture. Its birth was facilitated by the enormous wealth the merchants brought to the city of Florence. Living wealth of trade that was converted into death wealth: the Renaissance arts.

Figure 60 : Pazzi conspirator hanged from window (1478).

Source: Wikimedia Commons. Drawing by Leonardo da Vinci.

By that time, Florence had its ups and downs in economic, business, and civic life. The rich families built their urban villas which radiated wealth: the Palazzo Pazzi, the Palazzo Strozzi (1489–1538), the Palazzo Medici, (1444–1484), the Palazzo Pitti (1458–?). It was a visual display of the rivalries between the powerful families and their factions. These rivalries also led to conspiracies that included treason and murder, and led to the exile of families.

One of these conspiracies was the *Pazzi Conspiracy*, where the Pazzi family and the Salvati family planned to assassinate Lorenzo de' Medici and his brother Giuliano with the blessing of Pope Sixtus. On Sunday, 26 April 1478, during High Mass at the Duomo before a crowd of 10,000, the Medici brothers were assaulted. Giuliano de' Medici was stabbed 19 times by Bernardo Bandi and Francesco de' Pazzi. As he bled to death on the cathedral floor, his brother Lorenzo escaped with serious but non-life-threatening wounds. However, the coup d'état failed, and most of the conspirators were

---

[116] The migration waves of Byzantine Greek scholars and émigrés in the period following the crusaders sacking of Constantinople and the end of the Byzantine Empire in 1453 greatly assisted the revival of Greek and Roman literature and science via their greater familiarity with ancient languages and works.

[117] Humanists sought to create a citizenry able to speak and write with eloquence and clarity and thus capable of engaging in the civic life of their communities and persuading others to virtuous and prudent actions.

soon caught and summarily executed (Figure 60). The Pazzi were subsequently banished from Florence, and their lands and property confiscated.

Next to political tensions between the large families, by the fifteenth century the Florentine population enjoyed many pleasurable diversions from business and intellectual life. Their sponsor, Lorenzo de Medici 'The Magnificent' influenced the types of entertainment held and often sponsored the activities financially. Mystery plays, based on the theme of the Passion (the sufferings of Jesus), were regularly staged for the enjoyment and edification of the citizens. To celebrate the feast day of Saint John, Florence's patron saint, Florentines held a horse race that ran throughout the city.[118] And festivals held during the season before Lent— aka Carnival—were grand productions, especially in the late fifteenth century.

These celebrations assumed particular importance because Carnival represented the first occasion for public festivities following the return of the Medici to Florence after almost twenty years in exile. It provided an opportunity for the Medici to work out their place in the city following an eighteen-year absence and in light of the less-than-overwhelming support for their return. The richly layered, allegorical triumphs operated simultaneously on several levels, offering different messages to discrete groups within Florence. To the family's friends and supporters they presented an image of leadership and the restoration of the past glories of the fifteenth century. To their committed enemies, the carnival celebrations offered a triumphant declaration of return and victory: a ritual humiliation and assertion of defeat. But to the most important segment of the audience – the majority of the office-holding class who were neither ardent supporters nor avowed opponents of the family – the Medici presented

Figure 61: Cappella Pazzi at the Santa Croce Church, Florence.

Source: Wikimedia Commons. Artist C. Graeb Berlin.

---

[118] A similar tradition is still seen in Sienna today: the Palio di Sienna.

themselves as preservers and defenders of Florence's civic republican traditions. (Scott Baker, 2011, p. 491)

The richness was shown in the architecture of the town. A spirit of competition developed between the rich merchants and banking families (such as the Pazzi family, the Bardi family, the Strozzi family, and the Pitti family). They often competed with each other to see who could commission the grandest buildings and the finest works of art, like the many churches such as the *Basilica di Santa Croce* with its sixteen chapels, which was commissioned in 1294. The Pazzi family built the Pazzi Chapel (Figure 61), the Peruzzi family the Peruzzi Chapel, and the Bardi family the Bardi Chapel. Florence's wealth was also shown to the world in public buildings and churches such as its *Duomo;* the Cattedrale di Santa Maria del Fiore, designed by Filippo Brunelleschi and finished in 1436. With its magnificent dome, it was one of Italy's largest churches (Figure 62).

Renaissance Patronage

By the late fifteenth century the leading families in the city-states of Florence, Milan, and Sienna—to name a few—formed the wealthy elite. From their merchant forbearers, they had earned their mercantile wealth, and they were still expanding and investing it, obviously in their entrepreneurial activities such as banking (de Medici, Sforza), but now also in real estate: urban villas and country villas. Over time they became the

Figure 62: The Duomo of Florence.

Source; Wikimedia Commons.

patrons of artists. Lorenzo de' Medici was not only a shrewd banker and clever politician; he was also a scholar and a poet. Under Lorenzo's leadership, Florence became one of the most beautiful and prosperous cities on the Italian Peninsula, a city harbouring some famous artists: from Leonardo da Vinci to Michelangelo, Raphael, and Titian, to mention just a few of the many.

Leonardo Da Vinci (1452–1519), born in the nearby village of Vinci, had a broad interest; from painting to engineering, anatomy, and botany. He was educated as a painter— becoming a master of the Guild of Saint Luke, the guild of artists and doctors of medicine in 1503—and would produce some famous paintings: the *Mona Lisa* (Figure 63) and *The Last Supper*. In

Figure 63: *Mona Lisa* painted by Leonardo da Vinci.

Source: Wikimedia Commons, Louvre Museum

addition, his technological ingenuity let him also conceptualize and experiment with machines, including war-machines, flying machines, diving suits, and adding machines (Figure 11).

Although the available technology of his time often would not permit their development, he was a master of mechanical principles. Leonardo's approach to science was one of intense observation and detailed recording, his tools of investigation being almost exclusively his eyes. He kept a series of journals in which he drew his conceptual ideas, comments, and plans. For 17 years he lived in Milan in the service of its ruler, Ludovico Sforza. Later he went to Venice, where he was employed as a military architect and engineer, devising methods to defend the city from naval attack (Figure 11). From 1502 to 1503 he was briefly employed by Cesare Borgia, the son of Pope Alexander VI and commander of the papal army. He later worked in Rome, Bologna, and Venice, and he spent his last years in France at the home awarded him by King Francis I, who gave him the title 'Premier Painter and Engineer and Architect to the King'.[119]

---

[119] 'Leonardo Da Vinci, Biography'. Source: http://www.biography.com/people/leonardo-da-vinci-40396.

Michelangelo Buonarroti (1475–1564), born in Caprese, Tuscany, and raised in
Florence, was already, at young age, fascinated by painting. In 1488, at
the age of 13, he became an apprentice in the studio of Domenico
Ghirlandaio, a popular Florentine painter. In Florence, he found the
right environment, as it was greatest center of the arts in Italy by that
time. Michelangelo was invited by Lorenzo de' Medici 'Il Magnifico' to
come and live in the Medici palace, where he was welcomed and treated
as one of the family, learning from the famous and learned men who
surrounded Lorenzo. In 1496 he went to Rome, where he was
commissioned by a cardinal to make a statue for his grave tomb. At the
age of 22, from 1498 to 1499, he made from Carrara marble the
sculpture *Pieta*, showing Jesus on the lap of his mother, Mary, after the
crucifixion (Figure 65, top). Next, in Florence, from 1501 to 1503, he
created his sculpture *David*—the biblical figure killing the giant
Goliath—in white marble also from the Carrara area[120]—symbolizing the
defence of civil liberties embodied in the Republic of Florence (Figure
65, bottom). Back in Rome, in 1508 he demonstrated his genius for
painting by creating the frescoes of the Sistine Chapel ceiling (Figure
64). Along with being a painter, Michelangelo was also an accomplished
architect; the Dome of St. Peter's Basilica in Rome is an example of his
work. Michelangelo had a brooding, moody, yet sincere personality that
kept him from being as popular as his works. He died on February 18,
1564, at the age of 89.[121]

**Figure 64: The Sistine Chapel before (left) and after the painting of the
ceiling by Michelangelo (right).**

Source: Wikimedia Commons

---

[120] The giant block was quarried from the Fantiscritti quarries in Miseglia, Carrara. It
resulted in the 5.2 metre-high statue.
[121] 'Michelangelo Buonarroti, Biography'. Source;
http://pal.loswego.k12.or.us/Michelangelo-Info%20and%20Bio.pdf

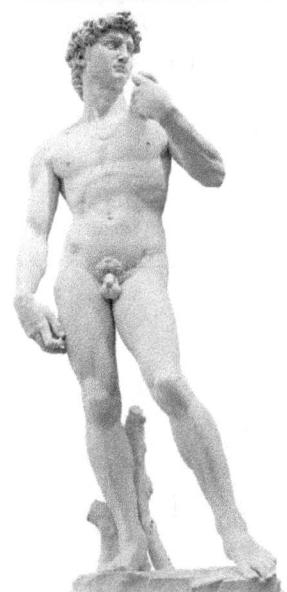

Figure 65: Michelangelo's *Pieta* (top) and *David* (bottom).

Source: www.arsmundi.com

These two well-known Italian artists were just a few of the many creating paintings, altarpieces, frescoes and sculptures for clerical masters. Their work was often, due to the background of their principals, of a religious nature. As many worked in, or came from, Florence, it became regarded as the birthplace of the Renaissance of art.

Papal Commission

From the preceding description, it becomes clear that the medieval merchants were not the only ones who had money to spend. In Rome, the Church had become very wealthy over time, and the successive popes showed their wealth by constructing magnificent buildings, for example, in the heart of Rome, the city that had become the center of the Catholic Church's religious power.

Already in 1477, Pope *Sixtius IV* had replaced the old chapel in the Apostolic Palace, the official residence of the pope. A team of Renaissance painters was commissioned to decorate it. The frescos on the walls of the new chapel—called the Sistine Chapel—were to contain, next to the portraits of popes, religious themes, such as the *Stories of the Life of Moses* and *Stories of the Life of Christ*. The paintings were completed in 1482. Next, between 1508 and 1512, under the patronage of Pope *Julius II*, Michelangelo painted the Sistine Chapel's ceiling, a masterpiece without precedent (Figure 64). Between 1535 and 1541, he painted *The Last Judgment* for Pope *Clement VII* and Pope *Paul III*, a painting that, due to its display of nudity, led to quite a few disputes.

In 1505, to replace the former basilica built in the fourth century on the historical site of the Circus of Nero by Pope *Constantine I* (272–337), it was decided that a new church, also called St Peter's Basilica, would be built. By then the basilica was already an important church, as papal coronations were held there and in 800 Charlemagne had been crowned emperor of the

Holy Roman Empire there. However, during the Avignon Papacy (1309–1377), the building had fallen into disrepair, and plans for either an entirely new basilica or an extreme modification of the old were made. Finally, in 1505 it was Pope *Julius II* (1443–1513)—the Warrior Pope—who made a decision to demolish the ancient basilica and replace it with a monumental structure to house his enormous tomb and 'aggrandize himself in the popular imagination'. Indeed, it became one of the largest churches in the world, leaving people in awe even in present times.

The work started in 1506, using the stones of the demolished Colosseum, and over the next 120 years, under the rule of many popes—each with an architect of his own preference (from Donate Bramante to Raphael and Michelangelo)—the new St. Peters Basilica was built (Figure 66). The costs were enormous, and all the cardinals and bishops of the Catholic realm had to contribute. One way of raising funds was selling 'indulgences' (where the sinners paid money to be forgiven their sins). This was done so enthusiastically that it was one of the reasons for Martin Luther to write his *Ninety-five Theses* that marked the start of the Protestant Reformation. All in all, the wealth of the Roman Catholic Church poured into Rome, and it was shown to the world in the papal architecture.

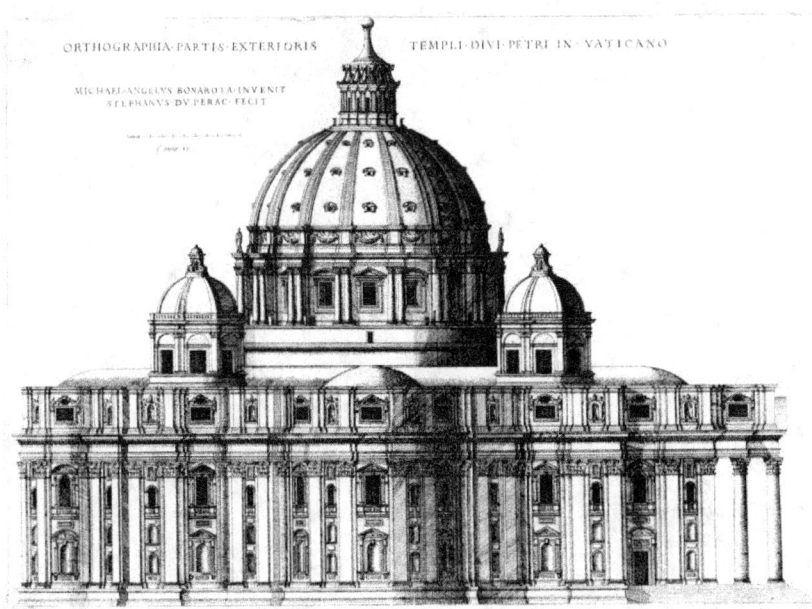

Figure 66: Elevation showing the exterior of Saint Peter's Basilica from the south as conceived by Michelangelo (c. 1560).

Source: Speculum Romanae Magnificentiae,
http://www.metmuseum.org/art/collection/search/364513. Artist: Etienne DuPérac.

Paintings and Sculptures

Florence boomed (economically) and bloomed (culturally). A side effect of all this economic prosperity spilling over in the arts and architecture was the presence of all these artists and their workshops, creating a sphere of modern times in Florence. As many of the artists travelled, they spread their art all over rich Italy, in the process becoming renowned. As a result, people from all over Europe flocked to Florence, the center of Renaissance arts.

In the refectory of the Convent of Santa Maria delle Grazi in Milan, Leonardo da Vinci, while working on his military designs, painted in 1495 to 1496 his *Last Supper* of Jesus and his disciples for his patron, Ludovico Sforza. In 1503 he started on his painting *Mona Lisa* (*La Gioconda*) (Figure 63). As part of his scientific studies, he later would create many drawings of the human body and its anatomy (Figure 11), his inventive designs (Figure 12), and models in his notebooks.

Another artist of that time, *Raffaello Sanzio da Urbino* (aka Raphael, 1483–1520) created sculptures and paintings while working in Florence, Sienna, and Rome. There, he was commissioned by the popes Julius II and Leo X

Figure 67: The Raphael Room *Stanza della Segnatura* in the Vatican, painted by Raphael.

Source: Wikimedia Commons.

Figure 68: Raphael's cartoon The Miraculous Draught of Fishes (1515).

Source: Wikimedia Commons.

to paint in the reception rooms of the Vatican, the 'Stanze' or 'Raphael Rooms' (Figure 67). Another papal commission was the Raphael Cartoons, a series of ten cartoons created as design for tapestries—to be made in Brussels—with scenes of the lives of Saint Paul and Saint Peter, for use in the Sistine Chapel (Figure 68).

From 1509 to 1511 he created the painting *School of Athens*, representing philosophy as a distinct branch of knowledge. Many of the pupils in his workshop became renowned artists themselves.

Overview of Italy's Early Times

The contours of present-day Italy originated as the heart and soul of the mighty Roman Empire, with Rome as its capital. The Romans had dominated the peninsula for centuries. Next, over time, barbaric invasions and massive migrations led to their decline and, ultimately, the collapse of their empire. In the fifth century new rulers established the Lombard Kingdom of Italy. By the ninth century that Kingdom of the Lombards was

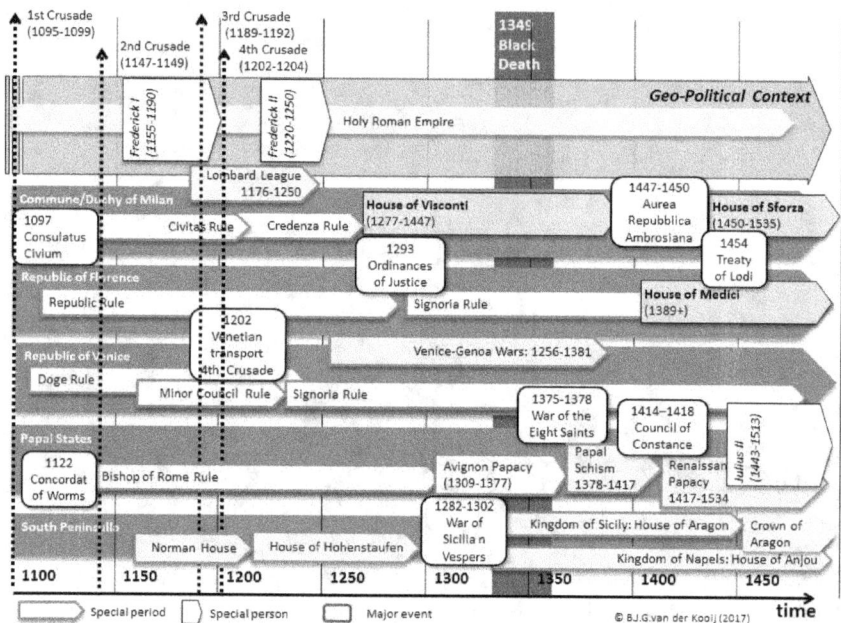

Figure 69: Overview of the development of Italian city-states (1100–1500).

Figure created by author.

incorporated in the Carolingian Empire under Carolingian rulers. They were succeeded by the different rulers of the Holy Roman Empire and the Papacy.

Then, the northern and central part of the peninsula saw—in the twelfth and thirteenth centuries—the rise of the city-states. Some merchant republics, such as Venice, Genoa, and Pisa rose to great wealth due to their Mediterranean trading activities. The fourteenth century was a time of incessant local conflict in which the prosperous city-states enlarged their territories at the cost of their neighbours. Decades of fighting saw Florence, Milan, and Venice emerge as the dominant players, and these three powers finally set aside their differences and agreed to the *Peace of Lodi* in 1454, a peace that brought calm to the region for the first time in centuries. This peace lasted till the start of the Italian Wars (1494–1559) at the end of the fifteenth century. That, in a nutshell, was the path of development of the Italian Peninsula from the early ages up till the 1500s that created the Age of the City-states (Figure 69). Over time, the enormous accumulation of wealth by the merchants and bankers of those days became a matter of serious interest to the rulers around the peninsula and the Po delta.

The Age of the City-states

The described city-states were just a few of those that dotted the Italian Peninsula. Next to Milan, Florence, Venice, and Genoa, numerous other city-states—Pisa, Lucca, Cremona, Sienna, Spoleto, Todi, Terni—grew and flourished, making Italy the most highly urbanized area in Western Europe in the fourteenth century without any form of central government. These city-states saw a mix of competing political powers—from feudal aristocratic to religious papal powers and international imperial powers— influencing their local governmental affairs. Often the growing cities expanded into the surrounding countryside by subduing the nobles and annexing their land. Land-owning nobles took up residence in the towns and became a part of city life. Within these towns and cities, political power belonged to the possessors of urban wealth, that is, the 'popolo' of bankers, merchants, and businessmen.

Within the context of that constant warfare between the different cities, competing for land, wealth, and influence, many of the city-states under the rule of urban elites grew and prospered. It made them rich, and they showed it in their architecture and arts. Other state-cities like the Republic of Pisa and the Republic of Sienna showed similar characteristics of popular government, military turmoil, trade dominance, and wealth accumulation.

The rise of the middle class of 'popolo': In contrast to many other European regions, which saw the remnants of feudal structures dominate society long after medieval times, Italy saw early the rise of the popular influence in civil administration. It was the rise of the new class of the 'popolo', who were neither the old feudal aristocracy nor the common serfs of people working as peasants or artisan. This 'bourgeoisie' was the new class of industrious people organized in merchant guilds. With increasing wealth and prosperity, the members of the guilds, commoners by birth but often rich and powerful, demanded a greater share in government. The old ruling class might be called the 'grandi', or great men, while the guildsmen took the (misleading) name of 'popolo', the people. It was the 'popolo grasso' of the larger guilds that ruled together with the 'popolo minuto' of the lesser guilds, but the majority of the common workers had no influence at all.

A state of constant military turmoil: As described, constant conflict was the fate of these vigorous urban societies. In addition to the internal struggle between families and the struggles for independence from emperor and pope, there was the constant regional fighting of the cities against one another. They fought for commercial supremacy, control of trade routes, access to seaports, territorial expansion (eg for the cereal production), and possession of natural resources (such as salt from

Salsomaggiore[122] and Cervia[123]). Like their internal conflicts, these intercity wars and rivalries were likely to be prolonged, bitter, and ruthless. It was the constant rivalry between the maritime republics of Genoa, Venice and Pisa and between Florence, Sienna, and Pisa, and Milan, Cremona, and Padua that resulted in the domination of five of the great states: Venice, Florence, Milan, the Papal States, and, in the south, the Kingdom of Naples. By the middle of the fifteenth century, between them, a delicate equilibrium was reached. Then the *Peace of Lodi* of 1454 became the basis of a league of all the Italian states, which maintained a precarious peace in Italy until 1494. Though there were conflicts during these forty years, they were all contained without foreign intervention on a large scale.

Trade dominance and wealth accumulation: The Po delta, with its rich soil, was the cereals granary of Northern Italy. Agricultural activities in the delta created grain surpluses that were traded by the grain merchants. Wool was transformed by early industrious activities in Florence and created the textile industry. Fine Florentine cloth was sold everywhere. The harbours of Genoa, Venice, and Pisa saw a massive influx from goods coming from the Orient, the Crimean area, and coastal Mediterranean areas. From there, goods were transported into Europe, Milan being the pivot point of logistics for the mainland routes up to the north, east, and west. From Genoa and Venice, the sea trading routes reached all over Europe. Their merchant activities, covering the Mediterranean regions up to the Middle East and Adriatic Sea in the East, and Flanders and England in the West, resulted in the mercantile economy that saw the transfer of wealth on a large scale to these republics (Figure 70).

Directly related to the trading activities, as the goods had to be paid for, was the development of financing activities. It was the rich merchants (like the Medici, Bardi, and Peruzzi families in Florence, and the communal Venetian Bank) who developed the first Lombardian banking networks all over Europe using 'bills of exchange'.[124] In addition, the

---

[122] The brine wells of Salsomaggiore (near Parma) were crucial to the development of the food economy in the Parma area—hence the famous (salted) Parma ham and cheese. The salt from this area was preferred to sea salt because of its high content of iodine, bromine, and sulphur, which could —especially iodine and bromine— block the growth of bacteria and, therefore, facilitate the preservation of meats.

[123] In Cervia, along the coast of the Adriatic Sea, salt was collected from seawater. The salt, collected in saltpans, was called 'sweet salt' due to its low content of potassium, magnesium chloride, calcium, and other substances that usually give sodium chloride a slightly bitter aftertaste.

[124] The bills of exchange certified that a particular person or company had paid a particular Medici branch a certain sum of money, and instructed the recipient Medici branch to pay back that sum in local currency.

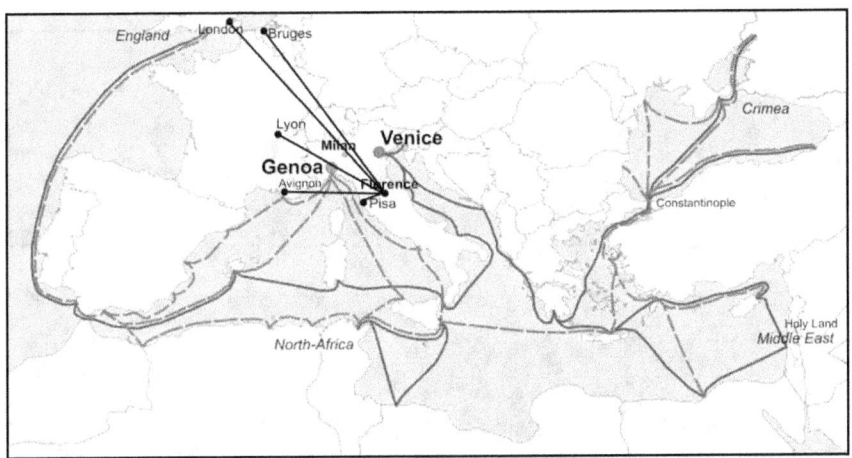

Figure 70: Medieval Venetian and Genoese trading routes and the Medicis' banking sphere.

Source: Unknown. (Graph adapted by author)

rich religious military orders of the Templers and Hospitallers, heavily involved in the crusades, offered banking services. This was not only the (virtual) transfer of payment between commercial parties; part of it was merchant banking in which the (agricultural, textile) production was pre-financed.

One might wonder what the dominant driving forces were behind this early development of Italy in the period of time from 1000 to 1500. It seems that among the major forces were population growth and urbanization, early technological change, and the commercial economy.

Italy's population, though decimated by the pandemic Black Death that swept across Europe from 1348 to 1351 (Figure 71), grew in the period of 1000 to 1500 from some seven million to eleven million people. Most of them living in the northern and central part of the peninsula in city-states and fed by the (climate facilitated) rich harvests of the fertile Po delta. Medieval technology, based on the mechanical arts, showed a continuous range of improvements. Agricultural technology increased yields. Maritime technology improved shipbuilding and shipping. Military technology increased weaponry. Powered by wind, water, and animals, early sawmills developed.

All these developments contributed, but the most dominant force during this period was the 'commercial economy' created by Florentine merchants and bankers and spanning the Western world. Trading and

(Marcoccio, 2003, p. 7) (Figure 111). This was shortly after Edison had filed in the US for a patent on November 4, 1879, and was granted US Patent №. 223,898 on January 27, 1880, for his invention. As Edison's invention was getting a lot of attention, Cruto's work stayed out of the limelight, his contribution forgotten.

On February 25, 1892, he created a company, *Cruto & Compagnia*, to exploit his inventions. Demonstrating his lamp at expositions in Monaco at Bayern, Vienna, and Turin, he was able to license it to several countries. In 1886 he moved from the small facilities in the city of Piossasco to create a lamp factory, *Societa Italiana di Elettricita Sistema Cruto*, in Alpignano. There, he manufactured 1,000 carbon-filament lamps a day, the first to do so in Italy (Marcoccio, 2003, p. 16). He stayed with the company for a couple of years but then returned to his inventive activities after strong disagreements with the factory's new management. In 1927 the company was, after a bankruptcy, acquired by the then-expanding Philips Gloeilampen-fabriek from the Netherlands.[165]

A special and quite tragic case also involved an Italian, but this man had left Italy and sought his luck elsewhere: *Antonio Meucci* (1808–1889), born in Florence, educated at the Academy of Fine Arts. The 1830s were revolutionary times. Escaping the surveillance from Florence's police, he went in 1835 to Havana, Cuba, as an engineer in a theatre. There, he spent his free time studying the new phenomena around electricity: *electrochemistry* and *electrotherapy*. In 1848 he started experimenting with electrotherapy in cooperation with some local physicians, and this brought him to the speaking telegraph. After moving to New York in 1850, he developed an early telephone system he called the 'telettrophone' (Figure 112). However, due to misfortune, he tragically failed to exploit his creative work. In his place, it was Alexander Graham Bell who became well known and was granted US patent № 174.465, dated March 7, 1876, for his speaking telegraph.[166]

Figure 112: Replica of Meucci's telettrophone (1870).

Source: https://www.histoire-fr.com/mensonges_histoire_graham_bell_telephone.htm

---

[165] Sources: http://www.wikiwand.com/en/Alessandro_Cruto, http://ecomuseo.comune.alpignano.to.it/start.htm.
[166] See: B.J.G. van der Kooij: *The Invention of the Communication Engine 'Telephone'* (2015), pp. 138–144.

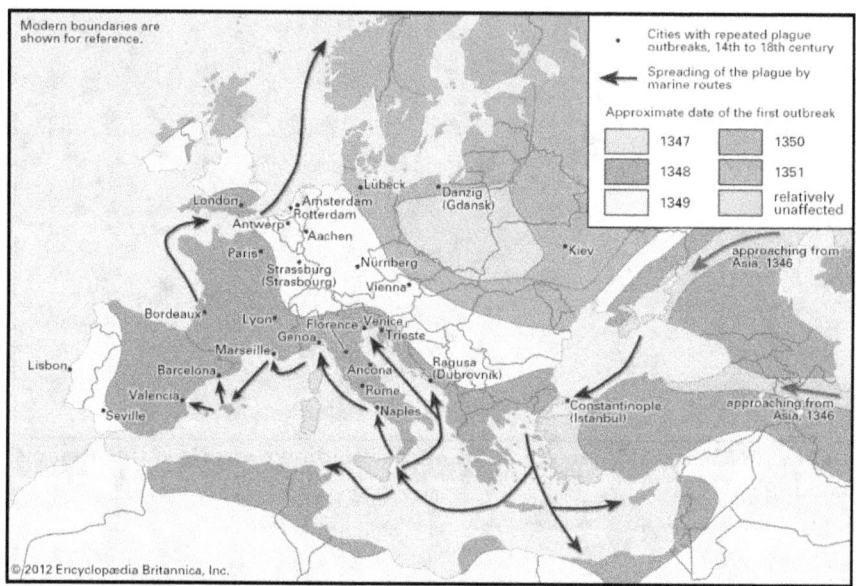

Figure 71: Black Death sweeping across Europe (1347–1351).
Source: Encyclopedia Brittanica.

transporting almost exclusively the riches from the Levant, shipping wool from England and Flanders to Florence,[125] financing agricultural and industrial activity, they transferred enormous wealth[126] to the republics governed by the powerful 'popolo'. Wealth was accumulated by the great Italian families, and it spread into local society. This wealth was used for trading but also political purposes, for financing military activities but also for enriching arts and architecture. On the Italian Peninsula, 'politics and popolo', 'trade and towns', 'war and wealth', they were all intertwined, creating the context for the times to come.

---

[125] Wool was one of the chief articles of trade and manufacture throughout Western Europe. The Florentine economy was largely based on it, and the textile manufactures of the Low Countries (Flanders) continued to be important.

[126] This Transfer of Wealth—next to the Transfer of Power and the Tranfer of Knowhow and Knowledge—is one of the major mechanisms that would create the context for innovation. See: B.J.G. van der Kooij: *Context for Innovation, British (R)evolutions in Perspective* (2017) pp 501-521.

Early Modern Italy

Italy by the early sixteenth century had grown from the medieval times into a politically divided society of powerful 'regional' city-states governed by urban merchant elites. Although foreign influences had been apparent—such as the efforts of the Holy Roman Emperors to dominate in the Lombardian *Kingdom of Italy* from Charlemagne's times—local powers had prevailed. As a result, the governmental and territorial struggles had been dominantly regional. True, the popes had mixed up their religious and secular ambitions by ruling the Church and expanding step by step the Papal States. This had resulted in clashes with the emerging republics and duchies. Massive riches had flown into the upper part of the peninsula, both from trade dominated by the maritime republics as well as from the revenues of the Church (crusades, indulgences, tithes). In addition, the different military conflicts had brought riches to Italy (eg the sack of Constantinople). In the period of a couple of centuries, a massive wealth, originating from loot, trade, and religion, had been transferred into Italy.

As a result, in the next centuries Italy was again faced with forces coming from outside the peninsula. Italy was to become, because of its wealth, regional divisions, and consequent weakness, the earliest battleground and victim of the emerging modern national states like the Kingdom of Spain, the Kingdom of France, and the Kingdom of Germany (later the Holy Roman Empire). Obviously, it was not for the nice climate, the beautiful woman, or the Italian cuisine that the foreign powers were interested in the wealthy Northern Italian states.[127] Their interest was in the accumulated wealth; there were riches to be obtained, not only just as loot of war. Both the rich population of the merchants as well as the peasants and artisans could be taxed permanently to the advantage of the occupiers. And tax, the rulers did...

To comprehend the complex history of Italy's development after 1500, one has to understand the complex geopolitical situation in the European theatre,[128] and its main players at that period of time. So, first, we will look at the European context, followed by the specific context created by a dominant player: the House of Habsburg.

---

[127] In the feudal system, it was about land possession (downward, from the monarch to the lords) and services rendered (upward, from the peasant and knights). Land expansion was by conquest, marriage, or alliance. Why would a ruler (from monarch to a local ruler) want to govern other people (countries, regions, cities) when there was not a reward for him. The reward was getting a grip of the wealth created by those people: natural wealth, commercial wealth, and industrial wealth.

[128] Geopolitics are concerned with the political powers in relation to geographic space. In these times the political forces were dominated by aristocracy and the elected monarchies.

In fact, it had started long time ago when Charlemagne, Charles the Great, King of the Franks, had died in 814. The division of his Carolingian Empire under his sons in three regions after the *Treaty of Verdun* in 843 (Figure 28) lay at the roots of the later struggles between the French rulers and the Habsburg rulers. At that time, his son *Louis the German* got the later Kingdom of Germany, *Charles the Bald* received the West that would become France, and *Lothair* got the middle region from the Low lands to the Kingdom of Italy. In the centuries to come Lothair's domains would become contested by both the French and German rulers.

European Context

At the start of the sixteenth century, on the European continent, the ruling Houses of several emerging nations fought each other, such as France warring with England and Spain in the *Hundred Years' War* (1337–1453), a range of succession conflicts over the last remnants of the 'Angevin Empire in France' (Figure 72, top).[129] England had several conflicts with the Dutch over mercantile affairs. Spain had problems with

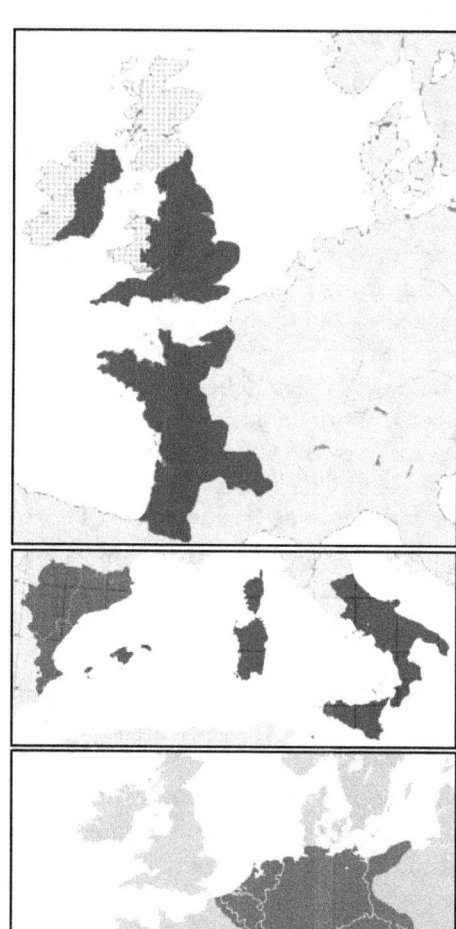

Figure 72: The Angevin Empire (top, c. 1172), the Crown of Aragon (middle, c. 1441) and the Holy Roman Empire (middle, c. 1600).

Source: Wikimedia Commons.

[129] The 'Angevin Empire' was the name for the possessions of the Angevin kings of England (1100–1200) in both England and France. In France, it consisted of the duchies of Normandy, Gascony, and Aquitaine as well as of the counties of Anjou, Poitou, Maine, Touraine, Saintonge, La Marche, Périgord, Limousin, Nantes, and Quercy.

England on religious matters. And in Central Europe, it was the Holy Roman Empire, a conglomerate of unions of different rulers, that was facing struggles of its own (Figure 72, bottom), for example, with France about the Burgundy region (the Rhinelands). And in the Mediterranean basin, the Spanish Crown of Aragon had considerable interests to defend (Figure 72, middle). All in all, at the start of the sixteenth century we find three dominant geopolitical players in relation to Italy: the House of France, the Holy Roman Empire, and the Spanish Crown of Aragon (Figure 72).

Next to those conflicting geopolitical interests, there was another force at play: the Roman Catholic Church with its religious monopoly in Europe, a dominance seriously challenged by the emerging Protestantism after Martin Luther started the *Reformation* in 1517. In addition, the temporal interest of the Church, the Papal States, caused continuous conflict. Take the 'Warrior Pope' Julius II (1503–1514) who, already as cardinal Della Rovere, constantly fought the power-hungry Borgia family ruling the Papal States. He had his eye on the Republic of Venice, and led an army to conquer Perugia and Bologna. The relationship between the emperors of the Holy Roman Empire and the popes of the Roman Catholic Church were ambivalent in nature, and the Papacy switched alliances regularly.

> For example, during the *League of Cambrai* (1508), an alliance of Pope Julius II, the Holy Roman Emperor Maximilian I, Louis XII of France, and Ferdinand II of Aragon, was officially created to stand against the Turkish expansion, but in reality, they were more occupied with the Republic of Venice. Concerned with the French presence in North Italy, Julius II first allied with Venice, hired Swiss mercenaries, and attacked the French-dominated Duchy of Ferrara. When that did not work out, he organized the *Holy League* (1511). Now Spain, the Holy Roman Empire, and Henry VIII of England joined forces. Clearly, the agenda of Pope Julius II—aka the 'Warrior Pope'—was about expanding the Papal States. In addition to an active military policy, he personally led troops into battle on at least two occasions. And to show off the magnificence of the Roman Catholic Church, he commissioned the rebuilding of St. Peter's Basilica.

House of France-House of Habsburg rivalry

At the end of the fifteenth century, France had emerged as a range of nation-states (Figure 73) with their own territories and identities. The emerging kingdom was ruled by different dynasties of the *House of France* (aka Capetian Dynasty with its many branches or Capetian Houses) from 987 up to 1328. The regions were controlled by local potentates: the Duke of Normandy, the Count of Blois, the Count of the Provence, etc. As usual

Figure 73: Territories of France, 1477.

Source: Wikimedia Commons.

between those rulers, on many levels, conflicting interests existed, conflicts about dominance, about territorial expansion, about religion.

One of those cadet branches was the *House of Valois*, which ruled large regions and had its specific territorial conflicts related to succession. Marriage was the tool for expansion for the rulers. For example, the House of Valois expanded in the west by acquiring the territory of the Duchy of Brittany (Figure 73, left top) through marriage.

Marriage united, but some marriages were the beginning of great conflicts. Take the expansion in the east, which was to be frustrated by the marriage of *Mary of Burgundy* (1457–1482), heiress of Charles the Bold, Duke of Burgundy, to *Maximilian of Austria* (1459–1519). When Maximillian was succeeded by his grandson Charles V—who inherited the Burgundian Netherlands and the Franche-Comté as heir of the House of Valois-Burgundy—the House of Habsburg entered the realm of the House of France. Even worse, it would encircle it.

By 1519 *Charles V* (1500–1558) had, from his own dynasty, the Habsburgs, inherited Austria and other lands in Central Europe such as the Habsburg Netherlands (1504) and the Spanish Empire (1516). In addition, he was elected emperor of the Holy Roman Empire in 1519. By then he was quite powerful, being the heir of three of Europe's leading dynasties: the *House of Valois-Burgundy* (ruling Burgundy and the Netherlands), the *House of Habsburg* (ruling the Holy Roman Empire), and the *House of Trastámara* (ruling Spain, which included the Crown of Aragon). Through these inheritances, he brought together under his rule extensive territories in Western, Central, and Southern Europe, and the Spanish colonies in the Americas and Asia. As a result, his domains spanned nearly four million square kilometres and were the first to be described as 'the empire on which the sun never sets'. And the colonies brought enormous riches to Spain that made the expansion foreign policies of its rulers possible.

After Columbus discovered the Americas in 1492, the enormous influx of wealth from the Peruvian, Bolivian, and Mexican gold and silver mines—in the 1550s transported to Spain in the yearly 'Silver Fleet'—had financed the territorial aspirations of Charles V and later his son Philip II, the self-acclaimed 'Defender of the Faith'. It was the *Golden Age of Spain* that made Habsburg Spain one of the most powerful states of its time. It brought Spain into many territorial conflicts, which culminated in the *Eighty Years War* (1568–1648) with the Dutch, the religious *Thirty Years' War* (1618–1648) between Protestant and Catholic states, and the *Franco-Spanish War* (1635–1659) between France and Spain.

So, not too surprisingly, the House of France felt entrapped by the Habsburgs. A situation in the early sixteenth century, where we see in the European theatre two dominant dynastic players:

On the one hand, there was the *House of Habsburg*. At the time of Habsburg Charles V's rise to power in 1521, the House of Habsburg surrounded the emerging nation of France: in the north by the Habsburg-controlled County of Flanders (Habsburg Netherlands), in the east by the region of Franche-Comte controlled by Habsburg's Holy Roman Empire, and in the south by the Habsburg-controlled Kingdom of Spain (Figure 74). And that Kingdom of Spain already, for a long time, had ruled over the Kingdom of Naples.

On the other hand, there was the *House of Valois*. France was by then the territory of Charles VIII (1470–1498) of the House of Valois, a young boy under the regency of his sister Anne de Beaujeu. After the *Mad War* (aka Guerre Folle, 1485–1488), a conflict between the feudal lords and the French monarchy, the independent territories were centralized under one monarchy. By marriage, he had added the Duchy of Brittany, and later Charles VIII inherited former Angevin possessions, among them the claim to the throne of Naples.

This, in short, rough brushstrokes, paints a picture of the complex geopolitical background for the conflicts between the House of Habsburg and the House of Valois on the Italian Peninsula. The struggles between these houses and, thus, the nations of France and Spain, carried on in many places. It was the first great problem of international relations in the modern sense, and the fight for control of Italy was only one part of it. However, the outcome of that territorial conflict was to be the loss of Italian independence until the second half of the nineteenth century. It began, just as in centuries before with the barbarians, with an invasion of Italy by a foreign power: French armies invading the Italy claimed and ruled by the branches of the Habsburgs.

The House of Habsburg

Thus, one of those foreign powers was the increasing dominance of the Habsburg dynasty, which left its mark on the developments in Europe, and whose behavior was characterized by diplomatic alliances, intermarriage with inbreeding problems,[130] and succession problems due to the lack of male heirs.[131] As a consequence of this all, many of the dominant rulers in the European geopolitical conflicts came from the House of Habsburg. Originating from the Swiss regions, the Habsburgs became, in 1282, rulers of the Duchy of Austria. As a result of their marriage policy, they expanded

---

[130] Historically, family line relations were often formed in royal houses to secure political alliances, strengthen the lines of succession, and ensure the noble purity of the bloodline. In the Habsburg dynasty, it manifested in the Habsburg Jaw. Through generations of royals marrying each other and therefore closing the ranks of their gene pools, the Habsburg Jaw manifested itself just about everywhere in medieval Europe.

[131] In an age when rulers led their armies into battle, there was a fairly high mortality rate amongst male rulers. Adding to the problem was a newly virulent syphilis, which seems to have returned to Europe with Columbus in 1492 and which caused infertility when it didn't kill. So, the result was that girls fairly often were to inherit when their parents failed to produce a male heir.

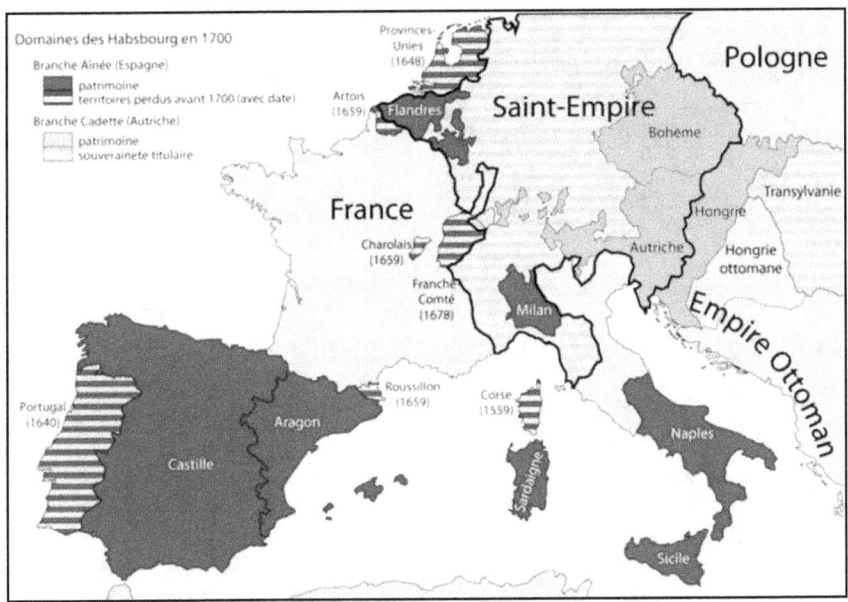

Figure 74: Habsburg's sphere of influence—Spanish Branch and Austrian Branch (c. 1700).

Portugal and the United Provinces were lost to the senior Spanish branch. Transylvania was a vasal and various de facto sovereign states of the Holy Roman Empire were controlled the junior Austrian) branch.

Source; Adapted from Wikimedia Commons

their sphere of influence and became, in 1437, rulers of Bohemia and Hungary (Figure 74).

Jumping ahead of time, we see that by Charles V's abdication in 1555, the House of Habsburg had been divided into the Spanish branch and the Austrian branch. The *Spanish Habsburg* ruled a large part of Western Europe, from Spain up to the Spanish Netherlands. They—chiefly Charles V and Philip II—reached the zenith of their influence and power, controlling territory that included the Americas, the East Indies, the Low Countries, and territories in France and Germany. The *Austrian Branch* ruled over the middle European regions of Austria, Slovenia, Western Germany (called Bohemia), and Hungary. And both branches looked eagerly at the riches to be obtained on the Italian Peninsula.

Over time the Eastern Habsburgs played important roles as elected rulers of the Holy Roman Empire, first as the elected kings of the Romans,

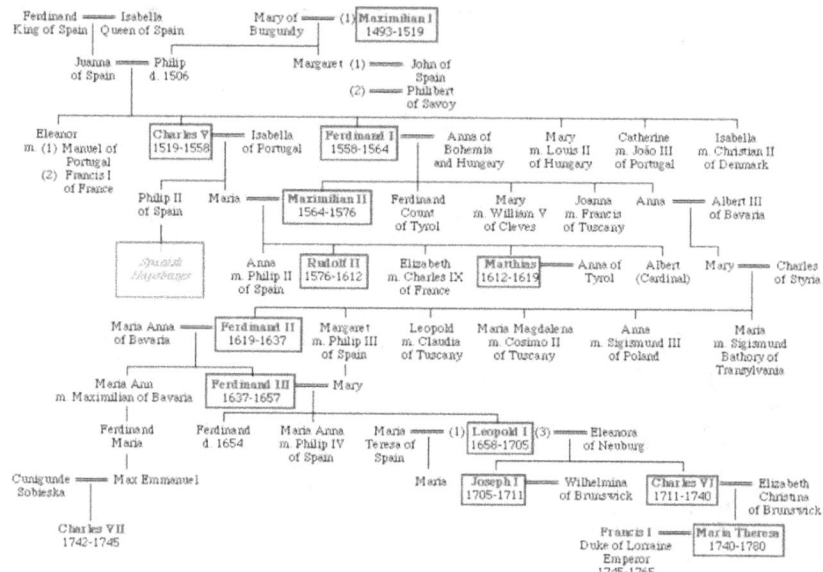

Figure 75: The Austrian Branch of rulers of the House of Habsburg.

Rulers are indicated in boxes

Source: edstephan.org. Artist Ed Stephan

later as Holy Roman Emperors after having received the Imperial Regalia—the Imperial Crown, the Holy Lance, and the Imperial Sword—and title of emperor' from the hands of the pope. This was the case with Frederick III (1424–1493), who was anointed Holy Roman Emperor in 1452 in Rome. He was succeeded by his son Maximillian I (1459–1519), who proclaimed himself the 'chosen emperor' in 1493. Maximillian expanded the Habsburg territory through war and marriage. In 1477 he married Mary of Burgundy, adding the Duchy of Burgundy. His son Philip married Joanna of Castille, adding Spain. Thus, his grandson Charles V eventually held the thrones of Castille and Aragon. This created the Spanish Branch of the House of Habsburg (Figure 75).

It took some marriages, but the extending Habsburg monarchy soon created a massive sphere of influence in mid-Europe. In 1521 Charles V assigned the Austrian hereditary lands of the Habsburgs (Austria and Slovenia) to his brother Ferdinand I, who married Anne of Bohemia and Hungary. After the death of his brother-in-law Louis II, Ferdinand ruled as King of Bohemia and Hungary (1526–1564). In 1558 he was elected king of the Romans. He was succeeded by his son Maximillian II (1527–1576), who became emperor in 1564. Till the end of the forty-year reign of Maria

Theresa (1717–1780)—the last ruler of the House of Habsburg—the Habsburg ruled Austria.

Thus, the both branches of the House of Habsburg controlled, by intermarriage and succession, a large part of Europe into the 1700s (Figure 74). The Spanish Habsburgs died out in 1700 (prompting the *War of the Spanish Succession*), as did the last male of the Austrian Habsburg line in 1740 (prompting the *War of the Austrian Succession*), and finally the last female of the Habsburg line died in 1780.

As it was not only by marriage that the Habsburgs expanded their territory, many military conflicts were also fought out by increasingly powerful armies, so it is time to explore shortly that dimension of the military context.

Military Context

As much of the geopolitical confrontations resulted in military conflicts, medieval warfare 'by bow, sword, and lance' dominated social conflict-solving. Military strategies and tactics were based on siege warfare and battlefield confrontations of a *cavalry* of mounted knights complemented by *infantry*, or foot soldiers (Figure 76). Over time the lightly armoured infantry, being cheaper to raise, became more important and increased the size of armies. Their deployment was in different forms, and the combat was dominated by 'firepower' (archers with crossbows) and 'mass' (pike men with lances) in tightly packed phalanxes of men in square formation.

Figure 76: Landsknecht mercenaries from Swabia (c.1450).

Source: Unknown.

Then the evolutionary use of firearms and cannons influenced warfare heavily and created the period of *gunpowder warfare*. After the arquebus and musket, in the late medieval period infantry musketeers were using 'matchlock' firearms. Infantry formations included a mix of troops armed with both firearms to provide striking power and pikes to allow for the close defence of the arquebusiers or musketeers from a cavalry charge (Figure 77).

One has to realize that the composition of armies then was different. Long ago, the Roman armies had already used, alongside the legionnaires, the warriors of the nearby tribes—the Perengrini—as 'auxillia'. Later, in feudal times, knights were soldiering on a part-time basis, when their ruler demanded their assistance. However, veterans from later wars such as the *Hundred Years War* (1337–1453) offered their services as professionals. Thus, soldiering, after feudal times, had become professionalized. Armies often contained contingents of mercenary soldiers who fought for money (and loot), such as the 'highlander' Swiss mercenaries and the 'lowlander' German Landsknechts from Swabia (Figure 76).

Figure 77: Matchlock musketeer (c. 1650).
Source: Unknown.

The impenetrable Pike squares (adopted from the Swiss), rows of deadly new Arquebuses, and a new mobile artillery system were to become the standard of this new deadly mercenary army. The Landsknecht army, as it was, was a fearsome sight to behold. Thousands of men wearing what would appear to be brightly colored rags and hats with many gaudy, brightly colored feathers. … All of that constant fighting would require lots of travel. Just imagine the logistics of moving the thousands of mercenaries, their gear, all of the camp followers and other laborers, all of the supplies and equipment necessary for the fighting force to wage war. Miles of walking Landsknechte, still more miles of the baggage train and required equipment. [132]

The use of Swiss mercenaries by others started at the end of the fifteenth century because entire ready-made Swiss mercenary contingents of skilled soldiers could be obtained by simply contracting with their local governments; the various Swiss cantons.[133] The Landsknechts, mainly coming from the south of Germany, served any paymaster. In the Great Italian Wars, the Swiss and the Landsknechts became bitter rivals.

Salvation for the paymaster came from the Andes; for recruiting offices from the Alps. A new race of hill-folk was coming to market, the Swiss, for two centuries the mercenary soldiery par excellence. … A nation of armed peasants, afflicted with the chronic overpopulation of the barren uplands, numerous enough to form massive columns of pikemen, they acquired by incessant practice the "extraordinary perfection of skill and discipline" demanded by their system of warfare, and they did this at home in their valleys, at no cost to their employers.

---

[132] Source: http://www.landsknecht.com/html/history.html
[133] The wealth—pay and loot—that was transferred to the Swiss Confederation in this way lies at the root of the Swiss banking system.

True, they were too democratic to be easily handled, and if the pay-chest gave out they changed sides or marched off home. Pas d'argent, pas de Suisse. While money lasted they fought like Trojans, local and professional pride giving them a high morale. Altogether they gave their employers the benefits without the drawbacks of a free citizen army. ... While the Swiss had a special link with France, Landsknechts served everywhere indiscriminately, and may have been preferred by some employers because, having less solidarity among themselves, they could be got cheaper and cheated more easily. However, Germany was industrially more advanced than Switzerland and therefore produced professionals who took to fire-arms more readily, such as the mounted arquebusier and the Schwartzreiter, a mounted pistoleer. ... Charles VIII started the Italian Wars in 1494 with 10,000 Swiss and Germans in his army. (Kiernan, 1957, pp. 69-70)

To hire the mercenaries, the rulers who employed them—these could be cities, local rulers, and monarchs—needed funds to pay for them. Those rulers that were wealthy, such as the Spanish ruler who had riches from the revenues of the Silver Fleet or the pope with Church capital, could afford to maintain large armies. Others, such as the rulers of Milan, had to get funds from the local population. That led to taxation of the population to pay for these high costs. And when the mercenaries did not get paid, they went out of control (of their paymaster) and took care of their own remuneration by looting.

It was not only the mercenaries who pillaged and looted. The by then emerging standing armies had to find their food supplies underway by pillaging local people, but there were also riches to be acquired by looting monasteries, cities, and estates. This was looting of monetary valuables (gold, gems) and looting of art and cultural valuables. The scale of plundering that took place under Napoleon's French Empire was unprecedented in modern history, such as the large-scale and systematic looting of Italy in 1796. But there was more, as the conquered people were taxed. During his rule Napoleon taxed Italy over 100 million livres tournoises in extraordinary contributions. He used 60 million to pay, feed, clothe, and reorganize the army of Italy deployed in every department. The rest had been sent to France.

Italy was practically destroyed by the years of war. If the French had stolen and robbed during the 1796–1799 period, their pillaging had been legally sanctioned after 1800. ... French rule cost the Cisalpine Republic alone—that is to say, to only a small part of Italy, from 1797–1804—288 million livres tournoises. ... Thus, once upon a time Italy was a rich country. After the Napoleonic era, it was truly one of the poorest regions of Europe. (Paoletti, 2008, pp. 85-86)

Later in time, part of the negotiations for peace treaties concerned the pay of 'war indemnities' as compensation. The Congress of Vienna (1815) addressed the issue after Napoleon's defeat at Waterloo and France paid some 40.5 million francs to the Italian States, a fraction of what they'd taken from it.

At that time armies consisted of formally organized colourful regiments armed with rifles, and bright coloured uniforms (Figure 78). In parallel with the move towards permanently employed soldiers, serving for several years, or indeed for life, this achieved several things: it encouraged loyalty, comradeship, and a vehicle for ensuring professionalism and a higher standard of training and skills.

Figure 78: 42nd Highland Regiment at the Battle of Waterloo (1815).

Source: Unknown.

Clearly, it was the military dominance that was often decisive in conflicts. A dominance created by ever-improving military technology, and Italy was on the receiving end of that dominance time after time, from Charles VIII's Italian Campaign (1494–1498) up to Napoleon's Italian Campaigns (1792–1802).

Italy's Stagnation and Reform (1500s–1815)

As we seen, in the early sixteenth century the textile industry flourished in Florence. Banking enjoyed a particularly dramatic growth. Genovese bankers became powerful, financing Spain's huge transatlantic fleets and the Spanish armies. Florence bankers financed the French crown. However, those same governments became their downfall, as the failed to repay their debts. Soon Italy began to experience an economic and social decline as the sixteenth century progressed. The Age of Discovery, with the discovery of the Americas by Christopher Columbus, had shifted the centre of trade in Europe from the Mediterranean to the Atlantic with the rise of the maritime peoples of the north: the English and the Dutch.

In addition, the Roman Catholic Church lost much of their former influence as the Protestant Reformation—starting in the 1520s with Martin Luther in the German state of Saxony—divided Europe into two camps: the Protestant rulers and the Catholic rulers. The first were—broadly speaking—to be found in the north: the Scandinavian territories, Northern Germany, and the Netherlands. The later were found in the south: France,

Spain, and Italy. The remaining Catholic princes increasingly sought to be the masters in their own houses and often clashed with the papacy over jurisdictional matters. Even in Italy itself, the political importance of the Papal States declined. So, one can observe on the Italian Peninsula a pattern of scattered regional city-states, all one way or the other under foreign rule. Only some regions stayed free of foreign domination, such as the Republic of Venice and the Papal States.

> The *Republic of Venice* continued to fight bitterly with the Ottoman Empire for control of outposts in the eastern Mediterranean. But by the eighteenth century, economic activity dwindled as the city withdrew in on itself and fell into stagnation, becoming easy pickings for the French revolutionary armies in 1796. Then it became a vassal state to France: part of the Bolognese Republic that soon was incorporated into the Cisalpine Republic.

> The *Papal States* also lost much of their former power as the Protestant Reformation saw the rise of countries that embraced the new religion. In England, with the rise of the Church of England—started after the quarrels with Henry VIII in the 1530s—and anti-Catholic sentiments related to succession issues and the French-British rivalry, the papal powers dwindled. The Church came also under attack as a result of the Enlightenment movement all over Europe. And in the 1790 the French Revolution toppled the Church's power in France.

Starting in the late fifteenth century, in the ages to come, Italy would be ruled—again—by foreign powers: the Spanish Rule, the Austrian Rule, and the French Rule.

> *The main cities had always been rich enough to put strong armies in the field. So long as other parts of Europe were weakened by rivalries between crown and feudal nobility, Italy was relatively safe; but once these disputes were resolved (as they began to be in France and Spain by the end of the fifteenth century, powerful monarchies suddenly appeared with an economic and military potential that outstripped anything in Italy.* (Duggan, 2014, p. 61)

It was in this period of time that the economic importance of Italy declined, as the Italian states played little part in the opening up of the New World. Also, the merchant entrepreneurs, who, as the urban middle class, were the backbone of earlier economic developments, turned away from commerce and industry and adopted an aristocratic lifestyle, transferring their money from trade to land, investing in arts and architecture.

> *From 1580 the slide into recession gathered pace, and in the decades that followed Italian trade and industry experienced a catastrophic slump. Woollen cloth manufacture disappeared almost completely, shipbuilding collapsed; and the*

principal ports, with the exception of Livorno … contracted sharply. Merchant banking also declined dramatically. By the end of the seventeenth century Italy had become an importer of large quantities of finished products from France, England, and Holland, and an exporter of primary or semi-finished goods: wheat, olive oil, wine, and above all silk. It had moved from a dominant to a subordinate position within the European economy. (Duggan, 2014, p. 69)

The economic decline hit the general population of the poor lower classes the hardest. Major epidemics of the plague returned in 1630–1631, killing high percentages of the population of Milan, Verona, Florence, and Venice. Bad harvests—which occurred everywhere in Europe in the mid-seventeenth century—also brought the lower classes to protest or starvation. But it was the burden of taxation that was favourable to the wealthy and discriminated against the poor.

Take as an example the south of Italy: the island of Sicily dominated by the still-active volcano Mount Etna. It was not only because of nature's whims that the common people there suffered.[134] In 1638, during the reign of Philip IV (Hapsburg) of Spain, the crown levied a 'head tax' to be paid by the feudatories of Sicily's feudal towns and the citizens of its demesnial cities to finance the cost of the Habsburgs' Thirty Year War. Also, climatic conditions contributed. In 1647 Sicily, normally an exporter of grain, saw its harvests failing due to drought, and food riots broke out in Palermo.

Meanwhile the local clergy organized processions to pray for rain and beg forgiveness for the sins that, they claimed, had brought God's punishment on the land. … Then a miracle occurred: it rained for two days and the crops began to grow again. … on May 1647 a second miracle occurred. A ship docked in Palermo harbour carrying several tons of grain. … The prospect of mass starvation, the outpouring of religious zeal and then two apparent miracles of deliverance created emotional overload in Palermo. … On May 20 about 200 people, many of them women and boys, gathered outside the city hall and shouted, "Long live the king and down with the evil government," "Big loaves, no excise" and, more simply "Bread, bread." (Parker, 2013, pp. 422-423)

A far more serious revolt broke out in Naples in July 1647. As in Palermo, it began with riots against high food prices caused in part by the excessive taxes that the almost-bankrupt Spanish government was exacting. The rebels sacked the houses of tax collectors and demanded lower taxes and local self-government.

---

[134] The period known as the Little Ice Age saw different cold intervals: one beginning in 1650, another about 1770, and the last in 1850. Many consequences were experienced locally. Farms and villages in the Swiss Alps were destroyed by encroaching glaciers during the mid-seventeenth century.

Some disasters were natural rather than economic. An eruption of Etna in 1669 seriously damaged several towns. Poor harvests and, more importantly, a long economic recession after 1671 had a particularly serious effect in Messina, where riots broke out in 1674 and lasted four tumultuous years.

These short brushstrokes show a painting of the early developments of the Italian peninsula after the Dark Ages. Let's go and look in more detail what happened.

Great Italian Wars (1494–1559) [135]

The Kingdom of Naples had always been in the attention of other nations such as the rulers of the Spanish House of Aragon and the French House of Valois. After the Spanish ruler of the Kingdom of Naples, Ferdinand I of Naples—aka Ferrante—died in 1494, the French king Charles VIII took his chance and invaded Italy.

This started the *First Italian War* (1494–1498), also called King Charles VIII's War. After crossing the Alps with a force of 25,000 men, sacking Lucca and entering Florence, passing the Papal States, and threatening Rome, he took Naples in February 1495 (Figure 79). There, he was crowned King of Naples. His rapid advance had alarmed other Italian powers, who formed the *League of Venice* (1495), aka the Holy League, to oppose him. This league included the pope, the Holy Roman Emperor Maximilian, Ferdinand and Isabella of Spain, Venice, and even Milan. After Charles was defeated by the league in July 1495 at the *Battle of Fornovo*, he left Italy without any success, and he died in 1498. This first effort had not been too successful.

Then came the *Second Italian War* (1499–1504), also called King Louis XII's War, when the French King Louis XII (1462–1515)—son of Charles VIII—after long diplomatic alliance-creating, invaded Italy with 17,000 infantry and 9,000 cavalry. He soon took Milan, Alessandria, Genoa, and the Papal States, thus becoming the dominant power in Northern Italy by 1500. Going south, he soon had also conquered Naples, and he became King of Naples from 1501–1504. However, by 1504 Louis XII, having been defeated at the *Battle of Cerignola* on April 28, 1503, and the *Battle of Garigliano* on December 29, 1503, was forced to withdraw from Naples, which was left under the control of the Spanish. This second effort had also not been successful.

---

[135] Text obtained from a range of sources, among them: Rickard, J (3 October 2015), 'The Italian Wars, 1494–1559,' http://www.historyofwar.org/articles/wars_italian_wars.html

A third French effort to establish its rule in Italy was made during the *War of the League of Cambrai* (1508–1516), aka the war of the Holy League. Territories were won, and territories were lost. The original anti-Venetian League of Cambrai collapsed, but it was partially restarted into the new Holy League. This time the French were less lucky, as the combined Venetian-papal armies drove the French out in 1512. Not for long, as soon the French were back, now facing the Old Swiss Confederacy (Figure 45), who had taken Milan in 1513. After the French crossed into Italy by the unknown roads over the Col d'Argintiere the French Landsknechts were

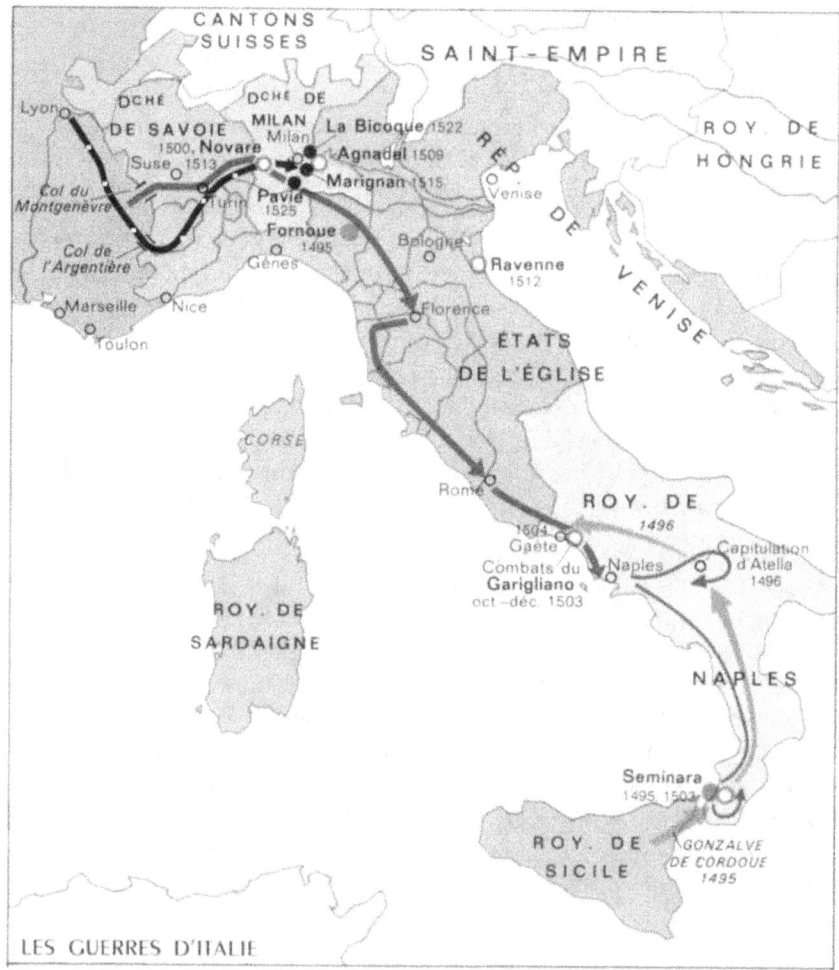

Figure 79: French campaigns during early Italians Wars (1494–1515).

Arrows indicate campaigns by French. Years indicate specific activity or occupation.

Source: Larousse http://www.larousse.fr/archives/grande-encyclopedie/page/7368

soon fighting the Swiss mercenaries in the *Battle of Marignano* (1515). The Swiss were beaten, and the *Treaty of Noyon* (1516) recognized French claims to Milan and Spanish claims to Naples.

This was the start of a range of battles in which France and Spain fought for the control of Italy: the several *Italian Wars* that lasted until 1559. Originally arising from dynastic disputes over the Duchy of Milan and the Kingdom of Naples, the wars rapidly became a general struggle for power and territory among their various participants and were marked with an increasing number of alliances, counter-alliances, and betrayals.

> *During the Italian wars, the weak, fragmented political system of the peninsula was confronted by the great European powers, which initiated the period of "foreign preponderance." The pressures and conquests of France, the Empire and then, more durably, of Spain had substantial effects and doubtless produced significant reshuffling: the definitive fall of the Florentine republic, the submission of the Sienese republic to the new grand duchy of the Medici, the substantial reduction of Bologna's and Perugia's "liberties" within the papal state.*
> (Chittolini, 1989, p. 701)

In the meantime France had also a range of other territorial problems. It became attacked from all sides. In 1521 imperial armies had invaded the northeast of France, in the southwest, the Spanish had obtained the Iberian part of Navarre in Spain, and in 1522 the English invaded the northwest of France, Brittany and Picardy, from their foothold Calais.

Then, in the *Italian War of 1521–1526*, the French control over Milan was lost again, this time to the coalition of the pope (Leo X), the Holy Roman Empire (Charles V), and the English (Henry VIII). First, in 1523 a French army of 18,000 men was defeated in the *Battle of Sessia*. Then the French tried again in 1524 with an army of some 40,000 men, but they lost the confrontations that followed, the final one between the House of Habsburg and the House of Valois being the French defeat at the bloody *Battle of Pavia* (1525). The French King Francis I surrendered and was imprisoned together with a large number of French nobles.[136] He was taken to Madrid, where he was forced to sign the *Treaty of Madrid* (1526), in which he surrendered his claims to Milan, Naples, Genoa, Flanders, Artois, Tournai, and the Duchy of Burgundy.

---

[136] Part of the deal for the return of the king was that his two sons were held in captivity to secure the agreement of cash and land cession. The ransom for the two princes was later set at the enormous sum of one million two hundred thousand gold crowns. It took some five years to raise this sum form the bourgeois, clergy, and nobles. When the transfer was made, 32 mules were packed to the hilt with gold.

The broken remnants of the French forces, aside from a small garrison left to hold the Castel Sforzesco in Milan, retreated across the Alps. By 1525 the French rule in Italy had totally collapsed. Now the states of Italy were reduced to second-rate powers, and the Duchy of Milan and the Kingdom Naples would come under the rule of the Spanish Habsburgs.

> *It was ironic (though not entirely a coincidence) that the political and military humiliation of Italy in the first half of the sixteenth century was accompanied by the emergence of a form of national culture. … As the sixteenth century progressed, Italian art and culture spread north and came to dominate much of Europe. Humanist education … became a hallmark of the rich from Scotland to Sicily. Italian dress, deportment, and even cooking set the standard at princely courts …* (Duggan, 2014, p. 66)

Spanish Rule over Italy (1530–1713)

Again, for Italy, it was to be a period of military turmoil between foreign powers and local powers. Active were several foreign powers: Charles V of the Holy Roman Empire, Francis I from the House of Valois, ruling in France, local city-states like Milan and Florence, and the Papal States. Dominating was Charles V, who inherited the *Crown of Aragon*,[137] which included the Kingdom of Naples, the Kingdom of Sicily, and the Kingdom of Sardinia (Figure 80).

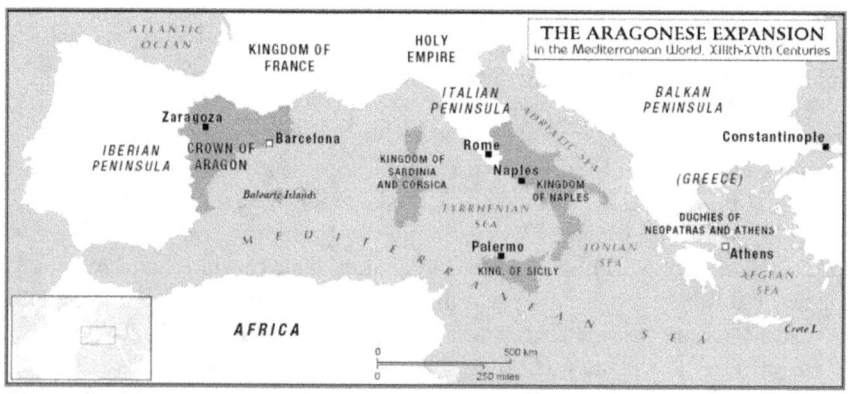

Figure 80: Territories of the Crown of Aragon (1450s).

Source: Edmaps.com, Christian Ionita.

---

[137] The Crown of Aragon was a state with primarily maritime realms, controlling a large portion of present-day eastern Spain, parts of what is now southern France, and a Mediterranean 'empire' which included the Balearic Islands, Sicily, Corsica, Sardinia, Malta, Southern Italy (from 1442), and parts of Greece (until 1388).

In response to the imperial dominance, the defeated French king, together with Pope Clement VII, in alliance with the Republic of Venice, Florence, and Milan, created the League of Cognac. It resulted in the *War of the League of Cognac* (1526–1530) with initial skirmishes in Lombardy. Then, to the pope's surprise and Charles embarrassment, Rome was sacked in 1527 by mutinous troops of Charles V (Figure 81). It was an event that was to have great consequences.

The mercenaries in the French army came from Spain (6,000 troops) and Germany (14,000 Landsknecht solders)—in an army of 34,000 soldiers—and had not been paid. That created discontent among the mercenaries. In addition, the protestant Landsknecht saw the capital of the Catholic Church from a different perspective. It

resulted in a mutiny in which Rome was taken in one day on May 6, 1527. The pope's Swiss Guard, also mercenaries, defended the Vatican and helped the pope to escape through the secret passage Passeto di Borgo to the fortress of the Castle Sant'Angelo, but they were soon massacred. After his surrender, Pope Clemence was held prisoner for seven months in the fortress. In exchange for his life, he paid a ransom of 400,000 ducati,[138] and he ceded Parma, Piacenza Civitavechhia, and Modena to the Holy Roman Empire. Meanwhile, in Florence, Republican enemies of the Medici took

Figure 81: Massacre during the Sack of Rome (1527).

Source; Wikimedia Commons.

---

[138] It was much more profitable to negotiate a ransom for people of higher status, then killing him.

advantage of the chaos to again expel the pope's family from the city. The population of Rome dropped from some 55,000 before the attack to 10,000. An estimated 6,000 to 12,000 people were murdered.

Charles was caught in a compromising situation, as his mercenary troops had challenged the authority of the Catholic Church and, by doing so, marked a considerable advance for Protestantism. The pope, born as Giulio of Giuliano de Medici and thus ruling member of Florence's de Medici dynasty, also had a problem, as he had chosen for the league, opposing the emperor. It created a complex political situation, intertwining Florence, the Church, the Holy Roman Empire, and France.

It was a confusing time, with the papacy switching alliances between the Holy Roman Empire and France, foreign armies crossing the peninsula, and local cities rebelling. However, Pope Clement VII and Emperor Charles V found each other in the *Peace of Barcelona* (1529), where the pope recognized Charles as king of Lombardy and the emperor, in return, helped overthrow the rebellious Florentine Republic, reinstating Medici rule.[139] In 1530 Charles was even crowned Holy Roman Emperor by Pope Clement VII in Bologna. It was to be the last imperial coronation by a pope. Taking Charles's side, the pope refused the annulment of the marriage of the English king Henry VIII from Charles's aunt, Catherina of Aragon, a decision that would come to have dear consequence for the position of the Catholic Church in England[140].

Then came final clash between France and the Holy Roman Empire in the last *Italian War* (1551–1559). Again Northern Italy saw many military confrontations between French and Spanish forces, such as the anti-Spanish revolt at Siena in 1552. Ultimately, the clashes resulted in a fifteen-month *Siege of Sienna* (1554–1555), one of the more turbulent Italian states, by French and Italians (ie Milanese under Sforza) troops. Siena was finally starved into surrender in April 1555. Florence eventually received control of the city, while the Spanish kept five sea ports.

The Italian struggles were just part of the larger conflict between France and Spain. The final peace came in 1559, after the *Peace of Cambrai*, a settlement between Henry II of France and Philip II of Spain. The Spanish were the big winners. As part of a greater deal concerning France's problems with Spanish dominance, Spain got control of Northern Italy except the Republic of Venice and the Duchy of Savoy. All other Italian

---

[139] Pope Clement VII was born as Guilio de Medici, which explains his interest in the rule of Florence.

[140] See: B.J.G. van der Kooij, *The Context of Innovation: British (R)evolutions in Perspective* (2016).

states were either ruled directly by Spain or were Spanish dependents. Naples, Sicily, and Sardinia (which had been Aragonese), as well as Milan, came under direct Spanish rule. Local councils and viceroys (in Naples, Palermo, and Cagliari) or governors (in Milan) controlled the internal affairs of these lands. Only a handful of independent republics—among them Lucca, Modena, Mantua, Genoa, and Florence—survived. The French failed to achieve any of their aims in Italy, ending up with no footholds on the peninsula. The Italians were obviously the big losers, with large parts of the country falling into foreign hands.

As Spanish powers declined in the sixteenth century, so did its Italian possessions in Naples, Sicily, Sardinia, and Milan. Southern Italy was impoverished, stagnant, and cut off from the mainstream of events in Europe. Naples was one of the continent's most overcrowded and unsanitary cities, with a crime-ridden and volatile populace. In 1647 a revolt broke out in Naples, resulting in full-scale war against Spanish rule. The Neapolitan aristocracy had long resented Spanish rule and welcomed the arrival of the Austrians in 1707. Hence, the *War of the Spanish Succession* saw control of much of Italy pass from Spain to Austria, culminating in the *Treaty of Utrecht* of 1713.

Austrian Rule over Northern Italy (1713–1792)

By 1700 Italy had grown into different spheres of influence. In the south, the Spanish ruled the Kingdom of Naples. The pope ruled the Papal States in the centre of Italy up to Bologna. The Duchy of Florence was ruled by the dynasty of the Medici. In the northwest the Duchy of Savoy and the Principality of Piedmont were under French control. In the northeast, the Republic of Venice was independent. And in between, the different duchies (eg Parma, Modena, Mantua) tried to stay clear from foreign domination (Figure 82), till they were confronted with the Austrian rulers from the Habsburg dynasty.

Going back in time, we see, by the end of the fifteenth century, Austria was part of the Holy Roman Empire (Figure 74). As the result of wars, marriages, and deaths, it became united with the Spanish Empire under Charles V (1500–1558) after the death of his grandfather Maximillian I (1459–1519). Following the dynastic tradition, the Habsburgs' hereditary territories were separated from this enormous empire at the *Diet of Worms* in 1521, when Charles left them to the rule of his younger brother Ferdinand I (1521–1564). He was confronted with the Protestant Reformation, which threw his lands in religious turmoil, internal divisions, and Ottoman Turkish expansionism. The Turks were even laying siege to Vienna (1528–1532). Ferdinand's election to emperor in 1558 reunited the Austrian lands, but after his death, the Habsburg tradition left his lands to his three sons,

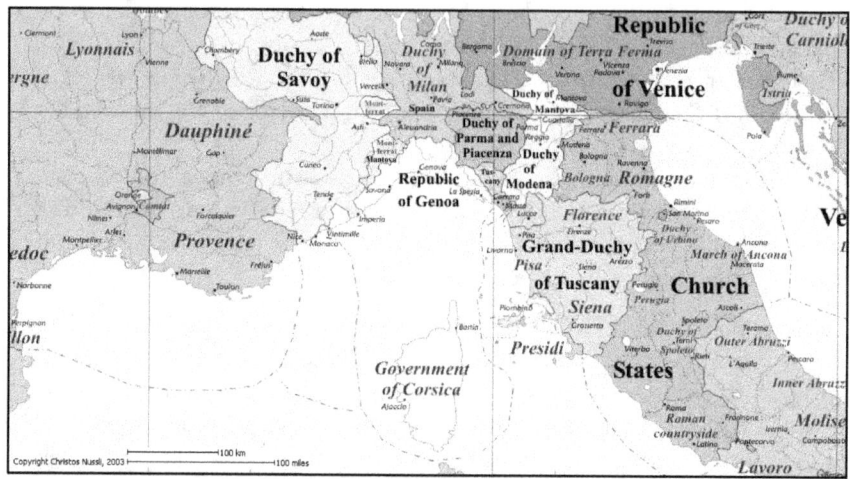

Figure 82: Political map of Northern Italy (1700).

Source: www.kingscollge.net. Artist Chritos Nussli

dividing Austria again into 'Lower Austria', 'Inner Austria', and 'Upper Austria'. In contrast to the authority of kings of Western Europe, where feudal structures were already in decline, Ferdinand's authority continued to rest on the consent of the nobles as expressed in the local diets, which successfully resisted administrative centralization.

In the seventeenth century Austria became involved in the disastrous religious *Thirty Years War* (1618–1648)[141] that swept over Europe, with the religion-inspired *Revolt of Bohemia* (1618–1620). The Austrian lands finally came under one archduchy in 1620, but Ferdinand III (1578–1637) re-divided them again in 1623 in the Habsburg tradition by parcelling out 'Upper Austria' (the later Austria and Tirol). It was not until after 1665 that Austria was reunited in one archduchy under Leopold I (1640–1705).

Despite the setbacks of the Thirty Years' War, Austria was able to recover economically and demographically, re-establish the Roman Catholic Church in a Counter-Reformation, and consolidate the Habsburg hegemony. It showed in architecture and arts, often referred to as the period of *Austrian Baroque* (1685–1740). Militarily, it was a time of problems on the western front—with the Habsburg Netherlands and the French

---

[141] The Thirty Years' War was a religious war. It was between countries that, after the Protestant Reformation, adhered to Protestantism, and the countries that under influence of the Counter Reformation adhered to the Roman Catholic Church. It developed in a war between the great powers of that time: France and the Habsburg Holy Roman Empire.

Burgundy—the northern front—Prussia and Sweden—and the southeastern front with the Ottoman Empire. By 1683, the Ottoman Turks were back at the gates of Vienna. Consequently, they were defeated at the *Battle of Vienna* (1683) with a decisive Austrian victory that saved Western civilization and began the fall of the Ottoman Empire.

In the eighteenth century it was Charles VI (1685–1740) who, in 1711, succeeded his elder brother Joseph (Figure 75) as Holy Roman Emperor, King of Bohemia, Hungary and Croatia, and Serbia, and Archduke of Austria. Being the sole male survivor of this line of the House of Habsburg, in an effort to solve the problem of lacking male offspring, he issued the edict *Pragmatic Sanction* (1713) to allow female heritage. Getting the other European powers' approval was complicated, as, in the European theatre, it was a time of massive confrontations due to several succession wars, for example, the *War of Spanish Succession* (1701–1715) that took place after the death of the Spanish Charles II without heirs. Louis XIV of France was to inherit, but France faced opposition from a coalition of England, the Netherlands, the Austrian Empire, and Prussia. Hence, by the *Treaty of Utrecht* (1713), Spain lost all its possessions in Italy and the Low Countries. The Spanish Netherlands, Duchy of Milan, Kingdom of Naples, and Sardinia were given to Emperor Charles VI, emperor of the Habsburg-ruled Austria, while Sicily was awarded to the duke of Savoy.

So, Spanish rule having ended, large parts of Italy came under Austrian rule but not for long, as in 1717 Sardinia was reconquered by the Spanish, followed by Sicily in 1718. It was at the *Treaty of Vienna* (1738) that France accepted the Pragmatic Sanction. Again entire kingdoms and duchies were merrily traded. Austria gave up Naples and Sicily, and an independent Kingdom of the Two Sicilies was formed with a secondary line of the Spanish Bourbon family as rulers, who had to pledge never to unite with Spain. In compensation, the House of Habsburg received the duchies of Modena, Mantua, Milan, Parma, and Piacenza in Northern Italy. Austrian rule was now established in an important portion of Italy. By the mid-eighteenth century, Austrians controlled Lombardy and Tuscany.

After the death of Charles III in 1740, his daughter Maria Theresa became, in 1745, ruler of Austria thanks to the Pragmatic Sanction now being accepted by the European powers. She also became empress consort, when her husband, Francis of Lorraine, grand duke of Tuscany, was elected as Emperor of the Holy Roman Empire. In Lombardy, the cautious reforms of Maria Theresa had enjoyed support from local reformers.

After Francis died in 1765 their son Joseph II (1741–1790)—brother of Marie Antoinette, who married the last French king before the French Revolution, Louis XIV[142]—became elected as Holy Roman Emperor and was co-regent with his mother in the Austrian dominions till her death in 1780. Joseph, ruling alone from 1780 to 1790, became an enlightened despot and started issuing edicts, over 6,000 in all, plus 11,000 new laws designed to regulate and reorder every aspect of the empire. He inspired a complete reform of the legal system and the administrative system, abolished brutal punishments and the death penalty in most instances, and imposed the principle of complete equality of treatment for all offenders. He ended censorship of the press and theatre.

With his *Imperial Patent* of 1785, he abolished serfdom, giving the serfs personal freedom by abolishing the lords' control over their marriage, freedom of movement, and choice of occupation, while keeping the landownership intact. He also sought to reform the Catholic Church, reduce ecclesiastical power, and abandon certain practises and abuses. In a way, his work was a precursor of the French Revolution.

The spirit of 'Josephinism' was benevolent and paternal. He intended to make his people happy, but strictly in accordance with his own criteria. His interferences with the feudal structure and its old customs began to produce unrest in all parts of his dominions, both among the nobility as well as ordinary people. By 1790 he was facing several rebellions in Belgium and Hungary. At his death in 1791 he was succeeded by his brother Leopold II (1747–1792), who had been ruling the Grand Duchy of Tuscany and who had brought the duchy to prosperity. He ruled for only a short period, as he died in 1792, the year the French Revolutionary Wars started. That event was going to be the dawn of the crumbling of feudal structures all over Europe.

Under Austrian rule, the period of 1748 to 1796 created a period of peace and stability without parallels since the sixteenth century (Figure 83). Many of the duchies—such as Tuscany and Modena—became private domains of the Habsburg family, separated from the Austrian State. In Northern Italy, the Austrians stayed a dominant political factor till the French showed up again. This time it was the Italian campaigns of Napoleon Bonaparte's revolutionary armies crossing the mighty barrier of the Alps.

---

[142] The outbreak of the French Revolution of 1789 led to Joseph II seeking to help the family of his estranged sister Queen Marie Antoinette of France and her husband King Louis XVI of France. Joseph, who kept an eye on the development of the revolution, became actively involved in the planning of a rescue attempt in 1791.

The Grand Duchy of Tuscany, which had replaced the Duchy of Florence, was ruled by the de Medici dynasty till its extinction in 1737. These masters of diplomacy, with their wealth and influence, developed Florence into a prominent city-state. The dynasty produced rulers, archbishops, and even popes: Leo X (1513–1521), Clement VII (1523–1534), Pius IV (1559–1565), and Leo XI (1605).

Figure 83: Political map of Italy (1798).

Situation before the French Revolutionary Wars. Insert top: Venetian Ionian Lands. Insert bottom: Fiefdoms of Lunigiana.

Source: Wikimedia Commons.

Their last ruler was Cosimo III who reigned from 1670–1723. Their wealth permitted them to pay enormous dues to the Holy Emperor. Cosimo III left a Tuscany one of the poorest nations in Europe; the treasury empty and the people weary of religious bigotry, the state itself was reduced to a gaming chip in European affairs. He was succeeded by his son Gian Gastone, the seventh and last Medici duke of Tuscany, who ruled from 1723 to 1737. His death ended some three hundred years of Medici rule over Florence. From then on the Grand Duchy of Tuscany was ruled by the House of Habsburgs-Loraine and Bourbon-Parma till 1807. Then Tuscany was annexed by France.

French Rule (1796–1815)

The period of the *French Revolution* (1789–1799) was a time in which France was occupied by internal affairs of quite some magnitude. After the storming of the Bastille on July 14, 1789, the absolute monarchy in France was under dispute, the ancient regime was toppled, and the Roman Catholic Church's influence minimalized. As soon as the revolution[143] was underway, France aspired to spread its ideas over Europe with a massive territorial revolution, in the process solving some international issues that had been bothering it for a long time. Political issues such as the Rhineland-Alsace situation in the northeast, the situation in the Savoy in the Southern Alps, and the conflict with the Spanish in the Pyrenees. Italy was soon to face the new French influence and the military capabilities of its rising star: General Napoleon Bonaparte.

The revolution in France greatly worried European monarchies, even more so after the revolutionary government executed King Louis XVI on January 21, 1793. They had reason to be, as, in the *French Revolutionary Wars* (1792–1802), the revolutionaries exported the revolution in a range of campaigns along the north, east, and south borders of France. The frightened European monarchies—and their aristocracies—reacted by creating coalitions. And those coalitions went to war.

In the *War of the First Coalition* (1792–1797)—a coalition of Austria, Prussia, Spain, and Portugal—France declared war on the Habsburg Monarchy of Austria. In the north, it started with the *Flanders Campaign* into the Austrian Netherlands, followed in the southwest by the *Roussillon Campaign* into Catalonia. Next, in the southeast, the Duchy of Savoy was annexed by France in 1792. Swiftly, the Nice region (part of the Duchy of Savoy) was taken by 30,000 troops.

---

[143] The French Revolution is covered in the case study The Invention of the Communication Engine 'Telegraph' (2015) pp. 38-167.

Then, in the *Saorgio Offensive* (1794), the French entered Piedmont over the crucial mountain pass of the Col de Tende (Figure 84, left lower corner). Along the coast of the Mediterranean Sea, in the *Montenotte Campaign* (1796), France again entered into the Piedmont region. This time it was an invasion under the command of the 26-year-old general Napoleon Bonaparte,[144] who reached Savona along the coast in April 1796. Next, the French entered the Po delta. After the *Battle of Lodi* (1796) (Figure 84, centre), the French captured Milan, followed by the *Battle of Castiglione* (1796) and the *Battle of Rivoli* (1797) with massive casualties. It was the *Siege of Mantua* (1796–1797), the primary Austrian fortress in northern Italy, which sealed the fate of the Austrians. Continuing their campaign, the

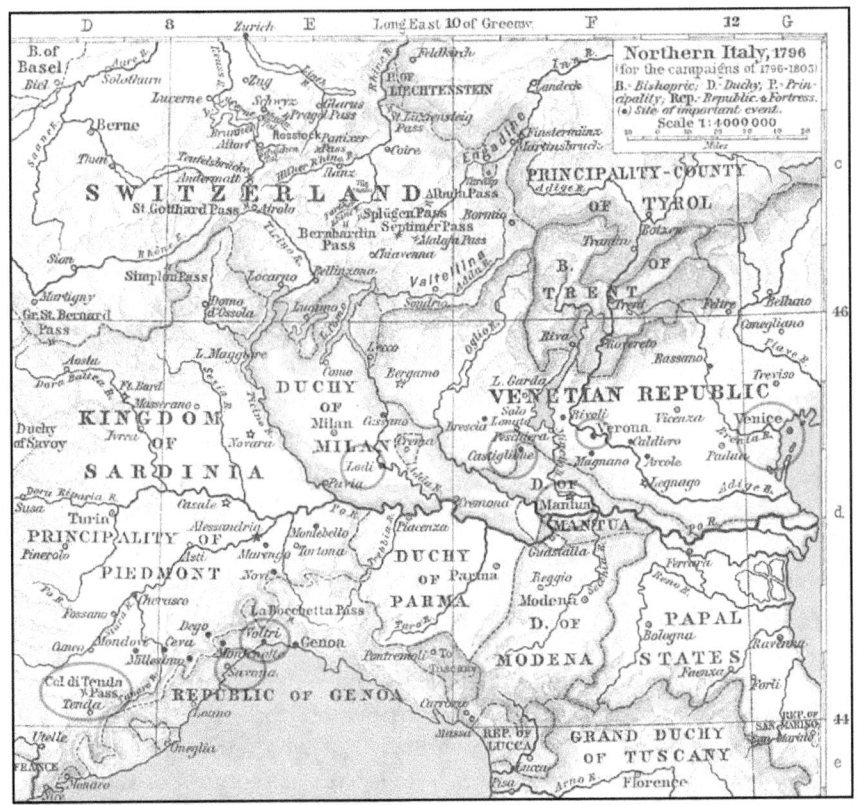

Figure 84: Northern Italy (1796).

Important places mentioned in text are circled.

Source: http://www.emersonkent.com/map_archive/northern_italy_1796.htm. Historical Atlas by William Shepherd (1911).

---

[144] It was Napoleon's first independent command, in which he lead the ragtag 'Army of Italy' to a series of astonishing victories over the more numerous and better-equipped Austrian and Piedmontese armies.

French troops arrived in Verona—part of the Republic of Venetia—in June 1796, soon occupying the military strongpoints and billeting troops in other buildings despite the Republic of Venice having declared its neutrality. After conflicts, the city rebelled against the French forces on the second day of Easter: the *Veronese Easter Rebellion.*

On the night of April 16, Easter Sunday, a manifesto was pinned up in Verona, inciting the population to rebel against the French and their local collaborators. Though it appeared to be signed by a Veronese, it was actually the work of a French collaborator. The purpose was to provide an excuse for the French to definitively occupy the town.[145] On April 17, 1797, brawls broke out between French soldiers and the local inhabitants. As a French soldier reported:

> [P]easants, taking advantage of the [Easter] festival, crowded into the town and mixed with the townsfolk and the Slav soldiers who still garrisoned the town, clogging the streets and the squares. Around midday, all of a sudden, upon a signal given by whistle blasts, this mob fell upon the French, attacked our isolated outposts, and massacred the guards. Our sick and wounded, who filled our hospitals, had their throats slashed with daggers. The bodies of the murdered French were thrown into the Adige River. The murderers spared neither women nor children. Some of the French were able to reach the forts we occupied. Others sought shelter in the palace of the Venetian magistrate, who granted them asylum, no doubt to preserve the appearance of neutrality should the assault fail, for he did nothing to stop or calm the insurgents. [146]

To control the situation, the French discharged cannons into the crowd gathered in the centre of the city. The Veronese population responded by raging through the streets. Armed with rifles, pistols, sabres, pitchforks, and staffs, the townsfolk rushed through the streets, killing, wounding and capturing several Frenchmen. The next days the rebellion intensified, and Verona came under continuous bombardment from the French. Finally, on April 23, the rebellion was suppressed by 15,000 French troops bombing the city day and night. Verona capitulated on April 25, 1797. As retribution, Verona had to pay massive reparations. The town was required to supply France with boots and clothing for 40,000 soldiers, a large amount of

---

[145] In reality, the actual history of such events is more complicated than as disclosed here. Many different factions did exist, those with pro-revolutionary ideas, those against the same ideas. They each used different means to reach their objective. Not everything reached the surface of history, as much is hidden in the caves of time.

[146] Sources: http://shannonselin.com/2016/03/napoleon-veronese-easter/; http://www.traditio.it/PASQUE%20VERONESI/pasque2007/luglio/30/Storia%20delle%20Pasque%20Veronesi%20in%20inglese..pdf

money equal to a month's pay for each soldier, and a hoard of famous paintings and sculptures, which were shipped off to Paris.

As soon as the town was recaptured, the French revolutionaries decided upon the immediate deportation to France, via Cisalpine and Milan, of the 2.500 men of the Venetian garrison which defended the town and particularly the Treviso infantry regiment. To lodge them, the country of the liberators of humanity set up the first modern concentration camp. Less than half or even one third came back from those prison and extermination camps after the Campoformio peace treaty. (Ibid)

By the start of the Veronese Easter Rebellion, the main body of the French army had passed already into the foothills of Austrian Tirol and advanced to within some 100 miles of Vienna. On April 18, 1797, the *Treaty of Loeben* was signed. Napoleon was ceded the Italian States in Lombardy. The whole of Northern Italy under his rule, Napoleon next turned his attention on the Republic of Venice (Figure 84, right) after peace talks had failed because Venice wanted to maintain a perfect neutrality.

Meanwhile the Venetian Senate, on hearing of these Austrian successes, had plucked up courage to throw aside their flimsy neutrality, and not only declared war against France, but hired 10,000 Slavonian mercenary soldiers, sending 6,000 of them to Verona to open the contest with the massacre of 400 French wounded in the city's hospital. The vindictive Italians, wherever the French party was inferior in numbers, resorted to similar atrocities. The few troops left in

Figure 85: Napoleon's arrival in Venice (1797).
Source: Wikimedia Commons, artist Luigi Querena

Lombardy by Napoleon were obliged to shut themselves up in garrisons. Napoleon's rear was for the time being, cut off from receiving any supplies because of the Venetian army. (Blair, 2002)

Soon Napoleon ranged his heavy artillery around Venice and blockaded the harbour with his warships (Figure 85). On May 12, 1797, the Great Council of Venice voted to dissolve the thousand-year-old republic and surrender the city to Napoleon. By that time the Venetians had panicked. Popular tumults filled the streets and canals of Venice.

Bonaparte appeared, while the confusion was at its height on the opposite coast of the Lagoon. Some of his troops were already in the heart of the city, when on the 31st of May, a hasty message reached him, announcing that the Senate submitted wholly. He exacted severe revenge. The leaders who had aided the Lombard insurgents were delivered to him. The oligarchy ceased to rule, and a democratic government was formed provisionally, on the model of France. Venice consented to surrender to the victor large territories on the mainland of Italy; five ships of war; 3,000,000 francs in gold, and as many more in naval stores; twenty of their best pictures, and 500 manuscripts. ... It was a glorious reform for the Venetian nation; it was a terrible downfall for the Venetian aristocracy. The banner of the new Republic now floated from the windows of the Doge's palace, and as it waved exultingly in the breeze, it was greeted with the most enthusiastic acclamations by the people, who had been trampled under the foot of oppression for fifteen hundred years. (ibid)

Figure 86: The four horses of St. Marc's Basilica (top) and *The Wedding at Cana* by Paolo Veronese (bottom, 1563).

Source: Wikimedia Commons.

Clearly, the rich Venetian merchants had to pay for their opposition. The city was looted. Among the many treasures sent to Paris were the four bronze horses of St. Mark's Basilica (Figure 86, top), the same horses the

Venetians had brought to Venice after the Sacking of Constantinople in 1204. Ultimately, they would be returned to Venice in 1815. But many paintings, among which was *The Wedding at Cana* by Paolo Veronese (Figure 86, bottom), an expansive canvas (67.29 m²), remained in the Louvre.

With the *Treaty of Campo Formio* (1797), the Holy Roman Empire ceded the Austrian Netherlands—creating the Batavian Republic after the

Figure 87: French client States in Italy: Cisalpine Republic and Parthenopean Republic (1799).

Source: http://etc.usf.edu/maps/pages/7500/7506/7506.htm

Flanders Campaign (1792–1795)—to France's rule. Also, Northern Italy was ceded to France. The region was divided into different sister republics: the *Cisalpine Republic* (1797–1802)—that incorporated the Bolognese Republic, the *Cispadene Republic* (south of the river Po), and the *Transpedane Republic* (north of the river Po)—the *Ligurian Republic* (1797–1805), the short-lived *Roman Republic* (1798–1799), and the *Parthenopean Republic* (1799) on the territory of the Kingdom of Napels (Figure 87). The revolutionary ideas of the French Revolution were transferred into reality.

However, the treaty was short lived, and soon the *War of the Second Coalition* (1798–1802) erupted. Now a coalition of Britain, Austria, and Russia, and including the Ottoman Empire, Portugal, and Naples tried to resist France's revolutionary military conquests. After crossing the Alps, it was during the Mediterranean Campaign that Napoleon recaptured Milan and Turin. But after a crushing defeat in the *Battle of Marengo* in 1800 (Figure 88), Austria was prepared to honour the *Treaty of Campo Formio* at the *Convention of Allesandria* (June, 1800). By then Napoleon was, as the First Consul, the most powerful man of revolutionary France. Thanks to the victory at Marengo, Napoleon could finally set about reforming France according to his own vision.

In the south of Italy, the French were less successful. In January 1799, Napoleon Bonaparte had captured Naples from the Spanish King Ferdinand IV and proclaimed the short-lived *Parthenopaean Republic*, a

Figure 88: Victorious Napoleon at the Battle at Marengo (1800).

Source: Wikimedia Commons. Artist Alphonse Lalauze.

Figure 89: Napoleon's coronation as Emperor (1804).

Seated on the right is Pope Pius. Napoleon crowns himself.

Source: Wikimedia Commons, Louvre Museum, artist Jacques-Louis David.

French client state, as successor to the Kingdom of the Two Sicilies. However, Ferdinand returned after some months and, with English support, was put on the throne. It was not until 1806 that Napoleon, by that time Emperor of France, in a new campaign, again dethroned King Ferdinand and appointed his brother, Joseph Bonaparte, as King of Naples.

By 1802 Northern Italy had become under control of France. The newly created *Italian Republic* (1802–1805) was the successor of the Cisalpine Republic, now with Consul Napoleon as its president. In 1805 Napoleon, by now Emperor Napoleon I, claimed the crown of Italy for himself, putting the famous Iron Crown of Lombardy on his head at Milan on May 26, 1805 (Figure 89). Seated upon a throne, he was invested with the usual insignia of royalty by the Cardinal Archbishop of Milan, and ascending the altar, he took the iron crown and, placing it on his head, exclaimed, being part of the ceremony used at the enthronement of the Lombard kings: "Dieu me la donne, gare à qui la touche" ("God gives it to me. Beware, whoever touches it"). Napoleon was now King of Italy, although the territory that was included extended only across Lombardy and the Emilia Romagna. The rest of the peninsula at that time did not acknowledge Napoleon as its king.

In France, the de-Christianisation by the French Revolution had ended the dominance of the Roman Catholic Church, as the state seized Catholic property, purged the clergy, and confiscated tithe gathering in 1789–1794.

Hence, for the popes in Rome and ruling the Papal States, the French revolution was a disaster. Already, in 1796, papal territory (ie Ancona and Loreto) had been occupied by the French. In 1798 Rome was occupied, and the *Roman Republic* (1798–1799) was proclaimed. Pope Pius VI was held in prison in Valence, France, where he died in 1799.

After the *Concordat* of 1801 relations between the French Republic and the Church improved. The Papal States were restored under Pope Pius VII, who also undertook measures to further improve the relations with Napoleon. The pope travelled to Paris in 1804 to officiate at Napoleon's imperial coronation, thus sanctioning 'higher' approval to Napoleon's rise (Figure 89). Nevertheless, in the course of the years 1806 to 1807, Napoleon came into sharp collision with the pope on various matters both political and religious.

By 1808 relations had deteriorated again, as the pope did not want to sanction the annulment of Napoleon's brother Jerome's marriage and

Figure 90: Extent of Napoleon's empire (1812).

Source: Encyclopaedia Britannica

refused to bring the ports of the Papal States into the Continental System.[147] This all resulted in a French occupation of Rome in February 1808, and in the following month another section of the Papal States (the Marches) was annexed by the Napoleonic Kingdom of Italy. By 1809 Napoleon had annexed all that remained of the Papal States, including the city of Rome, and announced that the pope no longer had any form of temporal authority. Pope Pius VII responded with an immediate use of his spiritual authority, excommunicating Napoleon himself and everyone else connected with this outrage. He was immediately arrested and removed to imprisonment in France. He remained there until 1814, when, after the French were defeated, he was permitted to return to Rome, where he was greeted warmly as a hero and defender of the faith.

By 1809 the whole of the Italian Peninsula was under French control. Napoleon, being the ruler of the French Empire, controlled a large part of Europe (Figure 90). His kingship would only last till 1814, but the spirit of the revolution that characterized the Age of Reform would leave its mark on Europe and spark the Independence Wars in Italy.

> *By 1815 Napoleon and his armies were defeated, but impact of their presence in the Italian states would endure long afterwards. Ideas from another relatively new movement called the Enlightenment had permeated Italian thought due to the French occupation. Napoleon and his armies were a product of the French Revolution, a movement heavily influenced by Enlightenment ideas. These ideas included a strong confidence in human reason, happiness, liberty, and the idea that people had rights unto themselves. The areas that were under French occupation or influence had their traditional political systems removed and replaced by systems that reflected Enlightenment ideals. These systems allowed for a measure of freedom previously unknown to most of Europe.* (Hill, 2005, p. 18)

After the *Treaty of Paris* of 1814/5, and when Napoleon was finally imprisoned and shipped off to the remote island of St. Helena, the rulers of Europe—the large nations like Austria, Great Britain, Russia, and Prussia, but also the other smaller states and the Bourbon representative—congregated at the *Congress of Vienna* (November 1814–June 1815) to decide on Europe's future. The Congress of Vienna settlement, despite later changes, formed the framework for European international politics until the outbreak of the First World War in 1914, and it decided on Italy's quite revolutionary future...

---

[147] The Continental System was the blockade of French territory for British trade.

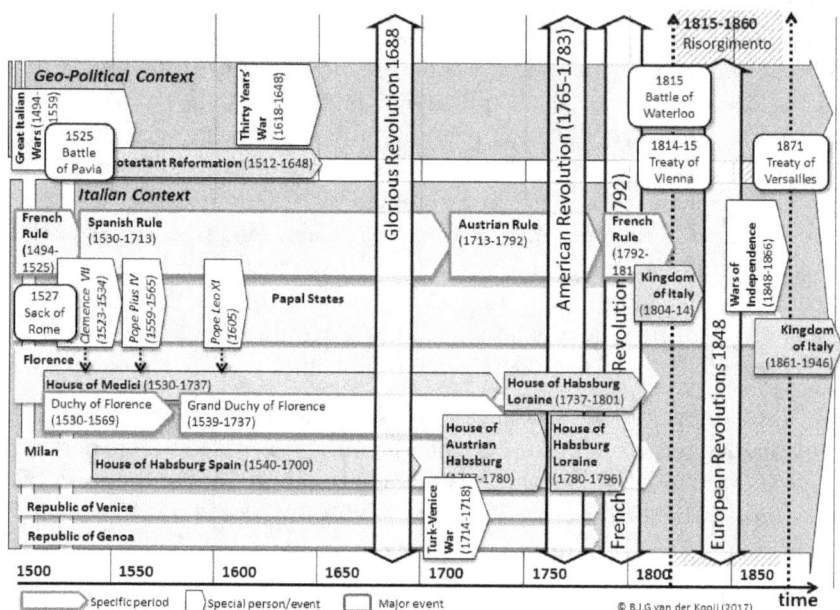

Figure 91: Overview of Italy's Age of Stagnation and Reform (1494–1815).
Figure created by author.

From the 1500s Italy had seen a continuous stream of foreign powers invading and ruling major parts of the country (Figure 91). After the French Rule, the Spanish Rule, and the Austrian Rule, the French, under Napoleon, had ruled again. In fact, Italy had been at the whim of the great European powers, which considered, especially the northern parts of the peninsula, an attractive source of taxation. The only 'region' that escaped foreign dominance for a lon time was that of the Papal States ruled by the Papacy. Several popes came from the Medici family ruling Florence and Tuscany. Over time Venice's power declined, as had Genovese power. The French Revolution and its usurpation of Italy—creating the different republics like the Cisalpine Republic and, later, the Kingdom of Italy under Napoleon— also revolutionized the governmental and institutional structures. It made Italy ready for the revolution to come after Napoleon's defeat at Waterloo. It would take some time, though, before the European-wide revolutions of 1848 would start the Italian Revolution and the unification of Italy.

Italy's Revolt in the European Context (1815–1850s)

At the start of the nineteenth century large parts of Europe were, as a result of Napoleon's military campaigns in the Revolutionary Wars (1792–1802), in turmoil. The governing classes were especially faced with the revolutionary ideas that Napoleons invading armies had spread all over Europe. In reaction, the former powers united in several coalitions, fighting their common enemy, Napoleon. It was the War of the Sixth Coalition (1813–1814) that defeated the French armies and sent Napoleon off to Elba. But by then the revolutionary ideas of the French had already sparked an interest among the peoples of Europe.

Congress of Vienna (1815)

The members of the coalition were united in their wish to restore the balance of power in Europe and eliminate the democratic and liberal ideas of the French Revolution. In addition, the big powers wanted to gain some advantage for themselves by remapping the boundaries of their territories. The objective was to return Europe to 'status quo ante bellum',[148] an objective based on three basic political philosophies: *compensation*, the idea that states should get back what they had lost, *legitimacy*, laws which established the legal government's right to return to the old regime, and *balance*, the idea that if all parts are equal within the whole, no one part can upset the balance.

Figure 92: The diplomats of the Congress of Vienna (1815).

Source: https://www.napoleon.org/histoire-des-2-empires/tableaux/le-congres-de-vienne/ Artist: Jaen-Baptiste Isabey

---

[148] Status quo ante bellum: the pre-war situation.

The monarchs had not forgotten the Reign of Terror that had befallen France and its king Louis XIV. Neither had the nobility or the Church. Just as the aristocrats had lost their heads, lands, and riches, the Church had lost its clergy, abbeys, and monasteries. The old order was to be restored, so the diplomats representing their countries of Russia, Great Britain, Prussia, Austria, and the French House of Bourbon—assembled in November 1814 in Vienna. However, to the surprise of many, Napoleon came back from Elba after 300 days of exile, assembled an army, and fought the Seventh Coalition to be beaten at Waterloo in 1815. After this interruption the diplomats continued the *Congress of Vienna* (Figure 92) and reached an agreement in June 1815.

Each power gained (Figure 93): Russia got most of Poland (aka Congress Poland); Prussia got parts of Poland, Saxony, and the Rhinelands/Westphalia; Austria regained control over Tirol and Salzburg; the Northern and Southern Netherlands were united; Britain was to hold former Dutch Cape Colony and various other colonies in Africa and Asia; and so on, every member of the great powers getting its share. In the process, the German Confederation of 38 states was created from the previous 360 states of the Holy Roman Empire.

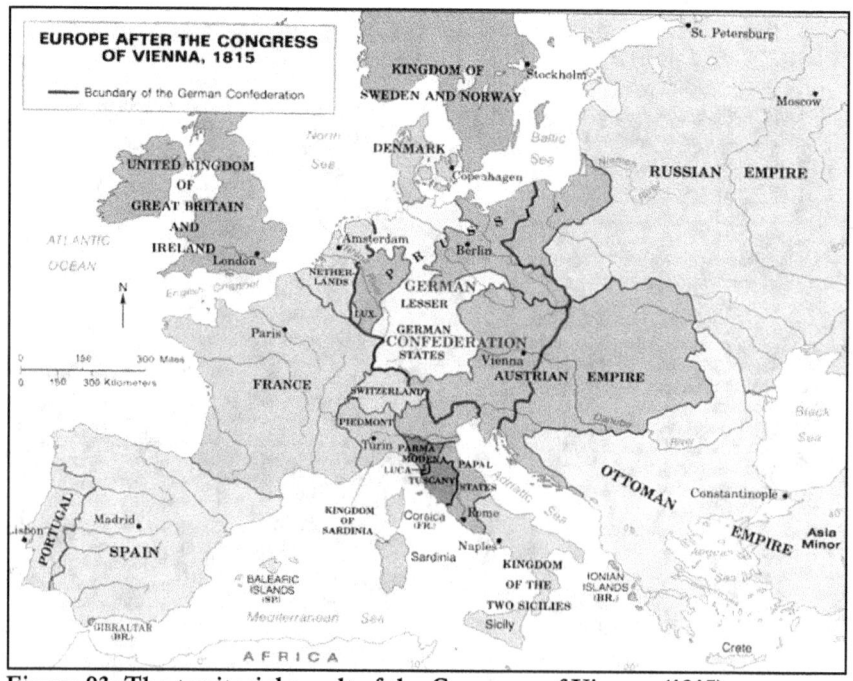

Figure 93: The territorial result of the Congress of Vienna (1815).

Source: Wikimedia Commons.

Figure 94: Italy before the Unification (1843).

Map shows the different nation-states (aka kingdoms) and territories.

Source: Wikimedia Commons.

In addition, in the Mediterranean area, they had (re)created the states on the Italian Peninsula (Figure 94). In the north, it was the Kingdom of Sardinia (where the king now ruled Piedmont, Nice, and Savoy) and the Kingdom of Lombardy-Venetia (under the rule of the Austrian Empire). In the center, the Papal States (under temporal papal rule) and the duchies of Tuscany, Lucca, and Modena (under House of Hapsburg-Loraine rule) were

restored. The Duchies of Parma, Piacenza, and Guastalla were given
to Marie Louise, Napoleon's wife and daughter of the Habsburg Emperor
Francis II of Austria. Also in the South, the Kingdom of the Two Sicilies
(under French Bourbon rule) was created.

Thus, the congress created a range of nation-states in Italy under the
rule of the former foreign powers. Again Italy was divided into over a
dozen states large and small, where foreign powers like Spain, France, and
Austria—and the Catholic Church—were the dominating political powers.
The old interests were served, and the interests of those who inhabited the
lands, the non-aristocratic people, were totally neglected.

Restoration of the Old Powers (1815–1850s)

The old powers that ruled society before the French revolutionary wars
were back in the saddle, so life seemed to be back to normal for the ruling
class. Austria was again dominating a large part of Northern Italy; it was to
be the 'Age of Metternich'[149] with his secret police. But the population
had—during the revolutionary Napoleonic rule—tasted the seeds of
democratic freedom, not only people in the countries like France, the
Netherlands, and the German states, but also in Italy.

> *Little attention* [in the Treaty of Vienna] *was given to the desires of the people
> in these territories. Northern Italy was now under the control of a conservative
> foreign power that was determined to quash the twin threats of democracy and
> liberalism. The rest of the Italian peninsula was divided up into its pre 1789
> kingdoms. To ensure control over the territory many members of the Habsburg
> family, the same family which ruled in Austria, were given control over many of
> these kingdoms. The Papal States were restored to their former border and Pope
> Pius VII was given back temporal power over them. In the north the states of
> Lombardy and Venetia were left under the control of Austria to provide it with a
> foothold to meddle in Italian affairs.* (Hill, 2005, p. 20)

Remarkably, the result of the Congress was the geopolitical restoration
of the former powers: what constituted as the *Conservative Order* of the
former ruling aristocracies. However, the Congress of Vienna was only the
beginning of a conservative reaction bent on containing the liberal and
nationalist forces unleashed by the French Revolution. After 1815 the
political philosophy of conservatism was supported by hereditary monarchs,
government bureaucracies, landowning aristocracies, and revived
churches. It also created some new nations that would not last for a long
time, such as the United Kingdom of the Netherlands, where the former

---

[149] Prince Klemens von Metternich (1773–1859) was the most important politician in
Austria. He was foreign minister in 1809–1821, and Chancellor from 1821–1848. He was a
staunch follower of the Ancient Regime and a bitter enemy of liberalism.

United Provinces (the Dutch Republic) of the north were combined with the formerly Austrian-ruled territories of the south (aka Belgium). It was not to last very long, as in 1830 this kingdom split up in two parts: the Kingdom of Belgium and the Kingdom of the Netherlands.

Not too surprisingly, there was the after-war economic depression, which added to the growing unrest. The climate also certainly played a role, for example, in 1816, when 'the year without a summer' after the eruption of the Tambora volcano had enormous consequences. Cool temperatures and heavy rains resulted in failed harvests, leading to food shortages all over the world, and thus in Europe. The consequence was discontent, which translated to popular revolt, turmoil, and uprisings. Over the next decades those early seeds of revolution would slowly grow.

Prelude: Early Italian Revolt (1820–1830)

Not surprisingly, very soon after the Congress of Vienna, in 1820, on Sicily and Naples, a military revolt broke out against the rule of the Spanish Habsburg King Ferdinand VII. Stimulated by revolt in Spain, the demand was for a constitutional monarchy. It was led by the secret societies like the 'Carbonari',[150] who wanted an equivalent of the Spanish Constitution of 1812 and who had already made several attempts at revolution between 1812 and 1820. However, the Neapolitan Revolution of 1820 was put down by the Austrian forces. The same happened in 1821 with the revolution in Piedmont. A decade later, all over Europe again saw revolutionary action.

> *The second wave of revolutionism occurred in 1829–34, and affected all Europe west of Russia and the North American continent; for the great reforming age of President Andrew Jackson (1829–37), though not directly connected with the European upheavals, must count as part of it. In Europe the overthrow of the Bourbons in France stimulated various other risings. Belgium (1830) won independence from Holland, Poland (1830–1) was suppressed only after considerable military operations, various parts of Italy and Germany were agitated, liberalism prevailed in Switzerland—a much less pacific country then than now—while a period of civil war between liberals and clericals opened in Spain and Portugal.* (Hobsbawm, 2010c, p. 110)

In Italy, by 1830 revolutionary turmoil had broken out in several duchies in central Italy and in some of the territories of the Papal States. It was again instigated by the secret societies like the Carbonari, with their anti-clerical and anti-imperial philosophy originating from Napoleonic rule, who aimed at the creation of a republic or constitutional monarchy. These societies were comprised of nobles, officers of the army, small landlords,

---

[150] The Carbonari (Italian for "charcoal makers") was an informal network of covert revolutionary societies active in Italy from about 1800 to 1831.

government officials, peasants, and priests, along with members from the small urban middle class. The turmoil was similar to those in France; the July Revolution of 1830 opposing royalist restoration in France.

However, after the failed uprisings of 1831 the governments of the various Italian states cracked down on the Carbonari, who now virtually ceased to exist. The revolutionaries created a new movement, *La Giovine Italia* ('Young Italy') that also was repressed. But they prepared the scene for changes to come.

Among those revolutionaries, we find *Guiseppe Garibaldi* (1807–1882), born in (the French-annexed city of) Nice. Coming from a family of fishermen and coastal traders, he became a sailor himself. In 1833 he served in the navy of Piedmont-Sardinia. During a voyage to Russia, Garibaldi became acquainted with the republican ideas of Giuseppe Mazzini. Garibaldi enthusiastically supported the aims of Mazzini's 'Young Italy' and became a lifelong supporter of Italian unification under a democratic republican government.

After participating in a failed insurrection in Piedmont, he fled to Marseille, and from there to Brazil. In 1842 Garibaldi was commander of the Uruguayan fleet. There, by 1846 the election of Pope Pius IX caused a sensation among Italian patriots, both at home and in exile. Pius's initial reforms seemed to identify him as the liberal pope that would go on to lead the unification of Italy. So Garibaldi returned to Italy, offering his services to Charles Albert, king of Sardinia. However, after he had become involved in many uprisings he had to flee abroad again. He went to North America, and then to South American countries like Nicaragua and Peru, where he made many sea voyages as a ship's captain. After five years of exile he returned to Genoa in 1854. After years of activities in agriculture, he became major general in a volunteer military unit called the Hunters of the Alps, fighting the Austrians in the Second Italian War of Independence. Then, in 1860, he became involved in the Sicilian uprisings: his 'Expedition of the Thousand'.

Those early revolutionary actions of the 1820s and 1830s had not resulted in social change of importance, as the conservative foreign political powers were firmly in the saddle. It was the time of the Austrian *Conservative Order* and the *Age of Metternich*, trying to contain the liberal popular movements in favour of the previous ruling aristocracies by creating police states. The secret policy of the Habsburgs under Chancellor Klemens von Metternich's rule kept a watchful eye on the inhabitants of the Austrian Empire.

The secret police and its spying network was in full operation by the Congress of Vienna and its activity continued without limitations during the Vormärz. Regular espionage on individuals, upon whom there might be even the slightest suspicion of thinking or acting against the regime, became a systematic activity after 1815. ... Since the year 1811 ten thousand Naderer or secret policemen are at work. They are recruited from the lower classes of the merchants, of domestic servants, of workers, nay even of prostitutes, and they form a coalition which traverses the entire Viennese society as the red silk thread runs through the rope of the English navy. You can scarcely pronounce a word at Vienna which would escape them. You have no defence against them and if you take even your own servants, they become within fourteen days, even against their own will, your traitors (Aliprantis, n.d., pp. 3-4)

However, the continuous rise of liberalism would eventually be the downfall for Metternich and—the mentally handicapped—Ferdinand. Then, by the mid-nineteenth century, the still-divided Italy (Figure 94) became involved in the wave of revolutions that spread over Europe like a wildfire: the European Revolutions of 1848.

Revolutionary Turmoil in Europe (1848)

Between 1815 and 1848 the multinational Austrian Empire had entered a period of censorship and became a police state.[151] But it was also a period of growth, both of the population, the economy, and urbanization. *Klemens von Mettenich* (1773–1859), prince of Metternich-Winneburg-Beilstein, was as chancellor the dominant politician for decades. Metternich's policies were strongly against revolution and liberalism.

However, by 1840 Central Europe faced more and more rebellions due to economic recession and food shortages during 1845-1847 period, and political turmoil such as the uprisings against the nobility; the *Greater Poland Uprising* (1846) and the *Peasant Uprising* (1846)—aka the Galician Slaughter—the *Hungarian Revolution* (1848), and the *Slovak Uprising* (1848–1849). The Austrian Empire, ruling over ethnic Germans, Hungarians, Slovenes, Poles, Czechs, Slovaks, Ukrainians, Romanians, Croats, Venetians (Italians), and Serbs was confronted with quite some nationalist popular aspirations (Figure 95).

In addition, there were the *German Revolutions* (1848–1849). They were fuelled by the popular discontent with the autocratic political inheritance of the Holy Roman Empire: the Confederation of German States. It created

---

151 In 1804 the Holy Roman Emperor Francis II, who was also ruler of the lands of the Habsburg Monarchy, founded the Empire of Austria, in which all his lands were included. The fall and dissolution of the Holy Roman Empire was accelerated by French intervention in the Empire in September 1805.

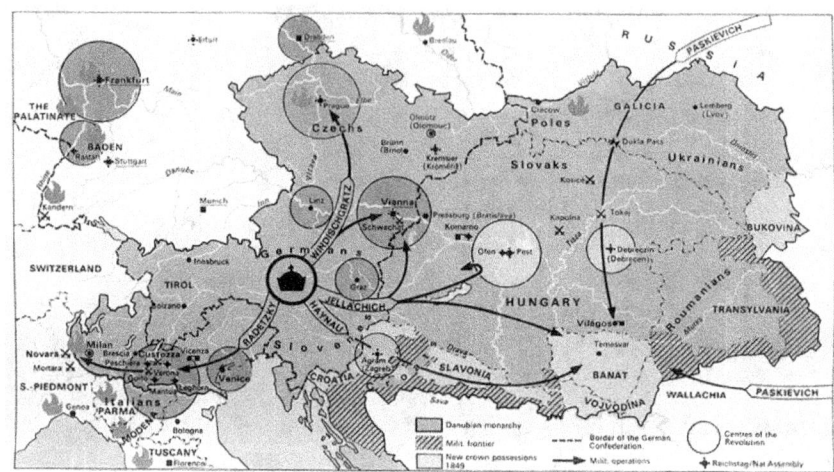

Figure 95: Austrian Revolts (1848–1849).

Map shows the areas that saw uprisings (circles) and the military commanders (arrows) sent to handle the riots.

Source: http://homepage.smc.edu/buckley_alan/Ps7/revolution_1848.htm. The Anchor Atlas of World History, Volume II. New York: Anchor Books. p. 58.

the revolutionary upsurge that resulted in the *Baden Revolution* (1848), the *Palatinate Uprising* (1849), and the *Greater Poland Uprising* (1848), and gave way to the demonstrations in Berlin (Prussia) and Dresden (Saxony). And there was the *February 1848 Revolution* in Paris, ending the restoration of the Orleans Monarchy and creating the Second French Republic.[152] All these uprisings were part of the *European Revolutions of 1848* (Figure 96), the 'Springtime of Peoples', the 'Völkerfrühling' that took place all over Europe. It was the dawn of the decline of the Austrian Empire.

All over Europe, fundamentally inspired by the American Revolution and the French Revolution and enhanced by the social consequences of the first Industrial Revolution, there was massive discontent among people. Again it was fuelled by an economic crisis: both an agrarian crisis and a financial crisis. As economic collapse breeds revolutions, a socio-political crisis emerged based on liberal ideas resulting from Enlightenment philosophers. Add to that a class conflict due to the new 'grande' bourgeoisie of capitalists and artisan classes emerging during the Industrial Revolution. This created the Spirit of the Times, which was dominated by constitutionalist, nationalist, and liberalist aspirations. And Italy was going to be part of it.

---

[152] See: B.J.G. van der Kooij, *The invention of the Communication Engine Telegraph* (2015), p.165–168.

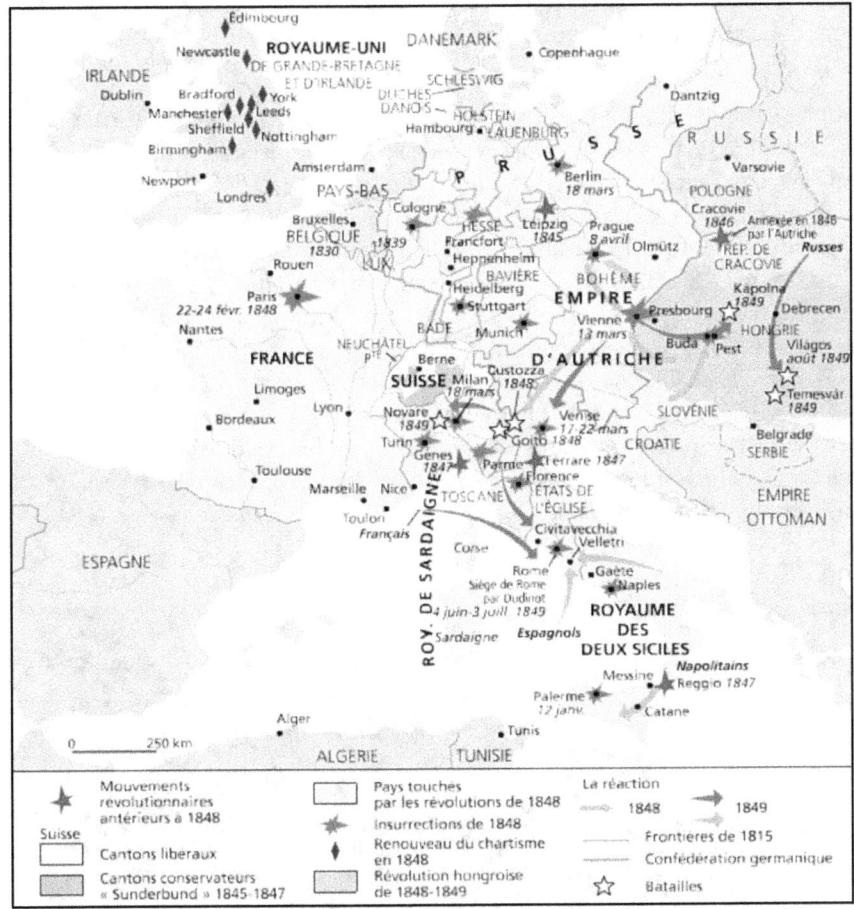

Figure 96: European Revolutions (1848).

Source: http://homepage.smc.edu/buckley_alan/Ps7/revolution_1848.htmThe Anchor Atlas of World History, Volume II. New York: Anchor Books. p. 58.

The Italian Revolution (1840s–1870s)

By the 1840s the revolutionary movement inspired by the French Revolution had also spread to Northern Italy. Along with all the other uprisings in its territory, the Austrian Empire now faced another rebellion as the Northern Italian states—led by the Kingdom of Sardinia and including Nice, Genoa, Savoy, mainland Piedmont, and the island of Sardinia—took part in the 1848 revolutions. It was the start of a revolution spreading over the peninsula—to be called the Italian Revolution[153]— that would consist of a range of interlinked revolutionary events. Before going into its details, let's have a look at the overall picture.

Over time the Italian people resented more and more Austrian rule, which was characterized by conservatism and the consequent heavy taxation. After several food riots all through the years 1840 to 1847, major rebellions took place in the two capitals of the Austrian-controlled Kingdom of Lombardy-Venetia: Milan (the rebellion of the 'Five Days of Milan', March 18–23, 1848) and Venice (the proclamation of the 'Republica di San Marco', August 12, 1848). This all started the *First Italian Independence War* (1848–1849), as the Kingdom of Sardinia, together with the Papal States, the Grand Duchy of Tuscany, and the Kingdom of the Two Sicilies, declared war against Austria. However, the poorly organized rebellion forces were no match for the Austrian forces under Field Marshall von Radetzky,[154] and soon the Austrians restored (their) order in Northern and Central Italy with the *Battle of Custoza* (1848) and the *Battle of Novara* (1849) (Figure 97). Austria was in power again.

Figure 97: Battle at Novarra (1849): Kaiserjaeger battling Piedmontese military.

Source: Wikimedia Commons. Artist unknown.

---

[153] A social revolution by definition is the change of regime in a country. The old regime is originating from the same people as the people who overthrow that existing power structure. Here in Italy, that existing power structure was of 'foreign' origin but included many of the local aristocratic rulers tied to the foreign rulers.

[154] This military commander was immortalized by Johann Strauss I's *Radetzky March*.

So, the northern revolutionary uprisings soon were crushed militarily and financially, and Piedmont was forced to pay an indemnity of 65 million francs to Austria. In Rome, the French—quite unexpectedly—supported the pope and crushed the rebellion there. Nevertheless, Austrian rule in the north was not to last long, as in the *Second Italian Independence War* (1859), Napoleon III joined in and defeated Austria in the *Battle of Magenta* and the *Battle of Solferino*. After some more conflicts in 1861, Victor Emmanuel was proclaimed king of Italy, and Rome was declared the capital of Italy. This was the start of an Italy that was to become a nation under the rule of one king—a herculean task when one considers the diverse bureaucratic structures and institutions (education, military, parliament) of the different regions. It would need the *Third Italian Independence War* (1866) to create a (more or less) unified Italy in 1870. In the period of some two decades (1848–1866), Italy saw a scattered revolutionary movement grow into a movement of unification under the slogan: 'Free from the Alps to the Adriatic.'

All this happened in a time that saw other major geopolitical events like the collapse of the Habsburg Empire with the 'Compromise of 1867' and the collapse of the Second French Empire in the Battle of Sedan (1870). Italy had the chance to become united because the foreign empires collapsed. In addition, there were, just like elsewhere in Europe, the emerging liberal ideas of the Enlightenment movement,[155] which had fuelled more liberal societal developments opposing the existing rule of monarchy and clergy. By now they had also reached Italy and stimulated local enlightenment scholars (aka 'illuminismo'), and that development gave room to some more liberal rulers such as Duke Leopold II of Tuscany (1797–1870), who was a not-too-successful model enlightenment monarch. The impact of the enlightenment ideas differed from region to region, with the Papal States, as a theocracy, lagging furthest behind.

This short overview shows that over the timeframe of two decades, the Italian peninsula freed itself from its former rulers. Let's look at this development more in detail.

First War of Independence (1848–1849)

Basically, the spark for independence was caused by the age-old phenomenon of taxation of the people who'd been conquered. Just as Napoleon had focussed on the riches of Northern Italy, so did the Austrians. Taken together, Lombardy and Venetia in 1848 were so populous as to be home to about one sixth of the persons then subject to

---

[155] See: B.J.G. van der Kooij: *The invention of the Communication Engine Telegraph.* (2015) p. 57–59, B.J.G. van der Kooij: *Context for Innovation: British (R)evolutions in Perspective.* (2016) p.42–48

Habsburg sovereignty. They were so easy to subject to taxation that they contributed almost one third of the Habsburg monarchy's overall tax revenues, but these revenues were taken from an Italian population that was already facing hunger and poverty due to very meagre seasonal harvests in 1846 and 1847. Hunger caused by dramatically inflated food prices led to many popular demonstrations and food revolts. As a result, the prosperous state of Piedmont-Sardinia, which featured a more modern, liberal government compared to other Italian states of that time, served as a driving force for unification in Italy, an early process that cumulated in 1848 when revolts broke out in Milan.

Five Days of Milan: In Milan, during the so-called 'tobacco riots' of January 1848, some 61 persons were seriously injured, resulting in a number of fatalities. Milan was by then the chief city of the Italian Peninsula in terms of economic activity. People demonstrated against such things as the high taxes imposed by Lombardy's Austrian authorities, who maintained a state monopoly on tobacco sales. Whilst taxes (on tobacco and salt) were prominent as the superficial focus of Milanese discontent, a broader political and cultural agenda underlay these protests. The boycott culminated in a bloody street battle on the third of January, when Austrian soldiers, in batches of three, were insulted and pelted with stones by an angry crowd. The whole city fought throughout the streets, raising barricades, firing from windows and roofs, and urging the rural population to join them (Figure 98). The populace was backed by the archbishop, and at least 100 priests joined in the fighting against the Austrians. After five days the turmoil cooled down, only to erupt two

Figure 98: Tobacco Riots in Milan in January 1848 (left) and the Bologna Uprising in August 1848.

Source: Wikimedia Commons,
Milan: British Library HMNTS 9315.h.12
Bologna: Archivio di Stato di Bologna, Artist Antonio Muzzi.

months later. By March 22 the Austrians left the city. In March 1849 a revolt broke out in Brescia. It lasted ten days, till the Austrian troops sacked the city, killing some thousand citizens. Also Bologna saw a massive uprising, expelling the Austrian garrisons (Figure 98).

Elsewhere on the Italian Peninsula, there were also uprisings, such as in Palermo on Sicily and Rome. It commenced on January 12, 1848, and therefore was the very first of the numerous revolutions to occur in the South that year.

Sicilian revolution of independence of 1848: Sicily had seen three popular revolts against the Bourbon rule before. In 1848 it became a quasi-independent state after rebellious insurrections, but not for long, as within sixteen months the Bourbon army took back control of the island.

Roman Republic: In Rome, in February 1849 the 'Roman Republic' was declared, aiming for religious freedom and liberal reforms. Pope Pius IX, disguised as a priest, fled from Rome. Guiseppe Garibaldi was appointed general by the provisional government of Milan in 1848 and general of the Roman Republic in 1849. On April 30, 1849, the Republican army, under Garibaldi's command, defeated a numerically far superior French army. Subsequently, French reinforcements arrived, and the Siege of Rome began on June 1. Despite the resistance of the Republican army, the French prevailed in June. On July 12, 1849, Pope Pius IX was escorted back into town, and he ruled under French protection until 1870.

All in all, the many uprising failed because Austrian's military power was too strong, and they crushed the revolutionaries in Piedmont. But the seeds of revolution were sown and sprouting revolution.

Second War of Independence (1859)

In 1859, again, a war broke out between Austria—governing the Lombardy-Venetia region—and the Second French Empire, which had joined forces with the Kingdom of Sardinia, ruling in Piedmont. The cooperation was triggered by an assassination attempt on Napoleon III: the Orsini attempt supported by Italian revolutionaries operating from Britain. The attempt brought widespread sympathy for the unification effort and resulted in secret treaty against Austria. France needed a weakened Austria, as did the revolutionaries.

Soon the French were facing their Austrian adversary when Austria demanded complete demobilization of the Sardinian army. The initial clash of the war was at the *Battle of Montebello* on May 20, 1859 (Figure 99), where the Austrian army—some 140,000 men—was defeated by the Sardinians

Figure 99: Battle of Montebello (1859).

Source: Wikimedia Commons. Artist Giovanni Fattori

with some 70,000 men. The same happened in a range of subsequent battles at Varese, Palestro, Turbigo, Magenta and Solferino. On July 11, 1859, an armistice was signed between France and Austria. The final arrangement was ironed out by 'backroom' deals instead of on the battlefield. This was because neither France, Austria, nor Sardinia wanted to risk another battle and could not handle further fighting.

As a result, most of Lombardy, with its capital Milan (excepting only the Austrian fortresses of Mantua and Legnago and the surrounding territory), was transferred from Austria to France, which would immediately cede these territories to Sardinia (keeping at the *Treaty of Turin* in 1860 the Duchy of Savoy and the County of Nice). The former rulers of Central Italy, who had been expelled by revolution shortly after the beginning of the war, were to be restored. However, this did not bring governmental stability, as Italy was still a divided country: the Austrians in Venetia, the Papal States under the Papacy, the Kingdom of the Two Sicilies ruled by the Bourbons, the small independent republic of San Marino, and the new expanded Kingdom of Piedmont-Sardinia. The dominant player in the developments that were to come was a ruler of the House of Savoy: Victor Emmanuelle II (1820–1878). He was supported by his prime minister Camillo Benso, count of Cavour (1810–1861), who had been politically masterminding the early events that led up to the unification.

The *Expedition of the Thousand* (Spedizione dei Mille) in 1860 was the next
effort to eliminate foreign rule (Figure 100). A corps of 1,089 volunteers
led by Giuseppe Garibaldi landed in Marsala, Sicily, in order to conquer
the Kingdom of the Two Sicilies, ruled by the French Bourbons.

*Garibaldi's whole Expedition was something of a charade. His volunteers were
poorly armed and clothed because Sardinia could not "officially" recognized them.
Yet the Sardinia government and their English co-conspirators spent thousands of*

Figure 100: Expedition of the Thousand (1860).

Source: https://storify.com/marialice01/ottone-i-di-sassonia

dollars bribing Neopolitan officiers to refrain from fighting and English ships accompanied Garibaldi's transport.[156]

In a range of battles—at Calatfimi, Palermo, Milazzo, Reggio Calabria, Volturnus—the revolutionary 'Southern Army' surprisingly easily defeated the Bourbon troops. After crossing the street of Messina between Sicilia and Calabria, by September 1860, the revolutionary troops were advancing northwards and everywhere received by the people as liberators (Figure 101). Finally, the 24,000 revolutionary troops reached the river Volturno near the capital Naples. They were met by some 25,000 Bourbon troops, and on September 30, 1860, the *Battle of Volturnus* commenced.

In the meantime another Piedmontese campaign, with 35,000 troops from the Kingdom of Sardinia, had invaded the Papal States (Figure 100) from the north, conquering Central Italy through battles such as the *Battle of Castelfidardo*, where the papal forces were beaten and the city of Ancona was taken. Arriving at the scene of the *Battle of Volurnus*, they joined Giuseppe Garibaldi's army. This resulted in the *Siege of Gaeta*, one of the strongest military fortresses in Europe. Despite heavy bombardments, an exploding powder depot, and severe conditions for all the people and soldiers packed

Figure 101: Garibaldi's entry into Naples (1860).

Source: Wikimedia Commons. Artist: Franz Wensel.

---

[156] Source: http://www.heritage-history.com/index.php?c=academy&s=war-dir&f=wars_italianunity#second

in the small fortress, King Francis II managed to hold out till February 13, 1861. Then the last Bourbon ruler of southern Italy surrendered the fortress to the Piedmontese. The Kingdom of Two Sicilies ended some days later when the last organized centre of resistance, Civitella del Tronto, surrendered on March 20, 1861.

So, in March 1861, the new *Kingdom of Italy* (Regno d'Italia)— the merger of the Kingdom of the Two Sicilies and the Kingdom of Sardinia— was formally established (Figure 102). Savoy's ruler, Victor Emmanuel II, was crowned king on March 17, 1861. The former Kingdom of Piedmont-Sardinia had grown into the Kingdom of Italy. And its ruler, Victor

Figure 102: Italy after the unification (1861).

Source: Wikimedia Commons.

Emmanuelle II, became the first king of a united Italy. The region of Veneto (part of the former Republic of Venice) was left out, and the French had gotten possession of Savoy and Nice as a reward for their cooperation.

Thus, the Kingdom of the Two Sicilies ceased to exist, and Bourbon rule ended. Only the region of Lazio, a remnant of the Papal States with the eternal city of Rome, was left out. The expedition had created the Kingdom of Italy and obtained the support of the powerful great landowners of southern Italy in exchange for the promise that their properties be left intact in the upcoming political settlement. Numerous Sicilian peasants, however, had joined the Expedition of the Mille hoping instead for a redistribution of the land to the people working it.

Third War of Independence (1866)

Over the years, in the south of the peninsula, brigandage had developed. Social unrest, especially among the lower classes, occurred due to poor conditions, and the fact that the 'Risorgimento'[157] benefited in the 'Mezzogiorno'—the regions in Southern Italy including the regions south of Rome, Sardinia, and Sicily—only the land-owning bourgeoisie classes.

They launched attacks against the Italian authorities (who were regarded as foreigners) and the land-owning upper-classes. Coming to be viewed by many of the common people as Robin Hood-like figures, some of them gained celebrity in the area via the spreading of their stories. The authorities repressed the brigands, and thousands were arrested and executed. It was the consequence of the feudal system breaking up, leaving many landless. The poverty also caused a large-scale emigration to Northern Italy, which started the first period of the *Italian Diaspora*. As millions of Italians also left for Europe and North and South America, their departure deprived Italy of its major economic source: their industrious people (Figure 103).

In the north of the peninsula, the situation was different. The reduced Papal States were under French protection, and the regions of Veneto and Trentino were still in Austrian hands. In Piedmont, King Victor Emmanuelle allied himself with the Kingdom of Prussia, which had just started the Austro-Prussian War of 1866 in the North and the Franco-Prussian War (1870–1871) in the West. And that turned the tables, both for Veneto and for the pope.

On French soil, the Franco-Prussian War of 1870–1871 had resulted in a victory for the Prussians after the *Treaty of Versailles* (1871). It would be the start of the modern Germany. But the war also created a side effect: the French withdrew their army back to France, leaving the Papal States

---

[157] Meaning resurgence or revival.

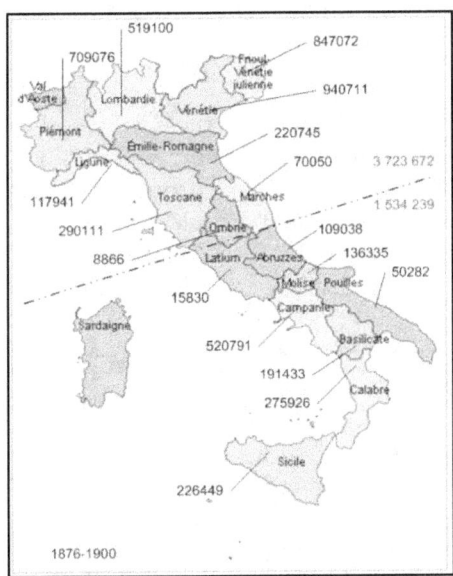

Figure 103: The first Italian Diaspora (1876–1900).

Figures show the estimates of the number of emigrants from 1876–1900 according to their region of origin.

Source: Wikimedia Commons

unprotected. Thus, the Piedmontese army moved in, and after the *Capture of Rome* the Papal States were occupied. The capture of Rome ended the 1,116-year reign (AD 754–1870) of the Papal States under the Holy See.

Venetia was added to the new kingdom in 1866 through an alliance with Prussia against Austria, but complete unification of the peninsula could not be achieved as long as Rome remained in the hands of the Pope. A French garrison stood between Victor Emmanuel and this final conquest. Napoleon III, needing the support of the clergy, did not wish to abandon the Pope, although he had been Victor Emmanuel's ally in the expulsion of Austria from northern Italy. But this last bulwark of the papal territories was withdrawn in 1870, when—under the threat of total defeat by Prussia—Napoleon ordered his soldiers out of Rome. On Sept. 20, 1870, the Italian army marched into the city, and on July 2, 1871, Victor Emmanuel himself entered Rome, from that time the capital of the kingdom of Italy. The Pope, who had lost the last vestiges of his temporal power although the Vatican and his freedom were guaranteed to him, refused to recognize the new kingdom, and Victor Emmanuel died on Jan. 9, 1878, unreconciled to the Church.[158]

After a brief stay in Florence the newly formed Italian government moved to Rome to make the Eternal City their capital. The pope withdrew behind the walls of the Vatican, declaring himself a prisoner of the Vatican and refusing to recognize the Kingdom of Italy in any way. In the *Law of Guarantees* (1871), the new king gave the pope the use of the Vatican area in Rome but denied him sovereignty over his territory. With the end of the Papal States in 1870, Pope Pius IX was the last pope to hold temporal

---

[158] Source: 'Victor Emmanuel II'. *Encyclopedia of World Biography*. Retrieved December 7, 2016 from Encyclopedia.com: http://www.encyclopedia.com/history/encyclopedias-almanacs-transcripts-and-maps/victor-emmanuel-ii

powers. Finally, Italy had gotten rid of foreign domination and was a kingdom on its own (Figure 104). It had taken some decades of revolutionary activities. Italy, by now, was unified under one crown, and Rome became its capital.

In reality, the central Italian states were annexed by Piedmont-Sardinia after referendums which the nationalists hailed as positive proof of popular support and which the Church condemned as totally fraudulent. Pope Pius IX and his successors refused to recognize the right of the Italian king to reign over what had formerly been the Papal States, or the right of the

Figure 104: The Italian unification resulting in the Kingdom of Italy (1870).

Source: Wikimedia Commons Author: Shepherd, William. *Historical Atlas*. New York: Henry Holt and Company, 1911.

Italian government to decide his prerogatives and make laws for him. It was the start of a standoff—the Roman Question—that would last well into the next century, until the *Lateran Treaty* of 1929 settled the Roman Question by establishing Vatican City as an independent state.

> *As King of Italy, Victor Emmanuel II toured the country, noting improvements and infrastructure projects that would have to be made. He also accepted friendly visits from princes from Great Britain, Prussia and even Austria. To the surprise of many he was invited to Vienna to be the special guest of Emperor Francis Joseph. The two former enemies met and embraced as friends in a touching moment of reconciliation between two ancient royal houses that had once been the closest of allies. (ibid)*

There was a new government, but it was 'Piedmontization all around'. The new Kingdom of Italy was structured by renaming the old Kingdom of Sardinia and annexing all the new provinces into its structures. The first King was Victor Emmanuel II, who kept his old title. National and regional officials were all appointed by Piedmont. Piedmontese tax rates and regulations, diplomats, and officials were imposed on all of Italy. The new constitution was Piedmont's old constitution; it was generally liberal and was welcomed by liberal elements. However, its anti-clerical provisions were resented in the pro-clerical regions around Venice, around Rome, in Sicily, and in the boot south of Naples. Italy was not politically unified that much, and many—representing the upcoming Italian irredentism[159]—would resent it in times to come.

That was the political part of Italy's revolutionary development into one nation by the end of the nineteenth century. In the meantime another kind of revolution had reached Italy; the Industrial Revolution.[160]

---

[159] Italian irredentism was a nationalist movement during the late nineteenth and early twentieth centuries in Italy.

[160] The concept of the Industrial Revolution describes the transition from agricultural to industrial societies under the influence of technological development. The evolutionary developments in Britain in the eighteenth century especially paved the way for similar developments in countries all over Europe later in time. See: B.J.G. van der Kooij, *Context for Innovation: British (R)evolutions in Perspective* (2016).

Industrial Revolution in Italy (1870s+)

As described, at the end of the nineteenth century Italy's scattered regional structure had become more or less unified as result of the events of the Italian Revolution. That is, only politically, as the economic diversification between the northern and southern regions was still considerable. The wealthy Po Valley in North Italy had the industrious, entrepreneurial heritage of the former city-states, from Turin and Milan in the northwest to Ferrera and Bologna in the southeast. The historic agricultural orientation—such as the cultivation of rice already started in Sforza times—stimulated the agro-mechanical industries. Another inheritance from medieval times was the early textile industry, such as the cottage industry organized in the mediaeval Florentine's 'Arte del Lane' (Wool Guilds). In addition, the central northern region had its cultural/economic/financial heritage, Florence, Lucca, and Pisa, and a considerable maritime heritage, Genova and Venice. In the early nineteenth century from those origins sprouted a related mechanical industry of small- and medium-sized companies manufacturing the textile and agricultural machinery. But then the effects of inventions in Britain—eg the steam engine—diffused to Italy.

Unlike other countries like England and Germany, Italy did not have considerable coal and iron deposits that could be mined. Neither had Italy a history of manufacturing like Britain, by that time the 'Workshop of the World'.[161] Italy's wealth over the ages was predominately from agriculture and trade. True, like England, early textile industry also had developed from cottage industries, and artisan's workshops organized in the different guilds had laid the basis for early industrialization. But unlike England, no massive factorization—aka the Industrial Revolution—had taken place by the early nineteenth century in Italy.

Nevertheless, the evolutionary ferment of the social French Revolutions and European Revolutions (1789–1792, 1830, 1848) creating the patriotic Risorgimento that swept across Lombardy in the 1830's also had other consequences. It resulted in both a growing industrialization and better educational system in the technical sciences.[162] And in 1839 the first eight-kilometre-long railroad line was built between Naples and Portici, followed by other local lines in the following years, such as the Milan-Venice railway, which was started in 1842 and, by 1846, reached Venice.

---

[161] See: B.J.G. van der Kooij, *Context for Innovation: British (R)evolutions in Perspective* (2016), p. 374.

[162] Much can be said about the Industrial Revolution in Italy. We will limit ourselves to those developments that are related to the GPT-Electricity.

Italy after the Scientific Revolution

Italy may have been slow in industrialising, but it was not lacking in scientific ardour. In the eleventh century universities had already emerged from mutual-aid societies protecting students and their teachers from the influence of local municipalities, such as the University of Bologna (1088) and its spin-off, the University of Padua (1222). Later in time other universities had grown out of the *Academia* (eg the Accademia Fiorentina, founded in 1540, and the Accademia dei Lincei, founded in 1603). The described work of the early Italian natural philosophers (eg Leonardo da Vinci, Galileo Galilei) ignited the early explorations in scientific fields. From that scientific heritage, in 1782 the *Accademia nazionale delle scienze* (National Academy of Sciences) was created.

One of the areas of attention of the later scholars was the phenomena of 'static electricity'. In Bologna, it was *Luigi Galvani* (1737–1798), professor at the Accademia delle Scienze dell 'Istituto di Bologna (Academy of Sciences of the Institute of Bologna, founded 1714) who discovered animal electricity in the 1780s. Then, at the University of Pavia, *Alessandro Volta* (1745–1827) discovered electrochemical electricity created by the Voltaic pile (Figure 121). A discovery that would spark a controversy, as well as a range of

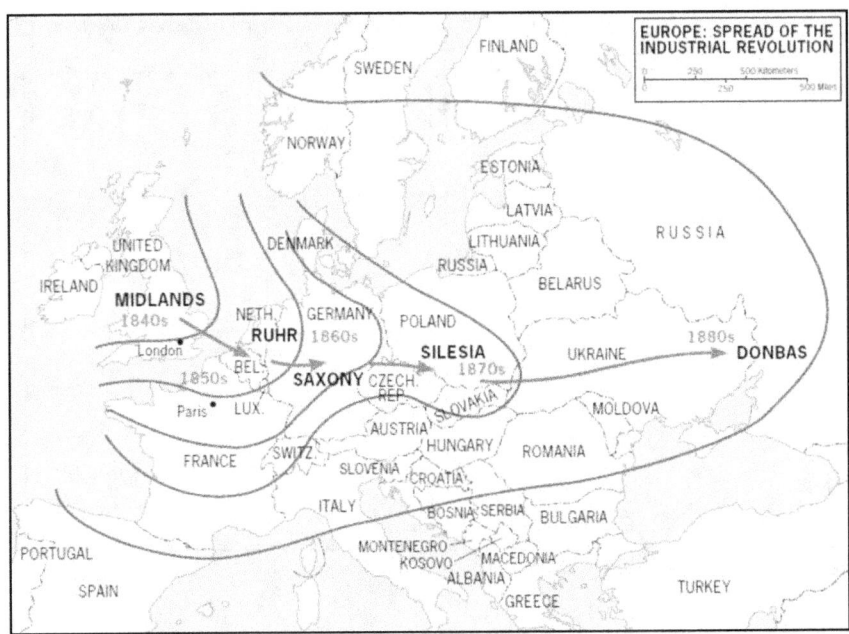

Figure 105: Spread of Industrial Revolution over Europe (1840–1880).

Source: http://imgur.com/JMif2ak

187

developments, ultimately creating electric light as well as electric motors and dynamos.

Not only in Italy but all over Europe, many tinkerers and tinkerers became fascinated by the new phenomenon of electricity and electroomagnetism; it became the Battery Mania. But first, there were the effects of the First Industrial Revolution that had sprouted in England and soon reached the continent (Figure 105).

The First Industrial Revolution Reaches Italy

As a result of the emerging *general purpose technology of steam*, the steam engine had been developed,[163] and it was powering Britain's First Industrial Revolution. It powered both the industrial activities—with stationary steam engines—as well as the logistical activities facilitated by the emerging railway infrastructure and mobile steam engines: the steam locomotives. From Britain, the Industrial Revolution spread over Europe in the nineteenth century.

Although some experimental rail road lines had already been created in the 1840s in Italy when Britain was in the midst of the *Railway Mania*, it took some time before the railroad system expanded considerably. By 1853 the Turin-Genova railway was built—using locomotives built in a locomotive factory in Genoa—and absorbed in 1865 by the *Società per le strade ferrate dell'Alta Italia* (Upper Italian Railway Company) that operated

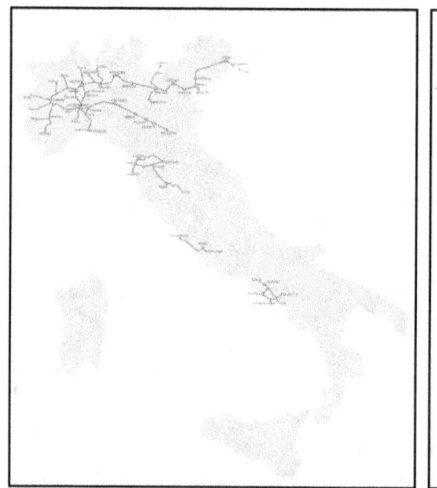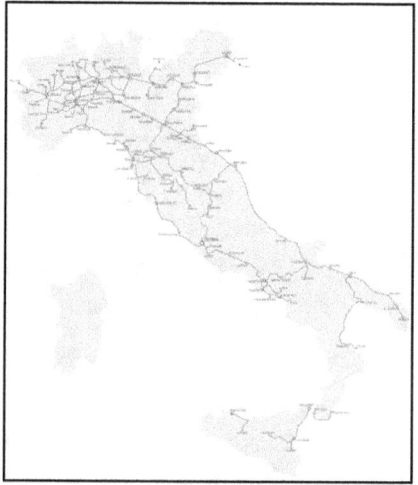

Figure 106: Railway network in Northern Italy in 1861 (left) and 1870 (right).

Source: Wikimedia Commons.

---

[163] B.J.G. van der Kooij, *The Invention of the Steam Engine* (2015)

lines mainly in the Po delta (Figure 106, left). After the third Italian War of Independence the railroads expanded considerably, and in 1872 there were in Italy about 7,000 km of railroads (Figure 106, right).

The same goes for the industrialization of Italy. Pre-industrial activities (aka crafts), the work of artisans in the organized form of the guilds, had already been present for centuries. However, some early industrial activities that characterized the First Industrial Revolution started in Italy in the 1840's. Then the quite-important silk-industry—a major export product—adopted the new source of steam power. Decades later, factory production in other industries developed mainly in Northern Italy, and this only intensified the economic backwardness of the south. In the north, it was the mechanical industries that flowered.

> *In Milan, during the 1870's, there were over 500 plants, both actual factories and large craftsmen's shops, which gave work to over 14,000 workers. ... Most of them worked in companies of the typographic industry, weaving companies, passementerie shops, mechanical business, in the silk industry and 1,400 were employed in the large tobacco plant. ... just ten years later the national census indicated that in Milan about 110,000 people were employed in all the sectors, ... It was, in fact, during those years, that the large factory, product of industrial capitalism, which had already existed over a century in the urban landscapes of the major cities of the Western advanced nations, was becoming also in Milan a presence that was no longer episodic but rather continuously spreading.* (Granata, 2015, pp. 13, 14)

In the Milan area, it was companies like the G.B. Pirelli Company (founded in 1872), started by Giovanni Baptista Pirelli, who'd graduated from de Politechnico of Milan. He produced rubber-related products, added tyres for velocipedes, followed by automobile tyres. In 1886 Ernesto Breda purchased an old mechanical company and converted it to a manufacturer of locomotives, till then a product of foreign companies. In Bologna, it was companies like Alessandro Calzoni S.P.A., which started as a mechanical workshop manufacturing cutlery in 1834 before expanding into agricultural machinery, drives for mills, oil presses, and hydraulic turbines.

The Second Industrial Revolution Reaches Italy

After Volta's discovery of the electric battery, many Italian scholars studied electromechanical phenomena, such as the professor in physics at the University of Padua *Salvatore dal Negro* (1768–1839), who built an early electromotor. Professor *Luigi Magrini* (1802–1868) and Professor *Guiseppe Botto* (1791–1865) also created (reciprocating) electric engines (Figure 107). In 1830 Botto described in a note a prototype electric motor on which he was working and published a description of it in a Memoria titled *Machine Loco-motive mise en mouvement par l'électro-magnétisme* to the Academy of Turin around 1836. However, the linear reciprocal electric motor proved a dead-end technology.

> *Salvatore dal Negro of the university at Padua reported in 1834 on an invention that he had worked out in 1831 of a permanent magnet pendulum kept in oscillation by an electromagnet that changed its polarity by a commutator switch. He added a linkage device so that he could raise a weight with it and found it lifted 60 grams, 5 centimeters in one second. A similar pendulum-instrument was made in 1834 by J. D. Botto in Turin.* (King, 1962, p. 261)

Figure 107: Botto's electrical motor (1834).

Source: Museo Gallileo, Florence

In the aftermath of the unification of Italy, in 1863, the *Politecnico di Milano* (University of Milan) was created with the departments of civil and industrial engineering. Later, in the 1890s, it was Galileo Ferraris, founder of the first electro-technical engineering school in Turin, who would contribute to the development of the induction motor.

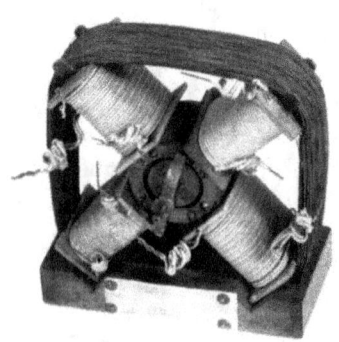

Figure 108: Ferraris's induction motor: fourth prototype (1886).

Source: Physics Museum of Sardinia, http://www.webalice.it/sergio.arien ti/immvite4/motoreeferRARIS_cm r_4.gif

Galileo Ferraris (1847–1897), born in Livorno Vercellesse, near Turin, created the foundation for the electric induction motor. Ferraris was teaching technical physics at the Museo Industrial of Turin (later the University of Turin). His research into electricity brought him in the mid-1880s to rotary electromagnetic fields discovered in the 1830s by the Frenchman Francois Arago. For demonstration purposes, he created in 1885 a new type of electric-magnetic motor: the induction motor (Figure 108). Next, he created the electric AC generator (a machine creating alternating currents from mechanical rotary power). On March 11, 1888, Ferraris published his research in a paper to the Royal Academy of Sciences in Turin. It looked like he was the first to invent the induction motor; however, he seemed to have been beaten. It was two months later, on May 1, 1888, that the American Nikola Tesla was granted US Patent № 381,968. He had his application filed on October 12, 1887 (Serial Number 252,132).[164]

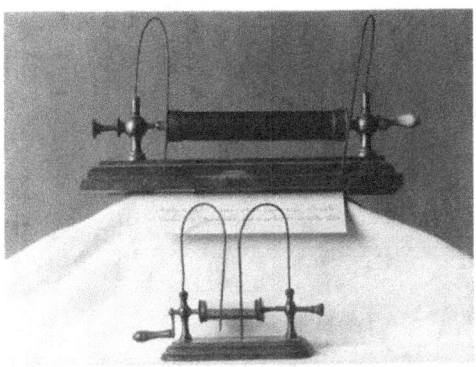

Figure 109: Coherer designed by Calzecchi-Onesti (1884).

Source: http://www.radiomarconi.com/marconi/onesti.html

It was *Temistocle Calzecchi-Onesti* (1853–1922), born in Lapedona (Ascoli Piceno), who became in 1879 professor of physics at Fermo/Monterubbiano. Based on earlier experiments with metal powders having a variable resistance/ impedance by many scientists, in 1884 he explored the possibilities of using metal powders for a microphone. He discovered that the electrical conductivity of his tubes with metal fillings changed due to various excitations such as lightning and electrostatic induction, and he published papers about them in *Il Nuovo Cimento* from 1884 to 1886. He saw his tubes (Figure 109) as detectors for lightning and seismic activity, but it was the Frenchman Eduard Branly (1890) and the Englishmen Oliver Lodge (1894) who applied them as detectors for Hertzian waves (more about that later on).

These Italian individuals, both the thinkers and the tinkerers, took part in creating the foundations of the general purpose technology of electricity of the Second Industrial Revolution in Italy. By the end of the nineteenth century Europe saw at the International Exposition of Electricity in Paris (1881) early inventors (such as Thomas Edison and Alexander Graham Bell) exhibiting the wonders of modern time, drawing visitors from all countries. It was also visited by Italians, and soon, on June 1883 the first electric power plant was built in Milan. A few months later it powered a mighty chandelier with some 2,450 electric lamps, hanging in the centre of the La Scala Theatre, baffling the elite of Milan on the opening night of the theatre season (Granata, 2015, p. 8). On January 6 the *Società Generale Italiana di Elettricità Sistema Edison* (General Italian Edison-System Electric Company) was founded in Milan. By 1892 Milan got its electrical trams,

---

[164] See: J.G. van der Kooij: *The invention of the Electro-motive Engine.* (2015) pp. 169–171, 193–197.

replacing the horse-powered ones. Not gifted with large coal deposits, Italy had erected its first hydroelectric power-generating facility in 1898, the Bertini Power Plant, built by the Italian Edison Company. Thus, by the end of the nineteenth century electric power and electric light had also reached Italy.

An interesting effort to improve upon the incandescent lamp was realized by the Italian *Alessandro Cruto* (1847–1908). It illustrates the rapid spread of the incandescent filament technology over the continents, as it took place in Italy. The young Cruto, an autodidact interested in

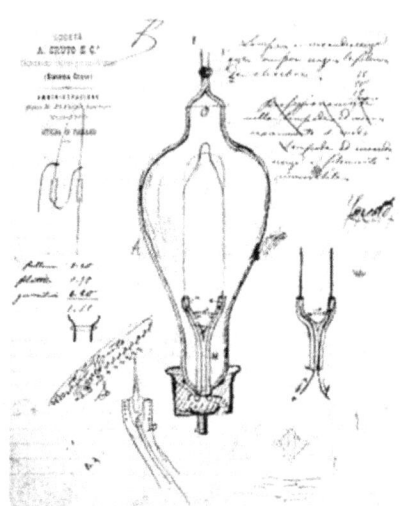

Figure 110: Alessandro Cruto's design for his incandescent lamp (ca. 1882).

Source: Bibliotheca Communale, Alpignano, 'Da Cruto a Philips 1886–2003,' p. 9.

Figure 111: Alessandro Cruto's incandescent lamp (1880–1882).

Courtesy photo (2014): Robert Martynse, ex-president, Philips Italia.

physics, originally experimenting in an effort to create artificial diamonds, succeeded in producing sheets of graphite from ethylene. He applied this knowhow of making pure carbon to create a carbon filament. He did this by depositing carbon on a platinum wire from ethylene under high pressure and temperature. As the platinum would evaporate during the process, it would result in a carbon filament of high purity (Figure 110).

During his efforts he was stimulated by professor Galileo Ferraris, who disputed the viability of Edison's bamboo filament. Then he managed to use Professor Naccari's laboratory at the University of Turin for his experiments. These experiments resulted in a pure carbon filament that gave a bright and white light on September 11, 1880

These are just a few of the numerous efforts by individual Italian thinkers and tinkerers participating in the development of the general purpose technology of electricity. Efforts that were converted into industrial activity, helping Italy into the Industrial Revolution. That industrialization of Italy was highly stimulated by unification, due to access to larger markets that created increasing demand, although it was also dependant on local entrepreneurial spirit, local financing, and local educational institutes, as the role of the state was nil at that time. Thus, again the dominance of foreign powers, now in the industrial sense, was unavoidable.

Overview of the Early Modern Times of Italy

Between the sixteenth century, when the Northern Italian regional city-states were at the peak of their powers, and the nineteenth century, the geopolitical forces of Europe came down over the Italian Peninsula. From the Great Italian Wars (1494–1559) until the Treaty of Versailles (1871), Italy was a divided and fragmented country, ruled—and subsequently heavily taxed—by foreign powers. From the times of the early French rule, followed by the Spanish rule and the Austrian rule, it was overrun by other

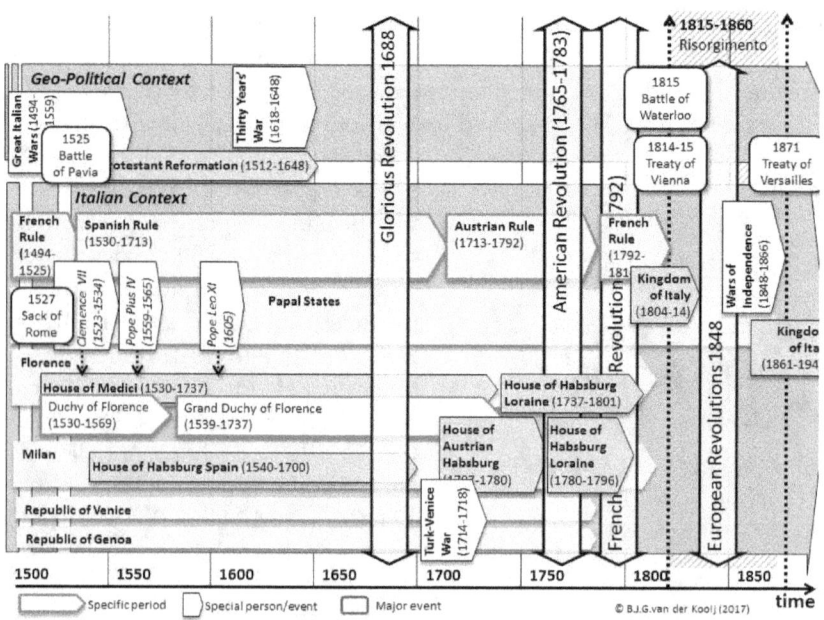

Figure 113: Overview of Italy's Age of Stagnation and Reform (1500s–1860s).

Figure created by author.

armies, ultimo to become the target of the French Revolutionary Wars (1792–1814).

These three centuries were marked by foreign domination, including the massive transfer of Italy's wealth to foreign rulers. Because, just like in feudal time the wealth went up in the temporal part of the feudal pyramid (Figure 29), now Italy's wealth went up to the conquering hordes and armies. Whether it was in the form of protection money, bounty or loot, war indemnity, tribute or taxation. The only who seemed to escape that fate was the Papacy. They, being on the top of the religious side of the feudal pyramid, collected their tithes, church taxes and indemnities and transferred it to Rome. Making in the process the Roman Catholic Church rich.

Within Italy, the remarkable rise of the Papal States and the temporal powers of the Roman Catholic Church were paralleled by the decline of the ecclesiastical powers of the Roman Catholic Church over Europe due to the *Protestant Reformation* (1512–1648). The Papacy originally had joined forces with the emperors of the Holy Roman Empire, but when that empire became religiously and politically divided in the sixteenth century they had to rely on French and Spanish support for their existence. And when the last French support withdrew in 1870, the Papal States were finished, and the pope—the prisoner of the Vatican—became the ruler of the Roman Catholic Church only.

During that period the ruling classes of the city-states such as Florence, which, with its Medici dynasty, had supplied popes, archbishops and cardinals, were pushed aside by the new rulers from the Habsburg dynasty, as its Spanish branch and its Austrian branch ruled in both the south and the north of the peninsula. The formerly mighty maritime Republic of Genova had become a vassal of Spain. Venice, which had already lost much of its Eastern Mediterranean influence to the Ottoman Empire after losing the Turk-Venice War (1712–1718), was crushed by Napoleon's revolutionary forces. But the seeds of revolution had also been sown on the Italian Peninsula, and the subsequent Risorgimento resulted in the Italian Unification under Piedmontese rule. The Kingdom of Italy was born, a nation that would soon display the same nationalist and imperialist tendencies as other European nations.

Europe and America at the End of the Nineteenth Century

We have been looking at Italy's development in depth over a time frame of several centuries. After its early development from Roman times, we described the growth of the powerful city-states and their influence on the Italian Peninsula, which was highly affected by the geopolitical European context of that time. Being the playball of the dominant European nations—each with its own geo-political power game—the Italian Peninsula was ruled by foreign powers until the 1860s. Powers that had, over time, even developed their own imperial aspirations.

The Age of Empires (1875–1914)

Europe at the end of the nineteenth century was the continent of the so-called imperial powers (Figure 114). It was the Age of Empires (Hobsbawm, 2010b). The former *First French Empire* of Emperor Napoleon Bonaparte (1804–1815), by now again a republic, had exercised in the early nineteenth century a profound socio-political influence on civil societies in Europe. The *British Empire* (1815–1914) under the rule of Queen Victoria had become the most powerful colonial nation outside the European continent. The young *German Empire* (1871–1918) was in the process of being shaped by the members of the German Confederation. The *Austro-*

Figure 114: Satirical map of Europe depicting the major powers (1870).
Source: http://www.crouchrarebooks.com/uploads/images/1642_1H.jpg. Artist Paul Hadol.

Hungarian Empire (1867–1918) had risen from the realms of the Habsburg dominions. The *Russian Empire* (1721-1917) lay dormant in the East heading at a revolution. All these empires, except Russia, were ruled by the remnants of absolute monarchies that had seen their 'divine rights' crumbled under rising parliamentary sovereignty during the preceding Age of Revolutions (Hobsbawm, 2010c).

It had, indeed, been the time of revolutions. The United States had become a powerful nation after their *American Revolution* (1765–1783) freed it from British dominance. France had tried to get rid of feudalism in the *French Revolution* (1789–1792). Central/Southern Europe had seen the *European Revolutions of 1848*. Italy had had its own revolution—the *Risorgimento*—in the 1850s to the 1870s. France had been beaten by the Prussians in the Franco-Prussian War of 1870–1871. However, the harsh conditions negotiated at the *Treaty of Versailles* (1871) gave way to the prelude of the First World War. An international war, one of the largest in history, where the different nations applied weaponry created with enormous technological and industrial sophistication. It was a time where the fruits of the Second Industrial Revolution would change the world. Just like the First Industrial Revolution, technical change would be the walking companion of political change and social change.

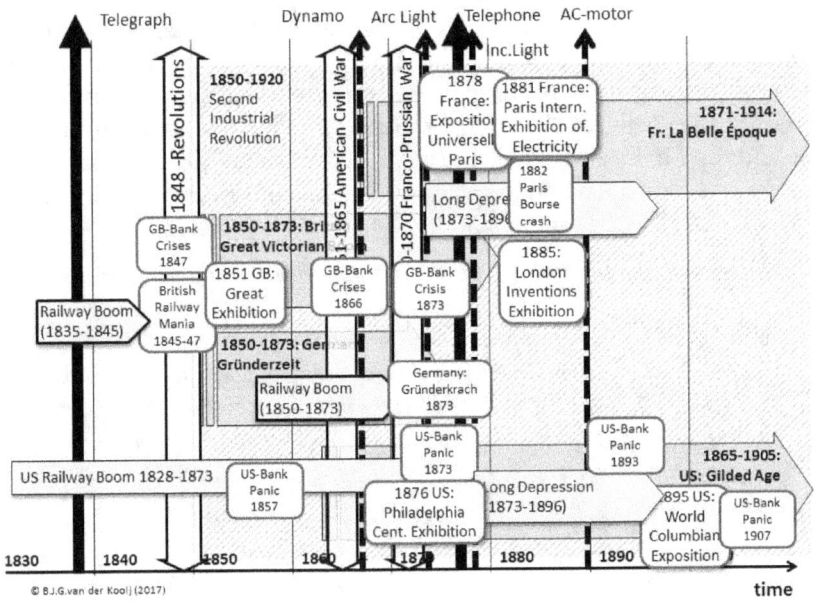

© B.J.G.van der Kooij (2017)   time

Figure 115: The economics consequences of the Second Industrial Revolution.

Figure created by author.

197

The European Revolution of 1848 was a watershed moment in the development of many European countries (Figure 96). In the second half of the nineteenth century Europe was characterized by technological change (Figure 105), by industrialization and urbanization, and by fast-growing economies. There was an explosion of urban housing, mass culture, and modernity. It was the time of England's' *Great Victorian Boom* (1850–1873) and America's *Gilded Age* (1870s–1900). In Germany and Austria, it was the *Gründerzeit* (1850–1873). It was also the time of the French *La Belle Époque*—with Paris being the City of Light—which had started after the Franco-Prussian War (1870–1871) and lasted till the beginning of the Great War (1914) (Figure 115).

The basic mechanism of the First and Second Industrial Revolutions—the transfer of power, wealth, and knowhow—was the driving forces for the further urbanization, industrialisation, manufacturing, and infrastructural development. The result was massive new entrepreneurial undertakings in construction (housing), transportation (steam-powered rail transport and shipping), and basic industries (like steel and textiles). That created the context for innovation in which new social classes rose to power and demanded their own place in their societies.[167]

Industrial modernity, in particular, really bit deep into societies and affected not thousands or hundreds of thousands of people, but tens of millions. It was also the time of (radical) feminism, due to the changing role of women in society that affected earlier patriarchal domination. It changed the relationship between men and women and created both the Spirit of the Time and the Madness of the Time prior to the Great War. All in all, it created a socio-political context that was marked by emancipation and socialism, as well as by militarism.

Many men also felt the pressure: the pressure of living in a modern society in which everything is up for grabs, in which everything is up for negotiation, and in which there is no longer any traditional set authority. This provoked not only a feminist revolution, but also a huge intensification and increase in what historians call "rituals of masculinity". There were more duels fought in civil society than ever before, men sported better waxed moustaches, people went around in uniforms, and it was the time of the first body-building craze.
(Blom, 2016, p. 50)

---

[167] See: B.J.G. van der Kooij, *Context for Innovation: British (R)evolutions in Perspective* (2016) for an in-depth analysis of the development of the Industrial Revolution in Britain.

Madness of the Time: Militarism, Imperialism, and Nationalism

The period from the late nineteenth to the early twentieth century was a one of polarization. The political tension arose due to the social climate of militarism,[168] nationalism, patriotism, and imperialism. In many countries, militarism influenced the political scene, and in the nineteenth-century European mind, politics and military power became inseparable:

> ... *governments were strongly influenced, if not dominated, by military leaders, their interests and priorities. Generals and admirals sometimes acted as de facto government ministers, advising political leaders, influencing domestic policy and demanding increases in defence and arms spending. This militarism fathered a dangerous child, the arms race, which gave rise to new military technologies and increased defence spending. Militarism affected more than policy; it also shaped culture, the media and public opinion.[169]*

By the late 1800s some European powers had grown almost drunk with patriotism and nationalism, not without some cause. Britain had enjoyed two centuries of mercantile, commercial, and naval dominance, and her empire now spanned one quarter of the globe. The lyrics of a popular patriotic song, 'Rule, Britannia!' trumpeted that 'Britons never will be slaves.' London had spent the nineteenth century advancing her imperial and commercial interests and avoiding wars—however, the unification of Germany, the speed of German armament, and the bellicosity of Kaiser Wilhelm II caused concern among British nationalists.[170].

> *British militarism, though more subdued than its German counterpart, was considered essential for maintaining the nation's imperial and trade interests. The Royal Navy, by far the world's largest naval force, protected shipping, trade routes and colonial ports. British land forces kept order and imposed imperial policies in India, Africa, Asia and the Pacific. (ibid)*

Along with that militarism, there was nationalism and patriotism.[171] Most pre-war Europeans believed in the cultural, economic, and military supremacy of their nation. In concert with its brothers, imperialism and militarism, nationalism contributed to a mass delusion that made a European war seem both necessary and winnable. It showed in the propaganda (Figure 116). Affecting the nations already established at that

---

[168] Described by Alfred Vagts as the 'domination of the military man over the civilian, an undue preponderance of military demands, and emphasis on military considerations'. Source: http://alphahistory.com/worldwar1/militarism/
[169] 'Militarism as a Cause of World War I'. Source: http://alphahistory.com/worldwar1/militarism/.
[170] See: B.J.G. van der Kooij, *Context for Innovation: British (R)evolutions in Perspective* (2016).
[171] Nationalism, patriotism: the social desire for national advancement or political independence combined with devotion and loyalty to one's own country.

Figure 116: World War I propaganda (1914): Britain (left), Germany (middle), and Russia (right).

Source: Wikimedia Common. Lawson Wood, Dobson, Molle and Co. Ltd (top).
http://www.ww1propaganda.com/world-war-1-posters

time, Russia, France, and Britain, nationalism also gave rise to new nations, such as the through the unification of Germany and, as described before, the unification of Italy. In Central Europe, by the 1860s it was the remnants of the former Holy Roman Empire—the 22 states of the Northern Confederation and the four southern states—that were trying to unify into one German State.

> Take the North German region of Prussia, the most powerful of the German States, considered to be the wellspring of European militarism. Prussia's crushing military defeat of France in 1871 revealed its army as the most dangerous and effective military force in Europe. This victory also secured German unification, allowing Prussian militarism and German nationalism to become closely intertwined. Prussian commanders, personnel, and methodology became the nucleus of the new German imperial army.

As the great powers beat their chests and filled their people with a sense of righteousness and superiority, another form of nationalism was on the rise in Southern Europe. This nationalism was not about supremacy or military power, but the right of ethnic groups to independence, autonomy, and self-government. With the world divided into large empires and spheres of influence, many different regions, races, and religious groups wanted freedom from their imperial masters.[172]

---

[172] 'Nationalism as a Cause of World War I'. Source: http://alphahistory.com/worldwar1/nationalism/.

These patriotic tendencies gave rise to a revolutionary climate that dominated the madness of time.

> *A new and aggressive nationalism, different from its predecessors in its thoughts, appeal and goals, emerged in Europe at the end of the 19th century... The new nationalism engaged the fierce us/them group emotions – loyalty inwards, aggression outwards – that characterise human relations at simpler sociological levels, like the family or the tribe. What was new was attaching these passions to the nation... In its outward-looking dimension, the new nationalism was fully a movement of the 'age of imperialism' – of the 'great game', the 'scramble for Africa', the enterprise of great powers.* (Rosenthal & Rodic, 2015)

It created turmoil, for example, in the *Powder Keg of Europe*: the Balkans, a region full of imperialistic ambitions and growing nationalism. No nationalist movement had a greater impact in the outbreak of war than Slavic groups in the Balkans. Pan-Slavism—the belief that the Slavic peoples of Eastern Europe should have their own nation—was a powerful force in the region. Slavic nationalism was strongest in Serbia, where it had risen significantly in the late nineteenth and early twentieth centuries. It was particularly opposed to the Austro-Hungarian Empire and its control and influence over the region. It was this imperialism, with the tensions between the Russian Empire, Ottoman Empire, and Austrian Empire, that brought regional conflicts: the *Bosnian Crisis* (1908) and the *Balkan Wars* (1912–1913).

Nationalistic ambitions had brought new players—such as the young German Empire that arose after the unification of German states in 1871—into the international political arena. The French Empire and the German Empire still had something to settle between them—the French *Revanchism*[173]—after Germans had conquered France in 1871, taken the Alsace-Lorraine region, and extracted huge war indemnities. Tensions between Germany and France arose during the *First and Second Moroccan Crises* (1905–1906, 1911). Also, the *Scramble for Africa* had created colonial issues and confrontations between the German Empire, the Belgian Colonial Empire, and the British Empire. Even Italy had started, in 1881, to take part in the Scramble for Africa by invading the Horn of Africa and colonizing Somalia and Eritrea. And there was the Anglo-German naval arms race (Figure 117). By that time the British Navy was the dominant naval power in the world. The Germans responded by building up their own mighty fleet.

---

[173] *Revanchism* is the manifestation of the political will to reverse territorial losses incurred by a country, often following a war or social movement. As a term, revanchism originated in 1870s France in the aftermath of the Franco-Prussian War among nationalists who wanted to avenge the French defeat and reclaim the lost territories of Alsace-Lorraine.

By the turn of the century Britain had begun to support Russia and France, creating together the great alliance of the *Triple Entente* in 1907 as a powerful counterweight to the *Triple Alliance* of 1882/1887 between Germany, Austria-Hungary, and the Kingdom of Italy. Although the naval arms race was over by 1912, when Britain and Germany signed an agreement over the African territories, the geopolitical situation was still tense.

And then came, in 1914, the explosion of the Powder Keg when Serbian irredentists assassinated Archduke Franz Ferdinand of the Austro-Hungarian Empire on June 28, 1914. This resulted in a mechanized war the likes of which had never been seen before.

Figure 117: Cartoon illustrating the naval arms race (Puck, 1909).

Source: Wikimedia Commons.

In the Franco-Prussian War of 1870–1, nine out of ten soldiers—90 percent— died from wounds received in direct hand-to-hand combat, from sabre, bayonet, or pistol. In the First World War, 3 percent of soldiers on the Western Front died in this way. Many died from illnesses, but an enormous number were simply blown to pieces while sitting in a trench and waiting for something else to happen. Now imagine what that meant psychologically and culturally. These men had gone in their millions to fight for something—for king and country, for the fatherland, for la république—and they found that they had been integrated into a machine of entirely random killing. The Western Front was by no means the only front of the First World War, it wasn't even the deadliest, but it is the one that left the deepest marks on European cultural history. (Blom, 2016, p. 51)

The Spirit of the Times: Heroes of Invention

Along with this madness of the times, with its militarism, nationalism, and patriotism, by the end of the nineteenth century, there certainly was a noticeable mental optimism—not so much in the political arena, as described, but in the conviction that technological development had created a different and better world. This was a period characterized by optimism, as, for the first time in a long time, there was regional peace in many parts of Europe, economic prosperity due to the Industrial Revolution, strength of colonial empires, and technological, scientific, and cultural novelty.

For many, life had changed, for example, with the improvement of the living standard due to the economic effects that the new technologies had created. People visited the popular exhibitions, from the *Great Exhibition of London* (1851) to the later world exhibitions in America and Europe (Figure 115). France saw the *International Exposition* (1867) in Paris, America the *Centennial International Exhibition* in Philadelphia (1876), and Belgium the *International Exposition* (1900) in Brussels—not to mention the many more-specialized exhibitions that were organized all over Europe, such as the *Great Emilian Exposition* (1888) in Bologna (Figure 157) and the *International Electrical Exhibition* (1891) in Frankfurt (Figure 118).

Figure 118: Poster for the International Electrical Exhibition in Frankfurt (1891).

Source: http://www.vde.com/ wiki/chronik_neu/Wiki- eiten/Ausstellungen _und_Museen.aspx

People flocked by the millions to gaze at the wonders of modernity: from the powerful steam machines to the electric power engines (dynamo and electric motor), from the arc lamps (in the hand of the woman in Figure 118) and incandescent lamps to the communication machines of the telegraph and telephone. They admired the fruits of the work of all those inventor-entrepreneurs of the nineteenth century: from the Americans Samuel Morse, Thomas Edison, and Alexander Graham Bell to the British Henry Bessemer, Isambard Brunel, and George and Robert Stephenson, as well as the Germans Werner Siemens and Georg von Arco. They became the 'heroes of invention' that emerged from the times of Industrial Enlightenment (Jones, 2008; Mokyr, 2011).

> *In Britain in the period 1850–1875, exemplified by the admiration for James Watt, the cult of the inventor reached its zenith, abetted by the lessons drawn from the Great Exhibition of 1851. Until approximately 1875, the industrial strength of the country, much of it due to technological innovation, caused society to celebrate the inventor as a greater national benefactor than the politician or*

military leader. The advent of such visible achievements as railways, bridges, and steamships reflected and reinforced the status of the inventor as national hero.[174]

The societal admiration was for the inventor-entrepreneur, like those who gave Britain the strength (as in the steam machine) and the power (as in guns, cannons, and battleships) to fight their wars. Their technologies enabled massive changes in society, as technical innovation transformed the conceptual artefacts and scientific instruments into practical products. It created the stimulus for the creative individuals to be entrepreneurial. These 'Times of Invention' were the dawn of the emerging 'Times of Innovation': the times that saw the transformation of a build-up scientific insights into reality using skills previously developed by entrepreneurs.[175] The individual contribution to innovation was enabled by the free society that did not impose religious, moral, or ethical restrictions on individual creativity the way it had done in the centuries before.

The Spirit of the Time also had an influence in another way. It was the dawn of female emancipation, in which women left their traditional roles and started to claim an equivalent social position in society: the right to vote, the right to work, the right to their own body. Gone were the days of masculine absolute power: the man as the master of the house, and the women to serve him. Victorian morality still restricted female sexuality, and the treatments for hysteria were still harsh, but increasing urbanization left middle-class women in charge of the residence while men went to work away from home. Depending of their status in society, men bought the favours of their mistresses (aristocracy) or went to numerous brothels. But the social undercurrent of female emancipation was there to stay. It was the Great War, with all those male casualties and the women taking part in the war effort in the factories and with the military at home, that would cement that development and allow women to truly start to gain their freedoms.

The Long Depression (1873-1896)

Society had seen many social revolutions by the middle of the nineteenth century. Along with the American Revolution and the French Revolution, Europe had gone through the 1848 Revolutions. Together with their evolutionary companions (the evolution of transportation, agriculture, urbanization) and the influence of imperialism and colonialism of the previously mentioned empires, these revolutions brought massive changes in the European and American societies of that time. Along with social

---

[174] Source: Cosgrove, R.A., *Heroes of Invention: Technology, Liberalism and British Identity 1750–1914* (review) https://muse.jhu.edu/article/458154

[175] As conceptualized by Schumpeter's description of innovation and the role of the entrepreneur creating 'business cycles'.

transformation, it brought changes later labeled as the First Industrial Revolution, a development that followed the British techno-economic evolutions.[176]

Part of that evolution had been the development of a capitalistic infrastructure. From early lending institutions like merchant banking (ie merchant capitalism), the industrial banking infrastructure (ie industrial capitalism) had emerged. The rise of private stock companies, with their stockholders providing the capital, had resulted in the stock exchange, where stocks in companies were traded. Parallel to the Industrial Revolution, capitalism and banking expanded, along with the inherent problems they possessed: bank crises and panics. While merchant capitalism had been a national affair with local markets, industrial capitalism had become an international affair.

Panics Crippling Economies

Industrialization, in general, brought progress, with its economic ups and downs depending on specific national situations. The driving force of technological innovation had made the world smaller. Railways and steamships brought people and goods everywhere faster than ever before, and the telegraph made long-distance communication with lightning speed possible. It was an early era of globalization. Nevertheless, America and Europe as a whole were experiencing the *Long Depression* (1873–1896), a period with several financial panics crippling the economies, hitting countries severely, such as the *Panic of 1857–1859* in the United States, which stayed in the US and only sent a ripple through world economies. It was related to the mechanism of money creation where banknotes were issued that could be converted to silver and gold at any time. The financial system was built on that trust that the bank would convert, and when that trust disappeared, the panic set in.

Before 1857, the US railroad industry was booming due to large migrations of people to the west, especially in Kansas. With the large influx of people, the railroads became a profitable industry, and the banks seized the opportunity and began to provide railroad companies with large loans. Triggered by a boom in railroad developments and speculation, prices of railroad stock speculation rose into a bubble. Then the failure of the bank Ohio Life Insurance and Trust Company—a financer of railroad companies—ignited a debt crisis. This was aggravated by agricultural crises, when prices of grain dropped and the farmers could not meet their obligations on their speculative land purchases. The ensuing debt crisis, and the

---

[176] See: B.J.G. van der Kooij, *Context for Innovation: British (R)evolutions in Perspective* (2016).

refusal of bank to convert money back into gold or silver when they ran out of their hard currency reserves, created a panic

The situation was different with the *Panic of 1873–1879*, which had serious worldwide effects (Figure 115). In Europe, it started with the crash of the Vienna Stock Exchange: the 'Gründerkrach' or 'Founders' Crash'. A series of Viennese bank failures ensued, causing a contraction of the money available for business lending, creating a debt crisis. The crash was related to the industrialization that had reached out over Europe from Great Britain (Figure 105) and where massive railroad development and industrial activity (eg steel and locomotive industry) had started. One aspect of this industrialization was related to urbanization, leading to growth of cities, especially in Germany and Austria, where private property developers created the infamous rental ghettos for the expanding urban lower classes. Another aspect was the rise of the wealthy merchant class and business class who—building their magnificent palaces of the nouveau-riche citizens—had the wealth to speculate. This was combined with the influx of capital after the 1871 War Reparations, in which the defeated France payed millions in war-indemnities to the Germans (5 billion francs or some $1,000 million).[177] This had pumped large amounts of money in the economies, creating a stock market boom and an asset bubble related to a real estate property boom. In short, it was the time of 'easy financing' fuelling economic expansion,[178] and when that failed, the stock market crashed.

> *Financial and industrial conditions soon showed the effects of this situation. A lively speculative movement, affecting both domestic and foreign securities, was followed by the organization of a host of new undertakings of all sorts, whose shares found ready purchasers … a strikingly large share of them (about a third) were in the field of banking. There was, in fact, a veritable mania for this form of investment, both in Germany and Austria. In 1870, the shares of forty-eight banks with a capital of 847,000,000 marks were quoted in Berlin; three years later ninety-five more banks with a paid-in capital of only 150,000,000 marks had been organized. These new institutions, though organized with scantier funds, undertook the financing of all kinds of enterprises, thereby adding to the general expansion.* (Monroe, 1919, p. 275)

---

[177] In 1807 Napoleon, as he often did after gaining victory, had charged the conquered German states with a similar indemnity. Now the Germans wanted an equal sum. The financing of the different instalments of the indemnity was by loans issued at the stock markets of Paris, London, Hamburg, and Amsterdam from the gold and silver reserves and from bills of exchange.

[178] In the beginning of the twenty-first century there was a similar development that resulted in the 2008 crash of the financial markets, creating financial turmoil all over the world, plunging economies into depressions that would take a decade to recover from.

The May 9, 1873, financial crash set a range of other crashes in motion. Not only private individuals but also companies like the German railway empire of Bethel Henry Strousberg had gone bankrupt. The crash spilled over to other countries like Germany, France, and Britain.

In America, industrialization had also resulted in a boom in railroad construction after the Civil War, attracting a large infusion of capital from speculators. When in September 1873 the Philadelphia bank Jay Cooke & Company forfeited and bankrupted, it ignited a chain reaction of US bank failures and bank runs (Figure 119), and consequently, companies

Figure 119: Run on the Fourth National Bank (1873).

Source: Wikimedia Commons. *Frank Leslie's Illustrated Newspaper*, 1873 Oct. 4, p. 67.

went bankrupt. By November 1873 some 55 of the nation's railroads had failed, and another 60 went bankrupt by the first anniversary of the crisis. In total, some 18,000 businesses failed between 1873 and 1875, sending the United States into a depression, which, in turn, had its effects on the European situation.

> *The breakdown in the United States, however, destroyed what confidence was left in German speculative circles, and the markets suffered a decline of vast proportions. Bank shares, which had been such favourites with investors, dropped more than 80 per cent in some cases, and the numerous companies which the banks had fostered suffered as badly. ... Railroads, banks, and even manufacturing concerns went down in a general collapse, which lasted fully two months and affected the whole of the country.* (Monroe, 1919, p. 278)

The Industrial Revolution also had other effects. Take for example, all those new invention using electricity. Such as the newly developed method of communication over long distances using electricity (Figure 120, top).

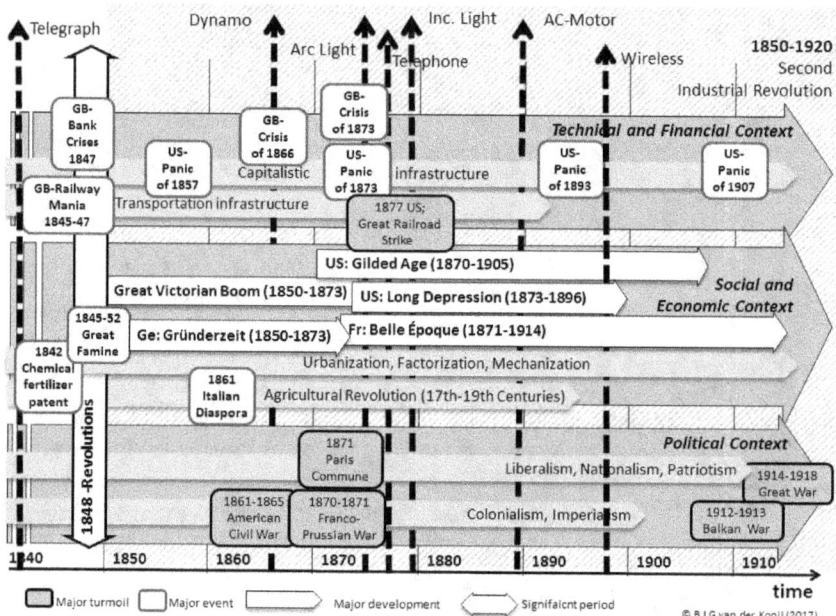

Figure 120: The overall context for the Communication Revolution (1840–1920).

Figure created by author.

After Morse created the Baltimore line of telegraph cables on poles in 1844, telegraph fever spread around, resulting in the telegraph boom of the 1850s. This boom changed the way people did business, made markets more transparent, facilitated the stock exchanges, and spread the 'news' all over the world. To an even larger extent, the telephone boom of the 1870s did the same. Technology was the walking companion of many of the social and economic revolutions of that time.

Next to these technology-based stimuli, there were social stimuli such as the massive immigration from the Old World: the Irish diaspora after the Great Famine (1845–1852) as well as the Italian Diaspora (1861). In the United States, the racial issue (slavery) resulted in a civil war between the North and the South, causing over a million casualties. After initial peace, much was left to be solved during the *Reconstruction Era* (1865–1877), such as the Reconstruction Amendments, which addressed social equality, discrimination, and the abolishment of slavery. The social problems caused by industrialization, urbanization, immigration, and corruption in government came to a head in the *Progressive Era* (1890–1919). Then a time of social activism and political reform began, of fighting the corruption of the 'political bosses' and of women's suffrage, but also of municipal reform and constitutional change.

In addition, that period of time also saw the *Panic of 1893–1897* in the United States. Similar to the Panic of 1857, this panic was marked by the collapse of railroad overbuilding and shaky railroad financing. Some fifty railroad companies went under in the chaos. And since this industry was one of the nation's largest and it supported other industries, those failures rippled outwards; more than 30 steel companies collapsed in the wake of the railroad failures. Additionally, there was a 15–20% decline in yields for the wheat crop due to severe drought conditions in 1892–93. The cotton-growing regions in the South also faced problems. Banks started to call in loans, setting off a series of regional bank failures: from some 600 in the first months to 4,000 by the end of 1893. As a result of other factors—among which was the mining of an overabundance of silver—people lost confidence and rushed to withdraw their money from the banks. Thus, the credit crunch rippled through the US economy (and soon also spread into Europe).

> *A series of bank failures followed, and the Northern Pacific Railway, the Union Pacific Railroad and the Atchison, Topeka & Santa Fe Railroad failed. This was followed by the bankruptcy of many other companies; in total over 15,000 companies and 500 banks failed (many in the west). According to high estimates, about 17%-19% of the workforce was unemployed at the Panic's peak. The huge spike in unemployment, combined with the loss of life savings kept in failed banks, meant that a once-secure middle-class could not meet their mortgage obligations. Many walked away from recently built homes as a result.[179]*

The result was an economic depression that created social turmoil; labor strikes, and massive unemployment (c. 20%). More than 1300 strikes, involving 750,000 workers, hit the nation's factories and mines in 1894. Jacob S. Coxey, an Ohio steel mill owner, protested and organized a march of an industrial army on Washington in 1894 that stirred panic among those who feared an insurrection of the unemployed. Coxly was arrested, but dozens of demonstrations like his broke out during this turbulent period. His ideas of government intervention (creating employment by public works) were by then considered to be revolutionary, but they were implemented later in time.

> *H. P. Robinson, editor of Railroad Age, wrote in January 1895: "It is probably safe to say that in no civilized country in this century, not actually in the throes of*

---

[179] 'Panic of 1893'. Source: https://www.saylor.org/site/wp-content/uploads/2011/08/HIST312-10.1.2-Panic-of-1893.pdf

war or open insurrection, has society been so disorganized as it was in the United States during the first half of 1894; never was human life held so cheap; never did the constituted authorities appear so incompetent to enforce respect for the law."[180]

Recovery came slowly, but by the middle of 1897 signs began indicating that the economy was stabilizing.

This description illustrates the overall context of some of the important countries of the industrialized world after the 1850s up to the twentieth century, also known as the Second Industrial Revolution (Figure 120). These short and rough brushstrokes paint a picture of a quite dynamic social-political and techno-economic context (Scheme 3) for the electric inventions we are studying here: the clusters of innovations during the different eras of communication.

---

[180] Source: http://www.encyclopedia.com/history/united-states-and-canada/us-history/panic-1893.

The Communication Revolution Expands

In our extensive analysis of the early developments in the nineteenth century into the field of long-distance communication elsewhere,[181] we observed how discoveries, inventions, and innovations resulted in the 'communication technology of telegraphy' that started the *Era of Telegraphic Communication*. The identification of 'era' seems to hint at quite relevant development. But relevant in what way? Obviously, it was relevant for communication over distance, but were those technical changes also relevant in other ways? What was their non-technical impact? Let's try and give some examples to elaborate this point:

Standardization of Time: The upcoming transportation facilities (train, coach) needed reliable timetables. In those early days every city more or less had its own time, which could differ considerably. The unreliable clock in the old bell tower might have been replaced by the individual mechanical pendulum clock, but times differed from city to city, as there was not a general reference. It didn't matter too much, as towns were mostly operating independently from each other. For naval operations, however, the situation was quite different, as the chronometer was used to establish the longitude of a position. Consequently, the need for some form of synchronization and standardization arose. That became possible when the telegraphic infrastructure in the 1850s started to develop seriously. As electric communication was so quick, it offered the chance to create a standard time: railway time was soon followed by the *Greenwich Standard Time*. Then, using the telegraphic infrastructure, a time-synchronizing signal could easily be sent over the wires. Even more, when wireless telegraphy was developed, synchronizing of the ships' clocks with radio time signals became possible. So, 'time' became standardized. This *synchronization of time* seems just a minor thing, as today's reality is that we live in a time-dominated society. While the feudal farm lacked any timepiece, the sun being its only reference to time, the industrialized society needed timekeeping to become organized. In today's society, timekeeping devices are everywhere, from smartphones to microwave ovens and dashboard clocks in cars.

Transparency over Space: Another example relates to the 'transparency in pricing and markets'. This is quite a mouthful, but just imagine those local communities with their local markets where the products of early industrialization (commodities like wheat and rice[182]) and cottage

---

[181] See: B.J.G. van der Kooij, *The Invention of the Communication Engine 'Telegraph'* (2015).
[182] Commodity: An economic good or service when the demand for it has no qualitative differentiation across a market. "From the taste of wheat, it is not possible to tell who produced it, a Russian serf, a French peasant, or an English capitalist."

industries (textiles, early iron products) were offered for sale. As the products were sold on a local market, from local producers to local customers, the pricing was local. This could result in regional price differences. With improved communication, without physically shipping the product, the price could be set regionally or nationally. Having the possibility to exchange price information for those commodities with other, distant markets, sellers could look for other buyers, thus creating regional, and even national, markets. Telegraphy thus facilitated the communication of market-sensitive information and made trading transparent over larger spaces. It is not surprising that the stock market embraced early telegraphy rapidly. It was all about *market-transparency*.

Incommunicado problem: A third example for the need of telecommunication would be in the maritime world of ships sailing the seas. After a ship left the harbor and sailed for its next destination, all communication with it was lost. The ships were without communication (they were 'incommunicado') until their return or arrival in the next port made any information about its well-being possible again. This was especially true in those early days when explorative voyages were long and when a ship could be incommunicado for months, even years. This was quite an unpleasant situation, as maritime dangers are so substantial. Also, later in time, when steamboats crossed the oceans, weeks went by without any knowledge about its positon or well-being. True, in the mid-1800s the use of semaphore signaling (with flags or panels) made communication within eyesight between ships and shore (aka ship-to-shore communication) possible.[183] Obviously, this short-distance communication was limited in its use due to the need for visibility. For all maritime activities, there always has been a strong need for a means of communication between shore and ships, and between a ship and other ships. It wasn't until wireless communication became available at the end of the nineteenth century that the *maritime incommunicado* problem was solved. Clearly, the sea—as we will go and see later on more in detail—proved to be an early push for innovation in this area.

These seemingly irrelevant examples of 'standard timekeeping', 'market transparency', and 'maritime incommunicado' were influenced by the technical developments of telegraphy. In turn, they influenced the development of society, both of the individual and the collective behavior of society's members. It was the First Industrial Revolution that saw the use of mechanical power (aka steam power) complementing the natural power

---

[183] In a similar way, navies, whose fleets had to be deployed in specific areas, were facing the 'incommunicado' problem. Just image the importance of naval communication when the Spanish Armada reached the English coast in 1588.

sources of wind, water, animals, and humans. The Fist Industrial Revolution freed human society from the limitations of natural power. It was followed by the Second Industrial Revolution, which saw the rise of electric power and its applications, such as electric light, which freed human society even more from things like the limitations of daylight. Those industrial revolutions were the result of technical change creating new technologies. One of those technologies was the *general purpose technology (GPT) of electricity*, and one of its early application areas was 'communication over distance'. The new phenomenon of electricity was applied to invent the telegraphic engines, fulfilling the basic human need of communication. These were the engines that resulted from the inventions of the Englishmen Cooke and Wheatstone, and the American Morse in the late 1830s.

In the first half of the nineteenth century the general purpose technology of electricity was in its infancy. It was battery powered, and that caused limitations on its use when power applications were explored (eg electric train/boat, electric DC motors). However, that did not seem to be much of a problem with communication technologies like the 'telegraph technology', which started to develop in the 1840s. To understand what caused the development of that technology, we have to go back to the discoveries that took place in the late eighteenth/early nineteenth century.

Discovering Electricity and Electromagnetism

As scientific thinking in the eighteen century developed, it focused on understanding the basic elements in nature. Among those was the understanding of the 'nature of heat' and the related 'power of fire' that resulted in the invention of the steam engine.[184] Not much later, this was followed by the understanding of the 'nature of lightning' and the related 'electromotive power'.[185] As the later developments of electricity (ie the generation, transport, and use of electricity) gave way to the ample availability of electric power, a range of other scientific curiosities became linked to electricity.

It was at the end of the eighteenth century and beginning of the nineteen century that the foundations for the understanding of the nature of electricity were being created. It started with *frictional electricity*, which is electrical charge caused by rubbing different materials together, as was shown by the experiments of *Benjamin Franklin* (1706–1790). The kite-carrying Franklin became famous for bringing lightning down to earth: his Philadelphia experiments. In these experiments, in 1750, he proved the existence of electricity by flying a kite in a thunderstorm. The kite twine

---

[184] See: B.J.G. van der Kooij, *The Invention of the Steam Engine* (2015).
[185] See: B.J.G. van der Kooij, *The Invention of the Electro-motive Engine* (2015).

conducted the 'electric fire' along the wire to a key at the bottom. Others, like *D'Alibard* in France in 1752 and *Georg Wilhelm Richmann* in 1753 in St. Petersburg, repeated his experiments. Their observations that lightning caused shocks was taken on by others, who were able to create electricity differently with simple frictional machines (Hauksbee, Faraday, Nolet, and others). The electricity-generating apparatus was born, and its medical application (aka early electrotherapy) was soon to follow.

It was the time of fluid theories, and electricity was explained as the result of electric fluids, for example, the two-fluid theory created by the Frenchmen *Cisternay du Fay* in the 1730s and the single-fluid theory proposed by Benjamin Franklin in the 1740s. Over time that fluid theory, just like the caloric theories of heat, were replaced by other views. Among those were the views of William Thomson, later known as Lord Kelvin (1824–1907), who synthesized the thinking on energy in the 1850s:

> *Thomson's worldview was based in part on the belief that all phenomena that caused force—such as electricity, magnetism, and heat—were the result of invisible material in motion. This belief placed him in the forefront of those scientists who opposed the view that forces were produced by imponderable fluids. ... He brought together disparate areas of physics—heat, thermodynamics, mechanics, hydrodynamics, magnetism, and electricity—and thus played a principal role in the great and final synthesis of 19th-century science, which viewed all physical change as energy-related phenomena.* (Sharlin, nd)

By that time many other scientists had looked for different forms of electricity. Soon electricity came in different flavours. For example, *Luigi Galvani*, while dissecting frogs, had encountered *animal electricity*, a new

phenomenon: a frog's leg in a nerve-muscle preparation contracted every time the muscle and the nerve were connected by a metal arc, which usually consisted of two different metals. The publication of his work got the attention of many scientists in those days. Among the scientists, we also find *Alessandra Volta* (1745–1827), a professor of experimental physics at the University of Pavia. He experimented with a pile of plates of silver and zinc

Figure 121: Alessandro Volta's chemical battery (ca 1800).

Source: http://alessandrovolta.it/wp-content/uploads/2011/07/144C.png

soaked in salt water, and his work resulted in another form of electricity, *voltaic electricity*. The 'wet' battery was born (Figure 121).

Over time these early scientists grasped, more or less, the nature of electricity. Based on the work of these early experimenters, others continued these explorations and added fundamental insight into the application of electricity. It was the Dane *Hans Christian Oersted* who, in 1820, observed during a lecture that a compass needle would move when an electric current passed through an electric cable located nearby (Figure 122). It was to be the discovery of electromagnetism. Its publication created uproar in the savant community of those days.

Figure 122: Hans Christian Oersted's needle experiment.

The voltaic battery is visible between the scientists, and the compass lies on the table.

Source: http://alessandrovolta.it/wp-content/uploads/2011/07/149A.png

Hearing of these experiments, the Frenchman *Andre-Marie Ampere* became excited by the discovery. After repeating Oersted's discovery he started experimenting himself, and eventually he explained the mechanism behind Oersted's discovery, where an electric current influenced the movement of a magnetic needle. But he could not explain the reverse action: magnetism influencing electric current. That was done by Michael Faraday, who was also intrigued with Oersted's discovery. He studied it extensively and experimented in 1831 with a soft iron ring with two sets of coils (as seen, more or less, in today's transformer). Connecting a battery to the first coil resulted in current in the second coil. He had found the induction effect and thus expanded the relationship between magnetism and electricity: electromagnetic induction. Next, it was *William Sturgeon* (1783–1850), who, in 1825, conceptualized that electricity and the properties of metal could create a magnetic force: the electromagnet was born (S. P. Thompson, 1890, p. 199). Its enormous power was soon demonstrated by the powerful magnets created by *Joseph Henry*.

In the firmament of knowledge, on the path covered with scientific contributions to the phenomenon of electricity, the efforts of these scientists created milestones (Figure 123). This was, quite some time later, recognized when the highly awarded *Elihu Thomson* (1853–1937), then acting president of the Massachusetts Institute of Technology, held a

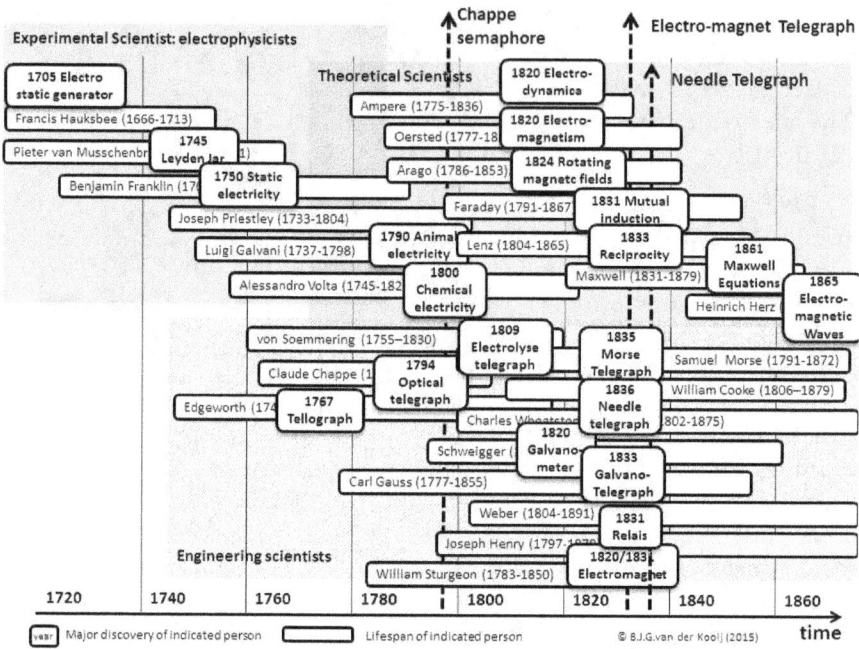

Figure 123: Overview of experimental, engineering, and theoretical scientists discovering telegraphy.

Source: B.J.G. van der Kooij, *The Invention of the Communication Engine 'Telegraph'* (2015), p. 447.

presentation for the *American Institute of Electrical Engineers* on October 8, 1920 in honour of Oersted's discovery a century before. He concluded, after describing in short the developments leading to the telegraph that changed the world of communication:

> *It is not necessary here to allude to the great developments in the field of electricity and electromagnetism as exemplified in generation and transmission of electrical energy. They have covered the past half century, but the foundation principles belong to those early years of upward of a century ago. Do we cause movement of iron masses by a current coil? It is the experiment of Oersted. Do we cause movement of coils, one with relation to another, as in our motors? It is the experiment of Ampère. Do we generate currents in a conducting mass in a magnetic field? It is the experiment of the Arago disk. When we measure current or energy by galvanometer, voltmeter, electrodynamometer, or wattmeter we have the work of Oersted, Ampère, Arago and Davy illustrated. But these early discoveries had a deeper significance still. They showed that electric currents and magnetism are inseparable – inseparable in practice, inseparable in theory. ...*

216

The discovery, then, of the relation between electricity and magnetism was in reality the discovery of a fundamental fact or principle lying that the foundation of the universe itself; the soul of energy, as also of matter, of electric waves from zero periodicity up to the most penetrating rays of the radium emanations. It is eminently fitting, then, that we celebrate the hundred anniversary of discoveries, the fruits of which have been of stupendous influence and value, and that the same time carry us to the very foundations of existence; but we meet also to do honor to the great men who first brought those discoveries to light. (Thomson, 1920, p. 1027)

Scientists ended up with knowledge about the fundamentals of electricity, and they managed to use its knowledge and apply electricity in real life. Basically electricity was used to create *linear movement* (ie the electromagnet) and *rotative movement* (ie the electromotive motor). In both cases, it was electricity that was used for the *transportation of energy*. This discovery would lead to an impressive range of inventions of its own. In addition to this power transmission capability, there proved to be another use for electricity: it would also become the medium for the *transportation of information* by means of telecommunication. And in those early days two artefacts proved to be fundamental to its development: the galvanometer and the electromechanical relay. At is foundations lay a famous controversy.

The Galvani-Volta Controversy

It happened in eighteenth-century Italy, where scientists—aka the electrophysiologists—at the University of Bologna and the University of Pavia were actively exploring the 'nature of lightning'. The 'loser' was *Luigi Galvani* (1737–1798), professor of anatomy at the University of Bologna, who became interested in 'medical electricity' (the medical application of static electricity generated by means of friction) and formulated his theory of 'biomedical electricity'. The 'winner' was Alessandro Volta (1745–1827), professor at the University of Pavia, originally a supporter of Galvani's theory, who—after repeating the experiments—developed his own theory of contact electricity. The controversy was born. (Pera & Mandelbaum, 1991; Soeiro, 2013)

The controversy occurred in the late eighteenth century. At this time France had invaded Italy, and Napoleon had created the Cisalpine Republic (1797–1802). Galvani was by that time deprived of his offices at the university and the Instituto delle Scienze because of his refusal to swear allegiance to Napoleon's Cisalpine Republic. As a result, he had to take refuge with his brother Giacomo and broke down completely from poverty and discouragement. Volta became world famous and was laurelled by Napoleon on several occasions. Their opposite fates were the consequences of investigating the nature of electricity from different points of view.

Nature of Lightning: Animal Electricity

At a given moment in the chain of discoveries that would lead to the 'discovery of electricity', scientists were exploring another form of electricity, later called 'medical electricity'. Among them was the physician and biologist *Luigi Galvani* (1737–1798), who, in 1786, discovered that a frog's legs would exhibit violent muscular contractions when its exposed nerves were touched with one metal and its muscles were touched with another metal while the two metals were connected.[186] Galvani explained this occurrence as the discharge of the 'nerveo-electrical fluid' previously accumulated in the muscle (Heilbron, 1979, p. 491). This was a biologic view that fit in with the 'fluid thinking'[187] that was common at that time.

His discovery was the result of a range of experiments with electricity. Aware of static electricity from lightning, Galvani started his experiments by observing the contractions due to atmospheric electricity during a weather storm.

He connected the frog nerve to a long metallic wire pointing toward the sky, in the highest place of his house and "…in correspondence of four thunders, contractions not small occurred in all muscles of the limbs, and, as a consequence, not small hops and movements of the limbs. These occurred just at the moment of the lightnings; they occurred well before the thunders when they were produced as a consequence of these ones" (Piccolino, 1998, p. 385).

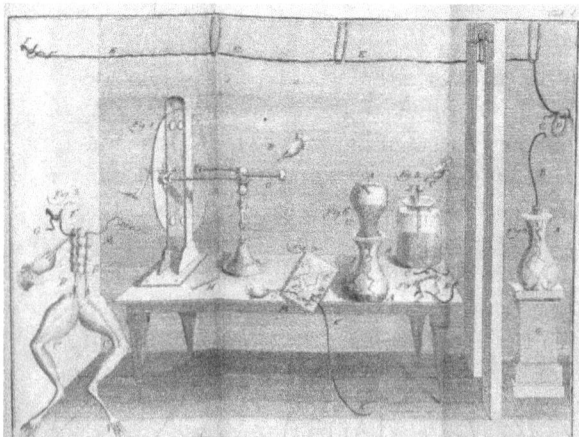

Next, he repeated the experiment during serene weather in order to investigate if the natural electricity present in the atmosphere of a calm day could succeed in evoking contractions:

Figure 124: Galvani's experiment with frog legs (1791).

Source: Wikipedia Commons

---

[186] Now we know that the effect was due to a (very small) electric current generated by a chemical reaction and acting with contractile effect on the muscles of the frog's legs.

[187] The 'two-fluid' theory, created by Charles François de Cisternay du Fay, postulated that electricity was the interaction between two electrical 'fluids'. An alternate simpler theory was proposed by Benjamin Franklin, called the unitary, or one-fluid, theory of electricity.

But nothing happened for a long time. Finally "tired of the vain waiting" he came near the railing and started manipulating the frogs. To his great surprise, the contractions appeared when he pushed and pressed, toward the iron bars of the railing, the metallic hooks inserted into the frog's spinal cord. (Ibid)

Galvani repeated his experiments indoors (Figure 124) and realized that in order to get contractions, it sufficed to connect the nervous structures (spinal cord) and the leg muscles (sural nerves) with a metallic conductor, therefore creating a circuit similar to that which develops in a Leyden jar when the internal and external plates are connected.

Galvani came to the conclusion that some form of intrinsic electricity was present in the animal, and that connective nerve and muscle together, by means of conductive materials, induced contractions by allowing for the flow of this internal electricity (Piccolino, 1998, p. 386).

This was a new phenomenon: a frog's leg in a nerve-muscle preparation contracted every time the muscle and the nerve were connected by a metal arc, which usually consisted of two different metals (Figure 125). Galvani concluded from his findings that an intrinsic presence of electricity existed: the "animal electricity" (Kipnis, 1987, pp. 114-116). He wrote about it:

And still we could never suppose that fortune were to be so friend to us, such as to allow us to be perhaps the first in handling, as it were, the electricity concealed in nerves, in extracting it from nerves, and, in some way, in putting it under everyone's eyes. (Piccolino, 1998, p. 381)

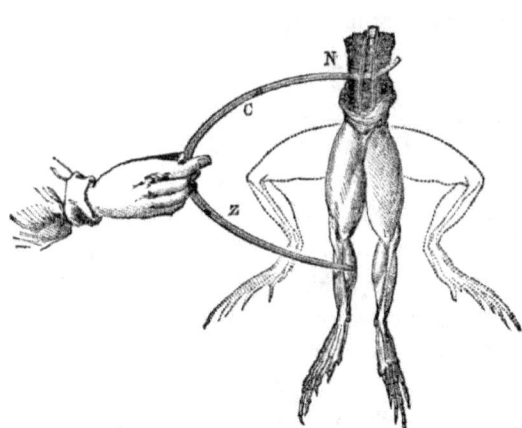

Figure 125: Principle of Galvani's experiment with frog legs (1790).

Source: Wikipedia, Wells, D.A (1859)

That 'animal electricity' had been a topic of interest for many scientists in the preceding decades who'd studied the relationship between electricity and life as observed in several species of fish like the electric eels (Gymnotus electricus). Thus, as it was apparent that animals could generate electricity in their bodies, they were looking for the 'neuro-electric fluid'. So, news about Galvani's discovery quickly spread from Italy to France, Germany, and England.

After reading about Galvani's theory as described in a 53-page Latin paper, *De Viribus Electricitatis in Motu Musculari Commentarius* that was written in 1792, many scientists started repeating Galvani's experiments with frogs.

> *The news of Galvani's discovery caused great feeling in scientific circles, and it produced a repercussion among physicians and the merely curious: astonishment, wonder, and immediate resolution to repeat Galvani's experiments, with the obvious consequence that frogs were decimated in great quantities everywhere, first in Italy and then in Europe.* (Bernardi, 2000a, p. 102)

An unusually large number of authors from all over Europe published their findings in the next couple of years (Kipnis, 1987, p. 117). At the University of Pavia, the feeling produced by Galvani's little Latin book, which soon became a best seller, was enormous. Mariano Fontana, a member of the Imperial Chancellery of Holy Roman Emperor Charles V, wrote to the author: "Now with endless pleasure I tell you that the result of your finest experiments is considered an original discovery, that the experiments have been repeated and found very exact… In short, here now all is animal electricity, and your name is famous in Pavia." (Bernardi, 2000)

Those were Galvani's finest days, but soon another Italian would come and spoil the fun.

Nature of Lightning: Voltaic Electricity

The work of Galvani also attracted the attention of Alessandra Volta (1745–1827), professor of experimental physics from the nearby University of Pavia. Volta was highly interested in the phenomenon of electric fire and had written treatises like *On the Attractive Force of the Electric Fire* (De Vi attractiva Ignis electrici, 1769) and *On the Method of Constructing the New Electrical Machine*. It led to his work on 'weak electricity' that complemented the works of others on 'strong electricity' (generated by the electrostatic machines of those days). Many scientists—such as the Englishmen Abraham Benett (1749–1799) and William Nicholson (1753–1815), and the Anglo-Italian Tiberius Cavallo (1749–1809)—took part in the hunt for weak electricity. Volta repeated—like so many scientists at that time—the Galvani experiments. After finding out the important role of the conductors used, and with his own experimenting with chemicals and metals resulting in a discovery he called 'metallic electricity' (Figure 126), he challenged Galvani's theory. In a letter he wrote in August 1796, he said:

> *One can consider this mutual contact of two different metals as the immediate cause that set the electric fluid in motion, instead of attributing this power to the double contact of these metals with the humid conductors…* (Kipnis, 1987, p. 122)

Some years later, on March 20, 1800, Volta wrote a letter to Joseph Banks, the president of the Royal Society (Volta, 1800). On June 26 this letter was read before the Royal Society in London. In this letter, Volta described his new source of energy: a pile of plates of silver (A) and zinc (Z) separated by cardboard (a) and soaked in salt water. The (electrochemical) battery from the spate groupings AZa, AZa, AZa, AZa, etc was born.

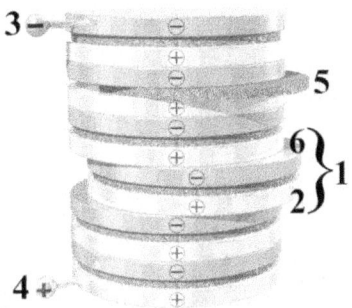

Figure 126: Principle voltaic pile.
The voltaic cell (1) is made up from metal disks (2, 6) isolated by a cardboard disk (5). Several cells make the battery, supplying electricity from their contacts (3, 4).

Source: http://simple.wikipedia.org/wiki/File:Voltaic_pile_3D_model.png

In this manner I continue coupling a plate of silver with one of zinc, and always in the same order, that is to say, the silver below and the zinc above it, or vice versa, according as I have begun, and interpose between each of those couples a moistened disk. I continue to form, of several of this stories, a column as high as possible without any danger of its falling.

The letter was distributed widely and stirred a massive interest in the scientific community of those days. The discussions between the adherents to Galvani's theory and those who believed in Volta's theory continued, and it even became a controversy. Scientists either adhered to Galvani's theory or to Volta's theory; it was basically the choice between the concepts of 'static electricity' (Galvani's theory) and 'contact electricity', the electricity created by contact potential (Volta's theory)[188] (Geddes & Hoff, 1971).

It was a controversy that had to been seen within the context of that revolutionary period in time:

Between 1791 and 1800, in the ten years from the publication of Galvani's Commentarius to the invention of the electric pile by Alessandro Volta, a scientific revolution occurred in Europe. It was not only a scientific controversy. The political problems and revolutionary events, which at the end of the eighteenth century changed French and Italian life, had a close relationship with the development and conclusion of the controversy between Galvani and Volta. (Bernardi, 2000b, p. 102)

---

[188] The definitive experimental clarification of the controversial issue was left to the noted German naturalist Alexander von Humboldt in 1797, who was able to uncover two separate and genuine phenomena, bimetal electricity and animal electrogenesis, and to point out where Galvani and Volta were right and where they were wrong.

Figure 127: Presentation of Volta's battery to Napoleon (November 1801).
Source: http://ppp.unipv.it/volta/pages/immagini/gv8_2.gif

These political problems and revolutionary events were related to the aftermath of the French Revolution, when Napoleon rose to power and conquered large parts of Northern Italy in his Italian Campaigns (1792–1802) and then created a vassal state, the Cisalpine Republic. Volta went with the flow of the times, and in 1801 he gave a demonstration in Paris of the research that had preceded his invention (Figure 127). The event was attended by Napoleon I, who awarded him a gold medal. Some years later Volta was even more heralded, as in 1805 the French emperor granted him an annuity and named him knight of the Legion of Honour; in 1809 Napoleon made Volta a senator of the Kingdom of Italy and, the following year, a count. Volta died in 1827 a famous man.

Whatever the controversy, the discovery of the 'voltaic pile' would have enormous consequences on the further development of electricity itself and its fields of application.

> *The pile was the last great discovery made with the instruments, concept, and methods of the eighteenth-century electricians. It opened up a limitless field. It was immediately applied to chemistry, notably to electrolysis, and soon brought forth the shy elements sodium and potassium from fused soda and potash. Its steady current provided the long sought means for establishing a relation between electricity and magnetism. The consequent study of electromagnetism transformed our civilization.* (Heilbron, 1979, p. 494)

The discovery of contact electricity changed the thinking about electricity. It resulted in 'battery mania', where scientists experimented with the new phenomenon. All that experimenting created understanding , and that resulted in another basis discovery: the existence of electromagnetic waves that could be used for distant communication. Let's have a look at how that came about.

Discovering Electromagnetic waves

One of the costlier components in the early communication systems of the telegraph or telephone, although technically less challenging, was the cable that transmitted the signal. Though the understanding of its behaviour over long distances took some scientific insight,[189] it was others factors that made it a problematic part of the system. One of them was the cluttering of the cities; another was its vulnerability (ie to sabotage, the aurora borealis, submarine problems) along the countryside or in oceans. In addition, there were places where the cable was not able to create a communication circuit, such as at sea. So, if a solution could be found to eliminate the cable, long-distance communication could realize its next leap.

As analysed elsewhere,[190] bridging the 'voltaic gap' created a bright spark (like the light created by a welding system). That the spark created an emission of something was clear: it was visible to the human eye. But there was more than only the visible emission. Obviously, the spark emitted light, that is—as we know today—an emission within a certain spectrum that our eyes can detect. Soon scientists became curious if that same spark would contain other (invisible) emissions as well.

Early Scientific Experimenting with Electromagnetism

Several scientists—too many to mention here—experimented with the amazing new phenomenon of electricity, like the Frenchman *Felix Savery* (1797–1841), professor of astronomy and geodesy at the École Polytechnique in Paris.

A highly respected mathematician and physicist, Savary also worked with scientists such as André-Marie Ampère, one of the main discoverers of electromagnetism and who the SI unit of current is named after. From his joint research with Ampère, Savary went on to write Mémoire sur l'application du calcul aux phénomènes élecro-dynamique (1823). Four years later, Savary's most famous work, Mémoire sur l'Aimantation, was published in the French journal Annales de Chimie et de Physique. It was Savary who was the first to

---

[189] Originally, as discovered by the German physicist George Simon Ohm in 1827, the cable was thought to have only 'resistance' defined by Ohm's law: $R=V/I$. Later it became clear, by the work of the Englishman Oliver Heaviside in 1886, that it was more complicated, as capacity and inductance also played a role over long distance. So, the 'impedance' extended the concept of resistance in an AC circuit.

[190] See: B.J.G. van der Kooij, *The invention of Electric Light* (2015).

describe in this paper his hypothesis of the oscillatory nature of the discharge of a Leyden jar connected to an inductor. [191]

The study of oscillations was also done by the Dutchman *Gerrit Moll* (1785–1838) in 1828 and—later—the experimental American scientist *Joseph Henry* (1797–1878), who showed in 1842 the existence of high-frequency oscillation as he discharged a 'Leyden jar'. He also observed the fact that, over a considerable distance, inductive coupling existed between a circuit with a discharge and a receiving circuit. It was the Englishmen *William Thomson* (1824–1907), the later Lord Kelvin, who provided a mathematical theory in 1853. In Germany, *Hermann Helmholtz* (1821–1894) and *Karl Wilhelm Knochenhauer* (1805–1875) did experiments of a similar nature. Knocherhauer postulated in 1842 a theory about the existence of electric oscillations. In the 1870s the German professor *Wilhelm von Bezold* (1837–1907) published in *Poggendorfs Annalen* (nr. 140, 1845, p. 541) about his experiments that a condenser discharge caused interference phenomena. These were what one could call the early scientific contributions towards explaining the 'phenomenon of electric oscillations' these scientists observed and theorized about.

In the nineteenth century it was already clear that electromagnetic radiation of wavelengths outside visible light existed (eg Hershel, Ritter, and others). Many scientists tried to create the high-frequency electrical oscillations of the 'transient electrical currents' (aka alternating currents, AC). There were quite a few tinkerers (aka engineers) who, each in his own way, contributed to the discovery of electromagnetic waves, as did the thinkers (aka scientists) towards the phenomenon of electromagnetic waves. Some even obtained patents for their work, as it was considered more than to be a contribution to scientific knowledge, and they wanted to get the honour of their discovery.

Their contributions followed different paths, as they were unaware of the basic physical properties of electricity. They were in the process of discovering them, as there was no 'theory of electromagnetic wave propagation' to guide experiments before Maxwell's treatise and its verification by Hertz and others. Their experiments followed different trajectories. To mention a few examples, as we do not have any intention to be complete in this overview, we will explore the following early trajectories and their contributors.

---

[191] Source: https://www.princeton.edu/ssp/joseph-henry-project/felix-savary-1827/ (Accessed November 2015)

Trajectory of Aetherial Telegraphy

On April 30, 1872, the patent issued to *William Henry Ward*, US Patent № 126,356, for a telegraphic tower contained the early concept of wireless transmission. Although it was seen as an 'improvement in collecting electricity for telegraphing' and was considered to be a source of aerial electricity, it conceptualized wireless transmission:

> *In this manner a message would be sent entirely by natural electricity in place of artificial. In the same manner a message may be sent across the ocean by having a high tower on each continent, each of which towers would have to be, - of course, through land-lines connected with the ear to enable the ground current with the aerial current to form a circuit. Different towers may be erected on the different continents, and if they are all what is technically called hooked on - that is to say, connected to the ear - a signal given that one tower will be repeated that all the towers, they being connected with each other by the aerial current.* (text of patent)

The idea behind his concept was quite simple and taken from telegraphy, where, in electric systems used, the two-wire system was replaced by a single wire and using grounding the circuit as a 'second wire'. Now the first wire was also eliminated, and the atmosphere was used as a conducting wire, an idea that seemed to have originated from the *aurora borealis* effect that had been observed on telegraph lines.[192]

This inspiration to transmit through the atmosphere—the ether, as it was called in those days—was picked up by *Mahlon Loomis* (1826–1886), a dentist in Washington DC. He experimented in 1866 with kites as antennas and received US Patent № 129,971 on July 30, 1872, for a wireless telegraph. The basic concept consisted of towers placed atop tall mountains, which supposedly would draw upon a perpetual source of electrical current from the upper troposphere, while at the same time achieving long-range signalling by using an electrical layer in the atmosphere as a substitute for the standard telegraph wire. Again it was—in a one-page patent—about improvement in telegraphing without cables, using the free electricity of the atmosphere. From the mid-1860s until his death in 1886 Loomis made numerous unsubstantiated claims that he had actually used this method for long-distance wireless communication, at first telegraphic and later by wireless telephone. Despite his bold claims, Loomis was unable to get financial support from either the US Congress or from private

---

[192] The *aurora borealis* effect is the result of the ionization of air particles when protons from the sun hit the atmosphere. The best know form is the Northern Lights, when the atmosphere lights up with a greenish glow. In long telegraph lines, acting as an antenna, these magnetic storms induce a current that comes in waves. In 1859 a large aurora took possession of all telegraph lines in New England, blocking all transmission between stations.

sources. The same Congress that refused the Loomis Aerial Telegraph Bill (May 21, 1872) had previously funded a test of Morse's wire telegraph (Thomas H. White, 2015).

At that time there was a great interest in the phenomenon of the 'ether', the medium in which people breathe and live. Their interest was in determining what the medium that transported heat and lightning was, the medium that allowed Oersted's electro-magnetic induction.

Also, the Serbian *Nicola Tesla* (1856–1943), well known for his work on electric induction motors,[193] was experimenting in the late 1890s with alternating currents of high frequencies. His work resulted in a stream of patents. His 'Tesla coils', covered by US Patent № 512,340, granted on January 9, 1894, made it possible to send and receive

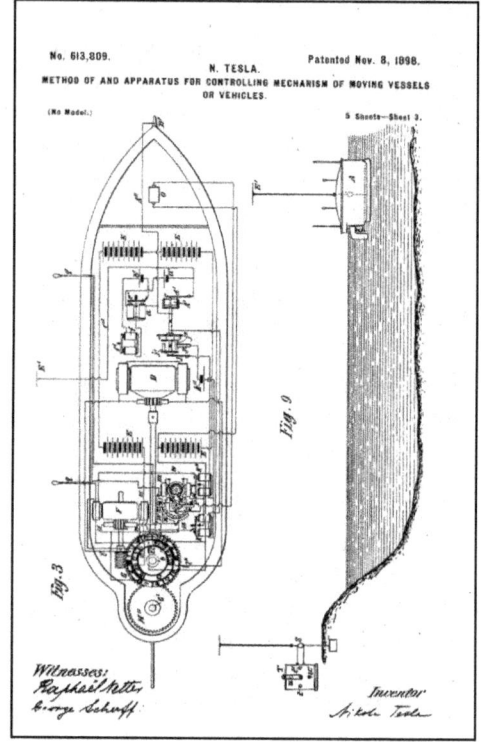

Figure 128: Figures from Tesla's US Patent № 613,809, dated November 8, 1898.

Source: USPTO.

high-frequency electromagnetic waves. He demonstrated this in 1898, when he showed a radio-controlled (model) boat for which he was granted US patent № 613,809 (Figure 128). He was also formulating ideas about wireless lighting and wireless electricity distribution (US Patent № 645,576, applied for on September 3, 1897, and granted on March 20, 1900).

Next, he tried to implement worldwide wireless telecommunication in his *Wardenclyffe Tower* project. This early wireless transmission station was built in Shoreham, New York, from 1901 to 1902, but it was not successful, and facing financing problems, the project was abandoned in 1906. But Tesla had his vision, much of which sounds close to what we experience today in the Internet Era:

---

[193] See: B.J.G. van der Kooij, *The Invention of the Electro-motive Engine* (2015), pp. 172–179.

The practical applications of this revolutionary principle [wireless art] have only begun. So far they have been confined to the use of oscillations which are quickly damped out in their passage through the medium. Still, even this has commanded universal attention. What will be achieved by waves which do not diminish with distance, baffles comprehension. ...

As soon as completed, it will be possible for a business man in New York to dictate instructions, and have them instantly appear in type at his office in London or elsewhere," Tesla said at the time. "He will be able to call up, from his desk, and talk to any telephone subscriber on the globe, without any change whatever in the existing equipment. An inexpensive instrument, not bigger than a watch, will enable its bearer to hear anywhere, on sea or land, music or song, the speech of a political leader, the address of an eminent man of science, or the sermon of an eloquent clergyman, delivered in some other place, however distant. In the same manner any picture, character, drawing or print can be transferred from one to another place. Millions of such instruments can be operated from but one plant of this kind. (Massie & Underhill, 1908, pp. 67-71)

Thomas Edison (1847–1931), in 1875, also spent some time contemplating the 'etheric forces' that were supposed to pervade the ether. He had observed that a rapidly vibrating spark created a spark in an adjacent relay. Edison announced the phenomenon to the press, and it was reported upon by *Scientific American* in 1875 that he had discovered a new force of nature—the etheric forces—that were subject to laws different from those of heat, light, electricity, or magnetism. He built a theory on it, but it was soon abandoned and not explored further on, mainly due to the opposition he faced among his peers (like the neurologist Georg Beard and the physicist Elihu Thompson) and due to his urgent work on telegraphy/telephony in preparation for the 1876 Philadelphia Centennial Exhibition (Wills, 2009).

However, he did pick up the idea of spark-induced transmission later in time. In 1886 Edison filed for a patent application (on December 29, 1891,

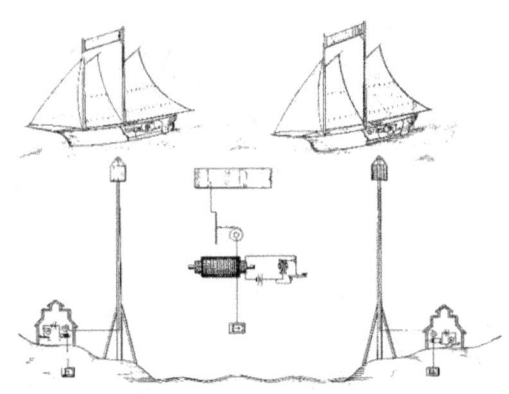

Figure 129: Part of Edison's patent № 465,971 (1891) showing wireless communication between ship and shore.

Source: USPTO. Wikimedia Commons

granted as US Patent № 465,971) on a system of electrical wireless communication between ships (Figure 129). The patent was based on mutual inductively coupled or magnetically coupled communication and was not related to the (high-frequency) electromagnetic waves caused by AC electricity. It caused a brief flurry of excitement in the maritime community, which saw it as a way to solve the maritime 'incommunicado' problems.

Trajectory of Electromagnetic Induction Telegraphy

Another development trajectory was followed by people like the Englishman *William H. Preece* (1834–1913), electrical engineer and chief engineer of the British Postal Telegraph. His experimenting into wireless telegraphy started in the early 1880s, when he was faced with a problem in telephone and telegraph communication in 1882: the mechanism of electromagnetic induction leading to the mixing of telephone and telegraph signals (aka crosstalk).

> *The discovery of the telephone has made us acquainted with many strange phenomena. It has enabled us, amongst other things, to establish beyond a doubt the fact that electric currents actually traverse the earth's crust. The theory that the earth acts as a great reservoir for electricity may be placed in the physicist's waste-paper basket, with phlogiston, the materiality of light, and other old-time hypotheses. Telephones have been fixed upon a wire passing from the ground floor to the top of a large building (the gas-pipes being used in place of a return wire), and Morse signals, sent from a telegraph office 250 yards distant, have been distinctly read. There are several cases on record of telephone circuits miles away from any telegraph wires, but in a line with the earth terminals, picking up telegraphic signals; and when an electric-light system uses the earth, it is stoppage to all telephonic communication in its neighbourhood.* (Fahie, 1899, p. 137)

He soon started experimenting with telegraphy based on his view of earth currents, trying to explain the strange phenomenon: it was the concept of 'electromagnetic induction' to be used for telegraphic communication.

> *In 1885 Preece arranged an exhaustive series of experiments in the neighbourhood of Newcastle, which were ably carried out by Mr A. W. Heaviside, to determine whether these disturbances were due to electro-magnetic induction, and were independent of earth conduction; and also to find out how far the distance between the wires could be extended before this influence ceased to be evident. Insulated squares of wire, each side being 440 yards long, were laid out horizontally on the ground one quarter of a mile apart, and distinct speech by telephones was carried on between them; while when removed 1000 yards apart inductive effects were still appreciable.* (Fahie, 1899, p. 145)

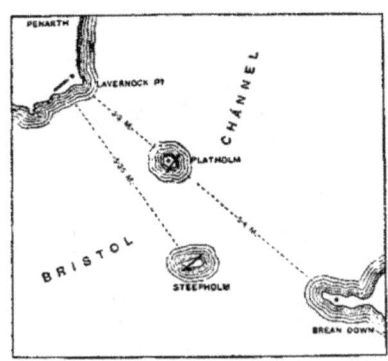

Fig. 20.

Figure 130: Preece's Bristol Channel experiments with wireless telegraphy (1892).

Source: Fahie, J.J., *A History of Wireless Telegraphy.*
http://earlyradiohistory.us/1901fa21.htm

Similar trials were held at other places over the coming years. The conclusion drawn from all these experiments was that the magnetic field extended uninterruptedly through the earth, as it did through the air. He then conducted the Bristol Chanel experiment in 1892 to find out if the same was the case in water (Figure 130).

The Bristol Channel proved a very convenient locality to test the practicability of communicating across distances of three and five miles without any intermediate conductors. Two islands, the Flat Holm and the Steep Holm, lie off Penarth and Lavernock Point, near Cardiff, the former having a lighthouse upon it. On the shore two thick copper wires combined in one circuit were suspended on poles for a distance of 1267 yards, the circuit being completed by the earth. On the sands at low-water mark, 600 yards from this primary circuit and parallel to it, two gutta-percha covered copper wires and one bare copper wire were laid down, their ends being buried in the ground by means of bars driven in the sand. ... The object of the experiments was not only to test the practicability of signalling between the shore and the lighthouse, but to differentiate the effects due to earth conduction from those due to electro-magnetic induction, and to determine the effects in water. ... The total absence of sound in the submerged cable was rather surprising, and led to the conclusion either that the electro-magnetic waves of energy are dissipated in the sea-water, which is a conductor, or else that they are reflected away from the surface of the water, like rays of light. (Fahie, 1899, pp. 148-149, 150, 151)

The induction-conduction concept, though working over short distances, was a dead-end technology for long-distance applications. Only specific circumstances—such as war—would enable systems based on this concept to be developed. However, it showed that wireless communication over distance was possible. Preece concluded:

Still, communication is possible even between England and France, across the Channel, and it may happen that between islands where the channels are rough and rugged, the bottom rocky, and the tides fierce, the system may be financially possible. It is, however, in time of war that it may become useful. It is possible to communicate with a beleaguered city either from the sea or on the land, or between armies separated by rivers, or even by enemies. As these waves are transmitted by the ether, they are independent of day or night, of fog, or snow, or rain, and therefore, if by any means a lighthouse can flash its indicating signals by electromagnetic disturbances through space, ships could find out their positions in spite of darkness and of weather. Fog would lose one of its terrors, and electricity become a great life-saving agency. (Fahie, 1899, pp. 158-159)

For a seagoing nation like Britain, the association of wireless communication and marine applications was something they could relate to, as ships, after leaving shore, were 'on their own without any means of communication' till they reached their destination port. It is not too surprising that he welcomed Marconi when he showed up in England in 1896 with another concept of wireless communication: the concept of electric waves.

Still, the concept of electromagnetic induction was experimented with in America later in time, but now it was not about telegraphy but about telephony. Another early pioneer in wireless communication, although he limited his experimentation to point-to-point telephony, was the self-taught tinkerer-experimenter *Nathan Stubblefield* (1860–1928). His public demonstrations in 1902 of a battery-operated wireless telephone received widespread attention (Figure 131). On March 8, 1898, Stubblefield was issued US Patent № 600.457 for an 'electric battery', which was an electrolytic coil of iron and insulated copper wire to be immersed in liquid or buried in the ground, where it could also serve as a ground terminal for wireless telephony. Using this system, he succeeded in 1902 in transmitting speech from a boat on the Potomac River to shore-based receivers (T. White, 1902, pp. 297-302).

Figure 131: Stubblefield using a ground-current wireless telephone (1902).

Source: Wikimedia Commons,

His invention resulted in the organization of the *Wireless Telephone Company of America*, with Mr. Stubblefield, participating with his invention, shortly as a director. The company was joined by A. Frederick Collins, owner of *Continental Wireless Tel. & Tel. Co.*, who had invented a similar system based on induction. However, the company proved to be one of the companies set up for selling stock (the wireless mania we will later touch upon). In 1911 some of its officials were charged with mail fraud. Four officers were convicted on all five counts, and three were fined and sentenced on January 10, 1913, to prison terms of up to four years. This was the end of the Continental Wireless Tel. & Tel. Co.[194]

Stubblefield, however, became distrustful of the promotors behind the venture and severed his connections with them. Continuing his experimenting, he received US Patent № 887,357 on May 12, 1908, for a wireless telephone system that used magnetic induction. This was soon followed by other patents; US Patent № 366,544, dated April 5, 1907, and Canadian Patent № 114,737, dated October 20, 1908. He did not succeed, however, in commercializing his latest invention, and he died in 1928. By that time other technologies for wireless communication had been developed, and his system based on induction, with a limited reach of three miles maximum, was outdated. (Fawcett, 1902, p. 363)

Those early experimenters all were trying to create a wireless communication system and thus avoid the problems of the cabled infrastructure. Then came the last decades of the nineteenth century. These were the years in which the theoretical knowhow about electricity increased rapidly. In their wake, the use of electricity as a carrier of information developed, both along experimental as well as theoretical ways. Let's, for a better understanding of the content of the efforts of all those curious men, try to explain some of the basic principles of electricity.

Discovering Wireless Signaling

The ideas of the tinkerers and thinkers might have been lovely, the experiments fun to execute, but they were not resulting in any practical development. Both the aerial and earth induction schemes passed into obscurity. Basically, this was because this form of induction communication had limited technical feasibility compared to the radiated high-frequency electromagnetic waves—aka 'electric waves'—as later developed by Herz/Marconi. But their efforts were exemplary for their time, when other inventive minds were also trying to get rid of the wires and developing ideas about what would be called 'radio waves' as they searched for them in the 1880s (Susskind, 1964).

It was *Amos Emerson Dolbear* (1837–1910), professor of physics at Tufts College near Boston, who had already conducted in the early 1880s experiments with induction coils in combination with electric sparks. And he had worked on a condenser 'microphone' receiver in a line system, using

---

[194] Source: 'The Collins Wireless Telephone'. http://www.sparkmuseum.com/collins.htm, (Accessed March 2015)

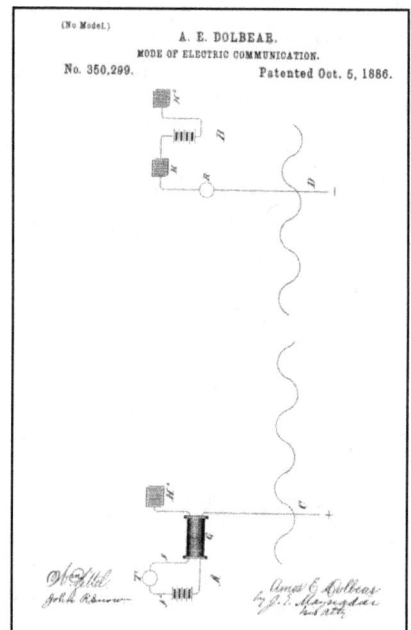

Figure 132: Figures from Dolbear's US Patent № 350,299, dated October 5, 1886.

Source: USPTO.

one line and the earth as the 'return line'. During his experiments he observed something he reported upon.

On March 23, 1882, Dolbear had read a paper in London "On the Development of a New Telephonic System." Many distinguished engineers were in the audience. He demonstrated his condenser microphone by setting up a dozen receivers in the lecture hall and putting the transmitter in an adjoining room, where another person counted to ten, whistled "God Save the Queen" and "Yankee Doodle," and performed on a cornet. As a climax of the demonstration, Dolbear said that he had found it unnecessary to have his device attached to the circuit at all; he disconnected it from the telephone wire and held it to the ear of the Chairman, who reported solemnly: "I hear the sound perfectly. It was about as loud in my ear as the cry of a new-born kitten." (Susskind, 1964, p. 38)

He applied for a patent on March 24, 1882, and was, some years later, granted US Patent № 350,299 on October 5, 1886, for a 'mode of electric communication' (Figure 132). In this patent, he claimed:

What I claim is – the art above described of communicating by electricity, consisting of first establishing a positive potential at one ground and a negative at another; secondly varying the potential of one ground by means of transmitting apparatus, whereby the potential of the other ground is varied; and lastly, operating receiving apparatus by the potential so varied, all substantially as described. (text of patent).

The concept was not implemented in any practical system, although this US Patent № 350,299 was later used in the priority battle with Marconi.

The Englishman *David Edward Hughes* (1831–1900), after returning from America to England in 1877 and experimenting with the microphone, also observed that sparks could create electromagnetic waves. When using his new induction balance across the room, he heard clicks in the telephone and correctly concluded they were produced by waves of electromagnetic energy as opposed to simple induction. To test this, he went from room to room in his house and then went outside into the London streets, listening to his battery-powered telephone receiver for what some call, in somewhat tongue-in-cheek fashion, 'the world's first mobile phone call'. He described it as follows in a letter to J.J. Fahie:

> *Further researches proved that an interrupted current in any coil gave out that each interruption such intense extra currents that the whole atmosphere in the room (or in several rooms distant) would have a momentary invisible charge, which became evident if a microphonic joint was used as a receiver with a telephone. This led me to experiment upon the best form of a receiver for these invisible electric waves, which evidently permeated great distances, and through all apparent obstacles, such as walls, &c. I found that all microphonic contacts or joints were extremely sensitive. Those formed of a hard carbon such as coke, or a combination of a piece of coke resting upon a bright steel contact, were very sensitive and self-restoring; whilst a loose contact between metals was equally sensitive, but would cohere, or remain in full contact, after the passage of an electric wave.* (Fahie, 1899, pp. 306-607)

These reports illustrate that—as a side effect of other research such as in the harmonious telegraphy and acoustic telegraphy—the existence of something like spark-generated electric waves had been observed. This was well before Hertz published in 1888 on his experiments on Maxwell's theory.

The preceding accounts provide an image of the 'electrical' scientists, who paid attention to the phenomena of the *wireless signalling*. As we see more in detail further on, within a decade many more would become interested in 'electric waves'. Some paid attention to the creation of the electromagnetic waves by a transmitter; others paid attention to the reception of the 'wireless' signal by the receiver, or they developed their own views on the phenomenon of electromagnetic waves that Maxwell had theorized about. For example, the American Amos Dolbear stated in 1893 after Herz's work on the 'Hertzian waves' became known:

> *A beam of Hertzian rays can be thus directed and not suffer so great loss, and being received by a proper electrical apparatus can be made visible; signals thus can be sent where the eye can see nothing in the space, and curiously enough such substances as wood and brick walls are transparent to such waves. There is no limit to this method but the curvation of the earth and the delicacy of the receiving apparatus.* (Dolbear, 1893, pp. 291-292)

Already in an early stage, much of the interest for the wireless communication was related to its maritime application, as identified by the

American Elihu Thompson in 1892 when he described a possible solution for many seagoing activities:

> *It may be remarked here, however, that electricians are not without some hope that signalling or telegraphing for moderate distances without wires, and even through dense fog may be an accomplished fact soon. Had we the means of obtaining electric oscillations of several millions per second, or waves similar to light waves, but of vastly lower rate of vibration, it might be possible by suitable reflectors to cause them to be carried a mile or so through a fog, and to recognize their presence by instruments constructed for the purpose. Many of the difficulties and dangers which now beset the navigator would, at least, be lessened, if not removed.* (E. Thompson, 1892, pp. 623-625)

Each scientist or engineer in his own way made his contributions, creating a scientific body of knowledge. All that was needed was the combination of all those elements into the *concept of wireless telegraphy* by spark-generated electric waves.

Principles of Wireless Signaling

We have seen by now some of the early contributors and their specific contributions in the different fields of wireless electricity that enhanced its body of knowledge. So, for a better understanding of their work, let us conclude by paying attention to some basics of those 'electric waves' that replaced the wires in the communication circuits like the telegraph.[195]

Firstly, one could ask, what is electricity? Alessandro Volta's discovery of electrochemical electricity resulted in the availability of the form of electricity called *direct current* (DC). Simply said, a direct current—of the free electrons in an atom—can be compared with a water current flowing from a high source through a pipe to a low point. The higher the source, the stronger the water flows. The same goes with the electrochemical battery: the higher its 'voltage', the stronger the current—of electrons—that flows directly to the other point (Figure 134, left). When the electromagnetic dynamo was developed, another form of electricity became available: *alternating current* (AC). As the result of a coil rotating in a magnetic field, the resulting current of electrons alternates pushing away from the source or retracting to the source (Figure 134, right). And the measure of the speed of the alternations as the result of the rotation speed of the dynamo is called the frequency.

---

[195] This segment is intended to increase the basic understanding of the non-technical reader of the phenomena the tinkerers and thinkers observed. As electricity is something invisible and hard to understand, we try to keep it as simple as possible.

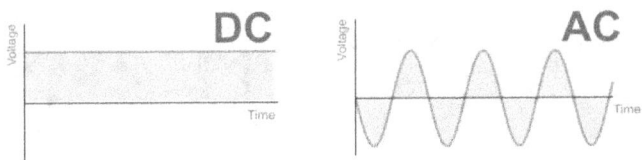

Figure 134: Principle of direct current (DC) and alternating current (AC).

Now, let's look at the use of DC electricity. As a starter, we take the example of an electric doorbell. Such an electrical circuit always has two wires connecting the battery with the components of the system (top of Figure 133). In simple terms, it works as follows. When the switch (A) is closing the electrical circuit, the electric current leaves the battery and activates the bell (B). The second wire returns the current to the battery. The principle is simple: pressing a button closes an electric circuit. Repeatedly closing the circuit creates a *pulse train* of electric currents.

This principle is the same for the basic wired telegraphic communication circuit. The transmitter (the telegraph key) is connected by two—sometimes quite long—wires to the receiver (such as the relay in a Morse telegraph apparatus). One wire is for the transmission of the pulse train of the electrical current; the other is the return wire. However, using the earth as the return wire makes it possible to eliminate one wire. So, the coded message is transmitted along *one* wire over a long distance as a pulse train (bottom of Figure 133).

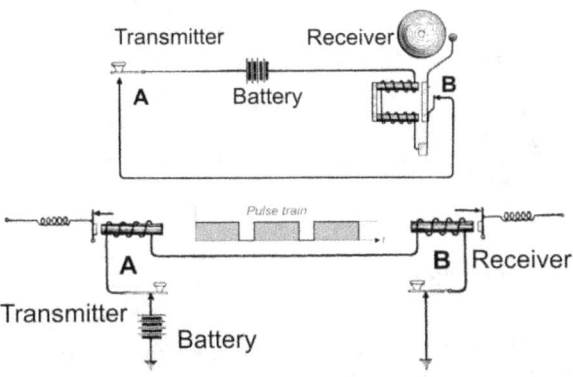

Figure 133: Principle of Distant Sounding.

System with a bell circuit (top) and a wired—two-way—telegraphic circuit using the earth as 'return line' (bottom). Pressing discontinuously the switch (A) creates a pulse train of electric currents activating the 'relay' (B)

Source: http://www.gutenberg.org/files/15617/15617-h/15617-h.htm

With wireless systems, this is not different. One can create a communication circuit with a transmitter and a receiver, and something in the 'ether' replacing the wire, to transmit the coded information. Both the transmitter and

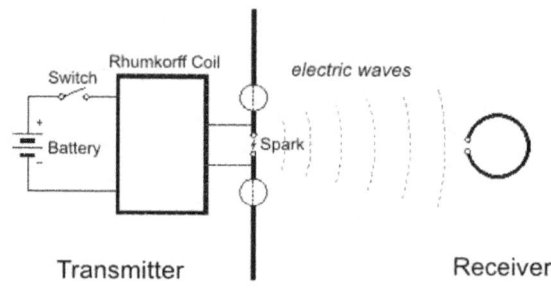

Figure 135: Principle of electric wave transmission.

Based on figures from Fillipovic, M., *Radio Receivers, From Crystal Set to Radio*.

the receiver are electrical circuits on their own, and the electromagnetic induction (like the forces of a magnet) or electromagnetic waves (like the rays of light) in the ether transmits the message (Figure 135).

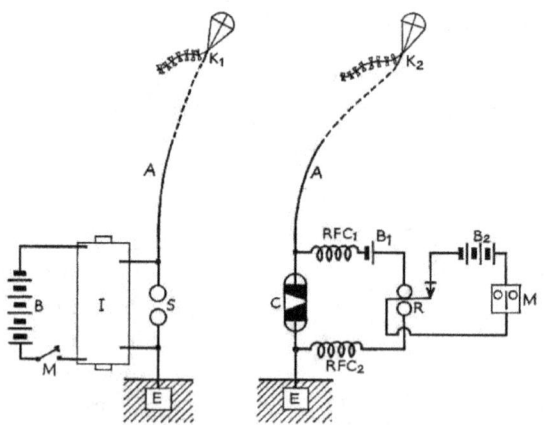

Figure 136: Principle of wireless communication circuit with transmitter (left) and receiver (right).

Pressing in the transmitter (left schematic), the button (M) closes the electrical circuit and creates a spark (S), which is converted by the antenna (A) into electromagnetic waves. In the receiver (right schematic), these waves, picked up by the antenna (A), are detected by the coherer (C) and processed in the receiver circuit. The ground connection (E) closes the circuit between the antennas (A) supported by the kites (K).

Source: http://www.newscotland1398.net/nfld1901/marconi-nfld.html

In the case of electromagnetic induction, the magnetic field created by a coil in an electric circuit executes a force on nearby metallic parts, such as in an electromechanical relay (Figure 133, item B), where the magnetic field attracts the metal part connected to a spring. However, electromagnetic induction also works over larger distances, but that requires large antennas (tens of kilometers in length) that can only be laid upon the terrain.

That is different in the case of electromagnetic waves. As shown in Figure 136, the transmitter (left) has a 'communication' circuit with the receiver (right) through a circuit between the antennas (A) and the earth (E). The circuit is E-S-A-K1-K2-A-C-E. At the transmitter side, the antenna is brought up vertically into the air by a kite (K1/K2). The sparks (S) emit the electric waves that are received in the antenna, also hanging from a kite, at the right. For creating the signal and for detecting the signal, separate electric circuits are applied. To detect the electromagnetic waves, one needs a *detector*, similar to the human eye, which is a detector of visible light waves. To emit the electromagnetic waves, one needs a *transmitter*, similar to a candle emitting visible light (Figure 138).

By their nature, wireless systems with a transmitter and a receiver are AC-based systems. The spark in the transmitter generates, from the DC current, the electric waves of a range of different frequencies emitted by the antenna (Figure 135). The receiver detects those electric waves with a decoder. The decoder changes its resistance—acting like a switch—and influences the current in the DC circuit. Between the transmitter and receiver, the electric waves are transmitted in the air. The farther they are away from each other, the better. Heinrich Hertz had a transmitter and receiver (Figure 146) that worked in one room. Marconi went outdoors (Figure 161), increasing the distance between transmitter and receiver by creating higher antennas, as we will see later on.

The Frequency Spectrum of Electromagnetic Waves

Another question is how the electric wave is created. In the early days of wireless communication a specific method of creating the electromagnetic waves was applied: the *spark generator* (Figure 137): a rather simple mechanism consisting of a battery, a capacitor, a coil (aka the Rhumkorff coil, after its inventor Heinrich Rhumkorff) and an interrupter. The combination of the coil and capacitor creates—when the circuit is interrupted—a spark emitting electromagnetic waves.

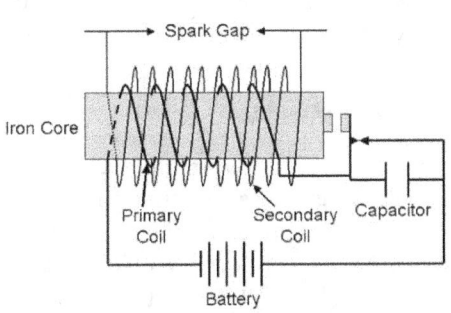

Figure 137: Rhumkorff Coil.

Electric waves are created from a spark gap in the secondary coil. In the primary coil the current is intermittently interrupted, causing magnetic changes in the iron core.

Source: unknown.

In order to create those electromagnetic waves, an interruption of an electric

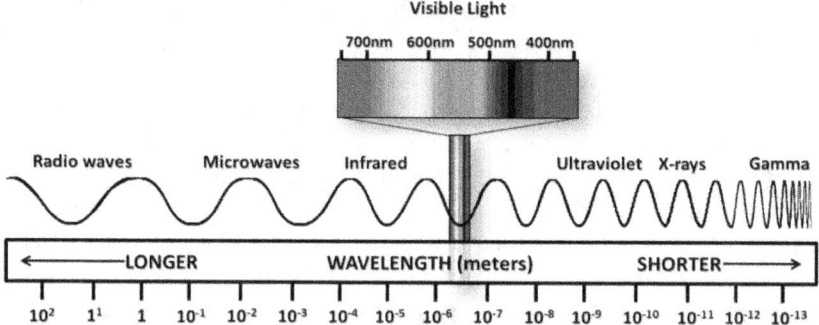

Figure 138: The Electromagnetic spectrum with the range of visible light and the range of radio waves.

Source: http://www.ces.fau.edu/nasa/images/Energy/VisibleLightSpectrum.jpg

current in a circuit was needed. That interruption of the 'force' (voltage) of a battery in the primary circuit of a transformer created in the secondary circuit a voltage high enough to ionize the air between two contacts (aka the spark gap) (Figure 137). The electric wave was created.

Just as white light is a mixture of many light frequencies—the well-known rainbow colors— the spark-generated 'electric waves' (aka 'radio waves') came in a mix of different frequencies. That spectrum of electric waves (aka radio waves) went from the long radio waves (Figure 138, left) to the—later applied—short waves (microwaves). The shorter waves (and thus of higher frequency) covered a different spectrum of electromagnetic waves, one that was discovered later (eg the X-rays in Figure 138, right).

Principle of Spark-induced Electromagnetic Waves

By the 1870s creating sparks was like controlling that mighty force of nature called 'lightning'. And using it to illuminate was even more magical. But now we had entered the unknown space with invisible electric waves. Just by pressing a switch, the result was the burst of electromagnetic waves that 'damped out' over their short lifetime (Figure 139, top). It was not a burst of a single frequency, but it was a spectrum of frequencies (Figure 139, bottom). Nevertheless, these undesirable characteristics, in the case of (Morse-) coded transmission, the short and long bursts of 'un-tuned, damped' electric waves—the dots and dashes—worked to create a transmission.

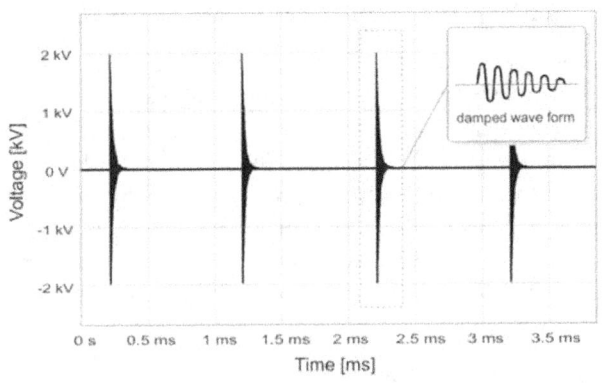

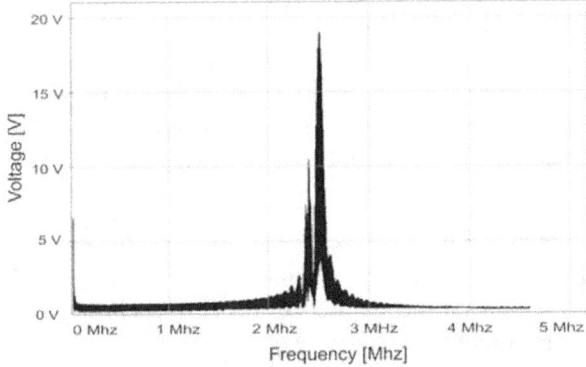

Figure 139: Multi-frequency (bottom) and damped electric waves (top) created by a spark gap generator.

Source: http://www-emt.tu-
ilmenau.de/ukolos/largerimage/sparkgap.php

The electric wave resulting from the spark could be transmitted into the air by capacitor plates working as an antenna. So, early 'directly coupled' transmitters consisted of a spark generator and an antenna created by those capacitor plates. Later, to control the burst-like emission of the electric waves, 'tuned' circuits appeared, the tuning being realized by adding resistance (R), capacity (C), and inductivity (L) to the circuit. The first development trajectory of wireless telegraphy after its invention was the improvement of the electric wave signal.

These spark-gap transmitters, emitting the *Hertzian waves*, were the dominant technology for the first three decades of radio (1887–1917), also called the early wireless telegraphy or 'spark' era. Because they generated damped waves, spark transmitters were electrically noisy; their energy was spread over a broad band of frequencies, creating radio noise which interfered with other transmitters. Later, a different transmission concept in regard to the generation of electromagnetic waves was found in the spark-less transmitter.

Principle of Continuous Electromagnetic Waves

The development of the spark-generated multi-frequency electric wave system resulted later in the constant-wave systems. Now the oscillations were generated by mechanical means: an *electric alternator* (a reciprocal electric motor that generated electricity). The specially designed alternators needed higher rotations and more coils to create the alternating high-frequency currents. The advantages of these electric alternators was that they created a continuous flow of 'un-damped' waves.[196] The transmission of information was realized, similar to the spark-based system, by switching the frequency on or off.

The mechanical problem, however, was to create an alternator that could create frequencies above 15 kHz with enough power to cover large distances. Due to their size and weight, their use was limited for long-wave communication by shore stations. It was these *continuous wave* transmitters that enabled the high-powered transmission stations. However, this mechanical technology did not survive the mainstream development of wireless transmission as it was replaced by the 'resonant oscillator'.

Basically, it was the 'sparks' that generated waves: in the visible spectrum (aka electric light) and invisible to the eye (aka electric waves). These sparks were noisy as they hummed, hissed, and howled, and they were smelly and uncomfortable to the eye. They were the smaller equivalents of those mighty lightning bolts that frightened people over the ages.

To surmount the damping effect, experiments with capacitors and coils were undertaken. This led to the 'arc oscillator': a combination of an arc transmitter with capacitors (C), coils (L), and resistances (R) across the carbon arc electrodes: the tuned spark systems (Figure 140).

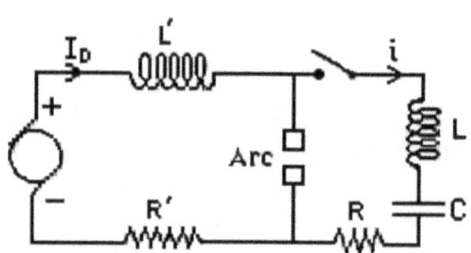

Figure 140: Principle of arc oscillator as used by Duddell.

Source: Unknown

These were the approaches to improve on the spark-generated wave by added components to the circuit. Another development focused on mechanical solutions. After early experimenting by Elihu Thomson (1853–1937) and later by William Duddell (1872–1917), it was the Dane Valdmar Poulsen (1869–1942) who

---

[196] Continuous waves are electromagnetic waves of a single constant frequency.

created the rotating arc transmitter; a continuous wave radio transmitter of often quite large mechanical construction. He was granted US Patent № 661,619 on November 13, 1900, for his invention.

The Principle of the Receiver

How was the electric wave detected? As we will see, for detection, an 'antenna' was needed. In Hertz's case, it was a simple coil-like device: the receiving loop (Figure 146). The antenna changed over time to a wire suspended in the air. The receiver was improved when a variable conductor (aka coherer) was applied in the circuit (Figure 136, right). When the electric wave created an electromagnetic field that was strong enough, the originally high conductivity of the coherer would drop. This change could trigger a circuit (Figure 141, top). The coherer (C) forms a circuit with the battery (B1), the coils (L), and the relay (R). When the coherer is triggered by the electric wave picked up by the antenna, the relay (R) is activated. Basically, this is the receiver function. Later that was improved when the detection action became the trigger for another action in the second circuit of battery (B2) and relay (S). And that circuit could include a Morse writing device to register the actions of the receiver.

The sensitivity of the receiver for electric waves was problematic, so a lot of effort was put into improving the receiver, improvements in the form of additional passive components such as a transformer (Figure 141, bottom).

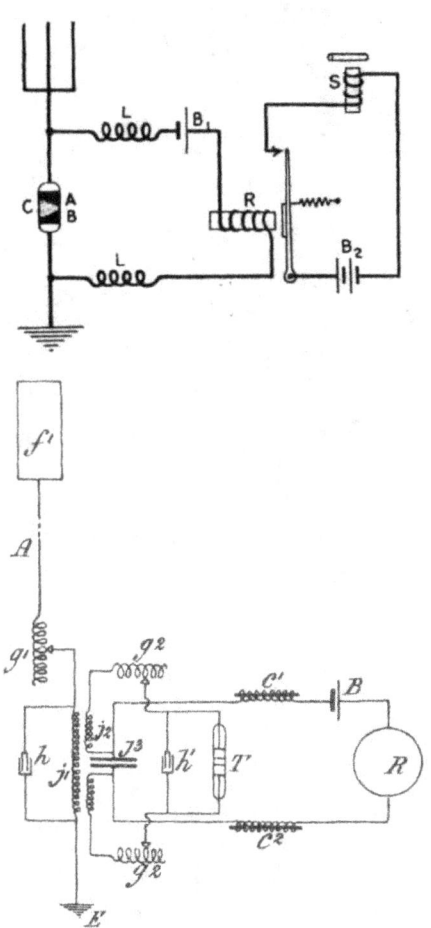

Figure 141: The receiver circuit.

Source: Unknown

The Electronic Oscillator

The coherer-based receivers were followed by other variations on the same theme. Then came the invention of the electronic oscillator. The word 'electronic' was applied because the receiver circuit used a new 'electronic' device: the diode (aka rectifier).[197]

This type of device originated from the cat's whisker diode: a thin wire touching a crystal (Figure 142, top). As a spin-off effect of the development of Edison's incandescent lamp, the vacuum tube rectifier (eg Fleming's thermionic valve) was developed. When this component was applied to a receiver circuit, it would act as the decoder—replacing the coherer—of an electric wave (Figure 142, middle). Soon the diode-based receiver was improved upon when Lee de Forest developed the 'triode' called 'Audion' (Figure 142, bottom). It was a leap forward in the development of wireless communication.

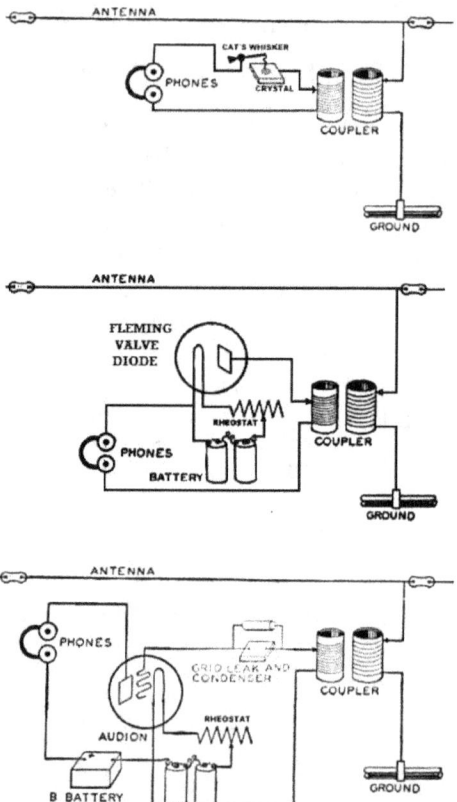

Figure 142: Principle of a receiver.

Receiver with a cat's whisker (top), with a diode (the Fleming valve, middle), and with a triode (Audion, bottom).

Source: http://www.geojohn.org/Radios/ MyRadios/Coherer/Coherer.html

Short Waves and Radio Broadcasting

Another giant leap in wireless communication would be the use of electric waves with a higher frequency: short-wave communication. Using the new vacuum tube, circuits were developed that could transmit electromagnetic waves of frequencies that followed the curvature of the

---

[197] The diode is a device that conducts an electric current primarily in one direction depending of the connection to a battery. This property was the basis for the name 'valve'.

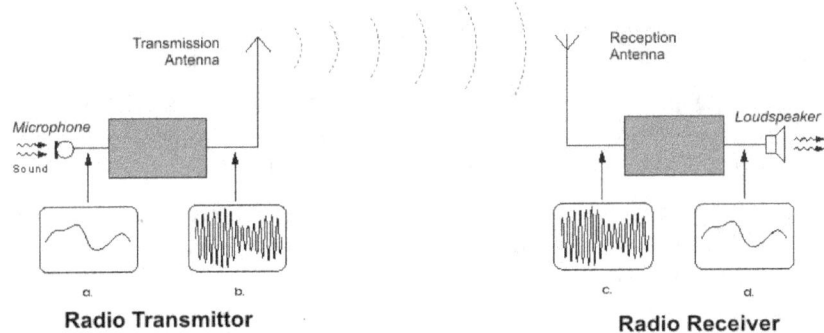

Radio Transmittor

Radio Receiver

Figure 143: Principle of Sound Transmission with Amplitude Modulation.

Based on figures from Fillipovic, M., *Radio Receivers, From Crystal Set to Radio.*

earth and consequently could be transmitted over large distances. Short-wave communication developed after the Great War and lay at the foundation of another communication development: the point-to-many communication that became known as radio broadcasting.

Quite logically, the next step after wireless telegraphy (aka radio telegraphy) was the wireless broadcasting of sound (aka radio telephony). The 'single wave' electric waves would be used very well for the transmission of sound (Figure 143). In the radio transmitter, the sound signal from the microphone (a) would be mixed—modulated—with a continuous wave. It resulted in an amplitude modulated signal (b) that was transmitted by the antenna. In the receiver circuit, this signal would be decoded—or 'demodulated'—back to the original sound.

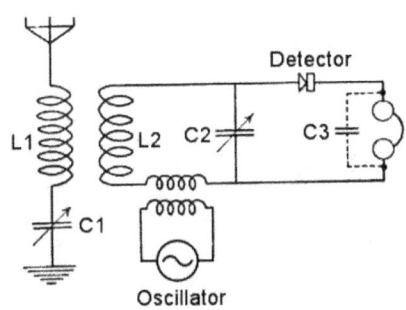

Figure 144: Principle of heterodyne receiver.

Source: Wikimedia Commons

The mixing of frequencies, as realized by the so-called 'heterodyne' circuit (Figure 144), was discovered by Reginal Fessenden. It was the start of radio broadcasting and wireless telephony. However, to get momentum, this development needed the invention of other components, such as Fleming's thermionic valve and Lee de Forest's Audion.

Later in time, after these discoveries, other systems for creating and detecting electromagnetic waves became available, but their origin was simple. Just as the incandescent lamp (an electric-powered red-hot wire enclosed in a glass vessel emitting visible frequencies) succeeded the spark light, a glass-enclosed vessel (the vacuum tube) would succeed the spark-based electric wave generator. But that is another story...

This concludes our short technical (and quite incomplete) overview of basics of electric waves and their circuits for transmission and detection. It is time to look at those who contributed to the invention of wireless telegraphy.

The Invention of Wireless Telegraphy

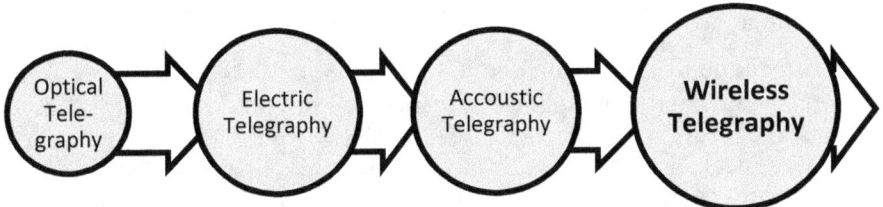

Development of communication over long distances started with *optical telegraphy* (the semaphore system of Claude Chappe) at the beginning of the nineteenth century. His mechanical system, when electricity became available was followed by galvano-electric telegraphy (the needle telegraph of Cook and Wheatstone) and electromagnetic telegraphy (the relay telegraph of Samuel Morse) in the 1840s.[198] It took a few more decades, but by the 1870s the coded signal of dots and dashes was replaced by electric speech (ie the human voice) by Alexander Graham Bell: acoustic telegraphy (aka telephony) was born.[199] By that time communication over long distances—telecommunication—had changed society, and electricity had played a crucial role in its development. One could say that the Era of Telegraphy and the Era of Telephony had started a Communication Revolution. Technical developments had been accompanied by developments in societies and their economies, that created the context for the innovation to come.

---

[198] Basically the galvanometer is a coil that moves in a stationary field between to magnets. And a relay is a magnetised iron object that moves under the influence of a magnetic field created by a stationary coil. See: B.J.G. van der Kooij, *The invention of the Communication Engine 'Telegraph'* (2015).
[199] See: B.J.G. van der Kooij, *The invention of the Communication Engine 'Telephone'* (2015).

Then came the Era of Wireless Telegraphy. The early development of the wireless telegraph is the story of electromagnetic waves: how to create them, how to use them, and how to receive them. As we have analysed elsewhere, it is a story that found its origins in the work of Hans Oersted in 1820,[200] when he published about his experiments with an electric current and a compass needle.[201] It was a sensation in the scientific world of that time, from Paris to Berlin and St. Petersburg, from London to Geneva and Turino. His original experiments were soon to be followed by the experiments of other curious scientists all over Europe. Among them was Michael Faraday, who by then was working on the phenomenon of magnetic induction. He concluded in the 1830s that a force between magnetism and electricity existed.[202] He observed the changes in that force, which, taking some time to interact, resulted in wave motions like 'the vibrations upon the surface of disturbed water'.[203] Faraday had created the 'lines of force' with a circular motion: the magnetic curves that were moving in an unknown medium called the 'ether'.

That was quite in contradiction to the then-popular Newtonian thinking of forces acting in a straight line (the 'action at a distance' concept with its universal gravitation). In the 1850s Faraday's work, published in his *Experimental Researches in Electricity*, inspired the Scot James Clerk Maxwell (1831–1879). He transformed the 'lines of force' into a mathematical approach that explained how they propagated.[204] But in what medium were the waves propagated? Was it one made up of the classical 'fluids', or was it the problematic medium called 'ether'? And, what was the nature of the medium in which those waves propagated? Maxwell developed from his mathematical explanations a theory published in a paper called 'A Dynamical Theory of the Electromagnetic Field' (1864) linking the phenomena of light with those of electricity and magnetism.[205]

---

[200] See: B.J.G. van der Kooij, *The Invention of the Electro-motive Engine* (2015).

[201] H.C. Oersted, *Experimenta circa effectum Conflictus Electrici in Acum Magneticam* (*Experiments in the Effect of a Current of Electricity on the Magnetic Needle*), Copenhagen, 1820.

[202] M. Faraday, *On the Induction of electrical currents and of the Evolution of Electricity from Magnetism* (Faraday, 1832).

[203] For reasons of his own, on March 12, 1832, Faraday deposited his observations from his 'Experimental Researches in Electricity', due to a controversy about priority with Humphry Davy, in a sealed document to be guarded in the strongbox of the Royal Society. When the document was opened more than a hundred years later—on June 20, 1937—this text was part of his original views expressed.

[204] J.C. Maxwell, 'On Faraday's Lines of Force' (read 1855/1856) (Clerk Maxwell, 1864). Part of *On Physical Lines of Force*.

[205] With the publication of *A Dynamical Theory of the Electromagnetic Field* in 1865, Maxwell demonstrated that electric and magnetic fields travel through space as waves moving at the speed of light. Maxwell proposed that light is an undulation in the same medium that is the

This linking was a breakthrough in scholarly thinking as he characterized the 'space that contains and surrounds bodies in electric or magnetic conditions'. His work became the foundation of electromagnetic theory, and it was later published in a classic textbook, *A Treatise on Electricity and Magnetism*, in 1873. His contributions to the unification of light and electrical phenomena are considered to be of the same magnitude as those of Isaac Newton (1642-1726) and—later in time—Albert Einstein (1879-1955).

His theory excited many scholars, who became known as the 'Maxwellians'. Although his mathematical approach was for many scientists hard to grasp, some more experimental British scientists like David Edward Hughes came close to creating electromagnetic waves, as did the more theoretical scientists like George Francis Fitzgerald (1851–1901),[206] Oliver Lodge (1851–1940),[207] and Oliver Heaviside (1850–1925),[208] and the Dutchman Hendrik Antoon Lorentz (1853–1928)[209]. Although they all expanded on Maxwell's theories, it was not till some years later that these theories were solidly confirmed. That was done by the German Heinrich Hertz (1857–1894) who, in the late 1880s, after studying Maxwell's theory, showed that electromagnetic waves could be created and detected.

In parallel to the proof of the experimental work of Heinrich Hertz—cut short by his premature death at the age of 36 in 1894—it was Oliver Lodge who played a vital role in the practical radio communication that followed, as we will see further on.

---

cause of electric and magnetic phenomena. The unification of light and electrical phenomena led to the prediction of the existence of radio waves.

[206] G.F. Fitzgerald, *On the Electro-Magnetic theory of Reflection and Refraction of light* (Fitzgerald, 1880).

[207] As early as 1879 Lodge became interested in generating (and detecting) electromagnetic waves, something Maxwell had never considered. Quite some time later he published a series of papers proving the existence of electromagnetic waves and their propagation in free space. On June 1, 1894, at a meeting of the British Association for the Advancement of Science at Oxford University, Lodge gave a memorial lecture on the work of Hertz, who had recently died.

[208] Oliver Heaviside published numerous articles in *The Electrician* about his electromagnetic theories from 1882 to 1902, such as 'Electromagnetic Induction and Its Propagation' in 1885, 1886, and 1887 and 'Electromagnetic Waves, the Propagation of Potential, and the Electromagnetic Effects of a Moving Charge' in 1888 and 1889.

[209] Lorentz was primarily interested in the theory of electromagnetism to explain the relationship of electricity, magnetism, and light. His work was the final point in the development of the classical aether theories, and he contributed to the thinking of special relativity. He received the Nobel Price in Physics in 1902 together with the Dutchman Pieter Zeeman.

Early Days of Wireless Communication

In its essence, the development of the wireless telegraph is a story about the replacement of the transmission carrier of wired telegraphy: the electric cables that by then spanned the globe and enabled communication around the world. It is also a story of how wireless transmission complemented cabled telegraphy. From the origins of the electric bell that could ring at a short distance, to the telegraphic writing at a large distance, it was the cable that conducted the information carried by electricity. Next to the transmitter at the beginning point and the receiver at the end point, the cable was the connecting medium in point-to-point communication. Except for the fundamental contributions of a few individuals—such as Morse and Cooke and Wheatstone[210]—its development over time, although rather important, was not one heralded in the scientific annals of invention. Its overall development was more the result of the practical works of the engineers whose thinking and tinkering solved the problems at hand.

So, basically, the telegraph is a device for the transmission of information from one point to another point—at a long distance—by means of a wire. It is a so-called *wired point-to-point communication* device. And those 'points' show its limitations, as they are both fixed in their position. That also goes for the cable that connects the points; it is fixed (often on poles). Cabled telegraph thus is a static system of communication that is limited to land application and that is not usable in mobile situations. So, ships, being mobile 'points' that cannot be connected by a cable, are excluded from cabled telegraphy.

It is not hard to image, however, that after the first cabled transmissions between two points, there soon appeared the next concept of sending messages from one point to many points: the *wired (one-way) point-to-many communication*, also called 'broadcasting', where information is distributed to many recipients. Take, for example, the 'financial' information of stock brokers and the stock markets that were distributed using special systems. Such as the ticker telegraph developed by Thomas Edison for the Gold and Stock Telegraph Company in 1867. Or imagine the distribution of news items by telegraph from one source to many editorial offices of local newspapers. Such as done by the news company called the Associated Press (1860s).

It was not only the telegraph that saw this development of 'broadcasting', as the telephone, which came on stage some decades later, underwent the change from the point-to-point concept to the point-to-many concept. Take the American *Telephone Herald*, where a range of

---

[210] See: B.J.G. van der Kooij, *The Invention of the Communication Engine 'Telegraph'* (2015).

'programs' was 'broadcasted' by telephone to a large audience of subscribers. There was also the Hungarian *Telephon Hirmondo* news service, as developed by Tividar Puskas, and the French *Theatrophone*, created by Clément Adler—both demonstrated at the Electrical Exhibition in Paris in 1881.[211] In the fall of 1876 experimental 'concerts' were already being transmitted over wire line by Bell from Paris to Brantford, Ontario, utilizing a triple mouthpiece telephone transmitter to accommodate several soloists. From 1876 through 1880 a variety of broadcasting transmissions were conducted, both in America and in Europe.

Surprisingly enough, this concept of broadcasting was not the focus of early wireless communication. When the use of electric waves came to its full existence in the first decade of the twentieth century, it was seen as a replacement for and/or addition to wired telegraphy. This was a replacement that came about due to a range of inherent properties, such as the fact that wireless telegraphy could service places unserviceable by cabled systems, among them the mobile maritime applications with their ship-to-shore and ship-to-ship communication. And those communications could be long-distance *point-to-point* transmissions, as in the case of the massive steamships crossing the Atlantic Ocean. Ultimately, after wireless communication had earned its place as an addition for cabled communication, it would create the novelty of broadcasting. But before that was going to happen, quite a few other events happened.

Experimenting to Prove a Theory

The preceding examples illuminate the development of the communication systems in different *application trajectories* and *technological trajectories*. They have one thing in common: they all were based on the wired transmission concepts that originated from the early days of the telegraph system and developed into the telephone system. As one realizes that a transmission system using wires (both in plain air and underground/ underwater) in its essence is limited in application to fixed points, expensive to install and to maintain, and vulnerable to disruptions caused by vandalism or weather conditions, then it is not hard to understand that inventive minds tried to circumvent the wire for communication purposes. They were looking for the ways and means of *wireless communication*.

As the electro-physicists of the late seventeenth century progressed with their experiments, discovering the nature of lightning (Figure 123), it became obvious that 'electricity' and 'electromagnetism' were closely related. This resulted in the 1840s in the telegraphic apparatus based on the

---

[211] See: B.J.G. van der Kooij, *The Invention of the Communication Engine 'Telephone'* (2015), pp. 283–287.

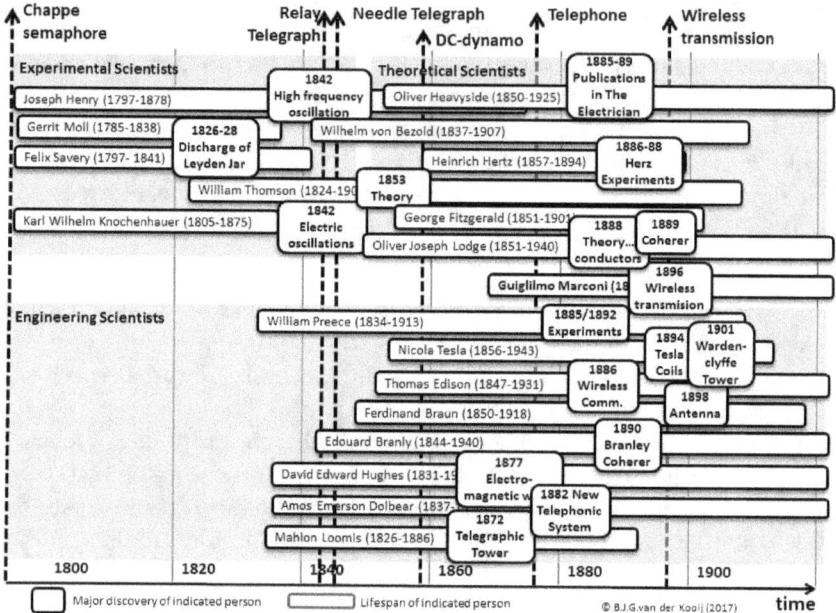

Figure 145: Overview of early contributors to the wireless transmission.
Figure created by author.

galvanometer principle (used by Cooke and Wheatstone) and the telegraphic apparatus based on the *electromagnet principle* (used by Morse). Further experimenting by many creative minds with transmitting musical tones over telegraph lines (Gray, Bell, and others) resulted in the acoustic telegraph using 'undulating currents'. Their systems were all wire-based connections.

After it also had become clear that electric sparks created more than just the waves of visible light, those invisible 'electromagnetic waves' were explored by a range of theoretical and practical scientists (Figure 145). They followed different approaches, from the more theoretical approach of the Maxwellian scholars to the practical development trajectories creating components for the creation and detection of the electric waves. Trajectories that paralleled the further development of the telegraphic and telephonic systems in the latter part of the nineteenth century.

We will address that further on, but let's first go and explore that fundamental experimental work of Heinrich Hertz.

Heinrich Hertz: Spark-generated Electromagnetic Waves

In the late 1880s it was the German physicists *Heinrich Hertz* (1857–1894) who proved the existence of electromagnetic waves as previously theorized by James Clerk Maxwell and published in his *Treatise on Electricity and Magnetism* in 1873.[212]

Heinz was the son of Dr. G.F. Hertz, a lawyer who became a senator of the Hanseatic city of Hamburg. His mother, Anna Elisabe Pfefferkorn, was the daughter of a physician, Dr. Johann Pfefferkorn. Not too surprisingly, their son 'Heinz' was a smart kid with a phenomenal memory, and he possessed an almost insatiable appetite for learning. After he had gotten a private education in his youth, he attended the Hamburg Gymnasium, spending a year as an apprentice engineer and a year in military service. In 1877 he was faced with a decision about his future: would it be natural sciences, or would it be civil engineering? His choice was for science, and he enrolled at the Polytechnicum in Munich, changing a year later to the University of Berlin. There, in 1880, Heinrich Hertz obtained his PhD, and for the next three years he remained for post-doctoral study under professor Herman von Helmholtz,[213] serving as his assistant. In 1883 Hertz took a post as a lecturer in theoretical physics that the University of Kiel. There, he studied Maxwell's theories and published about it: *On the relations between Maxwell's Fundamental Electromagnetic Equations and the Fundamental Equations of the Opposing Electromagnetics*. In 1885, at the age of 28, Hertz became a full professor at the Technical High School (ie University) of Karlsruhe.

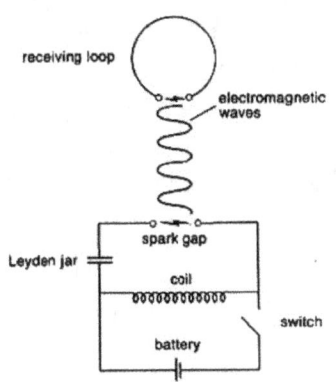

Figure 146: Principle of wireless transmission as discovered in the Hertz experiments.

Source: http://people.seas.harvard.edu/ ~jones/cscie129/nu_lectures/lecture6 /hertz/Hertz_exp.html

There, he observed in the autumn of 1886 that the discharge of a 'Leyden jar' (a capacitor, as we would call it today) over a 'Knockenhauer spiral' (an inductive coil) resulted in a spark that could cross a gap between two wires: the 'spark gap'. He had created a 'spark

---

[212] See: B.J.G. van der Kooij, *The Invention of the Electro-motive Engine* (2015).
[213] Helmholtz was interested in electromagnetic oscillations and created the Helmholtz resonator, a device creating tones. He wrote 'On the Sensation of Tone' that would contribute to the development of the telephone by Alexander Graham Bell. B.J.G. van der Kooij, *The Invention of the Communication Engine 'Telephone'* (2015), pp. 116–117.

oscillator' (aka 'transmitter') that could generate electromagnetic waves of a high frequency: it was to become the basic principle of the wireless transmitter (Figure 146). And in the same time he created the receiving loop (aka the 'receiver') that would be activated by receiving the electromagnetic waves (G. R. Garratt, 1994, pp. 34-50).

Between 1886 and 1889, being fully conversant by now with Maxwell's theory and dedicated to applying it, Hertz would conduct a series of experiments that would prove that the effects he was observing were results of Maxwell's predicted electromagnetic waves. Thus, in 1887 Hertz tested Maxwell's hypothesis. He used an oscillator made of polished brass knobs, each connected to an induction coil and separated by a tiny gap over which sparks could leap (Figure 147).

Hertz reasoned that if Maxwell's predictions were correct, electromagnetic waves would be transmitted during each series of sparks. To confirm this, Hertz made a simple receiver of looped wire. At the ends of the loop were small knobs separated by a tiny gap. The receiver was placed several yards from the oscillator. According to theory, if electromagnetic waves were spreading from the oscillator sparks, they would induce a current in the loop that would send sparks across the gap. This indeed occurred when Hertz turned on the oscillator, producing the first transmission and reception of electromagnetic waves.

Figure 147: Experimental Hertz oscillator (1888).

The battery (bottom right) supplies electricity to create a spark in the gap (a-b) in the electric circuit with the capacitors (A and B) and the Rhumkorff induction coil (bottom left).

Source: http://www.wdr5.de/sendungen/leonardo/
hertzheinrich108_v-ARDFotogalerie.jpg

Hertz published his findings in a paper called 'Ueber die Ausbreitungs-geschwindigkeit der elektrodynamischen Wirkungen', in early 1888 and in a book titled Untersuchungen Ueber die Ausbreitung der Elektrischen Kraft (Investigations on the Propagation of Electrical Energy) (Hertz, 1894). The scientific world became quite excited about his experiments on 'action at a distance': it originated the 'wireless mania' in science.

> *Great interest was excited by the experiments of Hertz, primarily on account of their immense scientific importance. It was not long, however, before several eminent scientists perceived that the property possessed by the Hertz waves of passing through fog and material obstacles made them particularly suitable for use for electric signaling.* (Fessenden, 1908, p. 555)

Oliver Lodge: Theoretical Understanding

It was the Englishman *Oliver Joseph Lodge* (1851–1940) who would contribute to the theoretical understanding of the new phenomenon discovered by Hertz. Lodge, born in Stoke-on-Trent, England, was the son of Oliver Lodge, a merchant who supplied pottery firms with raw materials. At the age of 14 he entered his father's business, but his interests lay elsewhere. Soon he was able to attend physics lectures in London, combining this with working for his father. After being enrolled at the University College in London in 1873, he became acquainted with Maxwell's theory. Lodge obtained a bachelor of science degree from the University of London in 1875 and a doctor of science in 1877. By that time he was fully conversant with Maxwell's theory. Soon he was appointed professor of physics and mathematics at University College, Liverpool, in 1881. It was in those years that he became acquainted with Heinrich Hertz.

> *While waiting to take up his Liverpool appointment, Lodge undertook an extensive European tour primarily to make personal contact with leading continental scientists to whom he carried letters of introduction. Calling without notice on the great professor Helmholtz in Berlin, Lodge fount him on the point of going out. After a few words, Helmholtz handed him over to his demonstrator, Dr Heinrich Hertz, then a young man of twenty-four, with whom Lodge quickly established a friendship which was to be closely maintained until Hertz untimely death in 1894.* (G. R. Garratt, 1994, p. 34)

Lodge became interested in electromagnetic waves in 1879. His first experiments were related to the phenomena of the destructive forces of 'electric lightning', which could be prevented by applying 'lighting conductors' to deflect the lightning to earth. Over the years he developed his scientific career, and by 1887, the 36-year-old Oliver Lodge was regarded in Great Britain as a highly accomplished scientist. He had become an expert on lightning and was invited by the Royal Society of Arts to present lectures on it, which he did in Bath in September 1888 at a meeting chaired by Fitzgerald.

In 1888 he also published 'On the Theory of Lightning Conductors' in *Philosophical Magazine* and concluded that a lightning flash consisted of an

Figure 148: Lodge coherer (1894).

Source: http://www.sciencemuseum.org.uk/

oscillating discharge (as in alternating current) and was not a simple discharge of current in one direction (as in a direct current). On March 8, 1889, Lodge held a lecture at the Royal Institution on the topic of 'The Discharge of a Leyden Jar' where he dwelled on the phenomenon of radiation and 'sympathetic resonance', comparing electrical radiation with the radiation of heat. He was even granted a British patent (GB Patent № 5,844 of 1889) for 'Protecting Electrical Instrument from Lightning'. Being aware of Hertz's work, he soon was experimenting with Hertz's discoveries. He devised a spark-gap detector he called the 'coherer',[214] to be used to detect electromagnetic waves generated by a spark discharge (Figure 148).[215] In 1890 he published a description of 'syntonic jars'; in fact, these were modified 'Leyden jars' (today called capacitors).

From all these public demonstrations, lectures, and publications, it is clear that he contributed before Marconi to the detection of electromagnetic waves by tuning a circuit ('syntony', as it was called), but he did not proceed with any commercial development and lost the opportunity to acquire popular acclaim as the 'inventor' of wireless telegraphy (G. R. Garratt, 1994, pp. 51-70).

On June 1, 1894, at a meeting of the British Association for the Advancement of Science at Oxford University, Lodge gave a memorial lecture, 'The Work by Hertz', on the ideas developed by Hertz and the German physicist's proof of the existence of electromagnetic waves six years earlier. Lodge set up a demonstration on the quasi-optical nature of

---

[214] The coherer is a frequency-controlled resistance where the conductive medium is metal filings contained in a glass tube. This early detection device predates vacuum tubes, magnetic detectors, and crystals as detectors. Their use even predates radio. It was the first detector used for radio reception.

[215] The basis for the operation of the coherer is that metal particles 'cohere' (cling together) and conduct electricity much better after being subjected to AC electricity. In rest, it has a high resistance, but when the high-frequency signal is detected, the coherer becomes conductive due to its low resistance. This attribute can be used to operate a switch. To mechanically restore the high conductivity, a decoherer mechanism was needed, often a tapper arm on a bell-like relay.

the 'Hertzian waves' and demonstrated their similarity to light and vision, including reflection and transmission over distances. On August 14, 1894, another demonstration showed that Hertzian waves had travelled a distance of 55 meters after being produced in another building by an assistant, and they could be detected by his receiving apparatus, called the 'coherer'.

> *Lodge clearly had all the necessary elements of an elementary wireless telegraphy system. While it could be argued successfully that Lodge did indeed achieve signalling of a sort in all three of these demonstrations, there is no indication that the sending of any true messages was accomplished or even attempted with this apparatus. It was not his intent to do so. Oliver Lodge never considered using his equipment for communicating ...* (Rybak, nd)

However, since this scientific demonstration was one year before Marconi's 1895 Pontecchio demonstration of a system for radio wireless telegraphy and contained many of the basic elements that would be used in Marconi's later wireless systems, Lodge's lecture became the focus of priority disputes with the *Marconi Company* a little over a decade later, when the priority discussion raged over invention of wireless telegraphy.

In May 1897 Lodge applied for a provisional patent protection in Great Britain (GB Patent № 11,575 of 1897) for 'Improvements in Syntonized Telegraphy without Line Wires'. This would become one of the most famous and fundamental patents in the history of wireless communication (G. R. Garratt, 1994, p. 67), Later he was granted US Patent № 609,154 on August 16, 1898, for his invention of syntonic tuning, in which he claimed:

> *What I claim, and desire to secure by Letters Patent of the United States, is—*
> *1. In a system of Hertzian-wave telegraphy, the combination, with a pair of capacity areas, of a self-inductance coil inserted between them electrically for the purpose of prolonging any electrical oscillations excited in the system and constituting such a system a radiator of definite frequency or pitch.* (text of patent)

Sometime later, at the time of the dispute, some people, including the physicist John Ambrose Fleming, pointed out that Lodge's Bath lecture in 1888 was a physics experiment, not a demonstration of telegraphic signalling. But in the British Court of Law, it was decided differently.

> *The final chapter of Lodge's pioneering work towards the development of radio-communication was not written until 1911 when an application for an extension of his master patent of 1897 came before Mr. Justice Parker. The validity of the patent was upheld, Lodge's claim to priority in the principles of resonance and selective tuning was established, and the exceptional extension of seven years was granted, mainly on the grounds that the inventor had been inadequately remunerated in consequence of the virtual monopoly created by the Marconi-*

Lloyds agreement and by the refusal of the Postmaster-General to grant a license for the working of the system in this country [England]. (G. R. Garratt, 1994, p. 69)

The British Controversy of the Practice versus Theory Debate

One of those scientists that attended the Lodge presentations was *George Francis Fitzgerald* (1851–1901*)*, professor in natural and experimental philosophy at the School of Engineering, Trinity College in Dublin. He had already written to Hertz on June 8, 1888:

> *I saw the other day that Prof. von Helmholtz announced your splendid verification of Maxwell's theory that electromagnetic disturbances are propagated with the velocity of light. You have been so kind as to send me copies of some of your former papers, for which I now thank you as I ought to have done before. Would it be too much to ask you to send me a copy of your paper describing how you have verified Maxwell's theory? I consider that no more important experiment has been made is century.* (O'Hara, 2007, p. 547)

Together with other British professors in physics like Oliver Lodge, Silvanus Thompson, and Oliver Heavyside, Fitzgerald had become a leading figure among the group of 'Maxwellians' who revised, extended, clarified, and confirmed James Clerk Maxwell's mathematical theories of the electromagnetic field during the late 1870s and the 1880s. And then came that contribution of the German Heinrich Hertz, published in 1888.

> *The British Maxwellians gave Hertz's experiments an enthusiastic welcome. Before 1888 their work had been considered quite esoteric and hypothetical; they were sometimes dismissed, according to Heaviside, as "working out a mere paper theory." Hertz's work gave the Maxwellians an answer to this reproach by putting the theory on a solid experimental foundation. As Heaviside put it, "the very slow influence of theoretical reasoning on conservative minds was enforced by the common-sense appeal to facts," and the Maxwellians could now attract attention and compel assent to the theory from those who were unswayed by its mathematical beauties. Hertz had given the Maxwellians a powerful tool with which to promote their theory, and they used it.* (Hunt, 1983, p. 344)

Clearly, the 'nature of electric waves' had now become a topic of scientific interests in the community of British scientists. It was the theoretical scientists who sought to understand and explain this aspect of the 'nature of lightning' and its properties by experiments. They became known as the 'scientific electricians' (Figure 145, top right).

The telegraph had been, since its invention, one of the areas of the application of electricity that was advancing rapidly. It was followed by the advance of telephony. The relatively simple digital 'dots and dashes' of the

(DC) telegraphic transmission were replaced by the multi-frequency (AC) currents of the telephonic transmission of electric speech. It brought many practical problems, and soon it was the practitioners of electricity—our engineering scientists, also known as the 'practical electricians' (Figure 145, bottom)—who were encountering problems in the field related to AC (alternating currents) and long transmission lines.

In England, the institution of the Post Office—as the monopolist for telegraphy and by that time the owner of the originally privately owned Telegraph Companies—made up the community of practitioners encountering these problems. Their spokesperson was William Preece (1834-1913), the chief electrician at the Post Office, in charge of the country's telegraphs and telephones. He was well known by his function and by his lectures, such as 'Recent Wonders of Electricity I/II' delivered before the Society of Arts, December 28, 1881, and January 4, 1882.[216]

> *The electriciens prided themselves with being 'practical men' of vast experience in the design and erection of lightning conductors over a period of many years and much of their criticism was based upon the fact that Lodge had developed his theory around a few simple laboratory experiments in which he employed sparks of only an inch or two in length. They argued that he was quite unjustified in extrapolating the results of such experiments to embrace a discharge with the dimensions of a lightning flash.* (G. R. M. Garratt, 1994, p. 59)

Preece, by that time, as we have seen and will explore later on in more detail, was experimenting with *magnetic induction* and *electric conduction* to realise 'inductive' wireless transmission (Figure 130). He experimented with a wireless system crossing the Solent between the UK mainland and the Isle of Wight in 1882 and continued those experiments until the 1890s.

> *To most "practical men" like Preece, a current in a wire was much like the flow of water in a pipe. There might be some modifications-picturing the pipe as elastic, to simulate capacitance, for instance, or filled with baffles to simulate resistance-but it was basically a simple and intuitive picture, and a remarkably useful one.* (Hunt, 1983, p. 345)

Thus, the electricians understood that electricity had to do with resistance and capacitance.[217] But next to the resistance and the capacitance

---

[216] Preece had also published on electricity, such as his contributions to the magazine *Popular Science Monthly* (1882).

[217] Electrical impedance is the measure of the opposition that an electrical circuit presents to an electric current when an electric voltage is applied: it consists of resistance, capacitance, and inductance. Resistance is denoted by the letter R, which represents the relation between electric current (I) and electrical voltage (V). Capacitance is denoted by the letter C and represents the behavior of current over time. Inductance, denoted by the letter L, is the opposition to a current over time.

of the transmission line, there was something else that influenced the behavior of electric currents over long distances: the self-inductance that acted like a flywheel to oppose any change in the current.

> *"Preece had long held that self-induction ("retardation") was the enemy of the telegrapher and the telephonist, and he asserted that long-distance telephony was only possible over lines of very low self-induction. He had experiments done, and from the quality of speaking on copper lines he argued that these must have a very low inductance. So, in his opinion, the transmission cables had to be thick."* (Hunt, 1983, pp. 346-347)

However, the more scientifically oriented professor Oliver Lodge did not agree with him and proposed thin wires.

> *In a Leyden jar or lightning discharge through a small resistance, the current oscillated, according to Lodge's calculations, with a frequency of about a million cycles per second. At such frequencies the self-induction would, in the Maxwellian view, not only become an important factor in opposing the rush of the current, but would completely overshadow the ordinary resistance of the wire. The contrast with Preece's view becomes clear: to him, a lightning conductor should simply be able to convey a large amount of electricity quickly, rather like a fat drainpipe. But according to Lodge, the real aim should be to minimize the self-induction. Providing a conductor of very low resistance was at best a waste of copper, Lodge said, since the self-induction would dominate the situation anyway.* (Hunt, 1983, p. 346)

Then it was the scientific hermit *Oliver Heavyside* (1850–1925), originally a rather obscure scientist of electromagnetism, who explained the phenomenon of 'self-induction': electrical energy. In his view, electricity was conveyed not by the current flowing in the wire, but by the ether in the space surrounding it (later called the 'skin effect'). The consequence being that the relatively high inductance of the copper lines, along with their low resistance, made them good for telephony. So, the cables should be thin. That view—to be expressed in a paper called 'The Bridge System of Telephony'—did it for the practical men; they blocked Heavyside's publications, and the controversy was born.

> He [Heavyside] *had found mathematically from Maxwell's theory that distortion could be lessened, even eliminated, by properly loading cables with extra self-induction. In an article sent to The Electrician in 1887, he described Preece's paper in characteristically blunt terms as "radically wrong . . . in methods, reasoning, results, and conclusions." But that article was not published. Preece had taken measures to keep Heaviside's work on self-induction from seeing print. . . . Preece and his "practical men" would not stand for a "mere mathematician"*

like Heaviside intruding self-induction and Maxwellian theory into the preserves of engineering practice. (Hunt, 1983, pp. 347-348)

Essentially, the explanation of the behavior of AC current due to resistance (R), capacitance (C), and inductance (L) seemed the basis of the controversy that arose between the two communities: the theorists versus the practitioners. However, the controversy went deeper, as it had been slumbering already for a long time.

In 1866 S. A. Varley was among the inventors of the self-exciting dynamo, work for which he felt he never received proper credit. He often expressed the belief that pioneers like himself were being pushed aside by glib-talking, mathematics-spouting professors of physics, many of whom, it seems, were Maxwellians. These professors, Varley claimed, were trying to establish a monopoly of authority; reaching deep into the bag of traditional British invective, he denounced Maxwellianism as "scientific Popery." Maxwell's field theory, Varley said, "rests solely on hypothesis," and Hertz's experiments "do not seem to bear very directly on practical electrodynamics." (Hunt, 1983, p. 350)

Quite different in rational standpoints but with a highly emotional undercurrent, the controversy got out of hand. It ended up in a total and ugly controversy between the theoretical scientists, who were not that much interested in developing a practical system, and the practical engineers of that time, who had practical implications (eg cost-effectiveness, reliability) as their objectives.

Practical men like himself knew that alternating current and related developments were bringing new phenomena into play, but they did not want to give up their old, common sense ways of dealing with electricity. They had been working with electric currents for years with great success, and they resented "the professors" telling them they had been all wrong. They did not, as Varley had said, want to take a back seat to the upstart Maxwellians. The practice versus theory debate was an attempt by the practical men to hold their own ground against the incursions of the scientists. (Hunt, 1983, p. 351)

As we will see further on, this controversy influenced also the later experiments on wireless transmission, both by Preece and by Lodge. It was by that time that a young Italian man called Guglielmo Marconi came to England with something quite new. His invention was met with ambivalence: by the scientific community with restraint and by the engineering community with enthusiasm.

Marconi had every reason to be suspicious of scientists and their complicated theories. Everything he had accomplished so far had been either ridiculed or at most considered of no particular interest by most mainstream scientists. No scientific theory was available to explain how or why his system worked at all but work it did. Marconi, in an extreme way, represented the triumph of the practical empiricist over the theoretical scientist, a characteristic that also certainly endeared him immediately to Preece who had been embroiled in such controversies long before Marconi entered the scene. (F. P. Marconi, nd)

But before examining how that contribution of Marconi was going to happen we have to go back in time to the foundations of wireless communication.

Development of Spark-based Signaling Systems

It had started in the early days of electricity, when all those curious people were investigating the 'nature of lightning'. The experimental scientists of those days investigated the phenomena of static electricity, natural electricity, and electrochemical electricity. The engineering scientists created working artefacts based on those phenomena, and it was not until quite sometime later that the theoretical theorists started developing their related theories (Figure 149).

In the second half of the nineteenth century, some 50 years after Volta invented his electrochemical battery around the 1800s, the *general purpose technology* (GPT) of electricity started outgrowing its early explorative years.

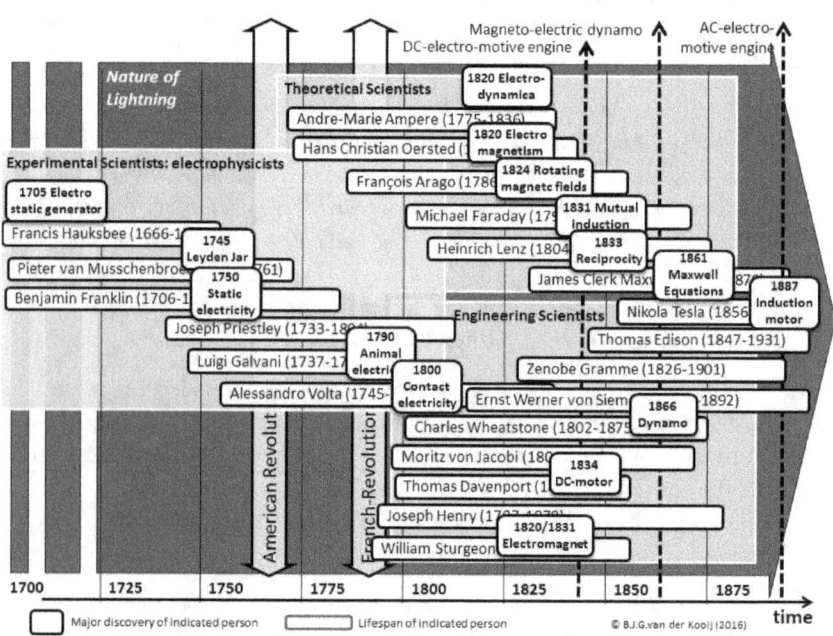

Figure 149: Overview of investigations in the nature of lightning.

Figure created by author.

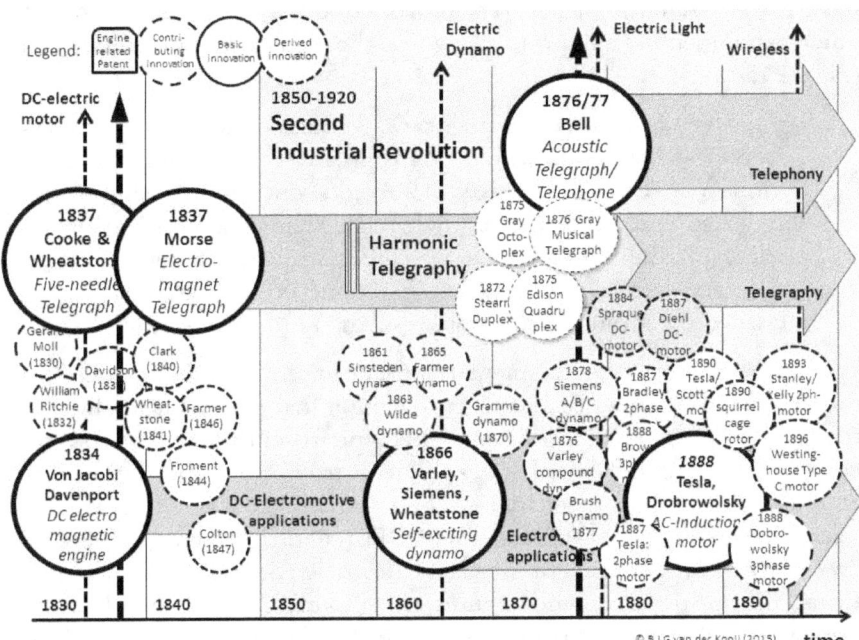

Figure 150: Overview of GPT-Electricity until the 1890s in relation to inventions in telegraphy and telephony.

Figure created by author.

As scientists began understanding better the fundamental mechanisms of electricity, there was a rush of renewed experimenting by practical engineers, resulting in a stream of new inventions (Figure 150). Among those inventions, we find the inventions related to the creation of electricity: the electromagnetic dynamos. First came the direct current versions (DC); later the self-exciting alternating current versions (AC). Soon an abundance of electricity became available, and the limiting properties of batteries were lifted. It fascinating application already being illustrated in London at the Great Exhibition of 1851, electricity was now becoming used in daily life, both professional as well as private. Soon the new dynamos were powering electric motors and creating electric light. It was the time of the Second Industrial Revolution of which the GPT-Electricity was such a great contributor.

Sure, theory development and engineering practice were important. Furthermore, it was the commercial activities of the inventor-entrepreneurs of that time that contributed highly to these developments, which were soon penetrating the daily life of common people. From their original ideas, these inventor-entrepreneurs created their concepts that materialized—after much experimenting and prototyping—as the archetypes of communication (Figure 150).

Many of them were Americans. Samuel Morse had created his relay-based telegraph with his transmission code, starting the monopoly of the *Morse system* that would conquer the technical world. In a similar way, Alexander Graham Bell had created the acoustic telegraph: the telephone that was the basis for the monopoly of his *Bell system*. And Thomas Edison had done similar activities and created the incandescent lamp, resulting in the *Edison system*.

Being aware of the fundamental impact of their inventions, foreseeing their commercial potential and eager to exploit that entrepreneurially, they focussed on the land many of them had come from: Britain. As a former colony of Britain, the United States still had strong cultural, emotional, and entrepreneurial bonds with Britain. There, experimental activities by scientists and engineers had, as we have illustrated elsewhere,[218] a massive awareness of the potential of the GPT-Electricity. Soon Edison, taking British patents on his invention and starting to exploit them in Britain, came in conflict with the British exploring electric light such as Joseph Swan and his incandescent lamp. Alexander Graham Bell, doing the same and patenting his inventions in Britain, was confronted with the monopoly of the British Post Office (Figure 210). All in all, it was in this dynamic context that the young Marconi entered on stage when he landed on the English shores in 1896. He would soon become entangled in a web of conflicting scientific aspirations, engineering frustrations, and political and economic interests.

The Wireless Mania in Science

When Marconi was busy in 1895 creating his early prototypes for wireless communication, he already was concerned that somebody else would doing the same and would be developing a working apparatus based on the increasing body of scientific knowledge on the Hertzian waves. Indeed, he was not the only one who had the idea to use Hertzian waves for distant wireless communication. Hertz had initiated a 'wireless mania' in both the scientific community of conceptual thinkers and in the engineering community of practical tinkerers with his ideas about spark-generated electric waves. The wireless mania was not restricted to one country. The telegraph having proved its importance worldwide, interest in wireless communication was all over Europe and the United States, even as far as the British colony of India.

---

[218] See the other publications in the *Invention Series*.

The search was basically a continuation of the exploration of the 'nature of lightning' (Figure 149) that had started with Benjamin Franklin's experiment into static electricity, a search in which curious people were looking for devices to trap the lightning and others were experimenting with creating 'sparks'. Sparks were interesting to scientists and engineers alike, as the idea to be able to control and use one of nature's most powerful—and frightening—forces fascinated many people. Its bright light, the massive sound, and the destructive force of lightning led their investigations into a range of research trajectories: obviously the trajectory of the visible effects such as light,[219] and soon also the trajectory of the effects resulting in movement at a distance.[220] That unsuspected side effect of 'distant action' was observed by many as they experimented with sparks. This became the early research trajectory of 'wireless signalling' and 'remote control'.[221]

Thus, the phenomenon of distant action could also be used for communication purposes. Just as an electromagnetic relay can be used as a distant detector of a wired transmission of coded information (eg Morse telegraph), the same applies for wireless transmission. Creating spark at one end would start the transmission, but how to detect the signal at the receiving end? Trying to create an answer to that problem created the research into the spark-gap detector called the cymoscope (ie the detecting telegraphic 'cymoscopes', later called 'coherers'[222]). The parallel search for devices to generate electric waves by 'lightning sparks' (later to be known as the Hertzian wave generators) would be its companion. In total it was the *Hunt for the Electric Waves* so many people were involved in. To illustrate this hunt, we will mention some of those curious people who, more or less, are related to the further developments around Marconi's monopoly to come and were active at the moment Marconi filed his provisional specification for a patent on March 5, 1896, and till he was granted the patent on June 2, 1897 (Figure 151).

---

[219] See: B.J.G. van der Kooij, *The Invention of the Electric Light* (2015).

[220] The interest originated from early observation that in experiments, 'sparks' had an effect on distant metal parts during their emissions of light. Next to the research trajectory of the spark light (Humphry et al), it created the research trajectory of electromagnets (Sturgeon et al) and the research trajectory of the electric waves (Hertz et al). Experimenting with electromagnets ultimately created wired telegraphy, while experimenting with electric waves ultimately created wireless telegraphy.

[221] As is often the case, the phenomenon of 'action at a distance', attracted people with a military background. They wondered what one could do with such a thing, like create new weaponry that could explode when commanded from a distance by remote control.

[222] Cymoscope is the name for any device detecting electric waves. Basically, a Cymoscope is a device that changes its conductivity under the influence of electric waves.

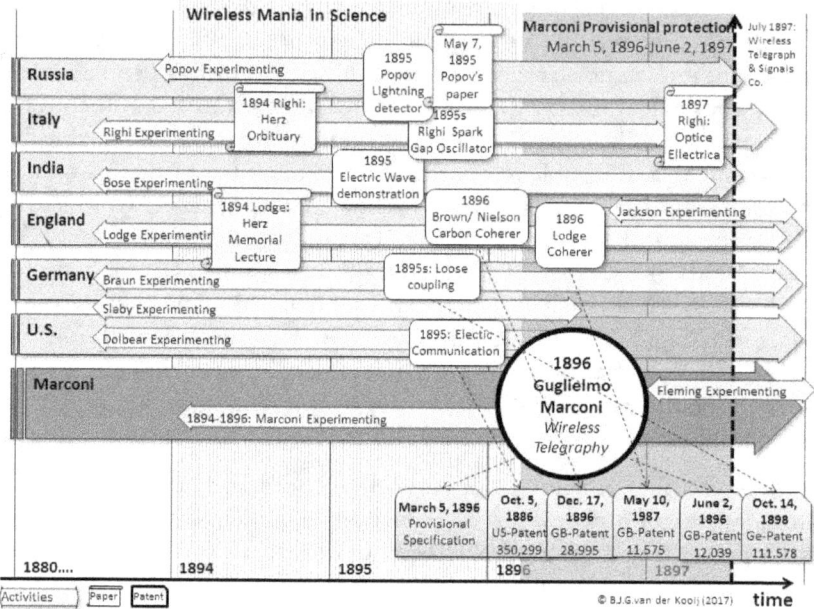

Figure 151: Contributors to the Wireless Mania in Science.

Figure created by author.

In England, *A.C. Brown* and *G.R. Nielson* worked on the early lightning detector using carbon as a filling for the detector tube.[223] They already had filed their provisional specification as early as December 17, 1896: GB Patent № 28,995. They did not realize the importance of the antennae and earth connections, but they expanded on the coherer already developed by Édouard Branly in 1890 (Figure 152).

In England, Captain *Henry Jackson* of the Naval Torpedo Training School was also experimenting with Hertzian waves and the way they were influenced by obstacles, creating apparatus for wireless transmission and demonstrating them on August 20, 1896. *Oliver Lodge*, after becoming interested in wireless telegraphy and developing his Lodge coherer, applied for a provisional patent protection, which he filed on May 10, 1897, as GB Patent № 11,575 for 'Improvement in Syntonized Telegraphy without Line Wires'.

---

[223] This was not so strange, as carbon was also used by Edison for his electric incandescent lamp. And later it was used in the development of the microphone by Hughes and Berliner.

In India, Professor *Jagdish Chandra Bose* was experimenting with electric waves. Around 1895 he demonstrated the existence of microwaves by ringing a bell at a distance. He also worked on a detector for electric waves: the Bose coherer. However, as a scientist, he was not interested in commercial telegraphy. Later he had a paper read at the Royal Society in London on April 7, 1889, about an 'iron-mercury-iron coherer with telephone detector'.

Those were the early beginnings in which experimenting with the cymoscope dominated. When 'distant communication' became a topic of interest, the mania expanded worldwide. It was not so much more about 'lightning', now it was about electric waves themselves. The experiments increased in number and in focus.

In Italy, professor *Augusto Righi* had already, in 1887, experimented with a spark-gap sphere oscillator (aka the 'Righi oscillator'), creating shorter electric waves, and published results of his work in a treatise, *Optice Elettrica*, in 1897.

In Germany, it was Professor *Carl F. Braun* of the University of Strasburg, who, in association with Siemens and Halske from Berlin, experimented with oscillating circuits. He introduced the closed-tuned circuit in the generating part of the transmitter. He added to the antennae a Leyden jar (capacitor) and an induction coil. Later he added inductive coupling in the transmitter to create the oscillations induced by the spark. In following patents, he expanded on this concept and filed for German Patent № 109,378 of January 26, 1899. It resulted in quite a complex apparatus, as in 1899 he created a spark-less antennae circuit. There was also *Adolf Slaby*, who had witnessed Marconi's demonstration for the Post Office, and his pupil *Georg von Arco*. Together, they worked on aerial arrangements and receiver improvements.

In America, where electric telegraphy was in full development, many scientists and engineers were aware of Hertz's work. For example, there was *Nicolas Tesla*, who worked with currents of high frequency and developed his grand scheme of space telegraphy: his World Wireless System for the transmission of electric energy without wires, as well as point-to-point wireless telecommunications. Similar work was touched upon by *Thomas Edison*, but he abandoned the trajectory, as he was much more interested in the development of electric light. They were not the only Americans; Amos Dolbear, John Stone Stone, Reginald Fessenden, and Lee de Forest were, each in their own way, involved in spark-generated wire telegraphy. Many of the work was done on the early detectors that had emerged from the lightning experimenting. *Amos Dolbear*, for example, was granted US Patent № 350,299 on October 5, 1886, for a 'mode of electric communication' (Figure 132). *Lee de Forest* worked on improving the early receivers and wave detectors.

There was also the New Zealander *Ernest Rutherford* (1871–1937), later First Baron Rutherford of Nelson, graduated from Canterbury College in Christchurch (NZ) in 1893. In 1894 he won a scholarship to the Cavendish Laboratory in Cambridge University in England, where he studied under J.J. Thompson. Being knowledgeable about Hertzian waves, he experimented with the magnetization of iron by the discharge of a Leyden jar, thus, by a spark with high-frequency emissions of electric waves. He found that when he changed the magnetic intensity (the cause), the changes of the magnetic induction (the effect) were lagging: the hysteresis effect.[224] From these experiments, he developed a device for the detection of electromagnetic waves—using the hysteresis property of iron—that would replace the unreliable coherer (Rutherford, 1897).

He also found that if a magnetized steel needle or a bundle of extremely thin iron wires were magnetized and placed in the interior of a small coil, the ends of which were connected to two long collecting wires, then an electric wave started from a Herz-oscillator at a distance caused an immediate demagnetization of the iron. Although Rutherford's detector was purely a laboratory demonstration, it created the basis for Marconi's radio wave detector (called 'Maggie') that was developed in the early 1900s.

These rough brushstrokes create a painting of the chaotic development of the 'wireless scene' as it emerged before, during, and after Marconi's invention in 1896. Clearly, many thinkers and tinkerers were actively trying to exploit this new application of the GPT-Electricity: the 'wireless mania' in the scientific community would start the third phase of the Communication Era. We will meet them again further on.

Early Contributors to Electric Wave Signaling

Thus, Hertz's experiments and confirmation of Maxwell's theory of electromagnetic waves in 1888 started a surge of interest in the scientific community, interest equivalent to what erupted after Oersted's discovery. So thinkers and tinkerers all over Europe started experimenting with

---

[224] The hysteresis property of iron, a complex concept, is used in electric systems. Basically, it is the lagging of an effect behind its cause: as when the change in magnetism of a body lags behind changes in the magnetic field.

electromagnetic waves within their specific domains of interests. Some of them became important contributors; others disappeared in the fog of history.

In 1890 French physicist *Edouard Branly* (1844–1940) published 'On the Changes in Resistance of Bodies under Different Electrical Conditions' in a French Journal, where he described his investigation of the effect of minute electrical charges on metal and many types of metal filings. He discovered that loose metal filings, which, in a normal state, have a high electrical resistance, lose this resistance in the presence of electric oscillations and become practically conductors of electricity. Based on this principle, he created a device he called the 'radioconductor' that would become known as the Branly coherer (Figure 152). It would be used as a receiver of electromagnetic waves.

In 1892 the English physicist *William Crookes* (1832–1919), later a pioneer of vacuum tubes, published in the *Fortnightly Review* an article entitled 'Some Possibilities of Electricity'. In this look into the future, he clearly discerned the coming of a new form of wireless telegraphy based on the application of Hertzian waves. He wrote:

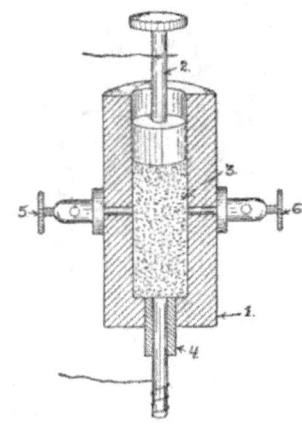

Figure 152: Branly radioconductor, aka coherer (1890).

It consisted of a tube (1) containing metal powder (3), in contact with two pairs of electrodes (2, 4 and 5, 6). The top and bottom electrodes were connected to a dipole antenna to pick up the radio waves, and also to a DC circuit consisting of a battery and a galvanometer. When a radio signal from a spark-gap radio transmitter was received by the antenna and applied across the electrodes, it caused the metal powder to become conductive. This allowed current from the battery to pass through the coherer and register on the galvanometer, indicating the presence of the radio wave. To restore the coherer to its high-resistance receptive condition, the metal particles had to be disturbed by tapping it.

Source: Wikimedia Commons.

Rays of light will not pierce through a wall, nor, as we know only too well, through a London fog. But the electrical vibrations of a yard or more in wavelength of which I have spoken will easily pierce such mediums, which to them will be transparent. Here, then, is revealed the bewildering possibility of telegraphy without wires, posts, cables, or any of our present costly appliances. Granted a few reasonable postulates, the whole thing comes well within the realms of possible

Figure 153: Replica of Popov's lightning detector (1895).

Source: www.imgur.com

fulfilment. At the present time experimentalists are able to generate electrical waves of any desired wave-length from a few feet upwards, and to keep up a succession of such waves radiating into space in all directions. ... Any two friends living within the radius of sensibility of their receiving instruments, having first decided on their special wave length and attuned their respective instruments to mutual receptivity, could thus communicate as long and as often as they pleased by timing the impulses to produce long and short intervals on the ordinary Morse code. ...

This is no mere dream of a visionary philosopher. All the requisites needed to bring it within the grasp of daily life are well within the possibilities of discovery, and are so reasonable and so clearly in the path of researches which are now being actively prosecuted in every capital of Europe that we may any day expect to hear that they have emerged from the realms of speculation into those of sober fact. Even now, indeed, telegraphing without wires is possible within a restricted radius of a few hundred yards, and some years ago I assisted at experiments where messages were transmitted from one part of a house to another without an intervening wire by almost the identical means here described. (Crookes, 1892, pp. 174-176)

The Russian professor *Alexander Popov* (1859–1905), professor in the Imperial Torpedo School in Cronstadt, Russia, experimented with electromagnetic waves in the 1890s. After reading about Oliver Lodge's work he improved

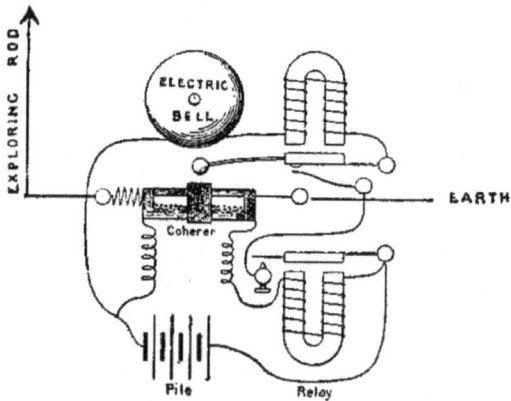

Figure 154: Popov's lightning detector (1895).

The schematic shows a remarkable similarity to Marconi's circuit but was only used as a detector for lightning.
Source: Fahie. J.J. (101)
http://earlyradiohistory.us/1901fa23.htm

268

upon his coherer, equipping it with an antenna. He used the device as a lightning detector and recorder but did not transmit messages with it (Figure 153, Figure 154). On May 7, 1895, he demonstrated it at a meeting of the Russian Physical and Chemical Society in St. Petersburg, but he did not claim a patent for it. Nor did he establish a wireless communication at that time. Nevertheless, he was later, due to political propaganda, acclaimed by Soviet authorities after 1945 as 'the true inventor of the radio', A claim that, after later investigation, was dismissed.

Figure 155: Captain Jackson's wireless transmitter and receiver (1897).

Source: http://www.g0akh.f2s.com/
SADARC/jackson.php Saltash Heritage

The Russians have good reason to be proud to have produced a pioneer of Popov's rank; but the officious Soviet campaign to designate him the "inventor of radio" and to enlarge his reputation out of proportion with his achievements amounts to a deviation from objectivity that must be deplored by all historians of technology who remain untouched by chauvinistic considerations. (Susskind, 1962, p. 2047)

People at the British Royal Navy had also been experimenting with alternatives for the then-traditional ship-to-ship communication: signal flags. It was captain *Henry Jackson*, who, at the HMS *Defiance* Torpedo School off Wearde Quay, Saltas, experimented with a wireless system based on the Hertzian waves (Figure 155). In August 1897 Jackson did start to send transmission between moored ships. Soon he was covering some miles.

Reports on Jackson's wireless system first appeared in the Annual Report of the Torpedo School of 1896 with a brief paragraph about 'transmitting electrical signals without connecting wires' allocating 'credit for the interesting experiments' to the staff of HMS Defiance. (Bruton, 2012, pp. 137-138)

As applying for a patent was strictly regulated by the Navy's policy's, Jackson did not apply for a patent on his invention of early ship-to-ship communication: "Furthermore the military application led to a veil of secrecy covering much of Jackson's work and contribution to early wireless history ...." (Bruton, 2012, pp. 138-139)

So, by the end of the nineteenth century, as a result of the numerous contributions of both the thinkers (the theoretical scientists) and the tinkerers (the engineering scientists), quite a bit of progress into understanding the fundamentals of wireless communication had been made (Figure 145). But as Fleming concluded, 'electric wave telegraphy' had yet to be born.

> *We are left, then, with the unquestionable fact that at the beginning of 1896, although the most eminent physicists had been occupied for nine years in labouring in the field of discovery laid open by Hertz, and although the noting of using Hertzian waves for telegraphy had been clearly suggested, no one had overcome the practical difficulties, or actually given any exhibition in public of the transmission of intelligence by alphabetic or telegraphic signals by this means. The appliances in certain forms existed, the advantages and possibilities of electric wave telegraphy had been pointed out, but no one had yet conquered the real practical difficulties, and exhibited the process in actual operation* (Fleming, 1919, pp. 517-518).

Remarkably enough, in all these years before and after Hertz's contribution in 1888, and with all those scientists contributing to the basic understanding, and all those engineers tinkering with parts of the *concept* of the wireless communication (Figure 156), no effective *system* of wireless communication had been developed yet by the mid-1890s. Their work had not resulted in an effective working system, as many of them were based of the electromagnetic field created by induction. This method worked maybe over short distances, but was too crude to be effective. It needed someone to come up with a totally different concept: electromagnetic waves.

On the positive side, all that experimenting had resulted in much theorizing and also experimental apparatus for either the generating (aka the 'oscillator') or the receiving (aka the 'coherer') of electromagnetic waves. And in parallel to that electromagnetic wave experimentation, there was the increasing development of the telegraphic and telephonic systems, which were based on cabled infrastructures, with their specific problems and limitations. These systems were waiting for a solution to those limitations: wireless telegraphy and wireless telephony.

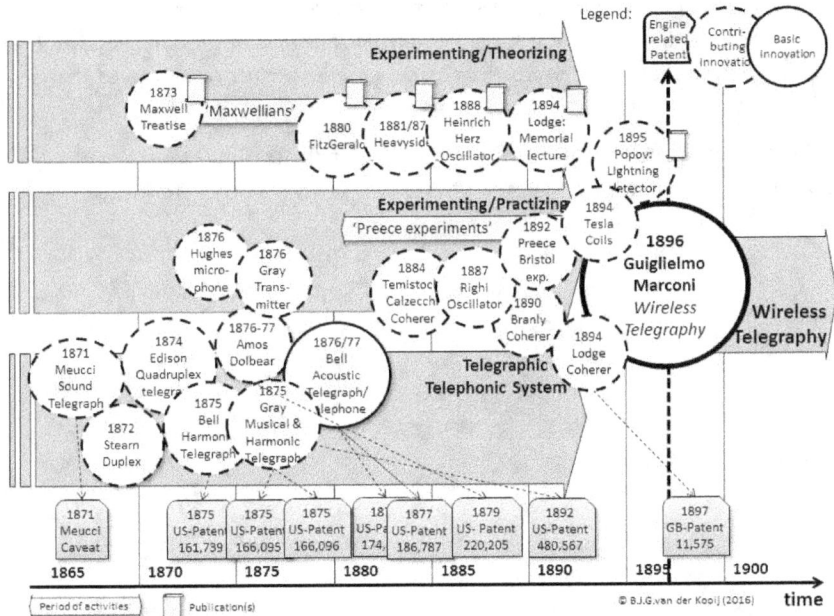

Figure 156: Overview of the early development trajectories leading to wireless telegraphy.

Figure created by author.

We have been exploring elsewhere early scientific development into the 'nature of light',[225] the 'nature of sound',[226] and here, for lack of a better word, the 'nature of electric waves in ether'. Now it is time to go and observe what the result was of all that accumulated knowledge and of all those experiments developing the technologies that were needed to create the early artefacts of telecommunication. As Fleming observed:

> *When a new field of discovery or invention is thus laid open, it invites the attention of two classes of minds. There are those who are chiefly drawn into its cultivation by a desire to increase purely scientific knowledge, and to explore the mysteries involved, regardless of any practical utility they may possess. On the other hand, there are others to whom the pursuit of novel facts or effects, or the unravelling of complex phenomena, or the construction of new theories, does not appeal. They are impelled to look at once for applications of the new knowledge which will minister to the convenience or mitigate the troubles of mankind.*

---

[225] See: B.J.G. van der Kooij, *The Invention of the Electric Light* (2015), pp. 37–41.
[226] See: B.J.G. van der Kooij, *The Invention of the Communication Engine 'Telephone'* (2015), pp. 112–118.

Probably in neither case is a more personal motive entirely absent, but whilst some minds regard the discovery of new physical facts or laws as an end in itself, others regard them only as a means to an end, and invent rather than discover or explore. (Fleming, 1919, p. 513)

Having explored the work of the early contributors to electric wave signalling, now we will zoom in on the work of that young man that would change the world.

Birth of Marconi's Wireless Telegraph (1894–1897)

The theoretical aspects of wireless communication (using electromagnetic waves) were understood better and better over the years, and some basic components like coherer and spark generators were developed, but this knowledge was not transformed into a usable system or apparatus for wireless communication. That would be done by a young Italian who, with an ordinary spark-induction coil, homemade coherers, and some other materials, realized a practical system for wireless telegraphy. After experimenting in the mid-1890s he patented his invention in 1896. It would become a milestone in the actual development of equipment, using the 'radio waves' that conquered the world in the time to come. And it started in Italy, in a small village near Bologna.

The 'Bologna' Context of Marconi's Younger Years

Italy at the end of the nineteenth century was—as we have seen in our analysis before—emerging as a young nation. The revolutionary unification into the Kingdom of Italy, starting within the context of the European-wide 1848 Revolutions, had taken some decades (1847–1870) of wars for independence. Originating from the reforms introduced by the French when they occupied Italy (1796–1815), after the 1820s the Risorgimento had been a period of continuous political upheaval, revolutionary movements, and insurrections. The Risorgimento also had reached Bologna.

> *Bologna in the late 1850s was at the centre of the revolutionary upheavals of the Risorgimento, the unification movement that defined and created the modern state of Italy. One of the prizes of papal power since the middle ages, the city changed hands frequently during the nineteenth century. Napoleon and the French army occupied Bologna in 1796, and for the next eighteen years, Bologna was a city marked by social, cultural, and economic change. Between 1796 and 1799 alone, more than forty religious institutions disappeared, their property and land distributed to the benefit of a new bourgeoisie. In July 1815, however, following Napoleon's defeat at Waterloo, Bologna was restored to the papal states, and it was soon a cauldron of political activity and intrigue, giving rise to a host of oppositional secret societies that drew their membership from the city's numerous university students. Bologna took part in the European turmoil of 1831 and 1848, and on June 12, 1859, Pope Pius IX finally gave up temporal power over the city. Less than a year later, in March 1860, Bologna joined the liberal constitutional monarchy of King Vittorio Emanuelle II ...* (Raboy, 2016, pp. 13-14)

That was Bologna's social-political context of that time, characterised by social unrest and political turmoil of the 1840s to the 1850s, turmoil that not only reached Bologna (Figure 98), but swept over the whole Italian Peninsula. That *socio-political revolution* that occurred in Italy needed the collapse of the Habsburg Empire with the Compromise of 1867 and the collapse of the Second French Empire in the Battle of Sedan (1870) to become a nation on its own.

Apart from this geopolitical development with all its socio-political conflicts, the *Scientific Revolution* had also left its traces in Italy. Italy's natural sciences had its roots in the mechanical experiments of Leonardo da Vinci and Galileo Galilei, part of the scientific awakening in this part of Europe. Early science academies like the *Accademia dei Lincei* (1603) aimed to understand all of the natural sciences, among which were the mechanical and biological philosophies. In the Bologna region, by the end of the eighteenth century it was the contributions in the bio-electromagnetics of Luigi Galvani, professor at the Academy of Sciences of the Institute of Bologna *(Accademia delle Scienze dell'Istituto di Bologna)* and Alessandro Volta, professor of experimental physics at the nearby University of Pavia *(Università degli Studi di Pavia)*, that created the foundations for the Electric Revolution to come in the nineteenth century.

During that century the secrets of the 'nature of lightning' were unravelled, creating a solid understanding of the phenomena of 'electricity'. It took the first half of the century for its rudimentary development in the form of the primitive battery-powered electromotor and the battery-powered telegraph. In Italy, inventors such as the previously mentioned Italians *Salvatore dal Negro* (1768–1839), *Luigi Magrini* (1802–1868), and *Guiseppe Botto* (1791–1865) who created (reciprocating) electric engines as early as the 1830s.[227] But when electricity became available in abundance by the mid-1800s due to the invention of electric dynamos, it saw its revolutionary breakthrough (Figure 150). The subsequent inventions of new DC and AC electro-motors, electric lighting, and telephony became part of the Second Industrial Revolution.

The phenomena of electricity fascinated many innovative and entrepreneurial tinkerers and thinkers worldwide. Their experiments and the resulting knowledge spread all over Europe, stimulating others to further progress and find alternative uses. These thinkers and tinkerers were also to be found in Italy, and one of them became world famous: a young adolescent with not too much formal schooling but with with great eagerness to learn.

---

[227] See: J.G. van der Kooij, *The invention of the Electro-motive Engine* (2015), p.75.

Figure 157: Poster of the Bologna Exhibition (1888).

Source:
http://www.bolognawelcome.com/en/home/live/events/complete
-calendar/expo-bologna-1888/

In 1888 in Bologna, the *Great Emilian Expositi*on was held on the occasion of the 800th anniversary of the University of Bologna and in the wake of the Expositions Universelles in France, Belgium, and England. The exhibition showed the fruits of technological progress as well as that of arts and music and was illuminated electrically. The most-visited pavilion—with a total of 500,000 visitors—was the one that showed the marvels of industry.

Marconi's Early Formative Years

Guglielmo Marconi was born on April 25, 1874, in Bologna, Italy, the son of an Italian aristocratic landowner, Guiseppe Marconi (1828–1904), and the Irish/Scottish Annie Fenwick Jameson (1840–1920). His father came from the Capugnano region of the Bolognese Apennine Mountains between Florence and Bologna. The Marconis, originally mountain men ('montanari'), by then were landed gentry, owning land in the Pontecchio region near Bologna. His mother was the daughter of John Jameson, a Scot who had gone to Ireland and into the business of distilling whiskey (Jameson & Sons) in the late 1780s. Jameson exported their distillation products all over the world, also to Italy, and much of their business was based in London, the heart of the British Empire.

Giuseppe Marconi belonged to the upper-middle class and was an eminently practical man intent on taking good care of the administration of his agricultural interests in the Pontecchio region. Amidst the political turmoil of 1848, he had inherited from his father, Domenico Marconi, an estate that had belonged to the Griffone family. It was an ancient house, large, plain, and nobly proportioned, called Villa Griffone, located in Pontecchio, nine miles south of Bologna. A house that made the Marconi's an important presence in the area, where mercantile traffic was growing steadily (the house still stands in rolling fields and vineyards looking out over a splendid view). Here, old Domenico began raising silkworms and made quite a success of it, while his son delighted in using his countryman's skill as an agriculturist to make Griffone's acres fruitful and beautiful. Guiseppe, meanwhile, spent most of his time in Bologna, where he became involved in various business ventures. One of his acquaintances was a banker, Giovanni Batista de 'Renoli, and Guiseppe was soon courting the Renolis' young daughter, Giulia. They married, but that marriage ended when Giulia Renoli died in 1858, leaving him with a son, Luigi. (D. Marconi, 2001, p. 5) (Raboy, 2016, p. 13)

The young *Annie Fenwick Jameson*, born in Ireland, raised in an old manor called Daphne Castle, educated in London, and, being a gifted singer, sent to Italy to extend her singing studies in bel canto at the Conservatorio in Bologna. This was compensation to the fact that her parents refused to agree with her ambitions to sing at the Covent Garden Opera House. In Bologna, she stayed with the family of a business relation, the Renolis. There, she met Guiseppe Marconi, and they fell in love. Guiseppe at that time was 17 years older than 23-year-old Annie. To prevent the love from developing into something serious, she was recalled by her parents to England. Nevertheless, through smuggled letters, their mutual affection was kept alive till they decided to elope and marry. Guiseppe travelled from Italy, and Anne sailed from England to Boulogne-sur-Mer in Northern France, where they were married on April 16, 1864. Returning to Italy together, they crossed the Alps by coach to Giuseppe's hometown of Bologna. (D. Marconi, 2001, pp. 1-7).

Annie, who was a devoted Protestant with a profound distaste for the Catholic Church, gave birth to two sons, Alphonse and Guglielmo. Alfonso came one year after their marriage, and Guglielmo came nine years later, being born on April 25, 1874, in a rented apartment in the massive, heavily shuttered Palazzo Marescalchi, a seventeenth-century baroque palace in the centre of Bologna. By then his father Giuseppe was in his mid-forties.

One year after their elopement, Giuseppe and Annie's first child was born in the country at Villa Griffone. They named the boy Alfonso. Nine years passed before Annie had another child. She came near dying the night that Guglielmo was born in Bologna. The windows were tightly shuttered to keep out the sharp crack of horses' hoofs and the rumble of carts over the cobbled street. Inside, the doctor came and went through the hushed, agonizing hours. At 9:15 on that Saturday morning, the doctor sent the nurse to tell distracted Giuseppe that he had another son. ...

Reading backward I can see which characteristics the younger Marconi son took from each of his utterly different parents. From his father, that independence of spirit which is the mark of mountain men, the aloneness that is often dour, the ability to make do with what is available, and the fortitude. From his mother, a will as stubborn as his father's, but matched with poetry and music and grace.

And to her radiant complexion and fair coloring he owed his blue eyes and the golden hair that darkened as he grew up. Ultimately he was his own man, an aggregate of opposites: patience and uncontrollable anger, courtesy and harshness, shyness and pleasure in adulation, devotion to purpose and thought thoughtlessness toward many who loved him. (D. Marconi, 2001, pp. 7-8)

Guglielmo and his brother, Alfonso, grew up in within a mixed English-Italian context and were bilingual. During his younger years Guglielmo was, like his brother, educated at a private school in Bedford, England. There, he learned to speak his upper-class English. The Italian language he learned while schooling in Italy. Travelling between England and Italy, and living in different locations in the Tuscany region, the boys and their mother, however, never spent an uninterrupted year in Villa Griffone.[228] The young Guglielmo combined three worlds: the Irish ancestry of the Jameson clan, the Italian ancestry of the Marconi family, and the English (Bedford)/Tuscan (Bologna) world where he grew up (Raboy, 2016, p. 20).

The inevitable ambivalence produced in his early life by is mixed Italian-Irish background was accentuated because his parents' interests diverged, and young Guglielmo spent long periods, sometimes years, away from his home on visits to Britain and homes of the English-speaking friends in Italy with his mother. During this roving infancy, the boy's formal education was fitfully managed by his mother or a private tutor. By the time he was eventually sent regularly to school in Florence he was neither amenable to the give and take of community life in the classroom, nor could he keep up with his fellow pupils, many of whom jeered at his poor Italian accent. (Jolly, 1974, p. 312)

As his parents were well-off, his upbringing was in the large English colonies of Bologna and Florence, and the Mediterranean seaport of Leghorn,[229] and as a child, he'd already traveled twice to England. As a result of this, and his mother's aversion to Catholicism, Marconi's education over the years was quite unorthodox.

Especially unorthodox, at least for the conservative and rather rigid educational establishment of most European countries of the time, was Annie's insistence that her son be given free rein to seriously study only what he enjoyed most and gave scant, if any, attention to his performance in those subjects in which he had no interest which, unfortunately, were those that most schools set great store in including grammar, literature, history, arithmetic etc. She made, instead, a great

---

[228] One's personality is created by nature and nurture, nature being unchangeable as the result of genetic formation and nurture being the result of one's upbringing. In Marconi's case, his formative years contributed highly to the way in which he would operate later as an inventor-entrepreneur.

[229] English name for Livorno, as it was a popular summer colony for upper-middle-class Brits.

effort to find for him the best possible tutors in his favorite subjects mainly in the sciences and music. Of particular significance was his apprenticeship with the Leghorn high school physics professor Vincenzo Rosa in whose laboratory Guglielmo picked up the first important notions and mastery of those experimental techniques that were to form such a crucial aspect of his later career. (F. P. Marconi, nd)

In Florence, Marconi briefly attended the Instituto Cavalerro, a private school he did not like too much, as, with his English-accented Italian speech, he was often the subject of teasing and bullying by his classmates. In the port of Livorno, only some 100 km from Florence on the west coast, but a world of difference from rural Pontecchio, between 1885 and 1889, he attended the Instituto Nationale, a technical school, where he was introduced to the formal studies of physics and electricity—and, to the joys of sailing in his sailboat in the waters of Livorno.

> *During these years he developed not only a deep passion for the sea… but also an immense interest in the electrical sciences, which among other things left him with little patience for his normal studies.* (Valotti, 2014, p. 262)

During the summer the family returned to Villa Griffone (Jolly, 1972, pp. 1-14) (Raboy, 2016, pp. 23-25). Marconi thrived at the institute and experimented with electricity and its application. He was even introduced to telegraphy by an ancient telegraphist he befriended, Nello Marchetti, who taught him Morse Code.

> *Marchetti had been a telegraphist at a time when it was a comparatively rare profession. When he discovered that the boy was intensely interested in electricity, he taught him, out of gratitude, the Morse code. On a telegraphist's key, set between pots of geraniums on a sunny window sill, the boy learned to tap out the*

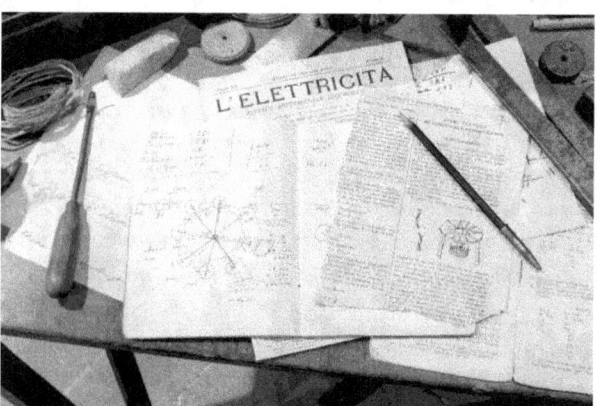

Figure 158: The magazine *L'Elettricita*.

Courtesy: Museo Marconi, Fondazione G. Marconi, Italy.

> *letters of the alphabet. Here, for the first time he sounded the three dots of the S, the signal that he would one day send around the world by wireless.* (D. Marconi, 2001, p. 17)

In 1891, Annie arranged for him to have private lesson from a high school teacher,

Vincenzo Rosa, who maintained a well-equipped laboratory. Marconi soon became absorbed in his informal studies, and encouraged by Rosa, he began doing all sorts of experiments on his own. The relationship with Rosa was intense, for they used to meet at least three to four times a day for lessons, conversations, and laboratory practice (Raboy, 2016, p. 25).

In 1892 Marconi, an avid reader of the magazine *L'Elettricita* (Figure 158) that had reported of an international competition for the development of a new type of battery, decided to participate in the competition. Soon he experimented with carbon electrodes, materials like antimony, aluminum, cobalt, nickel, and zinc. He worked on the project for several months, creating a three-anode battery, and in doing so, he enhanced his skills and knowhow. 'The laboratory work he did in the years 1891–1893, we can say now, gave him a solid and mainly practical basis in electricity, chemistry and metallurgy.' (Bigazzi, 1995, p. 4)

By 1893, the Marconis were back full time at Villa Griffone, and it was around this time that Marconi read of the experiments of a German physicist, Heinrich Hertz, on the generation and propagation of electromagnetic waves (Raboy, 2016, p. 27).

The Dawn of Marconi's Wireless Telegraph

The second half of the nineteen century was the time in which science and engineering were making fast progress all over the world. Physical transportation had been influenced greatly by the invention of the steam engine, and the massive need for communication had been met by the invention of the electric telegraph. It was the time that 'electricity' was maturing, both in the understanding of its electromagnetic properties as well as in its application. And there was the milestone of Heinrich Herz's discovery of electric waves. Guglielmo, reading about this and becoming fascinated by it, set up experiments at home and managed to get access to the Bologna laboratory of Professor Augusto Righi, their neighbour of the Villa Griffone, as he had a house close by, across the river Reno.

Augusto Righi (1850–1920) by that time was a well-known physicist who was focussing on the exploration of the secrets of electricity.[230] He published, for example, in 1887 *L'Ottica delle oscillazioni elettriche*: a treatise with the results of his experiments with a spark-gap sphere oscillator (aka the 'Righi oscillator') to check on Maxwell's theory. After Hertz announced

---

[230] Augusto Righi applied his talents to experimental work in electromagnetic studies and solid-state physics. He was the among the first scholars to demonstrate the hysteresis effect of iron and magnetic materials, and his contributions to the study of Hall and Kerr effects brought to light some basic aspects that were explained later by electronic theory.

his discovery of electromagnetic waves, Righi investigated them, especially their optical properties, and published the results in a treatise, *Optice Elettrica*, in 1897. He noticed that the smaller the spheres on the exciters of the spark, the shorter the waves, approaching those of light. Righi propagated electric waves as short as 2.5 centimetres, whereas Hertz had produced them at 30 centimetres in length, and Marconi's were originally meters long (Figure 138). Righi improved the Hertz oscillator

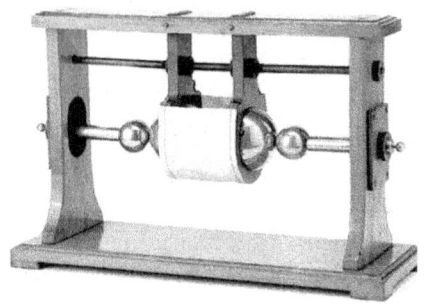

Figure 159: Spark-gap sphere oscillator by Augusto Righi (1890s).
Source: Museo First, Istituto Tecnico Toscano, Florence

that generated the waves; he placed the spark gap in Vaseline oil and made the waves more consistent and steady (Figure 159). He also experimented with the detectors of the Hertzian waves and tested the conductivity of different metallic powders and their use as a detector. All that experimenting interested Marconi highly.

> *Contacts with Professor Augusto Righi and his lectures certainly occurred, although not in a regular way. Righi seem to have suggested to the young Marconi to pursue an in-depth regular study before moving on with experiments. Marconi did not accept his advice because he was more concerned with the parallel work of other researchers and did not want to waste time. Nevertheless, he tried to take advantage of professor Righi's laboratory of Physics, in non formal way, in order to familiarize himself with apparatus and materials. This invaluable experience most certainly helped Marconi to develop his ability in shaping "the whole thing* [his idea] *into a practical, usable system", previously trained by Professor Rosa.*
> (Grandin, Mazzinghi, Olander, & Pelosi, 2012, p. 159)

It is not hard to understand that the environment of Righi's laboratory and his experimenting into the Hertzian waves fascinated and stimulated the young Marconi, who was by that time reaching the age of 20 years.

Within this timeframe, where the understanding of the phenomenon of electricity had matured, where early inventions like telegraphy were in full swing, where recent major inventions such as the telephone (1876/1877) and electric light (1879) proved its potency, and where the Hertzian waves were a hot topic, young Marconi's fascination, maybe even excitement, can be well understood. In addition, the peculiar way of his education, by his own wishes, interests, and commitments quite focussed on the physics of electrical phenomena (supported educationally by his teacher Vincenzo Rosa, and morally by his parents), created the environment for the birth of his concept of wireless telegraphy. Here was a young man with a vision, who was eager and capable of developing his ideas into tangible forms. As explained by Righi in a letter to Oliver Lodge on June 25, 1897.

> *The young Marconi has a strong desire to become an inventor. On several occasions he has questioned me on ideas and inventions that, though ingenious, would be difficult to realize. I have advised him to undertake formal studies and to follow a university course...* (Valotti, 2014, p. 266)

Conceptualization of Wireless Transmission[231]

When Hertz died on January 1, 1894, Marconi read the obituary written by professor Righi. It was by then that the idea of electric telegraphy using Hertzian waves started growing in his mind. This idea grew until the summer of 1895, when the family spent their summer holidays in a spa resort, *Andorno Bagni* in the Italian Alpine Biella-region in a remote and charming place called Sanctuario di Oropa. By then his first ideas seem to have been crystallized into his more solid concept for a system of wireless telegraphy.[232] "It was during his stay, more precisely during an excursion to the Oropa Sanctuary, that Marconi had the idea of the wireless telegraph" (Grandin et al., 2012, p. 153).

> *"It seemed to me," said Marconi in a lecture years later, recalling those thought-provoking vacation days in Biellese mountains, where he worked the idea out in his imagination, "that if the radiation could be increased, developed and controlled, it would be possible to signal across space for considerable distances. My chief trouble was that the idea was so elementary, so simple in logic, that it seemed difficult to believe no one else had thought of putting it into practice."* (Dunlap, 1941, p. 12)

It was the basic idea of a system for wireless telegraphic communication—ie Morse-coded information—using electrical waves. Today we call it *wireless one-way point-to-point communication*. This idea he later enhanced to wireless broadcasting, as he stated in 1937: "While in the Alps I got the idea that not merely signals but actual words and voices could be transmitted from one place to another without wires." (Raboy, 2016, p. 31). And this is *wireless one-way point-to-many communication*, aka 'radio'.

---

[231] I am indebted to the historian of technology Barbara Valotti, director of the Marconi Museum in Pontecchio, who was on October 3, 2016, graciously willing to share her insights with me and educate me in the early years of Marconi's life.

[232] Among historians, there is much debate on the exact year—1894 or 1895—when he developed his first ideas on wireless telegraphy (Grandin et al., 2012, pp. 150-151). His daughter mentions in her book the year 1894 (p.18). We follow the views of Barbara Valotti, historian at the Marconi museum (MM).

What set Marconi apart from the rest was that he saw wireless communication, in his mind's eye, quite literally as telegraphy without wires. Thus in his experiments he set out to reproduce as closely as possible the physical conditions condition of telegraphy, and use them to the same ends. ... It was Marconi who made the leap from Hertz's lab experiments to practical wireless telegraphy using electromagnetic waves as the medium of communication. This was his original contribution, and what caused him to be labelled a genius in the heroic version of his life story. Marconi's entire career was built on this one idea. ... Marconi not only invented mobile communication, it shaped and defined his world (Raboy, 2016, p. 52)

Experimenting with Hertzian Waves

Returning from that holiday in the Alps, burning to get to work, he started seriously with his experiments, shutting himself up in the attic of Villa Griffone, where the former silkworm room had been made available for his experiments (Figure 160).

The description of Hertz's transmission of electromagnetic waves across a room inspired Marconi with an all-consuming ambition to use them as the basis for a commercial telegraph communication system. He came back to Villa Griffone, shut himself up in two attics, which his mother had cleared to make a laboratory for him, and with single-minded dedication amounting to obsession he worked long, long hours to repeat Hertz's experiments. (Jolly, 1974, p. 312)

Subsequently, Guglielmo spent much time in his 'silkroom' laboratory in Villa Griffone, assembling instruments and copying experiments he read about. But his aging father was not too happy about his son's activities. He

Figure 160: Reconstruction of Marconi's attic laboratory in the Villa Griffone (1894–1895).

Courtesy Museo Marconi, Fondazione G. Marconi, Italy.
Source: http://www.fgm.it/en/museum-en/collection.html

blamed his wife for the fact that his son was wasting irreplaceable years and argued that the boy should have a proper education. He punished his son in every way he knew how, using money as his powerful weapon, as it was money that Guglielmo needed to buy long lengths of copper wire, batteries, induction coils and other electrical gadgets (D. Marconi, 2001, pp. 20-21).

> *His ambition to take up a recognised course leading to a scientific career was thwarted when he was unable to pass the matriculation examination of Bologna University. His father became even more disgruntled with the boy, and his irritation was increased by seeing his son, now nearly twenty, apparently aimlessly wasting his time and money on experimental dabbling—abetted by his mother.* (Jolly, 1974, p. 312)

To win over is father and to acquire his father's participation—and financial support—Guglielmo decided to show his father the results of his experiments. He demonstrated that he was able to ring a bell on the ground floor by pressing a button on the third floor without any connecting wires. And then he put forward his request.

Figure 161: Marconi's experimental spark-gap transmitter with tinfoil as antenna (1895).

Source: Wikimedia Commons.

Guglielmo's father treated his son request [for money] as he would have treated the request of any businessman. First, he had the boy explain the principles of his invention. His wife, fearful of a hitch, interceded with a plea that he be allowed to proceed on faith. Guiseppe rejected this approach out of hand on the grounds that if Guglielmo had, indeed, opened up a new scientific field he would have to know to get really substantial financing and he would have to get it from man who did not invest "on faith". This was dampening but also encouraging. Guiseppe had tacitly admitted that Guglielmo might have a future and that it might involve big business. (D. Marconi, 2001, p. 24)

Despite Giuseppe's reservations, with this demonstration, Guglielmo convinced his father of his invention. Guiseppe showed generosity and supplied his son with enough money—some 5,000 lire, equivalent to $1,000—to continue his experiments. Marconi later described him: "...as a cautious businessman, he was not so enthusiastic at first, as my mother; but he was not in the slightest averse to it." (Raboy, 2016, p. 26)

Figure 162: View from Villa Griffone towards the Celestini Hill (left, date unknown) and prototype standing in garden (recent photo).

Courtesy : Museo Marconi, Fondazione G. Marconi, Italy.
Source: http://www.radiomarconi.com/marconi/gm_diari.html

This money enabled Guglielmo to make his first prototype of the transmitter, a system consisting of a Ruhmkoff coil and Righi oscillator connected to an antenna made of tinfoil (Figure 161). For the receiver, he also made a new arrangement, replacing Herz's spark receiver with a Branly coherer. It was the archetype of a system for wireless transmission.

Celestini Hill Field Experiments

Till that moment Guglielmo had been meticulously copying the experiments of Hertz with coils, batteries, and wires. He had developed a system where pressing a key would result in the sounding a buzzer some thirty feet away. Now he started experimenting with bridging larger distances. Located on the gentle sloping hills bordering the river Reno, the Villa Griffone was surrounded by sloping large lands: the Celestini Hills. It was well situated when, by 1895, Guglielmo was ready for his first experiments in the open air (Figure 162).

Soon he set up experiments in the garden with his brother Alfonso, who assisted him with bridging increasingly longer distances between transmitter and receiver.

> However, there was much still to be done. Alfonso was Guglielmo's ally, dear Alfonso, so loving and admiring of his younger brother, so generous in his support! To Alfonso was assigned the receiving apparatus and with the help of a farmer on the place, he was to carry it to greater and greater distances from the sending mechanism. He was also entrusted with a tall pole at the top of which was affixed a white handkerchief. If Alfonso received a message sent by

Guglielmo, he was to lift the pole and waggle it so the handkerchief, waving in the breeze, could be seen from the house. (D. Marconi, 2001, p. 30)

Every time, they kept the transceiver and receiver within the line of sight (Figure 163). Then, in August 1895, he had discovered a new arrangement that enabled him to transmit even when the transmitter and receiver were not in the line of sight, such as was the case with a receiver placed behind a Celestini hill.

By change I held one of the metal slabs at a considerable height above the ground and set the other on the earth. With this arrangement the signals became so strong that they permitted me to increase the sending distance to a kilometer. ... I was sending waves through the air and getting signals that a distance of a mile, or

thereabouts, when I discovered that the wave which went to my receiver through the air was affecting another receiver which I had set up on the other side of the hill. It is my belief that they went through, but I do not wish to state that as a fact. (Jolly, 1974, pp. 27, 31)

This was an important discovery, as it was always assumed that the Hertzian waves travelled in a straight line and could not follow the curvature of the earth. So the receiver was deliberately placed at a further distance behind the hill. To communicate whether or not it worked, Alfonso used a gun.

The handkerchief was adequate to signal success from the fields in front of the Villa Griffone. It would not be seen if Alfonso went to the far side of the hill behind the house. For this he was armed with a hunting rifle and he marched sturdily off, up the narrow path past the farm buildings. It was the end of September now and the vines were heavy with purple grapes, the air golden. The walk over the rim of the hill took twenty minutes. Alfonso led, followed by the farmer Mignani and the carpenter

Figure 163: Marconi experimenting with his brother Alfonso in the garden of Villa Griffone (1895).

Marconi transmitting to his brother Alfonso on the family estate near Bologna, Italy, in 1895. His spark transmitting apparatus on the table was connected to the earth and to the metal sheet above him, which served as an antenna. Alfonso, in the background, had a similar antenna connected to his receiver.

Source:
http://www.newscotland1398.net/nfld1901/marconi-nfld.html

285

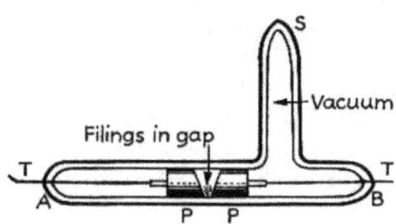

Figure 164: Marconi's coherer (1896).

When a Hertzian wave passes through the coherer (T_T), the filings in the gap P-P cling together ('cohere') reducing the original resistance.
Source: Wikimedia Commons.

Vornelli, lugging the antenna. Finally Guglielmo, watching tensely from a window, lost sight of the small procession as it dropped over the horizon. "After some minutes I started to send, manipulating the Morse key connected to the Ruhmkorff bobbin. In the distance a shot echoed down the valley." (D. Marconi, 2001, p. 30)

This was fundamental, as it contradicted the former belief that the electric waves, like light waves, travelled in a straight line:

The spark of a new era in communications had flashed. The scintilla was bewitching; Marconi saw no end to the dazzling beam that projected into the future. (Dunlap, 1941)

While experimenting with the several components of the system, he made a rather simple but significant contribution to the overall system by changing the antenna concept.

Instead of employing the Hertzian form of radiator, he connected one terminal of the secondary circuit of his induction coil to a metal plate or net laid on the ground, and the other by a wire to a metal can or cylinder placed on the summit of a pole. ... He then began systematically to examine the relation between the distance at which the spark could affect his coherer and the elevation of his cans or cylinders above the ground. (Fleming, 1919, p. 518)

So, what Marconi did was similar to wired telegraphy by one cable: the ground was used as a return 'wire'. The idea of changing the antennae concept was quite simple in nature; however, it proved fundament in creating his wireless transmission system. And increasing the height of the antennae would increase the distance to be bridged.

The same was the case for other parts of the system he improved on. Although, basically, they had been invented by others, step by step, Guglielmo added to their

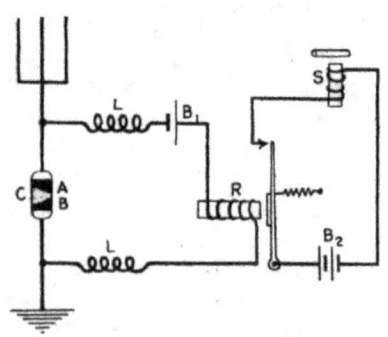

Figure 165: The electrical circuit of the improved receiver.

Source: Wikimedia Commons.

functionality and improved on their performance. He did that in several parts of the system:

- In the transmitter, he applied four instead of the usual two spark balls, just like Righi had done (Figure 159).
- In the transmitter circuit, he applied the ground as a return wire and created a half-wave dipole antenna (Figure 163).
- In the receiving circuit, he created his own nickel-silver sensitive 'coherer' tube based on the principle of the Branly coherer (Figure 164).
- In the receiver circuit (Figure 165), he added coils (L), a battery (B), and an additional relay (R) and battery (B2) to actuate a Morse printing device (S).
- In the secondary circuit of the receiver (Figure 166), to reset the coherer (A), he added a mechanical tapper (B) to disturb the cluttered mixture of metallic parts in order to restore the original state of non-conductivity.

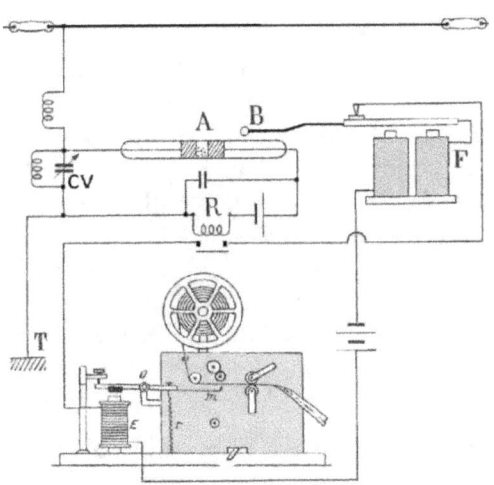

Figure 166: The receiver circuit of Marconi's wireless telegraph.

Below is shown the Morse receiver with paper roll. Its electromagnet (E) is activated by the electromagnet (F) in the receiving circuit. Nb.

Source: Wikimedia Commons.

It is said that Marconi tried several hundred combinations of metal filings of various sizes between metal plugs of different shapes and spacings before settling on undoubtedly a very refined version. Marconi's tube (which was evacuated) was much smaller, (about 2 inches [50 mm] long), the gap between the slightly tapered silver plugs was small (0.025 inches [0.635 mm]) and the faces had been treated with mercury. He used 95% nickel mixed with 5% silver.
(Simons, 1996, p. 43)

The Archetype of Wireless Telegraphy

Step by step, his experiments improved the total system. As a result the transmitter gave of a better signal due to the improved antenna, the receiver became more sensitive due to his coherer, the Morse telegraph functioned better due to the additional relay and battery, etc. And all that time he was concerned that somebody else would be developing a working apparatus based on the increasing body of scientific knowledge on Hertzian waves.

> *My chief trouble* [he said] *was that the idea was so elementary, so simple in logic, that it seemed difficult for me to believe that no one else had ought of putting it into practice. Surely, I argued, there must be much more mature scientists than myself who had followed the same line of ought and arrived that an almost similar conclusion.* (Jolly, 1974, p. 28)

Using the idea of applying Hertzian waves as the fundamental mode of operation of his system for wireless communication, the young 'practician' Marconi had managed to create a working prototype of his communication system and its improved components; it was to become his archetype of wireless communication.[233] In fact, by that time he had invented an apparatus for a 'wireless communication system'.

In a way, the apparent simplicity of his work—as we might consider it to be in hindsight—shows great similarity to the original invention of the telegraph. There, Cook and Wheatstone borrowed the idea of using a button, battery, and galvanometer, all connected by an increasingly longer wire, to create a simple system for 'distant writing': the telegraph. Samuel Morse even so just connected an electromechanical relay to a button and a battery by a ling wire, and he devised a quite simple code for transmitting the information.[234] Edison's invention of the incandescent lamp was also not that complicated. Basically, it was the result of electricity creating a hot wire that emitted visible light. But before the artefact itself was realized, it took a lot of experimenting and practical ingenuity.[235]

What great inventions seem to have in common is their elementary simplicity, where a basic phenomenon is transferred in a working artefact satisfying a basic (latent or apparent) need. The telegraph, in its simple form, played on the basic need for communication. The incandescent lamp played on the basic need for light. Marconi's system for wireless communication satisfied the need for reliable long-distance communication in its different forms: from marine communications of the early 1900s

---

[233] The word is used to denote the original system from which all later models were copied from. It was a collection of prototype components.
[234] See: B.J.G. van der Kooij, *The Invention of the Communication Engine 'Telegraph'* (2015).
[235] See: B.J.G. van der Kooij, *The Invention of the Electric Light* (2015).

through to our present day's mobile telephone networks and global broadcasting systems.

One has to realize also that, partly due to living and having done his experimenting in Italy, Marconi was an outsider to the British scientific community. A community that was interested in Maxwell's theory (the so-called Maxwellians), and by the many scholars preoccupied by the Hertzian waves (eg Lodge et al.). Although his technical understanding and engineering capabilities certainly were adequate, he was not like the theoretical and engineering scientists, who, with their experiments, were primarily focussing on the acquisition of knowledge. Remarkably, the painter Samuel Morse, the ex-military creator of medical wax models William Cooke, and the teacher of the deaf Alexander Graham Bell, all inventors of basic innovations in electricity, were similar outsiders.

Marconi Goes to England and Starts His Venture

By the 1895s, to the family, it was obvious that Guglielmo's experimenting had created something that could be of value. With his father's business background and broad experience in entrepreneurial activities, young Marconi's wish to explore further technical possibilities and maybe patent his invention, and considering the business background of the Annie's family of the Jameson clan, it is not hard to imagine the discussions that went on in the family. From collective brainstorming about where and how to use Guglielmo's invention, to the search for ways to realize some of those applications, maybe even how to make a business out of it.

> *That shot, the signal of victory from over the hill, changed the life of the Marconi family. Giuseppe was no longer required to take his youngest son's claims on faith. He was a shrewd man but he was a country man, so that though he knew that steps must be taken to release his son's invention to the public, he did not know what to do or how to do it.* (D. Marconi, 2001, p. 31)

To explore their first ideas about how to realize such a venture, there was much discussion with friends and family, and consultations with British relatives. Contacts with the Italian friends of the family even hinted at getting a patent. Quite a serious activity for such a young man to consider.

> *Yet another family friend, the physician Gardini, wrote confidentially on their behalf to Ferrero, who reportedly replied that the young Marconi should apply for a patent everywhere, using the "right of freedom of action" provided by the liberal Italian government, and do so before the secret of his important invention got out.* (Raboy, 2016, p. 32)

Obviously, it was clear that for converting the early prototypes into a saleable system, much more experimenting was needed. Thus, to expand his experimenting opportunities and find financial support to pay for it, his mother contacted her relatives of the Jameson Clan in England. Her sister's son, Henry Jameson-Davis, an engineer with contacts in scientific and financial circles, was willing to help. So, the 22-year-old Guglielmo Marconi and his mother, Anne Jameson, set off for England in February 12, 1896, carrying in his luggage and a black box with his apparatus.

> *Guglielmo had continued to cling to his black box but in the customs he was forced to surrender it to the inspectors. They found wires and batteries, strange shaped pieces of metal and dials and they reacted as officialdom is prone to. ... Incited by Guglielmo's replies to their questions, which, as they were not understood seemed evasive, and their own dutiful sense of importance, they manhandled the foreigners' mysterious box. Its contents were broken beyond repair.* (D. Marconi, 2001, p. 32)

The Context for the Decision

Since his mother was from Britain and her family and relatives lived there, going to England was obvious. Also, for her son, who spoke fluent upper-class English,[236] going to England was not a problem at all.

> *Several pretty obvious reasons prompted him or, maybe, one should say Annie to do this. First and foremost was the simple fact that his device had one unique and very simple application at the time: naval mobile telegraphy. What better place to market it than in England, then the pre-eminent naval power? Second, and, probably, just as important, the Jameson family to which Annie belonged was financially well off and politically and socially well connected enough in the English commercial if not the technical entrepreneurial class to provide the support that would be required to carry out the Marconi "grand design". Finally, their good knowledge of the language and customs of the country (assuredly better even than those of Guglielmo's native Italy) was another deciding factor.* (F. P. Marconi, nd)

Clearly, the choice of the Marconi family to take Guglielmo with his invention of the wireless telegraph to England has to be seen in the (political/technological/industrial) context of that time. Let us try and describe some of the major elements of that context.

---

[236] For people of all nations, it is easy to recognize social status by the type of mother language that is spoken. In the class-conscience England, the use of the upper-class English 'dialect' indicates one's aristocratic background. Marconi, however, was hard to place into the English class system, as he, along with his mastering of upper-class English, also could speak English with a distinctive Italian accent, indicating he was a foreigner.

Imperial expansion in Africa: It was
the time of colonial expansion
into Africa after the Berlin
Conference of 1884. The British
Empire, although in 'splendid
isolation' within the European
power game, was active on the
African continents, as were the
French, Germans, and Italians,
each with their own imperialistic
aspirations. In 1898, in Eastern
Africa, something had happened
between the two great powers of
that time, an imperial territorial
dispute between Britain and
France. It was an incident during
the so-called 'Scramble for
Africa' in which that continent
was colonized by the European
empires of that time. Who was
to control the Upper Nile Basin?
For the Brits, it was the desired
connection between their
possessions in South Africa,
East Africa, and Egypt (Figure

**Figure 167: The Rhodes Colossus
(1892).**

Caricature of Cecil John Rhodes after he
announced plans for a telegraph line and
railroad from Cape Town to Cairo.

Source: Wikimedia Commons. Artist: Edward
Sambourne.

167). France, pushing eastward from the Atlantic coast, wanted to
control the Sahel area from the Niger River to the Nile. The
confrontation came in Fashoda, where the two armies met. While
waiting for instruction from the home offices, due to the political
climate of the time, there was heated rhetoric on both sides. Ultimately,
the British won a diplomatic victory, but for a long time France
remembered the incident as an example of British brutality and injustice:
it became the 'Fashoda Syndrome'. And in this contest of power, Italy
hardly played a role of significance—except for the fact that it had
imperialistic aspirations of its own in abundance, aspirations that could
be fulfilled with the aid of one man: Guglielmo Marconi. However, the
young Italian nation was not yet familiar with the concept of state
support.

> *Events like the Fashoda incident of 1898 fed a sense of imminent European
> war, as did the Great Game with Russia. In the shadow of a global arms race
> and a growing conviction that new technologies conferred military and imperial
> advantages to whoever was the first in the field, the turn-of-the century British*

state invested more deeply in scientific research, and scientist, in turn, relied increasingly on state support. (Satia, 2010, p. 831)

Maritime incommunicado: It was also the time of intense seafaring. As described, earlier seagoing vessels were between their port of departure and their port of arrival 'incommunicado'. There was no way to communicate with them during the period at sea. On shore, it was not known what passed during voyages on those dangerous seas, voyages that could last months, even years, in the time of wind-powered sailing ships. Even with steam-powered ships, it could often take weeks to travel great distances.[237] And cabled telegraphy, in many European countries discovered by the military for its communication needs while deploying troops, was limited to land situations (eg the Crimean War of 1853). So, the early efforts to create wireless maritime communication (eg Edison's patent № 465,691 of 1891, Figure 129) could count on maritime and military interests: ship-to-shore communication and ship-to-ship communication. By the 1880s there had already been some testing of wireless solutions to communication by William H. Preece, the chief engineer of the British Post Office (such as in the Isle of Wight, 1882, Bristol Channel, 1892, Isle of Arran, 1894). However, that Post Office, an institution of the state, was functioning within a political context.

> *In England, where telegraphy had taken on commercial as well as strategic military importance, a Royal Commission on Electrical Communication with Lighthouses and Light-Vessels was appointed in 1892. The chief engineer of the British Post Office …, William H. Preece, was charged with developing a system for communicating with offshore lighthouses and ships ailing within British territorial waters.* (Raboy, 2016, p. 50)

The results of Preece's experiments were not too promising, and additional funding was difficult to get. These efforts were certainly discouraged by the fact that cabled telegraphy had such a solid position in society. Those with an investment in these cabled systems did not welcome any new system that might intrude and diminish their market share, expansion, and profitability. Nevertheless, facing this opposition, the maritime application of his system was quite clear to the young Marconi.

---

[237] In the 1850s crossing the Atlantic from Southampton to New York would take up to two weeks. The SS *Persia* was the fastest ship in 1856, able to make the crossing in 9 days, 1 hour, and 45 minutes. It was not until the 1910s that great steam-powered ships would make the transatlantic crossing within a week. From N. Y. to Southampton, the record of 5 days, 17 hours, 8 minutes was made by the *Kaiser Wilhelm der Grosse* of the North German Lloyd Line in 1897.

British ships covering the immense British empire scattered about all over the globe desperately needed, both from a commercial and military standpoint, a way to communicate rapidly with home base and amongst themselves. Wireless telegraphy offered the only practical way to do that. The Marconi Co. that Marconi founded in 1897 enjoyed for many years a virtual monopoly on this business. (F. P. Marconi, nd)

Industrial Revolution: It was the time of industrialization, as, by the end of the nineteenth century, Britain was in the midst of the Second Industrial Revolution. The Age of Mechanization had transformed Britain into the 'workshop of the world'. Its manufacturing capabilities—ie the factory system—combined with the development of mechanical technologies (eg the fine-mechanical industry, the arms industry) had given it huge economic advantages, and a high level of industrious (fine-)mechanical skills was widely available. Also, capital looking for investment was easily available, as the merchant class of England had seen many of its members—eg the sugar and tobacco barons—become quite wealthy. The previous railway mania of 1840 to 1850 and the economic Victorian Boom of the 1860s had brought a climate for industrial investment to the people. Furthermore, it was the time of the great inventions in electricity: the electric light and the telephone. Now it was the American entrepreneur-inventors with their capitalistic aspirations who flocked to the British Isles to try to exploit their inventions. Inventors such as Alexander Graham Bell and Thomas Edison came to London, applied for patents for their inventions, and started their companies to exploit those patents. It was the time in which the British inventor had to defend themselves through litigation when their patents were infringed upon by those foreigners. Americans, many of British origin (such as Bell), saw (quite optimistically) Britain as the springboard to access the European continent with their new products and communication systems.

Within that overall context of imperialism and colonialism, combined with maritime and industrial developments, Britain was facing some specific problems of its own as a consequence of these technological developments.

Telegraphic monopoly in Britain: In the decades following Cooke and Wheatstone's invention of their first cabled telegraph in 1834/1835, telegraphy had become part of the communication infrastructure in Britain. By 1868 the British public telegraph network consisted of 150,000 km of telegraph wires, 3,381 telegraph stations, and another 1,226 telegraph stations provided by the railway companies. At the maturity of the telegraph companies, during the early 1860s, there had been a technical consolidation into three wholly independent, incompatible national operating systems: Cooke and Wheatstone's single-needle telegraph with the Electric Telegraph Company, Bright's acoustic telegraph with the British & Irish Magnetic Telegraph Company,[238] and Hughes's printer with the United Kingdom Electric Telegraph Company. These systems were not technically compatible, which created some problems when telegrams had to be transmitted along lines using different systems. That created many user complaints. However, much of the telegraphy technology was in its infancy, much had to be discovered, experience had to be built up, and standards had to be created. But the complaints were still heard in the mid-1860s, when many political parties serviced the public interest and their own interests. The political question was whether the government should continue a laissez-fair attitude and leave the development to the market parties, as was the case in the United States, or if telegraphic communication could be treated the same as postal communication. Was this modern development, as part of the public interest, also to be controlled by the state? Should the Post Office, then, not be the organization to implement the telegraph infrastructure? These questions brought about years of political discussion and finally resulted in the introduction of the *Telegraph Bill* on April 1, 1868. The principle of nationalization received parliamentary approval in July 1868, and the following year a *Money Bill* was passed to implement the purchase. By February 1870 telegraphy had become a state monopoly. And the same happened to the telephone in Britain, the spoken word being another form of the written word. At the end of the nineteenth century that spoken word was on its way to also becoming a part of the state monopoly.

Colonial isolation: By the end of the nineteenth century Great Britain had become a dominant nation in the world. It created a region of influence—to put it mildly—spanning from the West to the East, and that empire had to be governed. Thus, the vast British Empire could very well use a new means of fast and cheap communication. So, in these times of the British Empire, with the British Navy 'ruling the world', governmental and merchant activity alike were facilitated by the worldwide cabled communication infrastructure that had emerged. British telegraph interests certainly dominated large parts of that

---

[238] This enterprise was led by the Scot John Pender (1816–1896), who became one of the *cable barons* of the age. He created the worldwide operating Pender Group of telegraph companies. It was later, during his American experiments, that Marconi was faced with their powers, as the Anglo-American Telegraph Company, part of the Pender Group, forbade any further experiments since they would infringe on the Pender Group's monopoly of communications in Newfoundland.

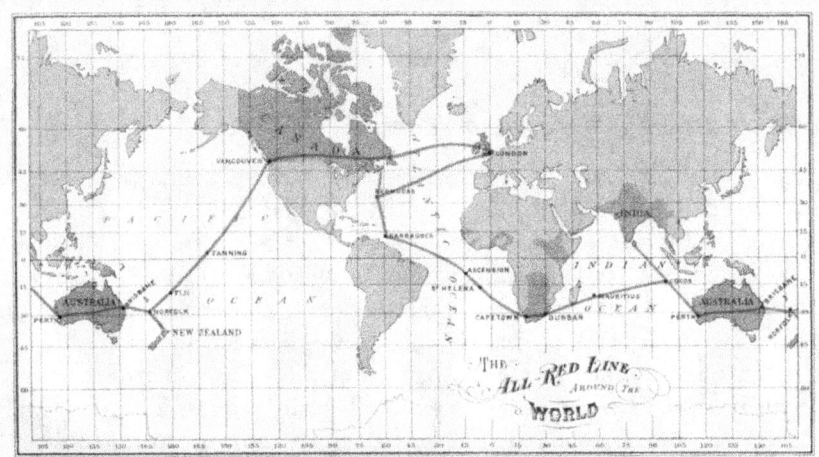

Figure 168: The British-controlled 'All Red Line' of telegraphic communication within the British Empire.

The cable network was interconnected through areas under British control (eg Canada).

Source: Wikimedia Commons. Artist: George Johnson.

worldwide telegraph infrastructure. However, over time that 'All Red Line'[239] of (maritime) telegraph cables spanning the globe through British-controlled regions (Figure 168)—enabling the government (and military) to communicate with the outposts of the empire—showed its limitations. The time was ready for an alternative to cabled communication.

The fact that everything related to telegraphy and telephony was part of the public interest—and thus subject to a state monopoly—would have great consequences, not only for Marconi and his entrepreneurial aspirations, but also for others who had entrepreneurial aspirations of their own. And within that context, young Marconi, hardly aware of these matters, arrived in England in 1896.

In this time of science and technology for and by the nation, Marconi was an interloper; despite his mother's British ancestry, he was a foreigner and, worse, a

---

[239] The All Red Line was an informal name for the system of cabled electrical telegraphs that linked much of the British Empire. It was inaugurated on 31 October 1902. However, some parts of the line were completed considerably earlier (eg the transatlantic cable as early as 1858).

tinkerer. Transmitting across the imperial map enabled him to prove his bona fides as a servant of the British state and style himself nostalgically as a "tinkerer-explorer" of the dark continent of space. (Satia, 2010, p. 831)

Start of Marconi's Pioneering Years

After arriving in London in February 1896, nephew Henry Jameson-Davis helped Guglielmo find accommodations where he could set up his experiments. But he did more than just support family members. He also became supportive of the specific activities of Guglielmo and his new invention and took him under his wing. Being a businessman of his own, he understood what it took to create business. And he knew that it was important to protect he invention. So, after helping Marconi assemble the materials he needed to replace his smashed instruments, Jameson-Davis brought Marconi contact a patent agent: Carpmeal & Company. For four months young Marconi devoted himself to the Herculean task of preparing the patent papers. Once that the legal claim to priority of the invention was staked, it was time to launch his invention and seek publicity, financial backing, and facilities for large-scale experiments. (Jolly, 1974, p. 35)

Marconi's Patent for Everything:
The First Wireless Telegraphy Patent[240]

It was obvious to Jameson-Davis and Marconi that he had to secure his invention by obtaining patent protection. Therefore, one of the first actions he undertook was to file an application for a patent on his invention. This first application, describing the novelty of his work, was important. If the novelty and the subsequent claims were well formulated, it would protect his invention. But what was there to claim as novelty? Was it his transmitter that could claim novelty? Or, alternatively, was it his coherer that had a patentable novelty?

> *In 1897, at which time the final specification of his first patent had not yet been submitted, Marconi was approached by people interested in the purchase of his invention. Moulton was engaged by the prospective buyers to assess the commercial value of the yet-to-be-obtained patent on the basis of the description provided by the inventor. Marconi was well aware not only of the potential value of his new system of wireless communication but also of the difficulty of obtaining a sound patent for a very complex invention—sound enough to withstand examination by those who had already declared their intention of challenging its validity.*
> (Guagnini, 2009, pp. 359-360)

---

[240] I am indebted to the professor Anna Guagnini, University of Bologna, author of many articles on Marconi, who was on November 18, 2016, graciously willing to share her insights with me and educate me in the patent affairs of Marconi's life.

The British Patent Law of 1883 had two important issues: the issue of specification and the issue of novelty.[241] The specification was checked by the clerks of the Patent Office, but the novelty requirement was left to be decided upon in court (if necessary).[242]

> *The core issue in Marconi's case was that he had not invented anything really new or at least easily recognizable as such but what was new was the use to which he put the old concepts and techniques in order to exploit them for a very practical purpose. His, then, would be a very difficult device to patent where priority of discovery of each component is emphasized as in the US and Italy, for example, especially since his predecessor's results (mainly Oliver Lodge) were of common knowledge in the world at the time. In England, on the other hand, what mattered most at the time was the possibility to claim the method itself as property and that was exactly what Marconi did with his crucial patents. These were used by him and his company to protect them, at least for the crucial first years, from potentially fatal attacks from powerful competitors. (F. P. Marconi, nd)*

Preparing a solid patent application was not a sinecure, and by the 1880s a whole new range of professions had emerged: the specialized patent agents, who supported inventors in their endeavours to obtain a patent, as well as the consulting engineers, who specialized in dealing with patent litigation. Obviously, Marconi—totally inexperienced in the legal matters of patenting—was advised to obtain some of this patent expertise. It was Edward Carpmeal of the chartered patent agents Carpmeal & Co. who originally assisted the young inventor. He was soon followed by others: Mr. J.C. Graham and Mr. Fletcher Moulton.

> *In the attempt to maximize his chances of obtaining such a patent, Marconi was already availing himself of the assistance of qualified patent agents and lawyers; he was also counselled by an experienced patent expert, John Cameron Graham. In the event, the prospective buyers abandoned the negotiation. But Graham suggested to his client that he should retain the services of the famous barrister who had been called as an examiner. So it was that Moulton became Marconi's*

---

[241] The specification should, where necessary, carefully distinguish what is new from what is old; otherwise, the patent may be held to include what is old and consequently void for want of novelty.

[242] Following a period of institutional reform, the *Patent Law Amendment Act* of 1852 established the modern-day Patent Office in October 1852. This resulted in a complete overhaul of the previously inefficient and aging British patent system and provided a simplified procedure for obtaining a patent. The Patent Office required a description of the invention be filed with the application and introduced the publication of applications. Moreover, separate patents for each of the countries in the UK were replaced with the issuing of a single UK patent, which significantly reduced legal fees by roughly three quarters. After a draft bill for revision of the British Patent Law ran into political problems in 1872, in 1873 the revised Patent Law passed the House of Lords.
Source: https://www.wilsongunn.com/history/history_patents.html

counsel in the preparation of the final specification of his first patent. (Guagnini, 2009, p. 360)

Marconi's patent lawyer did file the provisional specifications 'Improvement in telegraphy and in apparatus thereof' at the Patent Office on March 5, 1896. The application was entered in the *Official Journal* as № 5.028. A few days later it was accepted, and provisional protection was granted (Figure 169). In this provisional application, the invention was described as follows (Fahie, 1899, pp. 316-340):

> *According to this invention electrical actions or manifestations are transmitted through the air, earth, or water by means of electric oscillations of high frequency. At the transmitting station I employ a Ruhmkorff coil having in its primary circuit a Morse key, or other appliance for starting or interrupting the current, ... At the receiving instrument there is a local battery circuit containing an ordinary receiving telegraphic or signaling instrument, or other apparatus which may be necessary to work from a distance, and an appliance for closing the circuit, the latter being actuated by the oscillations from the transmitting instrument. ...* (text of patent application)[243]

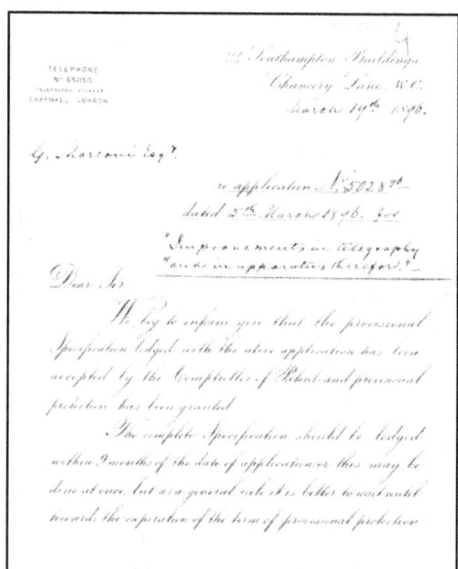

Figure 169: First page of Confirmation of the provisional application (1896).

Source: http://www.radiomarconi.com/marconi/ brevettomarconi.html

Next, while Marconi worked on improving his invention, the text for the final specification was worked on by his patent attorney. It became clear that the final specification would be quite different from the provisional specification. Changing a specification was a risky manoeuvre, as it could cause invalidation of the provisional protection.

In June 1896 Marconi abandoned the earlier provisional specification and—once again assisted by the Carpmeals—he submitted a new one, incorporating the results of new experiments. The title of his new applications was "Improvements in Transmitting Electrical Impulses and Signals, and in Apparatus

---

[243] J.J. Fahie, *A History of Wireless Telegraphy* (2nd edition, revised) (1901), pp. 316–340, APPENDIX E. Source: http://www.radiomarconi.com/marconi/popov/pat763772.html

therefor". The term 'telegraphy" that appeared in the former version was dropped, so as to avoid limiting the system to the transmission by traditional Morse's 'dot and dash' code. (Guagnini, 2002)

The new application contained nineteen claims. Of those, the first one was quite basic:

I declare that what I claim is—
1. The method of transmitting signals by means of electrical impulses to a receiver having a sensitive tube or other sensitive form of imperfect contact capable of being restored with certainty and regularity to its normal condition substantially as described. (text of patent application)

Thus, by changing the application, Marconi had broadened his 'Improvements in Telegraphy...' into 'Transmitting Electrical Impulses and Signals...' And now he claimed a 'method', and that is quite different from an apparatus.

Once the new provisional specification was submitted, Marconi would have protection for his invention. However, on November 8, 1896, he met again with his patent agents. They discussed the final complete specification, and they explored the possibilities of foreign patents. As the patent application required a description (the specification aspect) of the nature of the invention (the novelty aspect) and a precise statement of what was claimed (the protection aspect), this final specification was crucial—especially since the scientific community was already loudly expressing its reservations about the contributions of the foreign newcomer.

It was decided to hire the services of an expert, J. Fletcher Moulton, one of the most formidable lawyers practising in the London courts at the turn of nineteenth century, who was quite familiar with electric industrious activities and specialized in patent litigation in the emerging high-tech electrical industries of that time.[244]

Such expertise did not come cheaply, and although no detailed records of the payments to Graham and Moulton have survived, the fees must have been high. ... Costly as Moulton certainly was, Marconi and the Marconi Wireless Telegraph Co. retained him as counsel, to work (along with a highly qualified team of patent experts) not only on the preparation of subsequent patents but also on the assessment of rival patents and the fashioning of the company's patent policy. (Guagnini, 2009, p. 360)

---

[244] Such as the litigation for control of the telephone (Bell) and the incandescent lamp industries (Edison), and in the rapidly expanding field of the production and distribution of electricity.

N⁰ 12,039   A.D. 1896

Date of Application 2nd June, 1896
Complete Specification Left, 2nd Mar., 1897—Accepted, 2nd July, 1897

PROVISIONAL SPECIFICATION.

Improvements in Transmitting Electrical Impulses and Signals,
and in Apparatus therefor.

I, GUGLIELMO MARCONI, of 71 Hereford Road, Bayswater, in the County of Middlesex, do hereby declare the nature of this invention to be as follows:—

According to this invention electrical actions or manifestations are transmitted through the air, earth or water by means of electric oscillations of high frequency.

At the transmitting station I employ a Ruhmkorff coil having in the primary circuit a Morse key, or other appliance for starting or interrupting the current, and its pole appliances (such as insulated balls separated by small air spaces or high vacuum spaces, or compressed air or gas or insulating liquids kept in place by a suitable insulating material, or tubes separated by similar spaces, and carrying sliding discs) for producing the desired oscillations.

I find that a Ruhmkorff coil or other similar apparatus, works much better if one of its vibrating contacts or brakes, on the primary circuit, is caused to revolve which causes the secondary discharge to be more powerful and more regular, and keeps the platinum contacts of the vibrator cleaner and preserves them to good working order for an incomparably longer time than if they were not revolved. I cause them to revolve by means of a small electric motor actuated by the current which works the coil, or by another current, or in some cases I employ a mechanical (non electrical) motor.

The coil may however be replaced by any other source of high tension electricity.

At the receiving instrument there is a local battery circuit containing an ordinary receiving telegraphic or signalling instrument, or other apparatus which may be necessary to work from a distance and an appliance for closing the circuit, the latter being actuated by the oscillations from the transmitting instrument.

The appliance I employ consists of a tube containing conductive powder, or grains, or conductors in imperfect contact each end of the column of powder or the terminals of the imperfect contact being connected to a metallic plate, preferably of suitable length so as to cause the system to resonate electrically in unison with the electrical oscillations transmitted to it. In some cases I give these plates or conductors the shape of an ordinary Hertz resonator consisting of two semi-circular conductors, but with the difference that at the spark gap I place one of my sensitive tubes, whilst the other ends of the conductors are connected to small condensers.

I have found that the best rules for making the sensitive tubes are as follows:—

1st. The column of powder ought not to be long; the effects being better in sensitiveness, and regularity with tubes containing columns of powder or grains not exceeding two thirds of an inch in length.

2nd. The tube containing the powder ought to be sealed.

3rd. Each wire which passes through the tube, in order to establish electrical communication, ought to terminate with pieces or metal or small knobs of a comparatively large surface, or preferably with pieces of thicker wire, of a diameter equal to the internal diameter of the tube so as to oblige the powder or grains to be corked in between.

Figure 170: First page of the provisional specification of Marconi's GB patent № 12,039 (1897).

Source: http://www.radiomarconi.com/marconi/popov/pat763772.html

Moulton's assistance surprised the Maxwellians. Silvanus Thompson, professor of physics and working on electromagnetism, wrote to Oliver Lodge on June 30, 1897:

I happen to know that Moulton was called in to advise Marconi on the claim of his final specifications of patent, ... and he advised him to claim everything. I understand that as the claim was drawn, the claim, for telegraphy, not only the coherers, oscillators, & such like details, but even the Hertz waves! ... there is nothing new except the Hertz wave, the oscillator & the coherer, and these are not patented nor patentable.
(Hong, 2001, p. 43)

The complete final specification was filed on March 2, 1897, and accepted on June 2, 1897 (Figure 170). The specification showed the design of the system (Figure 171, Figure 172). The texts of the provisional and the complete specification were quite different. Emphasis was now put on the *system* for wireless transmission.

Above all, the nature of the invention was presented as consisting in the way in which the various components of the apparatus were assembled. ... the author made it clear that the invention ... "relates in great measure in which the above apparatus are made and connected together." (Guagnini, 2002)

On June 2, 1897, Marconi was granted the British patent № 12.039 for 'improvements in transmitting electrical impulses and signals, and in apparatus therefor'. It would become one of the first patents in wireless telegraphy. Everyone was surprised when the contents of Marconi's patent were publicized and they read his 19 claims. Most were related to the coherers and of various methods of connecting them, such as the ground connection. However, the claims were not limited to his improvements on the coherer, the tapper, and induction coil. In their totality, they covered the whole system of wireless transmission: it was the *combination of the elements* that created the *system of wireless transmission*. This was the new art of wireless telegraphy.

His first claim, indeed, was about the method:

> 1. *The method of transmitting signals by means of electrical impulses to a receiver having a sensitive tube or other sensitive form of imperfect contact capable of being restored with certainty and regularity to its normal condition substantially as described.* (text of application).

The Maxwellians were flabbergasted:

> *How could Marconi, thought of as a modest and open youth, dare to claim everything about Hertzian waves? How could he claim originality in regard to the Branly tube (which had been improved by Lodge) and the Righi-type ball transmitter?* (Hong, 2001, p. 44)

In the patent application for his invention, he described his improvements in the transmitting and receiving instruments in detail:

> *My first improvement consists in automatically tapping or disturbing the powder in the sensitive tube, or in shaking the imperfect contact, so that immediately the electrical stimulus from the transmitter has ceased the tube or imperfect contact regains its ordinary non-conductive state. ...*

> *A further improvement has for its object to prevent the electrical disturbances which are set up by the trembler and other apparatus in proximity or in circuit with the tube from themselves restoring the conductivity of the sensitive tube immediately after the trembler has destroyed it as has been described. ...*

> *Another improvement consists in a modified form of the plates connected to the sensitive tube, in order to make it possible to mount the receiver in an ordinary*

Fig. 5A.

Fig. 9.

Figure 171: Figure from the specification of Marconi's patent № 12,039 for a wireless telegraph system: Transmitter and receiver components (1897).

In the transmitter (Fig. 9), he applied four spark balls. In the receiver (Fig. 5a), the electric waves are detected by the sensitive tube (J).

Source: http://www.radiomarconi.com/marconi/popov/pat763772.html

circular parabolic reflector. … A further improvement has for its object to facilitate the focussing of the electric rays. (Text of patent)[245]

That being stated, the description of the claims followed. In the last few, he claimed to have invented the parts of the system:

Claim 15: A receiver consisting of a sensitive tube or other imperfect contact inserted in a circuit, one end of the sensitive tube or other imperfect contact being put to earth whilst the other end is connected to an insulated conductor.

Claim 16: The combination of a transmitter having one end of its sparking appliance or poles connected to earth, and the other to an insulated conductor, with a receiver as described in claim 15.

Claim 17: A receiver consisting of a sensitive tube or other imperfect contact inserted in a circuit, and earth connections to each end of the sensitive contact or tube through condensers or their equivalent.

Claim 18 The modifications in the transmitters and receivers in which the suspended plates are replaced by cylinders or the like placed hat-wise on poles, or by balloons or kites substantially as described.

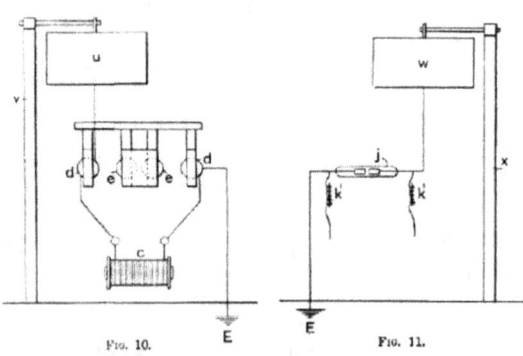

Fig. 10.   Fig. 11.

Figure 172: Figures from the specification of Marconi's patent № 12,039 for a wireless telegraph system: the total system (1897).

In the transmitter (Fig. 10), the electric waves created by the Rhumkorff coil (C) and spark balls (four balls, D, E) are transmitted by the tin foil antenna (U) on the pole (Y). In the receiver (Fig. 11), the electric waves captured by the tinfoil (W) on the pole (X) are detected by the sensitive tube (J).

Source: http://www.radiomarconi.com/marconi/popov/ pat763772.htmlhttp://earlyradiohistory.us/1897mar.htm

What Guglielmo had invented, and was to improve upon in the coming years, was a transmitter and a receiver in a communication system using electromagnetic waves. The transmitter generated an electrically coil-induced spark between discharge balls and was commanded by a telegraphic key. To generate the electromagnetic waves, he used a Rhumkorff coil in combination with tinfoil placed on a pole as an antenna, and

[245] Source: 'Marconi Telegraphy'. *The Electrician*, September 17, 1897, pp. 683–686. http://earlyradiohistory.us/1897mar.htm

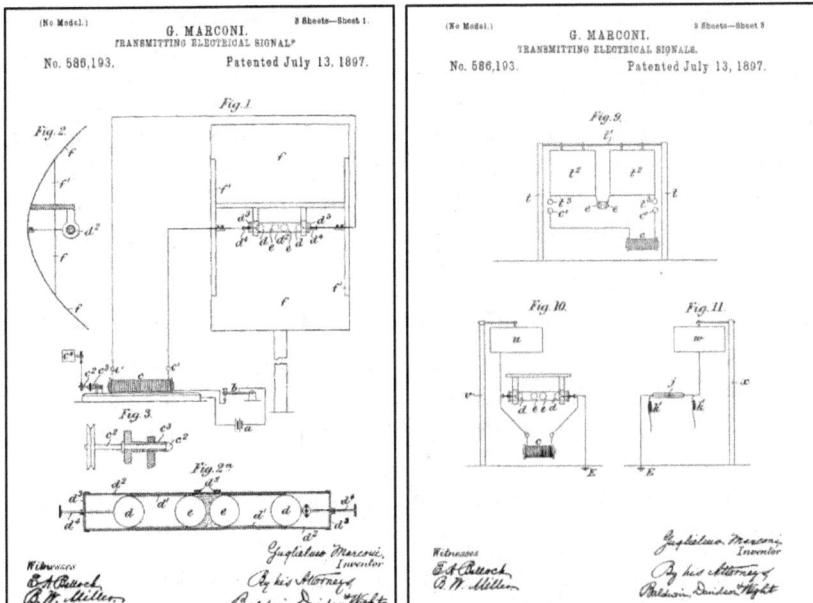

Figure 173: Marconi's US patent № 586,193 (1897), equivalent to the GB 12,039 patent.

Figure 2 shows the Hertzian reflector, and Figure 2a the four spark balls of the transmitter. Figure 9 and 10 show transmitter variations. Figure 11 shows the receiver.

Source: USPTO.

he grounded the other side with a metal plate to the earth (Figure 172, left). At the other end of the communication, the receiver used a metal-filled sensitive tube (aka the Branly-coherer) in combination with an antenna grounded with a metal plate to detect the waves (Figure 172, right). The patent for his invention was more than just the claim for wireless telegraphy; it was a blow to the Maxwellians.

> *Marconi was an Italian. The 'ether' had been discovered by great British scientists (Faraday, Kelvin, Maxwell). The Maxwellians, heirs to those scientists, had lost the priority for the discovery of electromagnetic waves to a German, Heinrich Hertz. … Marconi had opened up the possibilities of commercial use of the ether, and his comprehensive patent made matters worse. Furthermore, it was clear that wireless telegraphy would have naval and military uses. If his patent went unchallenged, Marconi not only would monopolize not only Hertzian waves but also important British national interests.* (Hong, 2001, pp. 44-45)

It was thus no accident that after Marconi got his patent, many British scientists and engineers would be challenging his novelty. With his broad claims, Marconi had violated the 'rules of the game', so there was no need for them to follow the rules. It was the start of an endless battle for priority (Hong, 2001, p. 45)

To secure his rights internationally, he also decided to apply for patent protection in the United States as soon as possible, so he also applied for the equivalent patent in the US on December 7, 1896. It was originally granted as US Patent № 586,193 on July 13, 1897. After some amendments it was reissued as 11,913, granted June 4, 1901 (Figure 173).

Testing with Some Help from Your Friends at the Post Office

The invention now legally protected, Henry Jameson-Davis brought Marconi—through his relationship with the electric engineer A.A. Campbell-Swinton—in contact with a dominant party in telegraphy in England: the Post Office. As a result, Marconi met with William H. Preece, Chief Engineer to the British Government Telegraph Service of the Post Office, whom we have met before in relation to Bell's telephone.[246]

Preece, 62 years old, met the 22-year-old Marconi. He recognized a fellow engineer in the young Italian tinkerer. Thus, Guglielmo demonstrated his invention in June 1896 successfully to Preece—the most powerful man in communications in England at the time—in the chief engineer's laboratory. The generously paternal matter in which Preece received young Marconi and the way in which, at least initially, he placed the Post Office facilities at Marconi's disposal was certainly related to the Preece-Lodge controversy mentioned before (G. R. M. Garratt, 1994, p. 59).

> *The only reason that this very important man was more than welcoming to an unknown entity such as Marconi was that he found himself at that exact moment in somewhat of a quandary, one that Marconi alone might help him to resolve. His problem, quite predictably, was to how to satisfy his obligation to his government to develop an effective, stable and secure system of communications within the British Empire especially with moving ships. He had pinned his hopes on an inductive method of wireless communication which he himself had devised but which he had reluctantly just come to realize would not work.* (F. P. Marconi, nd)

This visit opened for Marconi a first window of opportunity and resulted in the generous participation of the Post Office in setting up further tests on the rooftops of the General Post Office's building and a post office a mile away.

---

[246] See: B.J.G. van der Kooij, *The Invention of the Communication Engine 'Telegraph'* (2015) pp. 327–337.

The first experiments in England were from a room in the General Post Office, London, to an impromptu station on the roof, over 100 yards distant, with several walls, &c., intervening. Then, a little later, trials were made over Salisbury Plain for a clear open distance of nearly two miles. In these experiments roughly-made copper parabolic reflectors were employed, with resonance plates on each side of the detector. (Fahie, 1899)

The trials on July 27, 1896, set up to test the practicality of Marconi's system, impressed Preece greatly, and he stated: 'Young man, you have done something truly exceptional, I congratulate you on it' (Jolly, 1974, p. 37). It was the beginning of cooperation between the young Guglielmo Marconi and the much older William Preece.

Support for further experimenting with Marconi's wireless was taken up by the Post Office and other government departments (Admiralty and War Office). It was soon arranged to hold further trials such as on the Salisbury Plains in September 1896.[247] There, at Three Mile Hill, the trials were held under unfavourable weather conditions.

The tests were not just a technical affair among technicians; they were also a demonstration of the new invention to the numerous other people who were in attendance. It wasn't just Post Office officials who were present; representatives of the Navy and Army were also observing the work of the young Italian. Among them was a naval officer called Henry Jackson, who had been experimenting with a system of his own at the Naval Torpedo School at Plymouth. And there was a foreigner present: professor Slaby from Germany, also into the field of wireless telegraphy. In addition, the demonstrations were publicly reported in the papers, creating an early awareness in the general public.

Surprisingly, Marconi was able to send the messages at Salisbury in a more or less definite direction. By aiming the aerial reflectors he projected beam-like waves. The Hertzian energy was concentrated by a parabolic copper reflector or bowl, two or three feet in diameter, thereby shaping the waves into narrow strips, in much the same way that a searchlight stabs a streak of light through the darkness. (Dunlap, 1941, p. 55)

---

[247] The Salisbury Plains is an area that saw the early settlements of people in historic times, and places like Stonehenge and Old Sarum are located in the area (Rutherfurd, E.: 'Sarum, London, 1987'). By the time of the experiments the area was in the process of becoming a training ground for the military. Among the troops were those preparing for the Boer War in South Africa.

Figure 174: Marconi's black box containing the coherer receiver (1896).

Source: Museum of the History of Science, Oxford. Press pack
https://www.mhs.ox.ac.uk/marconi/presspack/images/receiver.jpg

Not long after the demonstration, on September 22, 1896, at a meeting of the British Association for the Advancement of Science in Liverpool, Preece 'told the world' of Marconi's invention. In addition, on December 12, 1896, a public lecture called 'Telegraphy without Wires' was held in Toynbee Hall, the educational and charitable institution in London's East End. The speaker, William Preece, operated the transmitter, and whenever he created an electric spark, a bell rang on a box that Marconi took to any part of the lecture room (Figure 174). There was no visible connection between the two. The demonstration caused a sensation and made the young Marconi a celebrity, as many interviews with journalist created a massive publicity.[248]

The audiences were astonished by the 22-year old inventor. He was always smartly dressed, spoke English perfectly in a low voice, and had a reserve which seemed more English than Italian. (Crowther, 1974)

By then the Post Office officials, practical engineers mostly, were convinced that that young Italian who spoke the English language so well had managed to create something new and usable. In the press, Marconi was heralded as the inventor of the wireless. This prompted a strong reaction from scientific circles. Oliver Lodge, who claimed he also had made valuable contributions prior to Marconi, was outraged.

The public announcement of his invention in September 1896 stirred strong feelings. The first reaction came immediately after Preece's intervention at the British Association for the Advancement of Science. The day after the communication, Professor Olivier Lodge invited members of Section A

---

[248] Source: https://www.mhs.ox.ac.uk/marconi/exhibition/marconiarrives.htm

Figure 175: Bristol Channel experiment (1897).

Drawing by George Kemp, which refers to the Bristol Channel experiments.

Source: http://www.carnetdevol.org/Wireless/marconi-transmanche.html

[Mathematics and Physical Sciences] *to witness in his laboratory an experiment, at an hour's notice, his assistant had arranged apparatus to do an exactly similar thing. He regretted that the resources of the Government should be opened to foreigners instead of encouragement being given to workers in the country.* (Guagnini, 2002)

The objective was to improve the apparatus and test what transmission distances could be bridged; the focus was on distance. So, next came the Bristol Channel tests between Lavernock Point and Flat Holm Island on May, 1897 (Figure 175). Testing was conducted over several days, with the trials on May 13 and 14, observed by Preece and other officials of the British Navy and Army (Figure 176). During these and later trials, George Kemp was active as Marconi's assistant.[249] On May 13, 1897, Marconi transmitted the first radio message in Morse code: 'CAN YOU HEAR ME'. Shortly afterwards, Kemp transmitted a reply: 'YES LOUD AND CLEAR'. They were a triumph for Marconi: he was able to send a signal across the water over a distance 8.7 miles.

Next to the previously mentioned demonstrations, which got huge publicity, Marconi also arranged to give lectures on his wireless system to specific interest groups of technical audiences.

Although Marconi did not like public speaking, he gave lectures at the Royal Institution, The Royal Society of Arts, the Institution of Electrical Engineers and many other venues, on the progress of his work, any or all of which could have been done by Preece, or many others, but none of them did. (Simons, 1996, p. 46)

---

[249] Georg Kemp was an ex-petty officer and had been one of Preece's laboratory assistants. July 1897, when the company was created, he joined Marconi from the Post Office, becoming his inseparable assistant and technician for more than thirty years.

Figure 176: Officials verifying the Bristol Channel experiment (1897).
British Post Office engineers inspect Marconi's wireless telegraphy (radio) equipment during a demonstration on Flat Holmisland, 13 May 1897.

Source: Wikipedia commons, Cardiff Council Flat Holm Project.

More Publicity of Marconi's Invention

Marconi, in the meantime, together with cousin Jameson-Davis, had developed plans to expand the business side of his activities. He did set up permanent wireless stations for continuous trials and public demonstrations. The first station was at the Needless Hotel, overlooking the Alum Bay in the western corner of the Isle of Wight, in November 1897. It was not by accident that he erected the station here.

> From the outset, Marconi shrewdly conducted his experiments in the presence of eminent personages, including ambassadors, military dignitaries, and illustrious scientists - an impressive roster of believers whose presence often proved more newsworthy than the experimental results themselves. In 1898, he landed the top customer of the empire, Queen Victoria, who had him erect a station at Osborne House to connect with the yacht of the convalescing Prince of Wales (Satia, 2010, pp. 846-847)

So, Marconi decided to create publicity at specific events. Both were related to a favourite sport of the royalty and nobility: the sailing regattas. The first was the Kingston Regatta of July, 1898, near Dublin that was reported upon minute by minute by the Dublin newspaper *Irish Daily*

Express, a newspaper sponsoring the event. Marconi hired a tug, which he fitted with a 75-foot antenna, and set up a receiving station under the control of George Kemp. The journal's sailing correspondent was able to communicate using Marconi's wireless system on their ship, the *Flying Huntress*, which followed in the wake of races at sea. Throughout the proceedings some 700 news flashes were sent, at ranges of 10 to 25 miles (16–40 km). For Marconi, this was exhilarating and secured him astonishing publicity. This event was the cause for an invitation by the *New York Herald* to do the same during the America's Cup race at the end of the year.

The second event was related to the son of Queen Victoria residing in Osborne House in the Isle of Wight. Edward, Prince of Wales, had injured his knee falling from the stairs at a Rothschild ball in Paris, but instead of staying at home in his mother's house, he decided to stay at the royal yacht *Osborne* that was moored some miles away in Cowes Bay, preparing for the Regatta of the Cowes Week (Figure 177). Marconi did set up a wireless connection between the Osborne House at the Isle of Wright and the royal yacht in the bay, enabling daily communication between Queen Victoria and Prince 'Bertie' about his health. "H.R.H. the Prince of Wales has passed another excellent night and is in very good spirits and health. The knee is most satisfactory" (Moffett, 1899, pp. 99-112). But other guests of the royal entourage also seized the opportunity to use the marvellous invention. They transmitted messages like: "Papa anxious to know if you and children coming today, 12.30, for trip round the island" and "Can you come to tea with us some day?" (Jolly, 1972, pp. 54-56)

Figure 177: Prince of Wales watching the races at Cowes from the royal yacht *Osborne.*

Source: The Illustrated London News. Artist: Melton Prior

As wireless telegraphy was a new phenomenon, Marconi was looking for applications where his system could be used effectively. At that time the Corporation of the Trinity House, responsible for British lighthouses, was looking for a reliable system of communicating with their lightships and offshore lighthouses. Ship-to-shore

communication seemed a promising market. In December 1898 test were held to see if wireless communication could be established with lighthouses and lightships. Marconi installed equipment for a wireless station in the South Foreland lighthouse, and his assistant, Kemp, went on board the East Goodwin lightship anchored near the dangerous Goodwin Sands off Dover. They successfully bridged a distance of 12 miles (Figure 178, top).

Marconi continued to improve and patent his system. The receiver modifications resulted in greater sensitivity by applying an 'oscillation transformer' between the antennae and the earth (Figure 180). This enabled him to cover greater distances. On March 27, 1899, the first wireless signals were transmitted across the English Channel (Figure 178, bottom).

Figure 178: Marconi's antennas at the South Foreland Lighthouse, Dover, England (top), and Chalet d'Artois, Wimereux, France (bottom), in 1899.

The mast supporting the vertical wire can be seen on the edge of the cliff.

Source:
http://ns1763.ca/marconi100/marconi1.html

At five o'clock on the afternoon of Monday, March 27th, everything being ready, Marconi pressed the sending-key for the first cross-channel message. There was nothing different in the transmission from the method grown familiar now through months at the Alum Bay and Poole stations. Transmitter and receiver were quite the same; and a seven-strand copper wire, well insulated and hung from the sprit of a mast 150 feet high, was used. The mast stood in the sand just at sea level, with no height of cliff or bank to give aid. "Brripp --- brripp --- brripp --- brripp --- brrrrrr," went the transmitter under Marconi's hand. The sparks flashed, and a dozen eyes looked out anxiously upon the sea as it broke fiercely over Napoleon's old fort that rose abandoned in the foreground.

Would the message carry all the way to England? Thirty-two miles seemed a long way. "Brripp --- brripp -- brrrrr -- brripp -- brrrrr -- brripp -- brripp." So he went, deliberately, with a short message telling them over there that he was using a two-centimeter spark, and signing three V's at the end. Then he stopped, and the room was silent, with a straining of ears for some sound from the receiver. A moment's pause, and then it came briskly, the usual clicking of dots and dashes as the tape rolled off its message. And there it was, short and commonplace enough, yet vastly important, since it was the first wireless message sent from England to the Continent: First "V," the call; then "M," meaning, "Your message is perfect;" then, "Same here 2 c m s. V V V," the last being an abbreviation for two centimeters and the conventional finishing signal. (Moffett, 1899, pp. 99-112)

Thus, the young Marconi—he was 25 at the time—had managed to establish a link over the English Channel between the South Foreland lighthouse near Dover (England) and the Chalet d'Artois at Wimereux, a small village near Boulogne-sur-Mer in France. By now he had crossed a distance of 50 km (32 miles) across the Channel.

The result [of bridging the Channel] *was to create an immense public interest in the achievement all over the world. Up to that moment wireless telegraphy by electric waves had attracted only a very limited attention: but the bridging of the English Channel by electric waves was one of those sensational facts which at once aroused the daily press to lively comment on the matter.* (Fleming, 1919, p. 529)

Figure 179: Marconi's experimental spark-gap transmitter (1897).

Early spark-gap radio transmitter used by Guglielmo Marconi during experiments at La Spezia, Italy, in July 1897.

All in all, the two and a half years between June, 1896, and December, 1898, were occupied by Marconi with numerous public demonstrations. As a result of continuous experimenting, he also improved his apparatus (Figure 179), for example, connecting his receiving relay to a Morse printing apparatus (Figure 180).

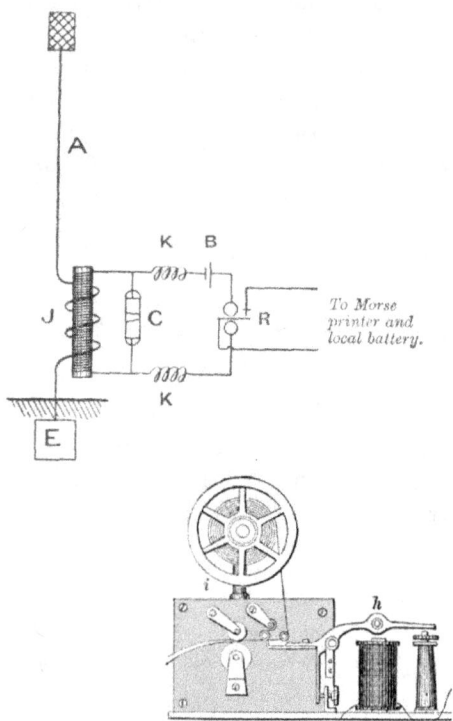

At the end of 1898 electric wave telegraphy had thus established by Marconi on a practical basis. He had demonstrated its utility, especially for communication between ship and ship and ship and shore, a work which could not be accomplished by any other system. (Fleming, 1919, p. 523)

The success of Marconi's demonstrations of his invention had also reached Italy. It didn't take long before he was invited to demonstrate his invention in Italy. There, after demonstrating his invention in La Spezia in July 15, 1897 (Figure 179), and meeting the Italian King and Queen, his fame was heralded by the Italian press (Jolly, 1972, pp. 38-41).

Figure 180: Marconi's receiver improvement with coils (1899, top) and the addition of a Morse printer (bottom).

In the antennae circuit (A-E) a transformer (J) is applied.

Source: http://www.newscotland1398.net/ nfld1901/marconi-nfld.html

Now that it was clear that transmission of telegraphic signals over land and water was feasible over increasing distance, Marconi focussed on shore-to-ship and ship-to-ship communication.

Marconi then concentrated on his original idea of communication with, and between ships at sea. He established a coastal station at the Needles Hotel, Alum Bay, Isle of Wight ..., and carried out tests with two steamers, achieving ranges of up to 18 miles (29 km) - always, it seems, in bad weather. Bad weather and the results of gales, continually appear in the records of Marconi's work. A second station was set up at the Madeira Hotel, Bournemouth. Lord Kelvin sent the first paid message (he insisted on paying), the first wireless telegram, from the Isle of Wight to Bournemouth in early 1898, thus creating a problem with the Post Office, whose monopoly covered all messages within the three mile limit. (Simons, 1996, p. 47)

In addition to the numerous demonstrations, experiments between naval ships were also held during the Naval Manoeuvres of 1899. Distances up to some 96 km were reached.

In October 1899 Marconi went to America and equipped two ships with his wireless system to report on the progress of the well-known sailing contest: the *America's Cup Race*. One of the participants was the wealthy British yachtsman Sir Thomas Lipton, founder of the tea company bearing his name, who challenged the holders of the America's Cup five times between 1899 and 1930, always with a boat named *Shamrock*. In 1899 he lost due to a broken mast, and Marconi witnessed the disaster from nearby (Figure 181). Within minutes the news was flashed ashore to the new papers *The New York Herald* and *The Francisco Call*. Both papers reported intensively on the Cup and thus contributed to Marconi's publicity.

Marconi was not the only one demonstrating his invention. For some other pioneers of the wireless communication—Lee de Forest and the American Wireless Telephone and Telegraph Company claiming ownership of Amos Dolbear's 1886 patent—the race proved to be a failure. None of the companies had effective tuning for their transmitters, so only one could transmit at a time to avoid causing mutual interference. Although an attempt was made to have the three systems avoid conflicts by rotating operations over five-minute intervals, the agreement broke down, resulting in chaos as the simultaneous transmissions clashed with each other.

SHAMROCK CRIPPLED AND COLUMBIA HAS A "WALKOVER."

TOPMAST OF THE CUP CHALLENGER CARRIED AWAY.

Figure 181: The *Shamrock* losing her topmast in the America's Cup Race (1899).

Source: http://cablecarguy.blogspot.fr/
2013_09_01_archive.html

From the day the races began on October 3, 1899, Marconi's fame in the United States was assured. The success of this demonstration of

Marconi's wireless system aroused worldwide excitement and led to the formation of the American Marconi Company.

> *During the progress of the America Cup Races, an agreement to conduct tests of his equipment aboard U.S. naval vessels was reached with Marconi. ... The installations were completed and the initial tests were commenced on 26 October. The trials were not fully completed when Marconi was requested to return to England to prepare apparatus for use in the Boer War. Always extremely anxious to do his utmost to be of service to the British Government, he made immediate arrangements to return. The remainder of the tests were cancelled.* [250]

However, when a request was later made by the US Navy to buy some sets of Marconi's equipment, problems arose, as Marconi only wanted to supply equipment under a lease and service contract. The US Navy refused, and that started their efforts to obtain wireless telegraphy from sources other than the British (ie the Germans and their own infant industry of the company Telefunken).

By that time both the British Navy, with officers present during the America's Cup race, and the British Army understood the importance of wireless communication. One of the areas of conflict in the British Empire was South Africa, where not everybody was that happy with the British imperial aspirations. The conflicts in South Africa with the descendants of former Dutch 'Boer' settlers culminated in the (Second) Boer War (1899–1902). As the military—after the tests on the 1896 Salisbury Plains—now were looking at wireless telegraphy as a strategic and tactical tool in warfare, the Boer War would see its first use.

> *Most notably, on declaration of the* [Boer] *war in October 1899, he immediately contacted the War Office to suggest that they use radio to control the flow of men and material into South-African ports and to "communicate across tracts of country where it would not be possible or prudent to carry ordinary telegraph wires." Indeed the war secured him his first contract. ... Within a month, he had dispatched a team of engineers and five wireless sets to the war zone, but they functioned so poorly that they cast fresh doubt on his claims about his system's immunity from "tapping" or disruption by the enemy. ...*
>
> *The director of Army Telegraphs retaliated by withdrawing the sets from service. By then, however, the navy was sufficiently impressed by the performance of his device in manoeuvres that it snapped up the rejected apparatuses for the naval blockade in Delagoa Bay, where they functioned well enough to prompt the purchase of some thirty sets in July 1900. ... Although this military contracts had immense publicity value, they did not generate further sales to the state, partly because the navy began building its own sets.* (Satia, 2010, pp. 836-837)

---

[250] Source: http://earlyradiohistory.us/1963hw03.htm

The Hero of Wireless Transmission

The period of 1896–1899 had been a time of constant experimenting and testing of Marconi's equipment for wireless communication in and around Britain (Figure 182). In 1899 America had also become an interesting testing ground, and the Atlantic crossing in 1901 cemented

Figure 182: Some of the test sites for Marconi's wireless experiments (1896–1901).

Dots indicate local experiments. Lines indicate long-distance experiments.

Source: based Maproom map https://maproom.net/wp-content/uploads/BIMC-MASTER-British-Isles-Mini-Countries.jpg

Marconi's public image. The main theme had been increasing the distances to be bridged by the Hertzian waves (Table 3).

> *With his pursuit of distance, Marconi played on a related set of security concerns that were more political-economic in nature: Britain's diplomatic isolation at a time of long-distance military conflict intensified calls for strengthening imperial ties, particularly among the "white" colonies of settlement, leading to Chamberlain's postwar calls for a tariff federation. While critics harped on the security weaknesses of his device for military use, Marconi traded on the multiple valences of the security concern as he explored other avenues for sustaining his commercial venture. Having failed to find contracts among state departments, he redirected his energies toward the creation of a wireless network that would capture the communication market of the empire itself. A sympathetic press continued to couch this application of the technology in terms of imperial security ... (Satia, 2010, p. 838)*

A—deliberately orchestrated—side effect was the publicity all the testing generated. The press loved to hear and report on Marconi's vision of the future of wireless, his experiments proving the validity of his ideas. That the highly visible sparks created the intangible, invisible waves carrying the message into the distance was a story worth telling. His story featured

Table 3: Increasing distances covered by Marconi apparatus at specific tests for specific interest groups (1896–1901).

Year	Location	Interest Group	Distance (miles/km)
March 1896	London, Post Office	Meeting with Preece	Local
June 1896	London, Post Office	Post Office officials	Local
Sept. 1896	Salisbury Plains	Army, Navy, Post Office	1.75/2.8
May 1897	Bristol Channel	Army, Navy, Post Office	8.75/
July 1897	La Spezia (Italy)	Italian Navy	11.8/18,9
Oct. 1897	Bath-Salisbury	-	34/54
Nov. 1897	Needles, Isle of Wight	Technical testing	14/22.4
July 1898	Ballycastle-Rathlin Island	Lloyds' insurers	18/28.8
July 1898	Kingston Regatta	General Public	25/40
August 1898	Cowes Race	Royalty/General Public	n.a.
Dec. 1898	East Goodwin Lightship	Trinity House	12/19
March 1899	Dover, English Channel	General Public	32/50
July 1899,	Navy experiments	Admiralty	45/72
Oct.1899	America's Cup	General Public	n.a.
Jan. 1901	Isle of Wight/Lizard	General Public	186/300
July 1901	Poldhu (England)/ Crookhaven (Ireland)	General Public	225/360
Dec. 1901	Poldhu (England)/Signal Hill (Canada)	General Public	2500/4000
Feb. 1902	US-Philadelphia crossing	General Public	2099/3358

Source: see text.

Figure 183 : Posters for the *International Exhibition* in Brussels (1897) and the *Greater Britain Exhibition* in London (1899).

Source: Wikimedia Commons. Artist: Henri Pivat Livemont (Brussels), https://imperialglobalexeter.com (London)

novelty in a time that was all about novelty; it was the time of the British Industrial Revolution. And electricity contributed a lot to all that novelty.

The fruits of technological progress were seen and admired by the public at large, such as at the *International Electro-Technical Exhibition* in Frankfurt (1891), the *Columbian Exposition* held in Chicago (1893), the *Exposition Internationale de Bruxelles* (1897), and the *Great Britain Exhibition* in London (1899) (Figure 183). Royalty, aristocrats, and politicians eagerly participated in creating the news events, as they also loved to be connected to the publicity of novelty. And thus Marconi became known as 'Mister Wireless' among the general public.

Marconi Gets Organized (1897–1903)

The young man who disembarked in early 1896 on the English shore brought with him the crude artefact of a system for wireless telegraphy. Despite its simplicity, in those days a device that could be called a high-tech invention. The artefact in the black box did mysterious things that were pushing the frontiers of existing knowledge and knowhow. It made communication at a distance possible through the dark space of the invisible ether. It was built with that new magical technology that was able to control lightning, and with the maturing electro-mechanical techniques of that time. It was operated on by the technical priests of that time. That all made wireless communication magical and fascinating for the general public.

As Britain at that time not only had become an industrious nation—being in the midst of the Second Industrial Revolution—it also was ruling the waves as a big maritime nation. The British Empire, with its colonies all over the globe, was connected and protected by the British Navy and traded by British merchant shipping. Realizing that this situation would offer him a stimulating environment for his invention, Guglielmo Marconi had gone to England to try and push his system. After being introduced to the people of the Post Office, he was soon engaged in technical affairs, demonstrating the feasibility of his system to an increasing number of people. By that time he was undoubtedly dreaming about the possibilities for introducing his system to a broader audience, maybe even capitalizing on his invention, but how that was to be realized was not that obvious at all. So, soon he was to face a fundamental issue and would have to make a decision that would determine the future of his invention, and of his personal life.

Quite a Decision to Make

His 1896 experiments drew the attention of the press, and his tests and demonstrations were reported worldwide, including in Italy, where the naval military had now become interested. In 1897 Marconi, busy in Britain with improving his equipment and being quite successful with his experiments and demonstrations, was faced with two issues. As an Italian citizen, he was liable to be drafted into the Army. The drafting problem was solved when, reluctant to renounce his Italian nationality, he was formally drafted as a cadet into the Italian Navy and became an assistant naval attaché of the Italian Embassy in London for a period of three years. It was just an administrative arrangement, as he could continue his experimenting as usual. In addition, there was the future of his invention. Being born in an entrepreneurial family, he had been thinking how to continue. It was clear that the moment approached when he had to make a decision on the course of action to take. Should he follow the entrepreneurial route and go into business, or should it be the civil servant route, selling his invention to the British state and joining the Post Office?

One has to realize that the 1890s were a period of time in which earlier technological advances had shown their commercial potential. Indeed, in the second half of the nineteenth century Britain was in the midst of massive industrialization, dominated the seas, and was at the zenith of its colonial empire. But that 'esprit' was not limited to Britain. The Spirit of the Times was broadly spread: the 'Gilded Age' in America, the Gründerzeit in Germany/Austria, and 'La Belle Époque' in France. In Britain, the Industrial Revolution had progressed rapidly. Many technical developments, such as the development of steam power, had resulted in a Transportation Revolution: people now travelled by steam-powered trains and ships. That industrialization also had created the Communication Revolution, where the communication engines of the telegraph and telephone had changed the way people (tele)communicated. The new means of communication had influenced greatly the business part of society, especially those areas where information was crucial (such as the stock markets and commodity markets).

Thus, for a keen businessman, it was not difficult to see from an investor's point of view that the next development in communication would offer quite a few business opportunities.[251] If those opportunities could be converted in profitable businesses strongly depended on the society in which they would develop: from the young Italian state to the mercantile-dominated Britain or the capitalistic America. And there was quite a distinction between those contexts to discern, such as in the business dynamics of America, which had loosened ties with its former colonial dominator, or in Britain, where the remnants of the feudal system still ruled in industrial times, or even in the newly minted kingdom of Italy, where the first steps of industrialization were emerging as the dust from the political turmoil was settling.[252]

In that context we find, coming from the Apennine Hills of Italy, a young man and his new communication engine. It was the end of his *Act of Invention* (Figure 232), in which an idea and concept had been transferred into an artefact (aka working prototype): a pioneering time in which he had

---

[251] The potential applications of electricity at that time were so overwhelming that it seemed like a Cambrian explosion of new developments: the electric light, the electric motors replacing horsepower, the new means of electric communication. It was a world of challenges for entrepreneurial people. The parallel with the Cambrian explosion of the information technologies at the end of the twentieth century comes to mind.

[252] See: B.J.G. van der Kooij, *Context for Innovation: British (R)evolutions in perspective* (2017), pp. 467–491.

been financially and morally supported by his parents. It was also the beginning of his *Act of Business* (Figure 233), where that same infant invention would be brought into maturity and made ready for the market of wireless telegraphy. The time of 'angels' supplying the seed funding was over.[253] Now was the time of business creation; with activities from financial investments to additional organizational, commercial, and technical development.[254]

Hectic Times for a Decision

Marconi had a nephew, the 20-years-older Henry Jameson-Davis, active as a milling engineer and well connected in the business circles of London. Coming from the Irish whiskey distiller's Jameson clan, he was well acquainted with many people in business. Realizing the great financial potential of his cousin's invention, he saw opportunities in its commercial development. Rather than simply selling the patent to, or tying into an agreement with, the Post Office, Davis proposed—in the tradition of British entrepreneurial thinking—to Marconi to create a 'limited company'.[255] A company that would be able to manufacture and technically improve the wireless equipment developed by Marconi, exploit his patent rights, and bring the total system (network and components) to the market, whatever that might be. A company that would be able to finance all those further resource-hungry experiments that were needed.

As such a company had to be financed, they needed investors with guts willing to invest in a novel thing as a 'wireless telegraphy', a technology in its infancy with initial prospects of replacing the cabled telegraphy. The first thing was to make a prospectus to inform potential investors of the new venture. That was done soon enough, and it contained the details of the future company's main asset: Marconi's invention as described by his lawyer, J. Fletcher Moulton, and supported by a favourable report from an outside expert. It stated optimistically: "the invention was exceedingly simple and reliable and capable of immediate industrial applications." Next,

---

[253] This is the early funding of the costs of operation by an informal investor or business angel. Sometimes it is family based, such as in Marconi's case with his father. In other cases, we find it 'friends and family' based.

[254] In today's terminology, this would be called venture capital, as it finances a high-risk activity.

[255] A company can be defined as a voluntary association of people formed and organized to carry on a business. The corporate laws provide for different legal forms, such as the limited company (Ltd). In a limited company, the liability of members or subscribers of the company is limited to what they have invested or guaranteed to the company. In a private limited company, the shares are held by private investors. In a public limited company, anyone can buy shares in that company on the stock market after its first public offering (IPO).

Jameson-Davis took this task to try and find people interested in investing in such a venture (Raboy, 2016, p. 71).

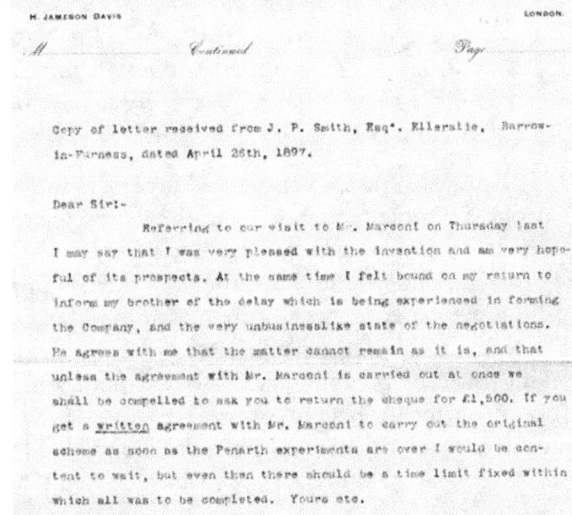

In addition to the financial issues, there was one other important activity to be considered. The provisional patent application might have been filed, but it still had to be finalized before it could be granted by the Patent Office. So, whether or not the invention could

Figure 184: Copied text of letter from investor J.P. Smith to Jameson-Davis.

Source: www.marconicalling.co.uk

obtain the protection of a patent was still uncertain. Marconi's lawyers took care of that part, as they filed the final specification of the patent application on March 2, 1897.

In early 1897 Jameson-Davis—doing the work without remuneration and footing the bills out of his own pocket—with his business contacts among corn merchants operating on the London market, had organized some potential investors for the prospective company that would own and exploit the Marconi patents. Marconi, by that time quite occupied with the technical, patenting, and early commercial activities but not experienced at all with these new aspects of setting up a company, was indecisive. It was on him to decide whether to try and sell his patent to the Post Office or to undertake the new business venture that was proposed. His lawyers advised him to be prudent before the tests were completed and the patent being obtained, and he stalled the proceedings. Marconi found himself 'in a difficulty', as he wrote to Preece explaining his dilemma. After laying out the situation, he stated: "I beg to state, however, that I have never sought these offers, or give encouragement to the promotors"(OX-1, 2017, Letter in Box 168)

Marconi was still uneasy about taking the entrepreneurial route rather than staying with a public agency like the Post Office.... Marconi was playing a high-

stakes game, trying to keep all the bases covered and seeking to alienate no one, and looking after himself first and foremost as he tried to figure out where his interests lay—as a private entrepreneur or a partner of a great public agency. The Post Office solved his dilemma by failing to move quickly enough. (Raboy, 2016, pp. 72-73)

By April 29, 1897, the potential investors were becoming impatient with the progress, and a certain J.P. Smith wanted his cheque of 1,500 pounds returned if 'this un-business like state of negotiations' continued (Figure 184). On the other hand, his father was advising prudence and advised first finishing the experiments. In the meantime Marconi himself, reflecting on his experience till then with the Post Office, was quite pessimistic about working with the bureaucracy of government. From the side of the Post Office, no offer for his patent was coming forward. And all that time Jameson-Davis was pressing for a decision.

Clearly relations between Marconi and his cousin were not quite as cordial as they seemed. In fact, the Marconis and their consigliere were close to a rupture. ... A meeting took place on March 31. ... They met again the following day ... The solicitors kept meeting separately with the two cousins... Marconi was playing a high-stakes game, trying to keep all bases covered and seeking to alienate no one, and looking after himself first and foremost as he tried to figure out where his interest lay—as a private entrepreneur or a partner of a great public agency. 72– 73 (Raboy, 2016, pp. 72-73)

Then, on April 30, 1897, Marconi agreed in principle with Davis's proposition and terms as outlined in the prospectus. He would go and sell his invention to a new, to-be-created company.

The new company's capital would be based on 100,000 stock options, each worth £1 par value at the issue of the stock (Figure 185). Marconi received for his patent rights 60,000 shares and in cash

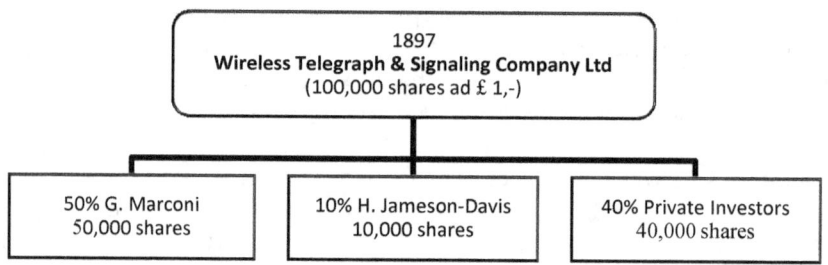

Figure 185: Shareholding in Wireless Telegraph & Signaling Company.
Capital in par value.

£15,000.[256] From his 60,000 shares, he was to give 10,000 shares to Davis: 'five thousand as "commission for your trouble" and five thousand "to pay all expenses in connection with the formation of the company".' And Marconi himself would receive a salary of £500 a year. The other 40,000 shares would be sold to seven unnamed private investors (Raboy, 2016, p. 74)

The deal may have been agreed between Jameson-Davis and Marconi, but at the same moment other pressing matters were at hand that required Marconi's attention. Such as the Bristol Channel tests (Figure 175) with the representatives from British government and Adolf Slaby present. At the same time others around him spread the word of the new invention, such as the influential presentations of Preece to the Royal Institution ('Signalling through Space without Wires', June 4, 1897) and Royal Society (June 16, 1897). On top of that, the patent for his invention was granted on July 2, 1897. And surprise…the news of his invention had reached Italy. Marconi was invited to demonstrate his invention to the Italian Navy (Figure 186).

As the agreement could not be finalized before his departure for Italy, a power of attorney from Marconi was arranged. So, during Marconi's stay in Italy to demonstrate to the admirals of the Italian Navy his invention in La

Figure 186: Marconi demonstrates his wireless system to the Italian Navy in La Spezia (July, 1897).

Sources: http://www.radiomarconi.com/marconi/bartolomeo2.jpg

---

[256] Equivalent to £1,535,000 in 2015; calculation based on historic standard of living. Source: http://www.measuringworthth.com/

Spezia, the *Wireless Telegraph and Signal Co. Ltd* was organized on July 20, 1897, by Jameson-Davis.[257] The rest of the stock—40,000 shares—was available to outside investors willing to invest in such a risky venture. And risky it was, as essentially this was a case of an 'unproven market' for an 'unproven product' with an 'infant technology'. In present day this would be high-risk start-up financing requiring a special kind of investor. In this case, the seven investors were corn and flour merchants who were, one way or the other, associated with the distilling business of the Jameson family. Their participation, although with a commercial motive, could become a long-term investment, as the funds they supplied were needed to develop both the technology as well as the market.

> *As far as Marconi was concerned, the company formed in 1897 was an instrument by which he was to continue his experiments. But by what lines? In what directions? ... For Marconi the company might have been a vehicle for technical experimentation; nevertheless, it was a company, a business enterprise that could survive only if it could sell products and services in the marketplace. Commercial problems and technological problems, therefore, were bound together in close interaction.* (Aitken, 2014b, p. 229)

Many of the first investors were from Ireland and related to the whisky-distilling activities of the Jameson clan. Such as the corn merchants: Cyril F. Bennet, Henry Obrè, Thomas Wiles, S.W. Ellerby, M.Y. Goodbody and Frank Wilson. In addition, James Fitzgerald Bannatyre and Robert A. Patterson put up some capital.

Thus, the new company was capitalized at £100,000,[258] which was quite a different value from the £10,000-estimate Preece had given in a confidential report entitled 'Marconi's Telegraph' sent to the secretary of the Post Office a few weeks earlier, in July 1897.

> *"My own view* [Preece's view] *is that subject to the system being made really practical ... and being favourably reported upon by the Admiralty and the War Department, the Government would be justified in acquiring the patent rights for £10,000 if the Attorney General pronounces in favour of the validity of the patent."* (Bruton, 2012, p. 72)

---

[257] The first name proposed—*Marconi's Patent Telegraph Ltd.*—was rejected by Marconi. However, the name of the company was two years later, on March 24, 1900, changed into *Marconi's Wireless Telegraph Co. Ltd.*
[258] Equivalent to £10,230,000 in 2015; calculation based on historic standard of living. Source: http://www.measuringworth.com/

The new business was administered by a managing director (Marconi's cousin Jameson-Davis, who did stay in management till 1899[259]) and a company secretary: H.W. Allen. Marconi not only became the major stockholder, which gave him control over the company; he also became its director and its employee. Marconi now, at the age of 23, was not only well known, but also well-off. Next to the £15,000[260] in cash and a nice salary of £500/year,[261] the potential value of his shares was enormous. But that was only the case if the company would grow and prosper. However, the signs seemed to be good, as Marconi wrote to his father in February 1898:

> *The company shares carry on increasing their value as our trials succeed over greater distances. In fact at about that time one of our shareholders sold all his shares which were more than 5,000 at £2 each, making a profit of £5,000; that is 160% in six months. I see today that our shares are quoted at the stock exchange at over £3 each, from their initial value of £1. By December 1898 the shares stood at £4 each.* (M. C. Marconi, 2002, p. 313)

One can try and understand the hectic character of this year 1897. On one hand, there were the technical developments and tests, the Post Office facilitating the technical efforts—a situation that saw a great personal involvement between Marconi and Preece, though in the bureaucratic context of government. Add to that the time-consuming work to be carried out obtaining patent protection and working with specialized patent lawyers. On the other hand, there were the explorations of an entrepreneurial route that could be followed. This needed again contacts with corporate lawyers and interesting prospective investors who wanted proof that the invention had potential before they could decide. Investors who also, businessmen that they were, pressed Marconi to make a decision. And last but not least, there were the 'commercial' contacts, where the new invention had to be demonstrated to interested parties like the navies of both Britain and Italy. For a young man—Marconi had reached the age of 23—this was quite a challenge, which he managed to solve with quite some result when one realizes that basically his invention had been nominally valued at some ten million pounds a year after he arrived in England.[262] And he owned—on paper—half of that value.

---

[259] In late 1899 Davis resigned as managing director, and when the company was restructured in February of the next year, Major Page Food became his successor. In 1902 Herbert Curt Hall became the managing director.

[260] Equivalent to £1,535,000 in 2015; calculation based on historic standard of living. Source: http://www.measuringworth.com/

[261] Equivalent to £51,160 in 2015; calculation based on historic standard of living. Source: http://www.measuringworth.com/

[262] The £100,000 value converted to 2015 value.

The moment the crucial decision was taken and working capital had become available, it was time to start the business. The board of the new venture had a lot on its agenda. They had a business to develop, one that was realizing something that had not been available before.

One has to realize that Marconi's invention was considered to be 'telegraphy without wires'. So, the business of telegraphy and the business of telephony—as they had developed after Samuel Morse's invention decades before and Alexander Graham Bell's invention not too long ago—were a source of inspiration. Being the businessmen that they were, they must have had a vision of what this thing 'telegraphy without wires' could become after seeing what had happened with its cabled predecessors, the telegraph and telephone.[263] Even without envisioning the new application of maritime communications, the business potential of just replacing the communications realized with cabled telegraphy and telephony was already mind boggling.

The Agenda: Patenting

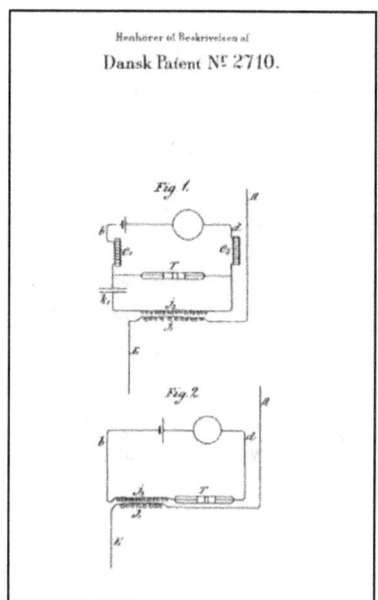

Figure 187: Marconi's Danish Patent 2,710 (1899).

Source: Espacenet

One of the first strategic considerations the Board of the newly formed Wireless Telegraph & Signal Company put on the agenda was the issue of patenting. Guglielmo Marconi might have obtained his first patent protection for his invention in Britain, but in other geographic regions, his invention also needed to be protected by the local patent laws.

At the first working meeting following the company's incorporation, the board of directors approved nineteen new patent applications and instructed agents to apply for patents in British Nord-Borneo, the Straits Settlement, Hong Kong, Gibraltar, Victoria (Australia), Ceylon and Canada. (Raboy, 2016, p. 87)

It was a lot of work, as patent laws in different countries had different rules and requirements.

---

[263] See: B.J.G. van der Kooij, *The Invention of the Communication Engine 'Telegraph'* (2015); B.J.G. van der Kooij, *The Invention of the Communication Engine 'Telephone'* (2015).

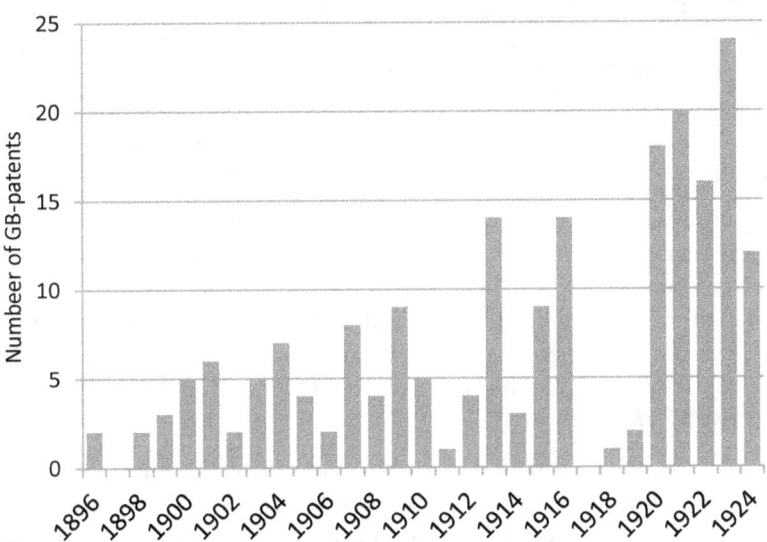

Figure 188: British Patents for Marconi and the Marconi Company (1896–1924).

Source: http://www.radiomarconi.com/marconi/brevettimarconi.html

Patent applications had to be prepared, corrected, and finalized according to all those different rules and protocols. And patents had to be maintained: either by paying an annual fee or by having someone actually doing some development or testing work on the device in the country itself.[264] So, as a rule, patenting on a broad, international scale was only done for something of a general nature worth patenting. As a consequence of this patent strategy, by 1900 over £75,000[265] had been invested in obtaining patents for Marconi's inventions in a range of countries, many of them in Europe, such as France and Denmark (Figure 187). Outside Europe, these were often countries related to the British Empire (eg Canada). It was the largest item on the balance sheet of that year (OX-3, 2017 Annual Report Sept. 30, 1900).

This patenting was the beginning of a strategy that would be fundamental to the rise of the Marconi monopoly. It resulted in a flood of

---

[264] Patenting in Britain was expensive in those days, as, after the fifth year, each year, a renewal fee had to be paid. Year 5: 5 pounds, year 6: 6 pounds, and so on up to year 14: 14 pounds.

[265] Equivalent to £7,318,000 in 2015; calculation based on historic standard of living. Source: http://www.measuringworth.com/

both British patents (Table 4) and US patents (Table 5) over the next decades (Figure 188). Among those many patents that Marconi was granted in the period 1896 to 1903 were the three patents that created the cornerstones for his monopoly: GB Patent № 12,039 from 1896; GB Patent № 7,777 from 1900; and GB Patent № 10,245 from 1902.

In addition, there were the patents for improvements. Some covered minor improvements in technical areas, eg the GB Patent № 401 of 1901 that was related to the adding of a condenser. Some were related to more important improvements, such as the syntony improvements in the receiver and transmitter. Again, in the period of 1904 to 1907 patent applications were filed and patents were granted in Britain, for example: GB Patent № 4,869 of October 13, 1904; GB Patent №14,788 of July 18, 1905; GB Patent № 16,655 of August 16, 1905.

And when the company became more commercially successful in the period 1908 to 1913 there was a constant stream of patent applications. After the wireless mania had started in the engineering community (Figure 230), Marconi continued to improve his system and patent his contributions far into the 1920s (Figure 188). Patenting was to become the cornerstone of the 'global' Marconi wireless monopoly.

Patenting was for Marconi an international affair. Take the example of his efforts to create the reliable magnetic receiver (aka Maggie). In the period 1902 to 1903 the Marconi Company undertook a massive effort to patent Marconi's invention for the magnetic receiver (as covered in GB Patent №10,245) worldwide: from Austria to France, Brazil, Rhodesia, Egypt, Peru, and Switzerland. This effort illustrates the importance of protecting the invention.

The Agenda: Internalization

Another important strategic item that stood at the agenda of the board was 'internalization'. As Marconi's inventions did not have geographical limitations, they could be used everywhere on the globe. By its very nature, wireless communication neglected geopolitical borders and was an international affair. And when it became clear that local, short-distance wireless telegraphy was facing the (British) monopoly, long-distance wireless telegraphy became even the more interesting. This was in addition to the rapid development of the marine markets for shore-to-ship communications.

Thus, by creating national subsidiaries all over the world that would be licensed to use the patents obtained in those countries, additional revenue could be created for the parent company as stockholder and as licensor.

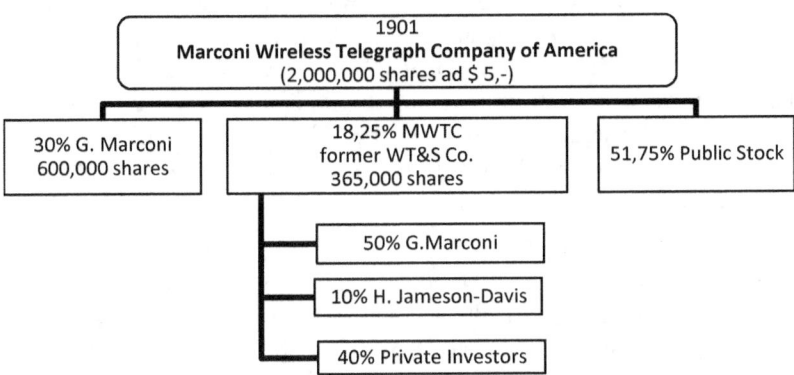

Figure 189: Shareholding in Marconi Wireless Telegraph Company of America.

Capital in nominal value.

The ensuing internationalization started with the Marconi Companies of America and Canada. But that was only the beginning.

> *When he was not working on a ship, Marconi spent much time in New York with the company director William Goodbody, meeting with potential investors. Goodbody had come over with Marconi to float the US and Canadian rights to the company's patents, which the company was willing to sell outright (£300,000[266] for both countries, £200,000 for the United States alone, or £150,000 just for Canada.)* (Raboy, 2016, p. 121)

In capitalistic America, in the wake of the wireless bubble, they did not have too many problems finding investors. This was even more true when Marconi was heralded so much as the hero of wireless telegraphy in the press. As a result, on November 22, 1899, the *Marconi Wireless Telegraph Company of America* (aka American Marconi) was incorporated and capitalized at $10 million (2,000,000 shares at a par value of $5).[267] The British parent firm held 365,000 shares, and Marconi was given 600,000 shares for American rights to his patents, present and future. The deal was quite advantageous to both the parent company in Britain and Marconi himself (Figure 189) and increased Marconi's personal wealth considerably.[268] He was in his mid-20s and already a multi-millionaire.

---

[266] Equivalent to £29,270,000 in 2015; calculation based on historic standard of living. Source: http://www.measuringworth.com/

[267] In 1919, after World War I, this company would become the Radio Corporation of America (RCA).

[268] Just imagine Guglielmo having a 50% share in the parent company now having thus a 50% part in the 365,000 MWTC of America shares: a nominal value of $912,500. Add to

That corporate strategy of internationalization was implemented rapidly. As we will see later on in more detail (Figure 195), to serve the international maritime market, the *Marconi International Marine Communication Co.* was created in 1900. Also, in 1900 they organized the *Compagnie de Télégraphie sans Fil* (C.T.S.F.) in Brussels. In 1903 the *Marconi Wireless Telegraph Company of Canada* was organized. They all contributed to the balance sheets of the parent company.

One has to see these corporate strategies of patenting and internationalization—that were based on a vision of wireless telegraphy being a dominant technology for the coming developments in worldwide telecommunication—in the context of that time. This was the time of the British Empire, where Britain was the dominant force in a lot of geopolitical theatres, and it exercised its influence in all its Eastern and African colonies, but also its dominance in the Western world of Europe and the Americas. It was the naval arms race (Figure 117) that had been a dominant 'tool of the empire' and resulted in the British Navy that 'ruled the waves'.

In addition, the telegraph race had created another 'tool of the empire' that had resulted in Britain's dominance of the worldwide telegraph networks. A network not only consisting of the governmental 'All Red Line' (Figure 168), but also the many commercial telegraph lines spanning the globe that were under British control. By now British government—eg the Admiralty—understood also that the wireless telegraphy would be the next 'tool of the empire' to support and maintain its dominance. Not surprisingly, Britain was also looking forward to ruling the electric waves.

That imperial aspiration was also the reason why the monopoly of a British company, especially in the American and European theatres, was so contested. For the Germans and the Americans, it was clear that foreign corporate dominance in such an important issue of distant communication was unacceptable, hence their support for local/national developments. In Germany, it resulted in the government-initiated creation of the Telefunken Company. In America, the Navy encouraged private enterprise to create alternative sources for their wireless aspirations. This resulted in all the efforts of the early American inventors (Fessenden, de Forest, Stone) to create their wireless companies.

---

that the 600,000 MWTC of America shares nominally worth $3,000,000 for a total value in the MWTC of America equivalent to $113 million in 2015. Calculation based on historic standard of living. Source: http://www.measuringworth.com/

Figure 190: The Hall Street factory (top, left), mounting shop (second, right), machine shop (bottom, left), and lacquering room (bottom, right) in Chelmsford (1898).

The dates of the photos are not known, so they might not represent the early pioneering days.

Source: http://www.marconicalling.co.uk/, www.marconiheritage.org

In Britain itself, the strategy of patenting created another form of opposition. In British politics, the so-called 'public interest' was often an issue of debate. This might have been due to the heritage of the democratic developments in British history, a heritage in which the Brits threw off the absolute monarchism and the dominance of aristocracy in order to establish a popular democracy.[269] The resulting democratic 'power to the people' translated into governmental policies where the old monopolies (like that of the Post Office emerging from the Royal Mail monopoly) that now were applied to the new technological developments in distant communication. First, the public interest had brought telegraphic communication into that monopoly, and then the same was going to happen with telephonic communication. Obviously, when wireless communication was considered to be in the public interest, similar developments were to be expected.

---

[269] See: B.J.G. van der Kooij, *Context for Innovation: British (R)evolutions in Perspective* (2016).

The Agenda: Manufacturing

In the pioneering years much of the production of the wireless equipment was done on an impromptu basis, such as the early assembling activities in the Haven Hotel on the Isle of Wright. But now the wireless system had matured, and more volume of sales was to be expected, and manufacturing of the equipment also had to be organized differently. Thus, the initial start of the company was soon followed by a lot of organizing and professionalizing its manufacturing activities.[270] In December 1898 the first factory at Chelmsford, Essex, in a former silk factory (Figure 190) was opened. After starting with a dozen men—like the engineer Georg Kemp, recruited from the Post Office—soon the company expanded to some fifty employees, with the production people making the early systems and the technical experts further developing the system.

Figure 191: Marconi coherer receiver Model 5A (top, c. 1900) and induction coil transmitter (bottom, 1898).

Source:
http://www.antiqueradios.com/forums/viewtopic.php?f=1&t=244082 (top),
collections.museumvictoria.com.au/content/media/17/597467-small.jpg (bottom)

From the beginning Marconi devised a working system that consisted of a team of technicians who reported directly to him. There were twenty in 1900 and by 1906 that number had increased to thirty-two. All were young men, more or less the same age as Marconi, who had studied electrical engineering in London. Some also had brief experiences in the telegraphic sector and in power plants. Many of them went on to hold important positions within the Company, whereas Kemp who was much older, kept his role as Marconi's irreplaceable right-hand man until he died in Southampton in 1933. Kemp was very efficient at solving all types of practical problems, kept a detailed work diary and was also even in charge of Marconi's personal diet.[271]

The factory was to manufacture the several transmitters and receivers Marconi developed over time (Figure 191). It was quite an ambitious undertaking.

---

[270] One reason for organizing a manufacturing process was the interchangeability of parts. When the military started buying wireless sets, their maintenance required that broken part could be replaced easily.

[271] Source: http://www.fgm.it/en/marconi-en/profiles/george-kemp.html

It was set up to manufacture wireless sets and receivers to Marconi's latest designs, but the new science was still in its infancy and the Company struggled, so had to diversify into manufacturing motor-car ignition coils, X-ray apparatus and other scientific equipment in order to balance the books. Although having a large factory for such a nascent industry was ambitious, previously the wireless equipment had been built by hand as required, using various modified apparatus bought from established scientific laboratory suppliers, a process that could never hope to cope with quantity production of commercial equipment. To fulfil any commercial order, especially for the Royal Navy, all equipment parts had to be interchangeable and all apparatus had to be built to a high quality and designed to be easily serviced and maintained. At the Hall Street works new departments responsible solely for their own specific areas of research, design and manufacture were established. In September 1899 a transmitting station was built directly across the road from the factory premises to test equipment as it came off the production line and the Hall Street mast soon became one of Chelmsford's landmarks, although now long gone.[272]

Consequences of the Decision

So, there we are. The decision was taken to go the entrepreneurial route. A company was organized, and a manufacturing plant established. It was an infant company with a Herculean task on its shoulders. But how to create a profitable business with an unproven 'product', and a wireless technology in its infancy, for a market that that did not yet exist (the marine applications) or that would show its teeth to the newcomer (the existing telegraph services)? In all that uncertainty, there was one certainty: this decision would have consequences.

The Post Office Withdraws

To Marconi's surprise, the creation of the company had a destructive influence on his relationship with the Post Office. In reply to a letter from Marconi—in which he informed Preece on August 6, 1897, of his reasons for the formation of the company—he received Preece's answer (obviously on instruction from his superiors):

I am very sorry to get your letter. You have taken a step that I fear is very inimical to your personal interests. I regret to say that I must stop all experiments and all action until I learn the conditions that are to determine the relations between your company and the Government Departments who have encourages and helped you so much. (Jolly, 1972, p. 48)

After the decision was made and the company created, Marconi soon

---

[272] Wander, T., Crosby T.: 'The Marconi Legacy – assessing the heritage of the Wireless Communication industry in Essex'. Source: http://marconiheritage.org/ej.html

realized that this letter announced that he was more or less being excluded by the Post Office from participating in further experimenting. The comradely pioneering spirit between Marconi and government departments like the Post Office and the War Office would soon change.

True, in the years to come there was further collaboration between Marconi and government officials, but during 1898 and 1899, when the Company was trying to build up interest, confidence, and income, it was more a properly liaison than active assistance. By the end of 1899—Preece had been retired by then—when there were prices and royalty claims to be settled, relations between the government and the company were anything but cordial and straightforward (Jolly, 1972, p. 50).

Moreover, the Post Office became one of Marconi's major antagonists and refused to purchase his equipment until around 1903 (Bruton, 2012, p. 73). Next to the early opposition that Marconi already faced from the scientific community, now the opposition from governmental institutions materialized. It soon became a serious matter that would hinder Marconi's entrepreneurial aspirations considerably.

> ... *the Admiralty challenged Marconi's patents, arguing that its own communication expert, Jackson, had developed equipment that preceded Marconi's. ...That was in 1899, the year that the Post Office moved against Marconi's Wireless Telegraph Company. As soon as Marconi had developed effective radio technology for communications between ship and shore, the Post Office widened its domestic monopoly to include three miles offshore. Marconi's company was unable to sell equipment without breaching the Post Office monopoly* [for paid telecommunication services]. (Hills, 2002, p. 95)

By the end of the century it became more and more clear to many in the British policymaking circuits that Marconi's patented concept for wireless communication was not just a nice idea, but was proving to be a game-changing invention. Now the battle of priority was getting into shape.

Competition Starts

The aforementioned opposition being the case, there was, not too surprisingly, interest from the entrepreneurial side in Marconi's discoveries, for example, from Professor A. Slaby of the Technical High School in Berlin, who was (reluctantly) invited by Preece to observe some of the early experiments at the Salisbury Plains after a request from the German Kaiser Wilhelm III.[273]

---

[273] In royal circles, the world is small. Kaiser Wilhelm III was the eldest grandchild of the British Queen Victoria. He visited the Cowes Racing Weeks on the Isle of Wright, competing with the Prince of Wales, Uncle Bertie.

Slaby had already suggested in an early stage in 1897 commercial arrangements between the Marconi Company and the German *Algemeine Elektricitats Geselschaft* (AEG: originally the *General Electric Company of Berlin*) whereby the later would manufacture and sell the Marconi apparatus. No agreement was reached, but in May 1903 the AEG, together with the company of *Siemens & Halske*, would create the *Geselschaft fur Drahtlose Telegraphie*. The parent companies would, being quite experienced in electric systems (telegraph/telephone), manufacture the equipment. Over time that company would become, with its Telefunken system, a formidable rival to the Marconi Company.[274] Telefunken, a systems manufacturer and not a service provider, was the first German company of many national industries that would further develop Marconi's idea, in the process circumventing his protective patents, as they were hindering their activities.[275] Clearly, on the international front, the opposition was preparing to battle the Marconi system.

A Market to Develop

Surely, further developing the technical capabilities of the system—and protecting them with patents—was a priority for Marconi. But he had also to look for buyers; the question was where would the market be for his system of wireless communication? Obviously, wireless communications could be used in a range of applications, like replacing the *public message services* of the cabled telegraphy. But that was not an accessible market due to the British state monopoly. Or it could be used for *military communication*, which was also not an easy market to open up, as it had, by definition, a secretive nature.

However, there was a potential market that was screaming for a solution: the *coastal communication market*. This was about the troublesome communication between lighthouses and lightships, or between ships and lighthouses.[276] This was the missing link that the (fixed) point-to-point communication of cabled telegraphy had not been able to solve. For a seagoing nation like Britain, ruling the waves of the British Empire, this was especially an issue that needed a solution.

---

[274] To illustrate this point: by 1905 the Marconi Company had 518 wireless installations in various parts of the world and dominated the market. However, that changed when in 1906 some 50% of the equipment in use by the American Navy was of German origin.

[275] Telefunken was based in Nauen, near Berlin. By 1908 they had their first transmission station ready and covered distances up to Tenerife, 3,600 km away.

[276] To transmit information from lightships, pigeons were used as well as semaphore flagging. Unfavorable weather conditions like fogs often frustrated this form of communication.

What all the applications had in common was that the wireless system needed to be able to bridge distance. Now that the relationship with the Post Office had cooled and their support was no longer available, Marconi had to organize the testing himself. So, in October 1897 he erected the world's first fixed-transmitting connection between the western tip of the Isle of Wight and the Royal Needles Hotel (aka Haven Hotel), near Poole overlooking the Alum Bay.

The owners of the hotel, a vacationing destination, were quite happy to have such a guest in the off-season. One of the ground-floor rooms at the Haven hotel served as the main laboratory, whilst other experimental work was carried out in huts on the grounds of the hotel. A month later, having enlisted coast guards to help erect a mast 120 feet high, Marconi was ready to transmit (Figure 192).

Figure 192: Haven Hotel near Poole, Dorset, with antenna (1903).

Photo taken at the occasion of a royal visit.

Source: http://www.marconicalling.co.uk/introsting.htm

> *Overhanging Alum Bay was the Needles Hotel alongside which towered Marconi masts braced against the winds as new symbols of safety for the men who went down to the sea in ships. From the halyard a wire dangled to a window of the wireless room, where seashore visitors caught their first glimpse of the flashing sparks as they enacted the mystery of talking through space.* (Dunlap, 1941, p. 62)

In December 1897, having fitted aerials and receivers to a pair of local ferry boats, he began tests of reception at sea. He was allowed to place receivers on the steamers *Solvent* and *Mary Flower*, which operated daily between Alum Bay and Bournemouth. Signal rates of about four words a minute were achieved, sometimes beyond the horizon and in atrocious weather, at ranges up to 18 miles (29 km) (W. J. Baker, 1970, p. 36).

> *A visitor to the main station in 1899 records that in the laboratory he found two of the earliest employees, the brothers Cave, at one table making coherers; at*

another, Paget was winding receiver chokes. At a third, Marconi himself was busy fitting V-gap plugs onto an experimental coherer. Outside, along the foreshore, Dr Eskine Murray was conducting parabolic mirror reflector tests, using centimetric wavelengths. At meal times, Marconi, his mother, his brother Alfonso, Dr and Mrs Erskine Murray, the rest of the staff and any visitors all shared a common table. Often for relaxation in the evenings, Murray would play his cello, Alfonso his violin, and Marconi would accompany them on piano.[277]

On the mainland, Marconi opened a station in the Madeira House at the seafront of Bournemouth, a station that would soon prove its worth when winter came and Victorian politics were dominated by the failing health of its former prime minister.

Marconi's fame was enhanced when he grabbed a unique opportunity to demonstrate the possibilities of his system: the illness and nearing death of former Prime Minister William Gladstone, the Grand Old Man of Victorian politics. That winter, by January 1898, heavy snowstorms had broken the telegraph lines in the Bournemouth region. So, the press gathered in Bournemouth—where Marconi had set up another station at Madeira House—was unable to report to Fleet Street on the progress of Gladstone's illness. Marconi heard of the reporters' predicament and offered the use of his Isle of Wight Station, which was telegraphically connected to London. Soon the reporters could transmit the news on the recovering Gladstone using Marconi's wireless connection. At the same time, as an additional promotional bonus, they enthusiastically reported about Marconi's invention. The incident was widely reported, not only giving Marconi some excellent publicity, but also making some good friends for him in the newspaper world (W. J. Baker, 1970, p. 37)

Gladstone died on May 19, 1898, and received a state funeral in Westminster Abbey.

Finding the First Customer

All the testing was draining the funds of the young company, and no real income had been produced by the sale of equipment. As the turn of the century drew near the Wireless Telegraph & Signal Company was close to financial collapse. That situation only changed after the he got in contact with the marine insurance company *Lloyd's of London*.[278] Lloyds, the marine underwriter's association, maintained over a thousand agents and subagents, who, in addition to other duties as representatives of that corporation, were

---

[277] Source: http://www.marconicalling.co.uk/introsting.htm

[278] Lloyd's of London (est. 1688) was in the business of marine insurance. It consisted of 'members', a collection of both corporations and private individuals, the latter being traditionally known as 'Names'.

especially charged with transmitting, immediately, all the latest maritime intelligence from their respective districts.[279] For Lloyd's, it was clear that wireless telegraphy could be an important development.

As the most extensive single system in the world for the collection, transmission, and dissemination of marine information, Lloyd's was naturally interested in any means which would facilitate communication with remote areas. And that new thing called wireless communication interested them highly, so they contacted Marconi. After some tests requested by Lloyd's at Ballycastle (Northern Ireland) with the lighthouse at Rathlin Island, the management of Lloyd's was convinced of the potential use of Marconi's invention, so the parties came to an agreement in September 1901. In the Lloyd's marine signal stations along the British coast, Marconi equipment was installed, and their keepers were trained to use it. The contract was to be in force for 14 years, a period that covered the life of the Marconi patents then in force.

The revenues from this first contract saved the company. Not only were the revenues of importance, but it also showed the investors that a commercial party believed in the system. It showed that Preece had been wrong when he'd stated earlier:

> *Two years after the practicability of Wireless Telegraphy was affirmed, and not a single independent commercial circuit exists. Marconi's operations are more concentrated on the stock exchange than on establishing useful circuits.*[280]

Further Experimenting

Experiments at the Needles continued throughout 1898 and 1899. Transmissions to ships at sea extended out to 64 km. Marconi's experiments were widely published, and scientist became curious, and they flocked to visit the Needles Wireless Telegraph Station. In June 1898 the scientist William Thomson, later Lord Kelvin, visited Alum Bay with his wife and the Second Lord Tennyson. Kelvin sent the following message to Glasgow University: "To Maclean, Physical Laboratory, University, Glasgow. Tell Blyth this is transmitted commercially through ether from Alum Bay to Bournemouth and by postal telegraph thence to Glasgow" (Figure 193).

---

[279] Rathlin in North Ireland was the most important point in this chain of information, as all shipping inward bound from America or Canada passed by Rathlin East lighthouse en route for Liverpool (then the largest port in the UK) or Glasgow or Belfast. The lighthouse keepers kept a record of all ships passing. The problem was to get this valuable information to London in the quickest possible time, by semaphore signals that were useless in bad or foggy weather or with pigeons that were under constant attack from falcons.

[280] Preece, W.,1899. 'Aetheric Telegraphy'. *Irish Times*, 4th May, p. 6. Source: https://irishmediaman.wordpress.com/category/ireland-history/

As he insisted on paying for the transmission of the message, it became a violation of the Post Office monopoly, just as he had intended. By now, not only had the scientific community showed interest, but members of Parliament also had become interested. In May 1898 a demonstration between the House of Commons and St. Thomas Hospital across the Thames was given.

In 1899 developments had moved rapidly, as new stations were established on lightships and on the English east coast. It created a lot of novelty in wireless telegraphy: the first ship-to-shore message, the first shipwreck rescue, the first use of the international distress signal, and the first transmissions to France. Marconi's name was equivalent to progress and novelty.

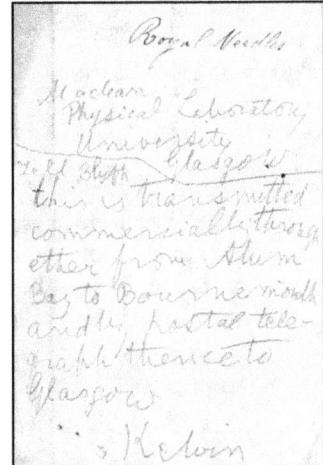

Figure 193: Text of message Kelvin did send on June 3, 1898.

Source: http://www.marconicalling.co.uk/

Preparing for the Next Phase

By 1898 Marconi had his activities well organized. Both his technical and experimental activities in developing his system were in full swing, as well as the early manufacturing and commercial activities. The new company, called the *Wireless Telegraph & Signalling Company*, was on solid footing. Many organisational activities had resulted in the transfer of the 'idea of a company' as envisioned in 1897 into the WTSC organisation's full-on action in 1899. It was clear that 'wireless was there to stay' and that there was a commercial market: the coastal communications market. The young Guglielmo Marconi had been at the centre of these developments that took place in less than half a decade. It had been a hectic period, but that would be true also in the times to come, as after 1900 business problems arose in abundancy.

Restructuring the Organization[281]

It was around 1899 that the Marconi board and the investors saw the need to restructure the company. This need arose partly due to the regulations set by the 1868/1869 *Telegraph Act* that granted the British state

---

[281] This part of our analysis is quite speculative, as not much information was available to explore this shareholding in detail.

a monopoly on all telegraphic communication. Another reason could have been that the growth of the activities (including financing its expansion and the growing risks of the business) had created a need to restructure the organisation. The company also needed additional funding as revenues were low.

> Henry Jameson Davis, till then the managing director—and largest shareholder next to Marconi himself—with a contract till August 1899, left the company. He was replaced as managing director by Samuel Flood-Page, a former manager of the Edison & Swan United Electric Company. Flood brought a wealth of experience in the 'electrical business' to the young company (Raboy, 2016, pp. 141-142).

The thinking about restructuring was stimulated by another of Annie's Irish business connections, John Erskine Jameson, who proposed to create a new venture by raising £1 million.[282] That massive plan never materialized, but it formed the context for the next business development: the creation of the *Marconi International Marine Communication Company*.

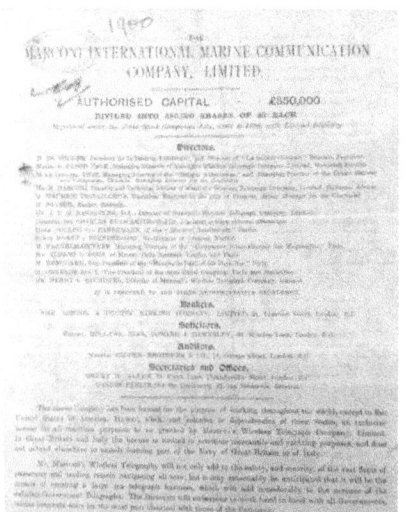

Figure 194: Allotment document of the International Marine Communications Company Ltd. (June 18, 1900).

Source: London Stock Exchange Archives in the Guildhall Library

As a result the Wireless Signal and Telegraph Company was restructured and split into two separate companies. The main company, named (against the wishes of Marconi) the Marconi Wireless Telegraph Company Limited was allocated responsibility for domestic wireless telegraphy in Britain. The second company, the Marconi International Marine Communication Company, was incorporated in April 1900 and was responsible for the potentially more profitable ship-to-shore activities and technologies. (Bruton, 2012, p. 35)

The documents for the newly formed operational organization (Figure 194), with its headquarters in Brussels and called the *Marconi International Marine Communications Company, Ltd* (MIMCC), stated that it would have all the rights for

---

[282] Equivalent to £97 million in 2015; calculation based on historic standard of living. Source: http://www.measuringworth.com/

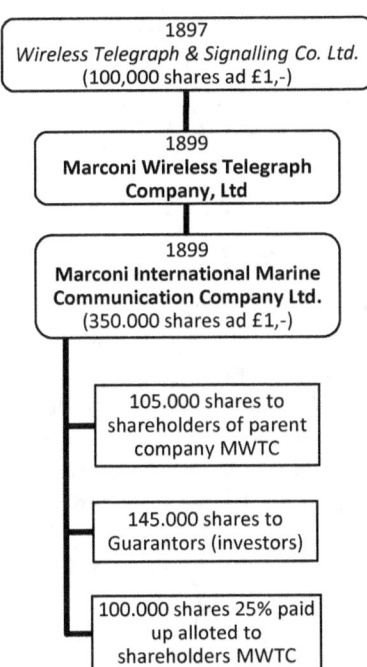

1897
Wireless Telegraph & Signalling Co. Ltd.
(100,000 shares ad £1,-)

1899
Marconi Wireless Telegraph Company, Ltd

1899
Marconi International Marine Communication Company Ltd.
(350.000 shares ad £1,-)

105.000 shares to shareholders of parent company MWTC

145.000 shares to Guarantors (investors)

100.000 shares 25% paid up alloted to shareholders MWTC

Figure 195: Shareholding in Marconi International Marine Communications Company, Ltd.

Capital in nominal value.

maritime communication, including the use of patents, all over the world except for some specified regions such as the US,[283] Hawaii, and Chili. This was to be an international company, organized in a similar way as its parent company MWTC. It would operate in the international maritime markets "creating a large sea telegraph business, which will add considerably to the existing Government Telegraphs." (text prospectus). Its clients would be the shipping companies to whom it would offer the tools and services for wireless communication on a lease basis, thus circumventing the British state monopoly on communication.

The capital of £350,000 was to be paid for the shares at one pound each (Figure 195). Some 145,000 shares went to Guarantors[284], among which investment bankers Govett & Sons, and Messrs Ochs Brothers.

From the total shares, 105,000 were for the shareholders of the MWTC; and another 100,000 shares were reserved to the shareholders of the parent company, now called the *Marconi Wireless Telegraph Company*. Davis, for example, was entitled to 3,040 shares, and Joseph Reed to 835 shares in the new company.

At the same time the MIMCC was created the parent company was also restructured. The *Marconi Wireless Telegraph Company Limited* (MWTC) was created as the successor of the WTSC, owning the

---

[283] With the creation of the Marconi Wireless Telegraph Company of America, this region was covered by that company. The same was the case with the Hawaiian Islands.
[284] A Guarantor on the Stock Exchange—often a bank acting as a middleman—is described as the seller's buyer and the buyer's seller.

patents obtained by Marconi. As it was also the owner of the manufacturing factory, MWTC would supply the equipment with a discount of 25% on the price charged to other customers.

The MIMCC was soon engaged in setting up its European affairs and created the Belgian *Companie de Telegraphie sans fil* in 1901. It was a wholly owned subsidiary without any local investors that would function of a vehicle for creating the national companies in Europe.

Simmering Tensions with a Potential Client[285]

The Post Office was not the only government branch that had experimented with different techniques for wireless communication before Marconi came to England. The Royal Navy was also conducting experiments based on Hertzian waves by 1895, such as those arranged by *Henry Bradwardine Jackson* at the Torpedo School in Portsmouth. After the initial test in 1896 covered distances of some hundred yards, by August 1897 they had bridged distances of some three miles with their own system. As a result of an earlier meeting between Marconi and the War Office, Jackson had also been invited to attend Marconi's demonstration for the Post Office in 1896—a few months after his arrival—at the Salisbury Plains.

> *At this War Office meeting at the end of August 1896, Jackson first met Marconi and they discovered they were both independently developing wireless systems along similar lines. At this early stage, the Marconi system and the Jackson system (as their respective wireless systems were referred to by the Admiralty) differed little technically but, according to a naval report, Marconi's system was more fully developed, and the instruments themselves were much more sensitive. Marconi's system also had a longer range but Jackson's apparatus was far better suited to maritime signalling.* (Bruton, 2012, p. 142)

Convinced of the importance of wireless communications for its operations, by July 1899 the Navy tested Marconi's system during naval operations, as was reported upon daily by correspondents from the *Times*, the *Daily Telegraph*, and the other British national newspapers:

> *The investigations into Military applications for Wireless Telegraphy were becoming a daily occurrence and Marconi is invariably present to clench any lucrative deal. At Bangor Bay near Belfast a number of battleships and cruisers are anchored and experimenting with Marconi's system. The cruiser Juno and the*

---

[285] I am indebted to Elizabeth Bruton, curator of technology and engineering at the Science Museum in London, who was, on March 22, 2017, graciously willing to share her insights with me and educate me in the early years of Marconi's stay in England. Much of the non-cited text here is inspired by her thesis.

flagship Alexandra are fitted with Wireless equipment and Marconi is on board
testing the usefulness of the invention in naval warfare. [286]

The tests were quite successful, bridging some sixty to seventy miles. Technically satisfied, the Admiralty, however, was quite unhappy with the proposed system of payment and royalties (£250[287] per wireless set). It was the start of a troublesome, ambivalent relationship between Marconi and the Admiralty.[288] Nevertheless, in 1901 a *deed of agreement* between the Marconi Company and the Admiralty was signed for the lease of 32 Marconi wireless sets for an annual royalty payment of £100/set.[289] It was an important contract, one that generated much-needed sales but also created future problems for the young company.

> *The agreement was one of the newly established company's earliest and most important contracts. However, the Admiralty continued to concurrently test two similar wireless systems, one designed and manufactured within the navy itself* [the Jackson System] *and the other manufactured by the Marconi Company. This continued beyond their initial full contract with the Marconi Company, signed in April 1903, until about 1904 when they switched to exclusively using Marconi wireless apparatus. While the wireless apparatus was supplied by the Marconi Company, wireless operators were naval serviceman and related training was controlled and determined by the navy and was integrated into the structure and working practices of this military institution. Henceforth the Admiralty continued to invest in wireless, both with land-based stations and on-board wireless sets. This process of installation, adaptation, and adoption continued up to and throughout World War One.* (Bruton, 2012, pp. 126-127)

Along with the previously mentioned deed of agreement, the Admiralty organized an *Interdepartmental Conference* in March and April 1901. There, the results of investigations into the position of Marconi's patents, the developments of other systems, and the need for regulation were presented.

In 1900 Professor Sylvanus Thompson was involved in the controversial Whitehall attack on Marconi's patents, when the Post Office commissioned both him and Professor Oliver Lodge to produce secret reports. The purpose was either to declare the Marconi Company patents invalid or to produce similar but

---

[286] Source: 'Naval Maneuvers'. *Irish Times*, July 24, 1899, p. 4.

[287] Equivalent to £24,480 in 2015; calculation based on historic standard of living. Source: http://www.measuringworth.com/

[288] The Admiralty invoked, for example, the provisions of the 1883 *Patents, Designs, and Trade Marks Act*, which permitted the Crown to manufacture a patented invention if it were urgently needed for defence purposes.

[289] Equivalent to £10,150 in 2015; calculation based on historic standard of living. Source: http://www.measuringworth.com/

technically different equipment: the latter involved Thompson. When the Admiralty received the two reports it was the pioneer of wireless telegraphy Captain (later, Admiral Sir) Henry Jackson, then commanding HMS *Vulcan*, whose opinion led a senior naval officer to report, "It would be unworthy to try to evade the Marconi Company's patent."

Jumping ahead in the future, the recommendations from the conference would lead to the *Wireless Telegraphy Act* of 1904, which regulated wireless telegraphy licensing by the postmaster general:

> *Under this Act no person is permitted to establish any Wireless Telegraphy station or install or work any Wireless Telegraphy apparatus in any place or on board any British ship except under a license from the Postmaster General.*

One clause was quite important, for it blocked Marconi's initial plans for ship-to-shore communications:

> *No person is permitted to work any apparatus for Wireless Telegraphy installed on a foreign ship while that ship is in territorial waters, otherwise then in accordance with regulations made by the Postmaster-General.*[290]

It was become clearer by the day that the market for paid telegraphic services (our present-day 'service provider') in Britain would not be available for a commercial party such as the Marconi Wireless Telegraph Company Ltd.

A Market Lost, a Market Gained

The prospects for the British coastal maritime communication market were bleak now that the British government had made it clear that the market was to come under the monopoly of the Post Office. However, a new market was under development, as in the meantime Marconi had continued increasing the transmission distances covered by technically improving his system. In 1902 the results from his long-distance test with the SS *Philadelphia*, covering some 1500 miles, had reached the Admiralty. It ignited a surprising action.

> *... in early 1902 wider concerns about long-distance wireless communication came to the fore and it was the Admiralty's desire to control and limit this application of wireless telegraph to military usage only that led to the Admiralty renegotiating their initial contract. In March 1902 Marconi sent a letter to Jackson describing long-distance signals received on SS Philadelphia in the mid-Atlantic when over 1500 miles from the sending station, Poldhu. This letter caused much consternation at the Admiralty, where senior officers had come to the*

---

conclusion that long-distance wireless telegraphy was in the best interests of the nation and of the Empire, especially in case of war. As a result, a letter was rapidly despatched to the Marconi Company offering to renegotiate their contract if their needs could be met. Their conditions boiled down to Britain having exclusive access to long-distance wireless telegraphy and furthermore to have access to details of other parties using Marconi wireless telegraphy systems. Essentially they were proposing that the Marconi Company, or at least the maritime arm, would become a state sponsored and state-controlled company. (Bruton, 2012, p. 159)

The new 'tool of the empire', wireless communication over a long distance, obviously had sparked the imagination of the Admiralty. This was an unforeseen effect of the concept of wireless communication that the young foreigner had demonstrated in 1896 on the Salisbury Plains. Thus, this new technology for long-distance communication had to be controlled in the 'public interest' also. But now by the military.

For the young Marconi Company, negotiating with the government of the British Empire that 'ruled the waves', this proposal would have been devastating. Under the contract, it could not license its patents in any foreign country, and long-distance communication between any foreign countries would be impossible. In addition, the establishment of long-distance stations in foreign nations, and transmitting or receiving any wireless messages over land in a foreign country over distances of greater than 50 miles, or over sea over distances above 150 miles, was not permitted. In short, Marconi had to give up, just as the national telegraphic service-business had been monopolized by the Post Office, the other essence of their business: the market for long distance communications (Bruton, 2012, p. 160).

Nevertheless, the board of the Marconi parent company was willing to consider the proposal—although excluding Italy and the US—for a renewable five-year contract with an annual payment of £75,000.[291] The Admiralty declined to pay that sum, as they refused a partial monopoly, with Italy and the US being excluded from the deal. The negotiations lasted until July 1903 without a conclusive result other than the Admiralty by signing an eleven-year contract quite similar to the earlier deed of agreement of 1901.

The Admiralty was required to pay an initial lump sum of £20,000 to cover patents and associated rights. Also within three months of the starting date, 31

---

[291] Equivalent to £7,343,000 in 2015; calculation based on historic standard of living. Source: http://www.measuringworth.com/. Calculated over the lifetime of the contract, this is equivalent to some £36.5 million in 2015.

March 1903, the Admiralty was required to pay £1,600 to cover royalties due from the initial 1901 Agreement. In return for meeting their obligations of the Agreement, the Marconi Company would be paid £5,000 annually. (Bruton, 2012, p. 162)

So, for its originally proposed state control over the Marconi Company—thus realizing a de facto state monopoly on worldwide long-distance wireless communication—the British government had not been willing to pay an annual sum of £75,000. Now they were content to conclude a deal for a minor part of what had been proposed. Less money indeed, but some would say, in hindsight, that this was an example of 'penny wise and pound foolish' behavior. By 1904 the Navy had converted all their sets to the Marconi standard. For the Marconi Company, this alliance with the Navy enabled the company to survive in the twentieth century. But for the moment, wireless communication had been lost as 'tool of empire'.

… the interests of the Admiralty and the Marconi Company converged for some of this period and so, when it was convenient to do so, the Admiralty closely aligned itself with the Marconi Company. But this was a temporary alliance at best and one that was questioned by upcoming domestic legislation and international regulations. (Bruton, 2012, p. 163)

The Admiralty was not the only governmental institution that proved to be a problem for the Marconi Company. After the 1897 letter from Preece cancelling the till-then amicable cooperation between the Post Office and Marconi, in 1903 the Post Office produced a highly controversial internal report called the 'General technical report on wireless telegraphy'.

This confidential report claimed historical priority and rights of exploitation of wireless for the Post Office… [the] report also included an open admission of intercepting signals sent between Marconi stations and the testing of multiple wireless sets quite possibly without the permission, license, or indeed knowledge of the sets' commercial owners. …Meanwhile, Preece (increasingly loudly) claimed historical priority and authority but clearly conflicted greatly with the official position of his former employers, the Post Office, and the government itself and their on-going and active negotiations with the Marconi Company. (Bruton, 2012, p. 76)

This illustrates the difficult business position for the Marconi enterprise. On the one hand, he needed the cooperation of the British government, both in terms of it being a giant customer with its mighty naval fleet and immense need for wireless long distance communication and as a part of a government that was also a powerful market party with the Post Office's monopoly on communication. On the other hand, the government, as it proved with the *Wireless Telegraph Act of 1904,* was the political regulator of telecommunications in Britain, a country which saw state-monopolistic behaviour prevailing over that of commercial monopolies like in America.

We will return to this topic later on, but let's first look at Marconi's efforts to improve upon his wireless system as the new possibilities of long-distance communication seemed to have opened a new market.

Technical Development: The Marconi System

As we have mentioned before, the electric spark was quite a rude mechanism to create electric waves. The sparks created electric waves that covered not a single frequency but a spectrum of frequencies. These bursts were creating electromagnetic interference and 'damped out' over their short lifetime (Figure 139). However, in the case of (Morse-)coded transmission, the short and long—the dots and dashes—bursts of 'untuned, damped' electric waves worked to create a transmission without wires.

Obviously, there was a range of problems related to the first prototypes of Marconi's transmitter and receiver in 1896, problems related to the distances that could be bridged, but improving the antenna—eg making the antennae higher—solved that partly. Also, the strength of the signal was important to the distance to be covered, but increasing the voltage to create bigger sparks solved that also. The sensitivity of the receiving apparatus was an issue, which was solved later with the 'reliable magnetic detector'. All in all, one could say that these issues were part of the (technical) 'growing pains' of the young system of wireless communication.

However, there were other (non-technical) problems in the application of the 'wireless', such as the topic of confidentially of the message transmitted. Everybody with a receiver could listen in to the Morse-coded transmission, a feature not much appreciated by everyone. For military and governmental applications, this would become a serious issue, as the nature of their communication was, by definition, confidential. If this topic could be solved, it would ease the (commercial) 'growing pains' of the young system of wireless communication considerably.

It would take some time before the more fundamental problem of damping of the spark-generated electric waves was solved. That solution would take the creation of 'continuous waves' created by a different technology. In addition, there was also another specific flaw in the system that took some time to solve: the 'interference problem'. The nature of spark-generated electric waves meant that in practice, it was impossible to tune a spark transmitter to a single frequency: "it was inherently a dirty radiator, polluting the spectrum with radiation that was unnecessary, wasteful of energy, and damaging the interest of the users." (Aitken, 2014b, p. 73).

The problem was the bursts of multi-frequency waves (Figure 137) emitted by the spark of the early 'untuned' systems. It was similar to the visible 'white' light—the combination of the individual rainbow colors— that a spark emitted. The result was the mixing up at the receiver station when several transmitters within reach of the receiver acted at the same time. This greatly restricted the number of spark-gap radio transmitters that could operate simultaneously in a specific geographical area without causing mutually disruptive interference. It was this 'interference problem' with its associated consequences that many scientists tried to solve.

> *As it turns out, imperial defense and secure communication were not the only conversations about security in which wireless was entangled; even the early hopes for ship-to-shore communication were about multiple kinds of security. .... At height of the naval arms race, wireless promised, but did not deliver, physical security at sea and the geopolitical security of a restored reputation as masters of the sea.* (Satia, 2010, pp. 838-839)

It was obvious that in order to solve the problem, the transmitter had to send electromagnetic waves of a specific frequency. And the receiver had to be sensitive to electromagnetic waves of a specific wavelength/frequency as sent by a specific transmitter.[292] The solution to these *problems of selective transmission and receiving* was to be found in the field of 'syntonic telegraphy'. The magazine *Scientific American* reported on June 6, 1901:

> *There can be no better evidence of the general utility of wireless telegraphy than that the time has already arrived when imperfections of the untuned system are making themselves felt. To quote Mr. Marconi again, "The ether about the English Channel is becoming exceedingly lively, and a non-tuned receiver keeps picking up messages from various sources which very often render unreadable the message one is trying to receive." ... Mr. Marconi is to be congratulated on the brilliant success of his efforts, and deserves the gratitude of all for having worked out so admirably a system for increasing the safety and convenience of those "that go down to the sea in ships, that do business in great waters.* (G. Marconi, 1901, pp. 130-131)

However, before that solution was created, something also had to be done about the (political) growing pains of the young system of wireless communication, as the company's future was at stake.

---

[292] Harmonic telegraphy: In the case of the wired telegraph, a similar problem had led to the use of different frequencies for each transmission signal in order to transmit several signals at the same time over one wire. See: B.J.G. van der Kooij, *The Invention of the Communication Engine 'Telephone'* (2015), p.131.

Syntony as a Cure for Many Problems

By that time there was a widely acknowledged understanding of the importance of wireless telegraphy in Britain, but also in Germany and America. Wired telegraphy had created the foundations for a considerable industrial development, the military had experienced the advantages of telecommunications, and many scientists were eager to explore further possibilities. Quite a few people were working on the understanding of the specific characteristics of wireless transmission. Some would play a more prominent role than others, such as Braun, Slaby, and Lodge. Let's have a look at their work in more detail.

The German *Ferdinand Braun* (1850–1918) was professor of physics at the Technische Hochschule (University of Technology) in Karlsruhe. In the early 1880s he became an electro-physicist. His early research was into the characteristics of electrolytes and crystals that conduct electricity.[293] In the late 1890s Braun, rather by chance, came to wireless telegraphy, stimulated by the entrepreneur Ludwig Stollwerck, an entrepreneur who had made his fortune with chocolate vending machines and who now was going to engage in a consortium dealing with the development of wireless telegraphy. Stollwerck had contacted Ferdinand Braun for expert advice. So, in 1898 Braun started to occupy himself with wireless telegraphy by attempting to transmit Morse signals through water by means of high-frequency currents. In addition, he developed a method of connecting an oscillary circuit indirectly (by means of inductive coupling) to an antenna.[294] This method of 'loose coupling" solved Marconi's problem of 'tight coupling'. (P. Russer, 2009)

> The tight coupling of the antenna to the oscillator in the Marconi arrangements yielded high damping and consequently the distribution of the transmitted power

---

[293] In 1874 he identified the rectification effect at the point of contact between metals and certain crystal materials with a semiconductor device—a discovery that helped bring about the invention of the radio a few decades later. He thus contributed to the birth of semiconductor electronics.

[294] Direct coupling (also called 'tight' coupling) of the antenna to the oscillator resulted in the production of bursts or pulses of highly damped oscillations. Much energy was dissipated in circuit losses. The highly damped pulses of oscillations were not effective for long-distance communicating. The radiated energy was spread over a wide range of frequencies, resulting in interference to other stations due to the inability to tune receivers to a particular frequency. Direct coupling of the antenna was limiting the range of Marconi's transmitter. The 'loose' coupling, which Braun now used between the spark-gap oscillator circuit and the antenna circuit, produced considerably less damping of the pulses of oscillations. The effect of low damping was highly beneficial in that much more energy was radiated and the energy was distributed over a much narrower range of frequencies. Making the two circuits resonant further increased the amount of energy transferred to the antenna.
Source: https://nitum.wordpress.com/2012/10/03/biography-of-karl-ferdinand-braun.

over a broad frequency band and therefore interference to other stations. The power losses were high and a tuning of the transmitter was not possible. … The loose coupling between the spark-gap oscillator and the antenna yielded lower damping, a narrower spectral width of the radiated pulses and higher power efficiency. The circuit could be tuned in frequency and using two resonant circuits further improved the efficiency. (P. Russer, 2009, pp. 242-243)

Subsequently, Braun introduced the closed circuit of oscillation into wireless telegraphy and was one of the first to send narrow frequency electric waves in definite directions. It was his 'closed circuit tuned oscillator' concept in the electric wave part of the transmitter, and its separation from the antenna (aka 'uncoupling'), that would enhance the quality of long-distance wireless telegraphy greatly. With his early experiments, he showed that he could bridge distances of 1600 m and later 3 km with electromagnetic waves. For his invention 'Telegraphier-system ohne fortlaufende Leitung', he had been granted German Patent № 115,081 on July 12, 1898; and for his 'Schaltungsweise des mit einer Luftleitung verbundenen Gebers für Funkentelegraphie', he had been granted the German Patent № 111,578 on October 14, 1898 (and, additionally, its British equivalent GB Patent № 1,862 on January 26, 1899).

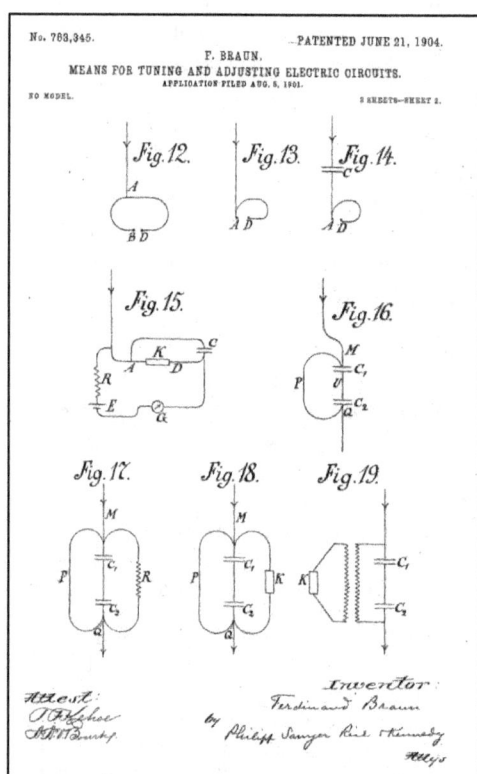

Figure 196: Braun's US Patent 763,345 (1904).

The patent shows figures with different use of resistors (R), condensators (C) and coils.

Source: USPTO.

On November 3, 1899, he applied for a British patent that was granted as GB Patent 5,104 on March 8, 1900. In addition, he applied on August 5, 1901, for the US Patent № 750,429, titled 'Wireless Electric Transmission of Signals over Surfaces', which was granted January 24, 1904. And on August 6, 1901, he would apply for the US Patent № 763,345, titled

'Means for Tuning and Adjusting Electric Circuits', which was granted June 21, 1904 (Figure 196). By 1904 the closed-circuit system of wireless telegraphy, connected with the name of Braun, was well known and generally adopted in principle.

Braun's fundamental modification in the layout of the transmitter circuit for the dispatch of electrical waves made it possible to produce powerful, narrow band waves with very little damping. Braun's papers on wireless telegraphy were published in 1901 in the form of a brochure under the title *Drahtlose Telegraphie durch Wasser und Luft* (Wireless Telegraphy Through Water and Air). He also widely published papers about his other experiments.[295]

Braun had experienced entrepreneurial activities when in 1882 his brother Wunibald Braun founded, with Eugene Hartmann, the company *Hartmann & Braun*, a manufacturing company of electrical measuring equipment. After his wireless experiments that were funded by Stolkwerck, Ferdinand founded the company *Professor Braun's Telegraphie Gmbh* (aka Telebraun) in May 1900 in Hamburg, also funded by Stollwerck. Soon transmission experiments were bridging a distance of 62 km in September 1900, from the island of Helgoland to Cuxhaven. This company merged in 1901 with Siemens & Halske,[296] and later with AEG,[297] and the company 'Gesellschaft fur Drahtlose Telegraphie' (Company for Wireless Telegraphy) was created. They created the 'Telefunken System' (Funken is the German word for spark). For his contribution to wireless telegraphy, he would receive, together with Marconi, the Noble Prize in Physics in 1909. (Peter Russer, 2012)

The German professor *Adolf Karl Heinrich Slaby* (1849–1913), scientific counsellor to the German Emperor Wilhelm II and working for the General Electric Company of Berlin (later AEG), had witnessed some of Marconi's early experiments in 1897, which he soon repeated after returning to Germany. He wrote:

Having returned to my home [after his visit to Marconi's demonstration], *I went to work at once to repeat the experiments with my own instruments, with*

---

[295] Braun also contributed to other fields of electromagnetism. He created the cathode ray tube, the forerunner of the CRT television tube. And as result of his research of conductivity of metal salt, he discovered the point-contact rectifier that would later result in the first point-contact transistor.

[296] See: B.J.G. van der Kooij, *The Invention of the Communication Engine 'Telegraph'* (2015), pp. 291–305.

[297] AEG was founded as the *Deutsche Edison-Gesellschaft für angewandte Elektricität* in 1883 in Berlin. It became the *Allgemeine Elektricitäts-Gesellschaft* (AEG) in 1887, iinitially producing electrical equipment (such as light bulbs, motors and generators).

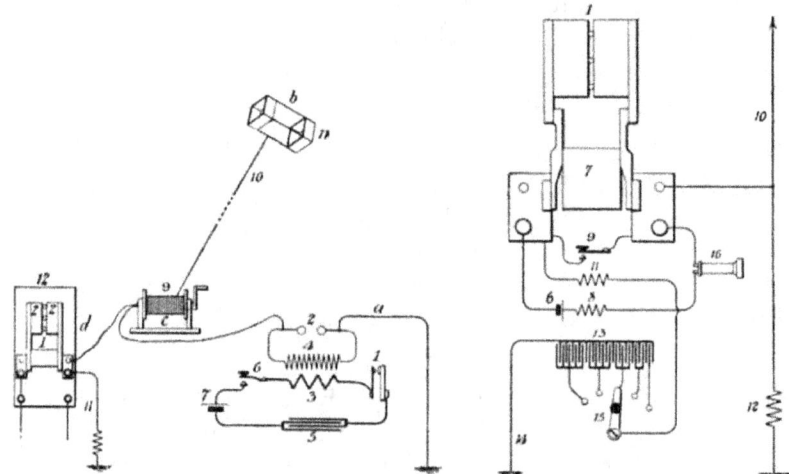

Figure 197: Slaby/Arco system for wireless telegraphy (1900).

Source: *Scientific American*, December 28, 1901, pp. 425–426.
http://earlyradiohistory.us/1901sla2.htm

the use of Marconi's wires. Success was instant. I set up telegraphic communication between my laboratory and a factory about two kilometers away, where a water-tower was placed at my disposal for the placing of the wire of transmission. … For carrying out extensive experiments, the waters of the Havel River near Potsdam were put at my disposal [by the Emperor], *as well as the surrounding royal parks—an actual laboratory of nature under a laughing sky, in surroundings of paradise! The imperial family delight to sail and row on the lakes formed by the Havel; therefore a detachment of sailors is stationed there during the summer, and I was permitted to employ the crews as helpers.* (Slaby, 1898)

With his assistant, the Count *Georg von Arco* (1869–1940) he devised variants on Marconi's scheme (Figure 197).

With generous support from Emperor Wilhelm II, and with the practical assistance of naval units, Slaby and his assistant Arco started further experiments in the summer of 1897. As soon as it was evident that they could supply useful arrangements, they got the order to develop wireless equipment and circumvent Marconi's patents that were not granted in Germany at that time. In the course of the years 1898/1899, Slaby and Arco filed five patents, which protected their whole telegraphing system, mainly in separation from Marconi. (Friedewald, 2000, p. 444)

Slaby and von Arco applied on October 16, 1900, for German Patent № 130,723, to be followed by German Patent № 131,585 of February 6, 1901, and German Patent № 131,586, applied for on November 9, 1900. The results of their works were presented in a lecture called 'Syntonic and Multiple Spark Telegraphy' on December 22, 1900, in the presence of the German emperor. Slaby's application for an American patent resulted in US Patent № 776,359 on November 29, 1904, for an 'Indicator for Electric Oscillations' and—together with Arco—US Patent № 785,276 on March 21,1905, for a 'System of Wireless Telegraphy with Tuned Microphone Receivers'.

It was the Englishman *Oliver Lodge*, whom we met before, who, based on his early experimenting with Leyden jars, had gotten a grasp on something he later called '*syntony*'. During these 'syntonic jar' experiments of 1889 he published 'Experiments on the Discharge of Leyden Jars' (1891). He demonstrated his findings on the 'resonant' Leyden jar in March 1889 at the Royal Institution, and performed a transmission on August 14, 1894, at a meeting of the British Association for the Advancement of Science at Oxford University.

In England, it took a while, but by 1897 Lodge, a busy man active in many scientific fields, picked up his experiments with wireless transmission again due to Marconi's growing popularity.

In principle, Lodge was extending the work of Preece, whose Post Office team had set up a system of inductive telegraphy between Lavernock Point near Cardiff and the island of Flat Holm. ... The work [his experiments with syntonic

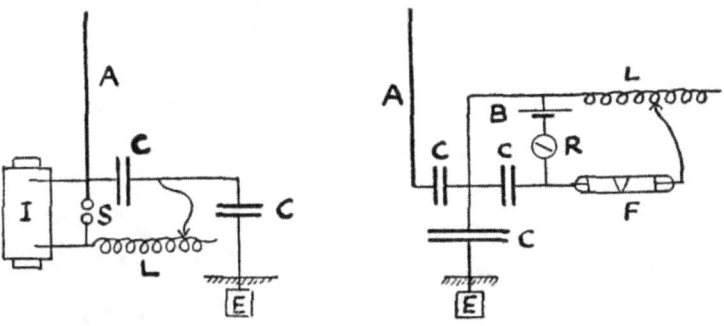

Figure 198: Principle of Lodge syntony experiments as illustrated by GB Patent № 11,348 (1901).

The use of capacitance (C) and inductance (L) to tune the transmitter circuit with the spark generator (I, S) (left) and the receiver circuit with the coherer (F) (right).

Source: Fleming, J.A.: 'Hertzian Wave Wireless Telegraphy. VI'.

circuits] *was carried out with a sense of urgency, even of desperation, and largely in secret. ... Resonant tuning was emphasized from the start.* (Rowlands, 1990, p. 176)

By 1897 Lodge had patented his design for a tuned circuit in GB Patent № 11,575, granted on May 10, 1897, and he was granted GB Patent № 11,348 in 1901 (Figure 198). By that time he had started working together with *Alexander Muirhead* (1848–1920), an entrepreneurial electrical engineer who had built up experience in the telegraph industry. On July 10, 1897, Lodge applied, together with Muirhead, for GB Patent № 16,405 and on August 11, 1897 for GB Patent № 18,644, followed by GB Patent № 29,069 on December 8, 1897 (Burns, 2004, p. 340).

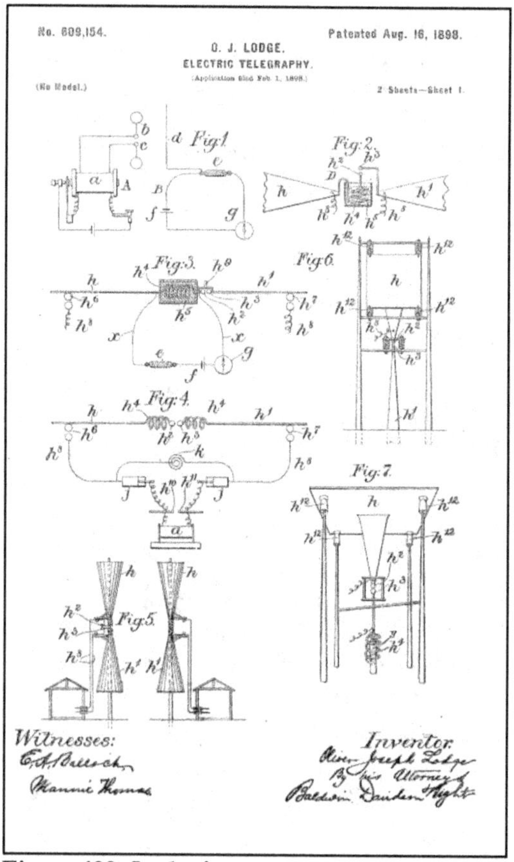

In 1901 Lodge and Muirhead would create a company: the *Lodge–Muirhead Wireless Syndicate*. However, their system—the Lodge-Muirhead system—was not widely adopted, as the monopoly of the British Postal Office also frustrated their commercialization attempts (Burns, 2004, p. 317).[298]

Lodge's experimenting resulted in August 1898 in US Patent № 609,154, titled 'Electric Telegraphy' (Figure 199). His system was still using a Ruhmkorff coil for the transmitter and Branly coherer for the detector. However, his novelty was that he

Figure 199: Lodge's syntony patent: US Patent № 609,154 (1898).

Source: USPTO.

---

[298] In 1911 the Marconi Wireless Telegraph Company bought the Lodge patents and the Lodge-Muirhead syndicate was dissolved.

utilized the concept of 'syntonic' tuning: it applied in a closed circuit capacitors and a variable inductance, both in the receiver and transmitter. In the patent, he claimed:

> *1. In a system of Hertzian-wave telegraphy, the combination of a pair of capacity areas, of a self-inductance coil inserted between them electrically for the purpose of prolonging any electrical oscillation in the system and constituting such a system a radiator of definite frequency and pitch.* (text of patent)

In January 1898, Lodge had made his invention public in a paper read before the Physical Society. In his lecture, he claimed to have created the means for electrical oscillations at a particular frequency of oscillation. Thus, individual messages could now be transmitted to individual stations without disturbing the receiving appliances of other stations, which were tuned at a different frequency (Aitken, 2014b, p. 131)

Already fascinated by the application of Hertzian waves, after this presentation, the scientific community became even more interested in 'syntony'. Among them was John Ambrose Fleming, who would be working later as scientific advisor in close cooperation with Marconi. He was busy with the new phenomenon of spark-generated electric waves. He developed a different method for generating sparks, for which he obtained British Patent № 18,865 on October 22, 1900, and British Patent № 22,126 on December 5, 1900, followed by a range of patents from 1902 to 1903.[299]

The military (Army, Navy, and War Office) became highly interested, as the advantages of tuned wireless telegraphy in military operations had by then become widely accepted. However, there still was the problem of secrecy. For security reasons, other parties should not be able to listen in to wireless transmissions. 'Syntony', by then not a scientific theory but more of a practical craft, seemed to be the solution to solve this problem.

The preceding analysis shows how the international scientific community was interested in the phenomenon of applying the Hertzian waves at the time, preceding Marconi's patent in 1896. It was the time of the wireless mania among the scientists (Figure 151), one that would soon be followed by the wireless mania in engineering.

---

[299] For Fleming, this was only the beginning, as later, in 1904, he invented the two-electrode vacuum-tube rectifier, which he called the oscillation valve, for which he received a patent on November 16, 1904 (US Patent № 803,684). This would become the start of a different technology for wireless communication, creating the 'continuous wave' wireless systems.

Marconi's Work on Tuned Systems

Obviously, Marconi was confronted with the same interference problem. But as Lodge's specific solution was patented, Marconi had to create another solution for the tuning problem. Again it took a lot of experimenting, as he reported in his paper 'Syntonic Wireless Telegraphy' presented at a meeting of the Society of Arts on May 15, 1901, and published by *The Electrician* (G. Marconi, Gill, & Barcroft, 1901).

In his presentation, he described the efforts made in order to tune (syntonize) the wireless system. He stressed that his work took place in a commercial context, and he underlined his limitations with regard to establishing the priority of his work by publishing:

> *A commercial concern, such as the one with which I am working, does not exist solely for the advancement of science, but especially for the purpose of securing a pecuniary return to those who have braved risks and undertaken sacrifices in assisting and forwarding the necessary experimental work. It is often considered possible that certain new methods and results may, if prematurely published before being fully patented, be utilized by persons whom I might call business rivals, thus preventing those who have born the initial cost of the first tests benefiting from a fair measure. I am therefore, frequently prevented from promptly publishing the*

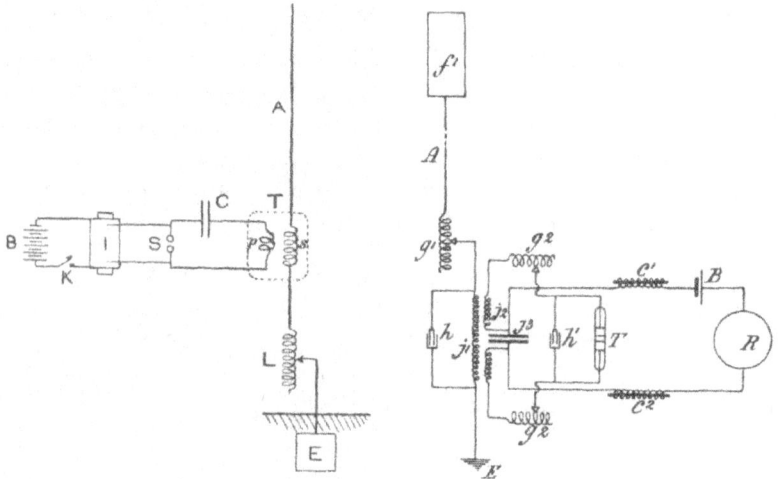

Figure 200: Marconi's improvements in the transmitter circuit (left) and receiver circuit of his syntonic wireless telegraph (1900).

Varying the induction (L) in the transmitter (left) brought the transmitter into resonance, creating an electric wave. By adapting the induction (g) and capacitor (c) in the receiver (right), the two circuits of the system were tuned to that resonance.

Source: (Fleming, 1919) p. 536 (left), 538.

methods by which I have obtained interesting results.[300]

In his presentation, he also underlined in detail that his work on a syntonic system certainly preceded the work of his competitors, scientists Lodge and Slaby/Arco.

Figure 201: Marconi's tuned transmitter (1899).

The Leyden jar (capacitance) and the coil (inductance) are clearly observable.

Source: Museum of History of Science. Oxford. Press pack https://www.mhs.ox.ac.uk/marconi/presspack/

> *Marconi borrowed Lodge's idea and modified it to suit his own requirements. This, however, is not the case; Lodge's syntonic jars' experiment of 1889 did not permit the radiation of electromagnetic waves for any significant distance; furthermore, Marconi's experiments with oscillation transformers began well before Lodge's patent of 1897, the constructional details of which, in any event, were not published at the time of the application.* (W. J. Baker, 1970, pp. 54-55)

Marconi, in his tuning experiments, also experimented with inductance coils and transformers, capacitors, and variable resistances (Figure 201). His initial solution, however, was just a partial solution, as he still could not receive two transmissions at the receiving station (Figure 200).

> *The first results were disappointing. Far from improving the sensitivity of the receiver it diminished it considerably. Other transformers were wound, using different ratios of turns, different wire diameters and different couplings: some improved the receiver sensitivity; other did not. Between March and December of 1897, several hundred transformers of different design were wound and tested, and Marc patents on the three most promising ones.*" (R. S. Baker, 1902, p. 54)

Next, he found the first solution by changing the antenna configuration and applying a cylindrical antenna. He patented his partial tuning solutions in GB Patents № 12,325 and № 12,326 in June of 1898, and № 5,387 on March 21, 1900 (for the US equivalent, see Figure 202). His improvements were later covered in several US patents: № 627,650 and №'s 647,007/8/9. These patents would protect the improvements he had made in the

[300] Published in *The Electrician*, May 14, 1901. Source: *The Tesla Collection.* http://teslacollection.com/tesla_articles/1901/ the_electrician/guglielmo_marconi/syntonic_wireless_telegraphy

important market of the Unites States. By 1900 Marconi had developed a syntonic wireless telegraph that differed considerably from other designs such as Lodge's.

Marconi's Syntony Patent:
The Second Wireless Telegraphy Patent

Marconi continued experimenting to obtain the best syntonic conditions for the transmitting jigger, the transmitting condensers, the receiving jigger, and the receiving condensers from July to September 1900. By then Marconi, the 'practician' and tinkerer, was tackling what Lodge and other famous physicists thought irreconcilable. Soon he introduced another revolutionary advance when he adapted a jigger (ie a high-frequency oscillation transformer) for the transmitter (Figure 203) (Hong, 2001, p. 98).

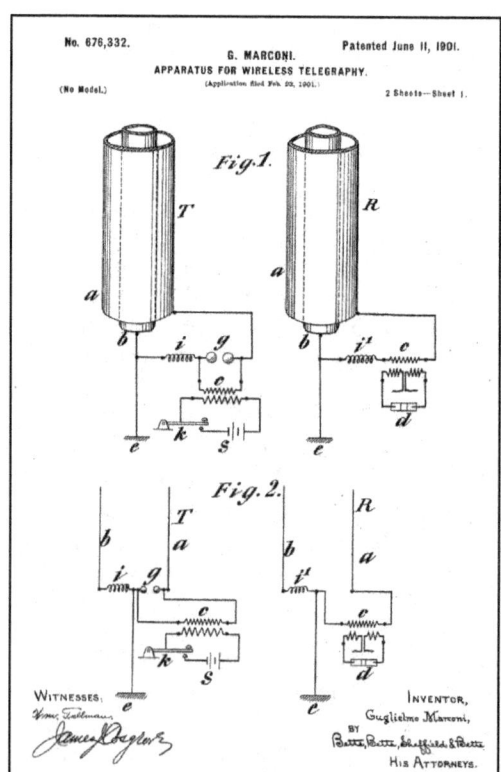

Figure 202: Marconi's US Patent 676,332 for the tuned circuit (1901).

This patent covers his invention as described in British Patent 5,387 (1900).

Source: USPTO.

What was new was that the antenna inductance was tapped so as to be able to adjust the periodicity of the oscillation [the frequency] *and also the Leyden Jar or fixed capacitor was replaced by one which could be varied in capacity. The receiving circuit used similar arrangements. This was it. The problem of syntony had been solved, and two major worries eliminated in one stroke, for the radiated power was nog longer dissipating over a broad area* [of many frequencies] *and consequently more important, adjacent stations could now conduct their business without interfering one with the other.* (W. J. Baker, 1970, pp. 56-57)

He filed the patent application on April 26, 1900, and obtained GB Patent 7,777 on April 13, 1901. The equivalent US patent filed on November 10, 1900, was granted on

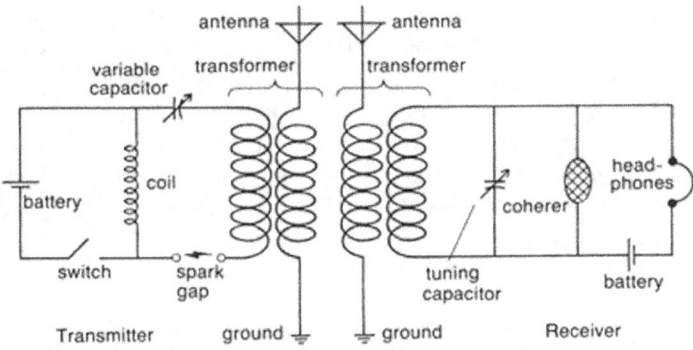

Figure 203: Principle of Marconi's tuned system.

Emitter circuit (left) and receiver circuit (right). By pressing the 'switch', the resulting spark creates an oscillation in the electrical circuit that causes the antenna to emit an electromechanical wave. That wave is received, causing the coherer to act and creating a sound in the headphones.

Source: http://iwcmediaecology.pbworks.com/w/page/ 8480806/Guglielmo%20Marconi

June 28, 1904, as US Patent № 763,772. The delay was caused by the US Patent Office, which had to conclude on his alleged infringements.

> *It was initially rejected based on it being anticipated by Lodge's tuning system, and refiled versions were rejected because of the prior patents by Braun, and Lodge. A further clarification and re-submission was rejected because it infringed on parts of two prior patents Tesla had obtained for his wireless power transmission system. Marconi's lawyers next managed to get a resubmitted patent reconsidered by another examiner who initially rejected it due to a pre-existing John Stone Stone tuning patent, but it was finally approved it in June 1904 based on it having a unique system of variable inductance tuning that was different from Stone who tuned by varying the length of the antenna.* (Wikipedia)

These patents would become the cornerstone of the Marconi Company's activities until they expired after 1914.

> *I believe that I am the first to discover and use any practical means for effective telegraphic transmission and intelligible reception of signals produced by artificially-formed Hertz oscillations. ... My invention relates in great measure to the manner in which the above apparatus is made and connected together.* (Text of patent)

Marconi's new syntonic scheme, which was also called the 'four-seven' system after its patent number № 7,777 (1900), was not, however, free of problems. Worst of all, it was vulnerable to patent disputes, since Lodge had also filed a patent on syntony and been granted GB Patent № 11,575 in 1897. In addition, Ferdinand Braun in Germany had also applied in 1899 for a patent on 'Improvements Relating to the Transmission of Electric Telegraph Signals without Connecting Wires', which was publicized in January 1900 as GB Patent № 1,862. All these systems used spark-induced electromagnetic waves as a carrier for the transmission, hence the name of 'tuned spark telegraphy'.

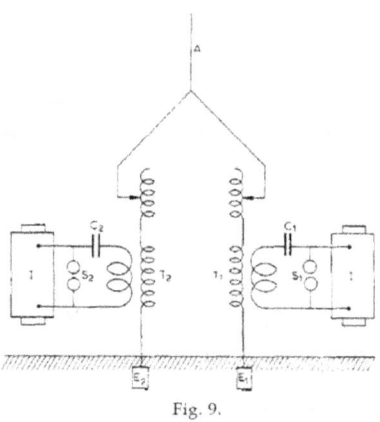

Fig. 9.

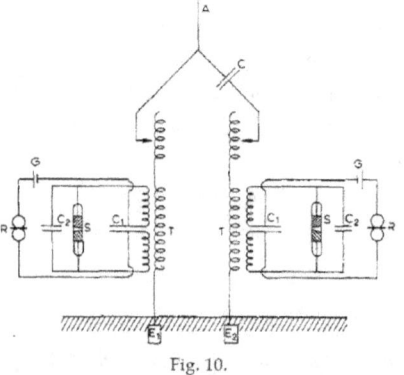

Fig. 10.

Figure 204: Figures from Marconi's US Patent № 676,332 for duplex wireless telegraph (1901).

Transmitter (top) and receiver (bottom) are tuneable.

Source: Marconi Nobel Lecture (1909)

By then, Marconi had addressed the tuning problem with a patent application for a much more sophisticated 'four-circuit' design, which featured two tuned circuits at both the transmitting and receiving antennas. Similar to the development in wired telegraphy, the invention of the tuned systems of wireless telegraphy allowed for the development of 'duplex wireless telegraphy', which enabled simultaneously sending and receiving messages from one and the same aerial (Figure 204).

Experiments with syntonic transmitters and receivers were even more successful than had been hoped for. In a series of demonstrations to influential people, including the Admiralty Commission, it was shown that that two differently tuned transmitter could be connected to a common antenna. A number of messages could also be received simultaneously on an antenna connected to differently tuned receivers. Duplex working had arrived, and other technical improvements which had been made enabled a keying speed of about twenty words per minute to be attained. This speed, although slow by later standards, was a considerable advance on what had been possible at the start of the Company's activities and made wireless telegraph competitive with the ordinary inland telegraph of the period. (W. J. Baker, 1970, p. 57)

The Reliable Magnetic Detector: The Third Wireless Telegraphy Patent

As we have seen, Marconi's wireless telegraph in its early versions was a 'spark transmitter system' that was quite basic in its operation (ie the use of electromagnetic emission due to the spark emitting waves and a coherer as detector). It had some inherent problems, such as the interference problem. These were solved when he improved his system into the 'syntonic wireless telegraphy' with a tuned transmitter (his previously described 7,777 patent). However, during his experimenting at sea with bridging increasingly large distances with his wireless transmissions, Marconi observed the limitations of the sensitivity of the coherer to electric waves. So he started working on that problem.

Marconi's third important patent was based on a phenomenon discovered by the New Zealander *Ernest Rutherford* (1871–1937), later First Baron Rutherford of Nelson, who graduated from Canterbury College in Christchurch (NZ) in 1893. In 1894 he won a scholarship to the Cavendish Laboratory in Cambridge University in England, where he studied under J.J. Thompson. Being knowledgeable about Hertzian waves, he experimented with the magnetization of iron by the discharge of a Leyden jar, thus by a spark with high-frequency emissions of electric waves. He found that when he changed the magnetic intensity (the cause), the changes of the magnetic induction (the effect)

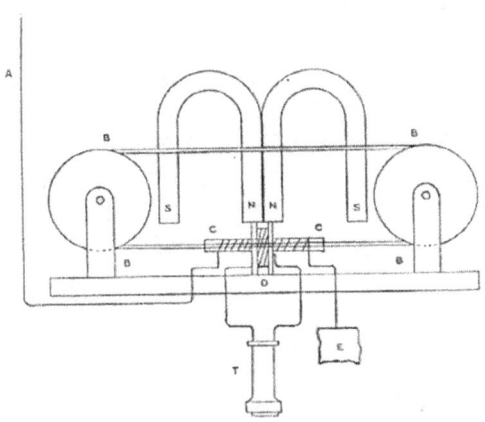

Figure 205: Principle of Marconi's reliable magnetic detector (1903).

A) Antenna wire; (B, B) Iron band around pulleys; (C, C) RF excitation winding on glass tube through which the iron band travels; (D) Audio pickup winding; (E) Ground plate; (S, N) Permanent magnets; (T) Telephone receiver.

Source: Wikimedia Commons

were lagging: the hysteresis effect.[301]

From these experiments, he developed a prototype of a device for the detection of electromagnetic waves—using the hysteresis property of iron—that would replace the unreliable coherer (Rutherford, 1897).

Although Rutherford's detector was purely a laboratory demonstration, it created the basis for Marconi's radio wave detector (called 'Maggie') that was developed in the early 1900s.

Figure 206: Replica of a prototype of Marconi's reliable detector.

Courtesy of Marconi Museum.

> *While installation work was in progress at Glace Bay, Marconi patented a new receiving set. The coherer was still used but other parts of the circuit were greatly improved. For example, he used earphones, a tuning transformer and variable condensers to vary the capacity of both the primary and secondary circuits of the tuner. This made tuning more selective; stations could be separated more easily to avoid overlapping.*
>
> *Mindful that the coherer was the weak link in the circuit he turned to develop a detector based on a discovery of Sir Ernest Rutherford in 1895. He had observed that a small, permanently magnetized needle, when suspended at the end of an electromagnet, was deflected by the rise and fall of the current in the coils of the electromagnet. In this principle Marconi saw an opportunity to sensitize his receiving set and he utilized it in designing a magnetic detector. He used a pair of horseshoe permanent magnets, slowly revolved over an electromagnet, the coils of which were connected to earphones. Fluctuations in the current caused by the incoming signals were audible in the headphones.* (Dunlap, 1941, pp. 131-132)

Marconi's apparatus, a hysteresis detector, consisted of two horseshoe magnets and a revolving wire (Figure 205). Marconi's invention of this detector was filed as British Patent № 10,245 in 1902 for two forms of magnetic detectors: one with revolving magnets and one with a moving

---

[301] The hysteresis property of iron, a complex concept, is used in electric systems. Basically, it is the lagging of an effect behind its cause: as when the change in magnetism of a body lags behind changes in the magnetic field.

Figure 207: Marconi's reliable magnetic detector (1903).

Source: www.liveauctioneers.com

wire (Figure 207). Although already filed on February 2, 1903, it was not before April 14, 1908, that he obtained the US Patents №'s 884,986, 84,987, 884,988, and 884,989 for the version with revolving magnets.

> *Marconi made initial sketches in January 1902 and on returning to England that April, spent long hours in his laboratory at the Haven Hotel near Poole. In May, he filed designs for the first of two patents in that year covering inventions based, as he explained, 'upon the discovery that a core or rod of magnetic material becomes sensitive* [to high-frequency oscillations] *when placed in a varying or moving magnetic field.'* [302]

It took several prototypes before he managed to create 'Maggie' (Figure 206). Due to the nature of the system, it had to be used with earphones. Maggie proved to be both more reliable and more sensitive than its predecessor, ensuring its commercial success as the standard detector on shipboard receivers from 1902 until about 1914, when it became replaced by the next generation of receivers: the more sensitive crystal and vacuum tube detectors.

All these efforts were part of a strategy to take the lead in the technical development of the wireless system. The consequence was that he was able to market equipment that was the most advanced of that time.

---

[302] Source: http://www.marconicalling.co.uk/introsting.htm

Marconi's Patents in Overview

Looking at the technical developments of Marconi's wireless system, one observes a continuous trajectory of improvement in the system, from the improvement in the transmitter (eg the tuning) and improvements in the receiver (eg Maggie) to the improvements in the overall system (eg syntony), it was a continuous path of technical development of the spark-based technology. This illustrates the emergence of a new technology that was rapidly leaving its infant state and in which Marconi's contributions kept him on the forefront of the technology, creating a commercial advantage over his competitors.

Table 4: Early patents granted to Guglielmo Marconi (1897–1902).

Patent №	Granted	Description
GB 5,028	March 5, 1896	Provisional Patent
GB 12.039	March 2, 1897	Improvements in transmitting electrical impulses and signals, and in apparatus therefor (filed June 2, 1896)
GB 29.306	Dec 10, 1897	Improvements in apparatus employed in wireless telegraphy.
GB 12.325 GB 12.326	June 1, 1898	Improvements in transmitting electrical impulses and signals, and in apparatus therefor (oscillation transformer, antenna-design with cylinders and jiggers).
GB 5.657	March 15, 1899	Improvements in apparatus employed in wireless telegraphy.
GB 6.982	April 1, 1899	Improvements in transmitting electrical impulses and signals, and in apparatus therefor (oscillation transformer).
GB 25,185 GB 25,186	Dec 19, 1899	Improvements in transmitting electrical impulses and signals, and in apparatus therefor (division of secondary circuit).
GB 5.387	Mar 21, 1900	Improvements in apparatus for wireless telegraphy.
Gb 18,865	Nov 14, 1900	Improvements in apparatus for the production of electrical oscillations. (Issued to Fleming and Marconi Company).
GB 7.777	April 13, 1901	System for tuned coupled circuits: Improvements in Apparatus for Wireless Telegraphy (filed April 26, 1900). Equivalent US Patent 763.722, 1904.
GB 409	Dec 14, 1901	Improvements in apparatus for wireless telegraphy: joining of primary and secondary windings of induction coil.
GB 410	Jan 7, 1901	Improvements in apparatus for wireless telegraphy.
GB 411	Jan 7, 1901	Improvements in apparatus for wireless telegraphy.
GB 10,245	May 3, 1902	Improvements in receivers suitable for wireless telegraphy.
GB 25,658	Nov 21, 1902	Improvements in apparatus for wireless telegraphy.

Source: (Trainer, 2007) Table 5, p.361.

http://www.radiomarconi.com/marconi/brevettimarconi.html

Patenting was important and well understood by those active in the new wireless technology. It often put pressure on the development, as there was always the element of time. Not only the moment of official granting of the patent but also the date of the initial patent application was important to establish the start of the priority right. And the wording of a patent could be crucial, as its claims would determine what was to be covered by the patent. And finally, patents had to be maintained, either by paying an annual fee (eg Britain) or by doing some actual work in the countries where the

Table 5: US patents granted to Guglielmo Marconi (1897–1910).

Patent №	Granted	Description
US 586,193	July 13, 1897	Transmitting electrical signals. This is the US equivalent of GB Patent № 12,039 for Wireless Telegraphy with 56 claims.
US 624,516	May 9, 1899	Apparatus employed in Wireless Telegraphy: protecting the receiver in a metal box.
US 627,650	June 27, 1899	Apparatus employed in Wireless Telegraphy: use of capacitors and coils.
US 647,007 /8/9	April 10, 1900	Apparatus employed in Wireless Telegraphy: Concerning the induction coil.
US 650,109 /110	May 22, 1900	Apparatus employed in Wireless Telegraphy: protecting from atmospheric electricity.
US 668,315	Feb 10, 1901	Receiver for electrical oscillations: improvements.
USRE 11913E	June 4, 1901	Transmitting electrical impulses and signals and apparatus therefor. Reissue of US 586,193 with 24 claims.
US 676,332	June 11, 1901	Marconi's second patent for Wireless Telegraphy: This is the US equivalent of GB Patent № 7,777 for Wireless Telegraphy with cylindrical aerial conductors.
US 757,559	April 19,1904	Wireless Telegraph System: combination of transmitter and receiver with one antenna.
US 760,463	May 24, 1904	Wireless Signalling System: lateral aerial
US 763,772	June 28, 1904	Apparatus for wireless telegraphy. Improvement in tuning between transmitter and receiver.
US 786,132	Mar 28, 1905	Wireless telegraphy: grounding of antenna with inductors and capacitors.
US 792,528	June 13, 1905	Wireless Telegraphy: tuning with variable inductor/capacitor combinations.
US 884,986	April 14, 1908	Marconi's third important patent for Wireless Telegraphy: the magnetic decoder.
US 884,987 /8/9	April 14, 1908	Wireless telegraphy: Improvements for detecting electrical oscillations.
US 935,381 /2/3	Sep 28, 1909	Transmitting Apparatus for Wireless telegraphy: producing alternating currents with rotating disks.
US 954,640	April, 12, 1910	Apparatus for Wireless telegraphy. Transmitter with rotating disks for the creation of sparks.

Source: USPTO.

Table 6: Some other patents granted to Guglielmo Marconi in Europe and Canada (1899–1910).

Patent №	Granted	Description
CH 18,393	Dec. 31, 1899	Récepteur perfectionné pour la télégraphie sans fil. (filed July 7, 1898)
DK 2,710	Nov 28, 1899	Anordninger ved Apparater til Telegrafering uden Traad. (filed Nov. 29, 1898)
DK 3,429	Oct. 8, 1900	Anordninger ved Apparater til Telegrafering uden Traad.
CA 68,941	Oct. 8, 1900	Wireless Telegraphy (filed May 25, 1899)
DK 4,158	Sept. 9, 1901	Anordninger ved Apparater til Telegrafering uden Traad.
CA 74,799	Feb. 18, 1902	Wireless Telegraphy Apparatus (filed Nov. 7, 1900)
CH 23,154	June 15, 1902	Installation pour la télégraphie sans fil (filed Nov. 7, 1900)
FR 326,064	May 15, 1903	Perfectionnements aux récepteurs pour la télégraphie sans fil (filed Nov. 3, 1902)
DK 5,967	Sept. 21, 1903	Modtagerapparat til traadløse Telegrafer.
FR 340,887	July 22, 1904	Perfectionnements aux récepteurs pour la télégraphie sans fil (filed Oct. 13, 1903)
CA 88,007	June, 28, 1904	Wireless Telegraphy Apparatus (filed May 26, 1900)
CA 110,966	Mar. 24, 1908	Wireless Telegraphy Apparatus (filed Jan. 10, 1907)
CA 112,784	July 7, 1908	Receiver for Wireless Telegraphy (filed July 18, 1907)
CA 113,312	Aug. 4, 1908	Wireless Telegraphy (Feb. 22, 1906)

DK: Denmark, FR: France, CA: Canada, CH: Switzerland
Source: Espacenet.

patent was established (eg France). All in all, patenting was an expensive and time-consuming activity, but it was an essential part of the business strategy.

So, the pillars supporting Marconi's entrepreneurial activities were formed by Marconi's patenting activities in Britain (Table 4). As a consequence of the policy to try and get worldwide coverage, these British patents were soon followed by their equivalent patents in other countries such as the United States of America (Table 5). Along with patenting in Europe and Canada (Table 6), many of the patents were acquired in countries that were part of the sphere of influence of the British Empire.

Of those many patents, three in particular became important. After his basic GB Patent №12,039 in 1896, it was GB Patent № 7,777 in 1900 and the GB patent №10,245 in 1902 that were to be the dominant ones, which became obvious when they were contested in several court cases. In addition, there were many more patents granted for the partial improvements to his system. For those improvements, equivalent patents were also applied for in different parts of the world, among them America, France, Italy, and Germany.

A Time for Technical Improvements (1897–1903)

The preceding analysis shows how the contributions of many people in America, Britain, and Germany had added to the early development of spark-based wireless telegraphy. This all happened in the time of the massive growth of (cabled) telegraphy, the introduction of duplexing (aka harmonious telegraphy), and acoustic telegraphy (aka telephony). In other words: it was the time of the second Industrial Revolution. And in this context, all those thinkers and tinkerers did their work. Much of that early explorative work with the Hertzian waves had the character of trial and error; the basic mechanism would take longer to be understood by the theorists. It was the practical engineers who, with their experiments, created artefacts that generated and received the electric waves. These apparatus, based on a range of basic components, each had its own trajectory of improvements. Then it had been young Marconi who'd combined those components into a *system of wireless communication* that could bridge increasing distances over time (Figure 156).

After Marconi's realization of a working system for wireless communication, both the components and the total system were soon to be

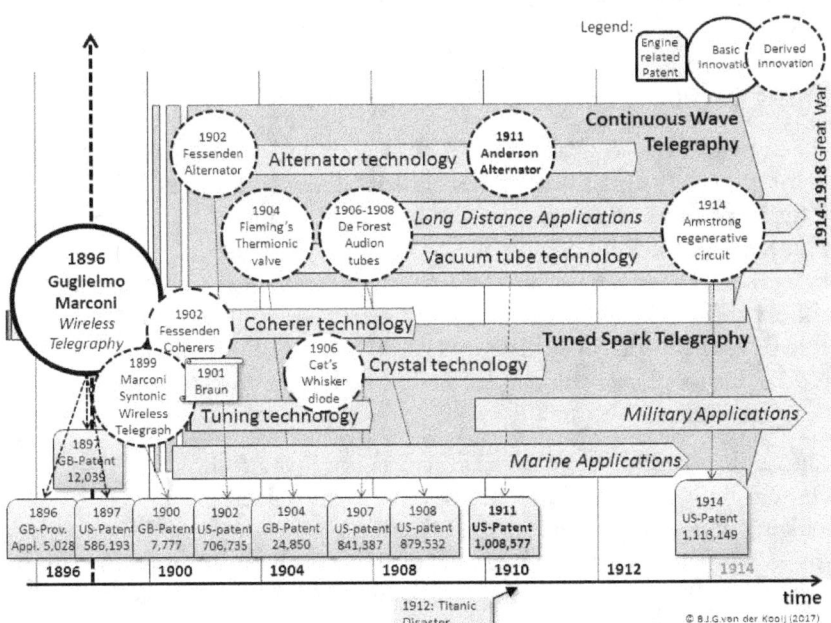

Figure 208: Overview of improvements after Marconi's invention.

Figure created by author.

improved upon in their separate trajectories. Such as there were (Figure 208):

Component improvement: The components of the system generating and emitting the electric waves (the transmitter), and the part that received the transmission (the receiver), were the first to be improved upon, for example, with Marconi's reliable detector (aka 'Maggie')—based on Rutherford's initial selective detector—which was patented in the early 1900s and would dominate early detection methods used in maritime applications for a long time.

System improvement: After the improvements in components, the method of transmission was improved upon. By applying additional components as inductors and transformers, capacitors, and resistances—thus uncoupling the antenna and tuning the system—the total system of wireless communication was improved, resulting in transmission over increasing distances. Marconi's contribution of his syntonic system would also dominate wireless telegraphy for a long time.

All these improvements were patented by their inventors. Marconi's patent protections—from his initial patent of 1897, GB Patent № 12,039—were also enhanced with the syntony patent GB № 7,777 of 1900, and the Maggie GB patent № 10,245 of 1902. With their equivalents in other countries, they would be the cornerstones of the monopoly of the Marconi System.

Business Development: The Marconi Monopoly

When he came to England in 1896, Marconi undoubtedly knew there was a lot to do to improve his archetype for wireless telegraphy. Hence, by 1900 Marconi had technically perfected his system, and he had proven that with his system, transmissions over considerable distances were possible. From a crude archetype, he had developed his *transmission system* into a functional and marketable system for wireless telegraphy, a system that was capable of transmitting over distances of up to a few hundred kilometers (Table 3). But there was still much to be done in that area, for example, crossing the Atlantic Ocean.

The choice for crossing the Atlantic was not just a technical choice in which he followed the evolutionary development trajectory of his invention by bridging increasingly larger distances. Nor was it just part of Marconi's efforts to convince the public—as he faced so much opposition from the scientific and technical community—that his wireless system was more than a technical gadget. His determination to cross the Atlantic seemed to have originated from different considerations. Wireless telegraphy was, in his vison, a tool for connecting, governing, and securing the British Empire at the turn of the century—a formidable business argument.

> *As the* [Boer] *war fed demands for tighter imperial bonds, his technology promised to put far-off places in instantaneous touch with "home" ... Marconi leaned on the wartime sense of vulnerability when he billed his transatlantic transmission as an event that reconnected a sundered people, "colonial cousins" whom punishing cable rates had long kept apart and whose cabled contact was susceptible to attack during war. ... Marconi promised to end colonial isolation. ... Besides strengthening the sentimental bonds of empire, wireless could also provide a more practical administrative security. The press promoted its uses in integrating neglected islands and transforming a fragmented geography into coherent units; viceroys and governors could consolidate control at the margins of their territories by signalling to outlying islands. ... The initial military application thus gave way to other kinds of security applications - sentimental, political, and commercial. This last was a vision of an empire engaged with renewed confidence in its traditional mercantile pursuits in an increasingly jealous world. Marconi's spectacular experimentation muffled the syntony issue and resolved the question of the precise nature of radio's service to national interests by associating it tightly with the symbolic security of a close-knit empire.* (Satia, 2010, pp. 843, 844, 845)

Ending colonial isolation was quite an ambitious objective for a 22-year-old man who in 1896 had stepped so anonymously from the ferry on the British shores, and who in just four years had become a celebrity, creating shockwaves among the British scientific elite. With his concept for wireless telegraphy, his experimental progress and tangible results, he had appalled the scientists also experimenting with wireless transmission. In the process, he—unwillingly—had alienated his former ally, the Post Office, when he'd decided to try following the entrepreneurial path to fund the cost of his experiments and demonstrations.

However, with those public demonstrations and the accompanying publicity, by now he had created awareness in the policymakers in England and in the United States. He had even become kind of famous, as his endeavours had been heralded in the British, Italian, and American press; the wireless telegraph was associated everywhere with the name Marconi, the Hero of Wireless. The spreading news of his experiments had even stimulated the rise of competition, such as in Germany, where Karl Ferdinand Braun had founded the company 'Professor Braun's Telegraphie Gmbh' (aka Telebraun) in May 1900 in Hamburg. Despite all this, his own company was still not a sound enterprise, especially as the British context for a communication enterprise was not that favourable at all.

Thus, to create new working capital, at the shareholder meeting on

October 8, 1898, the Marconi board decided on a restructuring of the organization and to issue new shares, increasing its share capital to £200.000.[303] The cost of improving the system and patenting these improvements had created a need for additional funding.

British Preference for State Monopolies

To understand that British entrepreneurial context for technology-based communication firms, we have to go back in time to a period when private 'communications' had been a monopoly of the Crown. Her/his majesty's mails were transported by horsepower and delivered by her/his majesty's servants: the Royal Mail. Ultimately, the king's own 'Royal Mail' had been transformed into the General Postal Office (GPO) in 1660.

With the arrival of the British telegraph, invented by Cook and Wheatstone, many entrepreneurial activities had started, among them Cooke and Wheatstone's *Electric Telegraph Company* established in 1846. Their first focus on the communication needs of rail companies had shifted towards the general market of telegraphic services when the business world

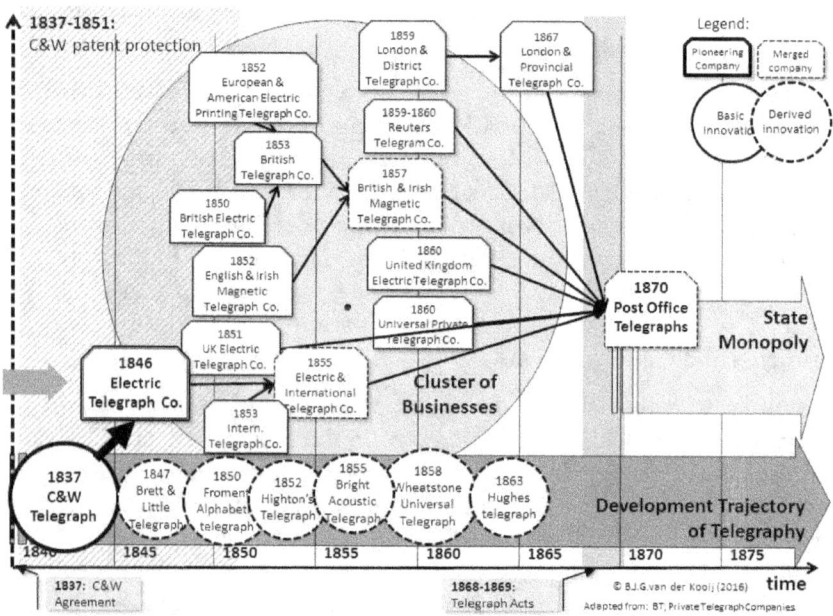

Figure 209: Cluster of businesses in British telegraphy.

Figure created by author.

---

[303] Equivalent to £20 million in 2015; calculation based on historic standard of living. Source: http://www.measuringworth.com/

saw the advantages of telegraphic communication. By the early 1850s a range of companies had been created for the purpose of delivering telegraphic messages with the speed of light. However, within two decades those entrepreneurial activities had been nationalized by the British government.[304] Then the *Telegraph Acts* of 1868 to 1869 had given the right to transmit paid telegraphic services to the British Post Office and the right to nationalize the existing telegraph 'service providers', thus creating a state monopoly on 'electric' communication (Figure 209).

The next important moment came after the telephone was invented by Alexander Graham Bell in 1876–1877. That resulted in an explosion of entrepreneurial activities in the United States. The Bell Company, realizing the importance of their business, set their focus on internationalization, and Britain was a logical choice. In 1878 they created *The Telephone Company Ltd* in Britain to exploit their (British) patents. They were soon followed by Edison, who also rushed to Britain and created the *Edison Telephone Company of London Ltd*. Their goal was supplying paid telephonic services, along with the sales of their equipment. By that time, however, the postmaster general, seeing their monopoly infringed upon, obtained in December 1880 a court judgement that telephone conversations were, technically within the remit of the *Telegraph Act*. Telephonic messages were considered similar to telegraphic messages.[305]

That being the case, the Post Office did not see telephony as that important, as became clear in a hearing of the Select Committee of the House of Commons on May 2, 1879. When asked by Lord Lindsay: "but do you consider that the telephone will be an instrument of the future which will be largely adopted by the public?" William Pierce of the Post Office Engineering staff replied:

> *I think not. It will not take the same position in this country as it has already done in America. I fancy that the descriptions we get of its use in America are a little exaggerated; but there are conditions in America which necessitate the use of instruments of this kind more than here. Here we have a superabundance of messengers, errand boys, and things of that kind. In America they are wanted, and one of the most striking things to an Englishman there is to see how the Americans have adopted in their houses call bells and telegraphs and telephones, and all kinds of aids to their domestic arrangements, which have been forced upon them by necessity. ... Few have worked at the telephone much more than I have. I have one in my office, but more for show, as I do not use it because I do not*

---

[304] See: B.J.G. van der Kooij, *The Invention of the Communication Engine 'Telegraph'* (2015), pp. 327–334.
[305] See: B.J.G. van der Kooij, *The Invention of the Communication Engine 'Telephone'* (2015), pp. 279–283.

want it. If I want to send a message to another room, I use a sounder or employ a boy to take it; and I have no doubt that is the case with many others, and that probably is the reason why the telephone has not been more adopted here.
(Kingsbury, 1915, pp. 208-209)

He was quite wrong. In Britain, the telephone market would also explode, but only sometime later than in America. For the moment, in the following decade, the development of the British telephone system would become a balancing act between government regulation and the development of private enterprise within a rapidly evolving telephone technology.

Having confirmed its monopoly, the Post Office could itself have developed and operated the telephone service as a unified system over the whole country. However, the government did not follow this course but arranged to license telephone companies to carry on business (for a 10% license fee of their gross income) or to start licensed telephone services, such as by the *United Telephone Company*. This 'liberalization' resulted in the fact that several new telephone services were started, like the *London and Globe Telephone and Maintenance Company Ltd.* In 1882. But again it did not have the desired effect.

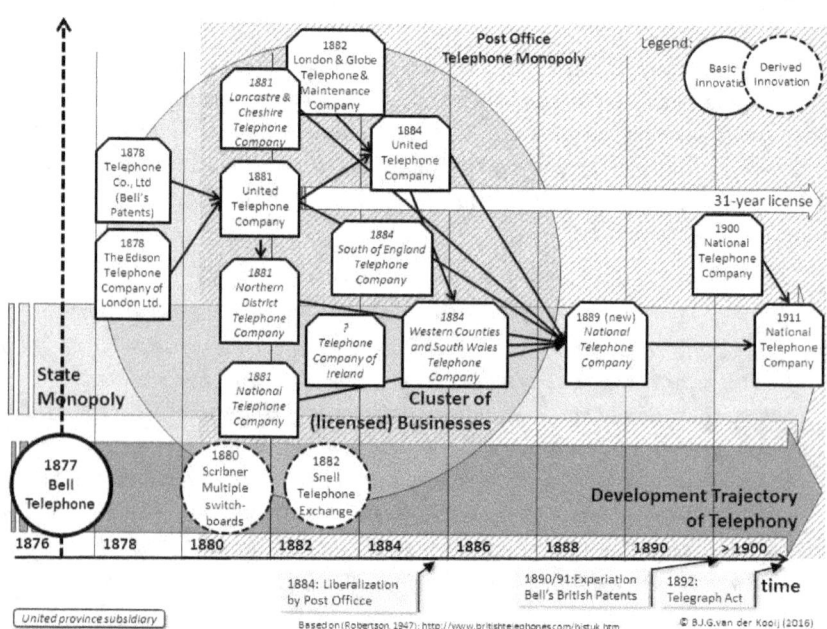

Figure 210: Cluster of businesses in British Postal Office telephony.
Figure created by author.

By 1884, as a result of this dour dualism of policy, it was clear that the telephone industry was being dragged to disaster. There were only 3,800 telephone subscribers in London, and 9,000 in the rest of the United Kingdom. At the same time there were 135,000 subscribers in the United States. (Robertson, 1947, p. 34)

In the meantime the Post Office development of telephone services was just enough to render it a troublesome competitor to the outside companies, yet it never attained vast proportions: 'The Post Office was still a competitor against private enterprise—though not, so far as local exchange facilities were concerned, a blazing successful one' (Robertson, 1947, p. 44). The private companies were restricted to specific exchange areas and to renting from the Post Office—limited by political constraints—the long-distance connections. This problematic situation from the start of British telephony led to political debates in which the words 'nationalization' and 'monopoly' could be heard. Within a few decades after the invention of the telephone, by 1911, the activities of the Postal Office had resulted in the complete nationalization of the telephone business (Figure 210).

Circumventing the British Post Office Monopoly

Although at the end of the nineteen century Marconi had received much publicity and a favourable reception from the elite circles of the British and Italian society all these tests and demonstrations had not resulted in much effective income for the company, and the short-term sales projections were not too good. Even worse, the British government, having recently realized the state monopoly for the telephone, saw wireless telegraphy as another candidate for their monopolistic aspirations. Within that context, Marconi was looking for a feasible business model, one of them being to copy the model of the—cabled—telegraphic service providers and compete in price and quality with them.

As would soon become clear, the selling of equipment was not the major problem, but realizing telegraph services with the equipment for wireless communication certainly was. The power invested in the Post Office by the *Telegraph Acts* of 1868 to 1869 prohibited the Marconi Company from instituting a competitive inland telegraph message-carrying service. And as the jurisdiction for a paid message service of the Postal Office stretched to three miles away from land, an offshore message service could not be started either. The crux was that telegraphic messages (wired or wireless) in the public services were paid for by the message. Marconi could not charge for a public (wireless) telegraphic service. And the company needed to create income if it were to survive.

Already in 1896 and 1897, one option would have been to sell the patent rights to the Post Office. However, the offer made by the Post Office to acquire the patent rights for a mere £10,000 (later upped to £15,000[306]) was not tempting. So, they—Marconi and his business investors—decided to develop the business on their own despite all the opposition—the Jamesons and their fellow corn merchants were astute businessmen, after all. The first few years, the first business model had been selling equipment for wireless telegraphy. Given the unknown functionality of the product, that required a lot of proving its value to the potential customers. Thus, the focus was on demonstrating to specific interest groups like the Army and Navy.

Along with selling the equipment, establishing paid telegraphic services was considered. However, given the Post Office monopoly, that was not feasible, and other possible sources of income had to be considered. Among them, the maritime market seemed to provide the best immediate prospect. But that market needed a business model other than the 'paid-message' model. So, by 1900 the company devised a strategy in which they leased their equipment to shipping companies with a Marconi operator on board the ship. The remuneration was for the lease and maintenance of the equipment and the seagoing operator, but no charges were made for the individual messages. This meant that messages could be exchanged with other ships fitted with Marconi equipment and with Marconi wireless stations, which were being erected along the British coastline, as no direct charge was made for the messages themselves. This way, the Post Office monopoly was circumvented, and the sales concept would become the keystone of Marconi's operations (and a source of many problems in markets elsewhere). It resulted in the formation of the *Marconi International Marine Communication Co. Ltd* on April 25, 1900. (W. J. Baker, 1970, p. 85)

The naval installations were to be the responsibility of the parent company, but the new company was to carry out the maritime application of the Marconi system.

> *Because the 1869 Act specifically exempted intracompany lines (ie private lines), Marconi's company could argue that renting equipment and wireless operators to ship owner and transmitting free of charge, the service represented intracompany communications and was not subject to the Post Office monopoly. Inasmuch as investment in ground stations had to be made up front, while rentals, primarily to passenger ships, produced little income, the strategy was expensive. But the policy had the advantage of allowing the company to specify that shore stations should only communicate with ships using its equipment. ... (Hills, 2002, p. 95)*

---

[306] Equivalent to £1,474,000 in 2015; calculation based on historic standard of living. Source: http://www.measuringworth.com/

For merchant vessels, the concept worked, and the first order came from the Beaver Line. Their vessel, the *Lake Champlain*, was equipped with a Marconi system for wireless telegraphy by May 1901. Soon orders came for the vessel *Luciana* from the Cunard Line, and the vessel *Kaiser Wilhelm der Grosse* from the North German Lloyd Company. Not everybody was that enthusiastic about the lease construction though, and the first efforts to circumvent Marconi's patent appeared.

> *The marine company demanded annual royalty of £100[307] and met with considerable resistance. The Admiralty regarded the price as preposterous. Instead in invoked the provisions of the Patents, Designs and Trade Marks Act of 1883, which gave the Crown the right to manufacture and employ an invention on terms settled by the Treasury, if the patent invention was needed urgently for national defence. Thus the admiralty itself began to manufacture their own equipment based as far as possible on the Marconi design, with Henry Jackson in overall charge.* (Hills, 2002, p. 95)

One sees here the first contours of the rising entrepreneur problems appearing. As a consequence of Marconi's grip on the wireless communication concept protected by his patent position, and the fact that his company was forced to choose a business model of leasing equipment, he was creating an executable monopoly on maritime wireless communication. Anyone who wanted to communicate wirelessly was confronted with the Marconi monopoly.

By that time many European navies started to regard the technology of wireless as a strategic weapon, and these strategic considerations vied with other commercial and regulatory priorities. The British Navy, suspicious of that 'Italian with his obvious connections with the Italian Navy', also realized the value of wireless telegraphy. It's not that strange that the British government was considering its options in relation to the Marconi patents.

> *In 1900 Whitehall* [ie the British Government] *laid plans for a determined attack on Marconi's Company by either having its patents declared invalid or by getting around them with the use of similar, but legally different apparatus. The Post Office had secretly commissioned Professors Oliver Lodge and Silvanus P. Thomson to examine these two possibilities, their reports and all related correspondence being treated as strictly confidential.* (Jolly, 1972, p. 87)

Although some sales for naval communication were made—for example, the sale of 32 wireless sets to the British Navy in 1901, for which

---

[307] Equivalent to £10,050 in 2015; calculation based on historic standard of living. Source: http://www.measuringworth.com/

an annual royalty of £100[308] per year during ten years would be paid—by that time the Marconi company was already cheated, as 50 copies were illegally made by the Admiralty without paying any royalty (Jolly, 1972, p. 92).

The American Commercial Telegraphy Monopoly

In the United States, Samuel Morse's invention of the telegraph in 1837 had resulted in the *telegraph boom* in the second half of the nineteenth century. Over the period of some decades a broad range of entrepreneurial activities all over America had resulted in a myriad of companies, active as either manufacturer of telegraphic equipment or as (telegraphic) service providers. Through a range of mergers and acquisitions, by 1870 a monopoly of the private company *Western Union Telegraph* had emerged as the dominant player in telegraphy (Figure 211), a situation not unlike the British monopoly of the Post Office (Figure 209) but with one big difference. In the capitalistic America that rejected governmental dominance, it was private enterprise that maintained the monopoly.[309]

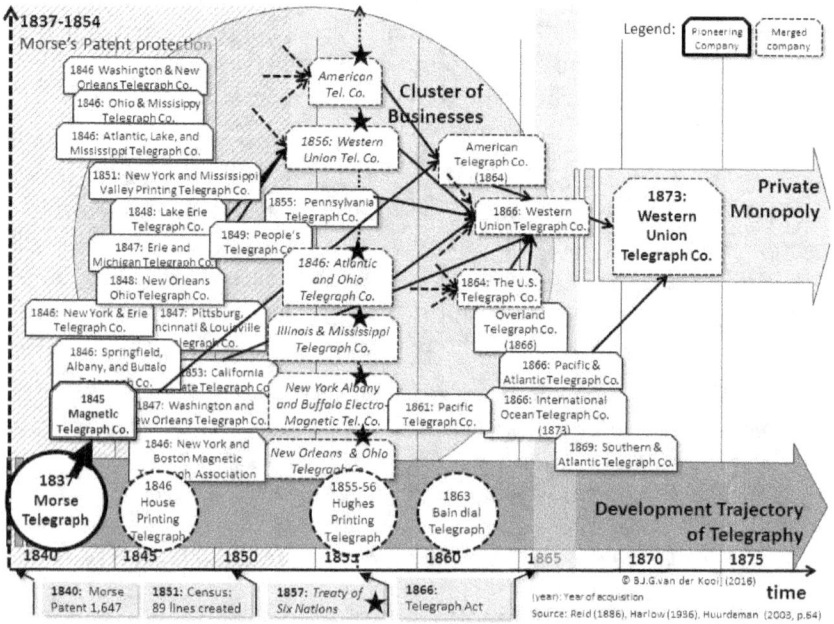

Figure 211: Cluster of businesses in American telegraphy.

Figure created by author.

---

[308] Equivalent to £9,696 in 2015; calculation based on historic standard of living. Source: http://www.measuringworth.com/

[309] See: B.J.G. van der Kooij, *The Invention of the Communication Engine 'Telegraph'* (2015), pp. 431–436.

Then came this young man Marconi to the United States, demonstrating his new concept for wireless communication, applying for a patent in America soon after he patented his invention in England (similar to GB Patent № 12,039 granted on June 2, 1897). He wanted to protect his concept for wireless telegraphy in the potential market of America and was granted US Patent №. 586,193 on July 13, 1897. After amendment, it was reissued as Patent № 11,913 and granted June 4, 1901. The patenting illustrates the vision and dedication the Marconi board had on the importance of Marconi's invention. Patenting it in America, the New World the British were quite familiar with, shows how they valued that potential market. It was the first step in a strategy of internationalization.

That move towards internationalisation, again, has to be seen within the context of the United States at that time. By the end of the nineteenth century the North American former colonies that had abolished British rule with the American Revolution (1765–1783) were not too happy with the British domination of the worldwide wired telegraphic system and the Atlantic submarine cables. Now, with this new invention, those Brits were trying to dominate business again. Soon both the US government as well as American enterprise were trying to break Marconi's monopoly. One obvious reason was that he was threatening the commercial position of the cable telegraph industry.

Marconi was not the first to confront the mighty American telegraph industry with revolutionary developments. Alexander Graham Bell's invention of the telephone (aka the speaking telegraph) had resulted in the David-Goliath contest between the Bell Company and Western Union that was won by the newcomer Bell. After a lot of patent litigation in which the Western Union lawyers contested Bell's priority—claiming that other inventors had already invented the telephone before him—Bell struck a deal with Western Union. Western Union agreed to stop the activities of their *American Speaking Telephone Co.* in the telephone business and assign all its 84 telephone patents to the Bell Company. In return, the Bell Telephone Company would refrain from entering the telegraph business.

> *This agreement, which was to remain in force for seventeen years, was a master-stroke of diplomacy on the part of the Bell Company. It was the Magna Carta of the telephone. It transformed a giant competitor into a friend. It added to the Bell System fifty-six thousand telephones in fifty-five cities. And it swung the valiant little company up to such a pinnacle of prosperity that its stock went skyrocketing until it touched one thousand dollars a share.* (H. N. Casson, 1910, p. 84)

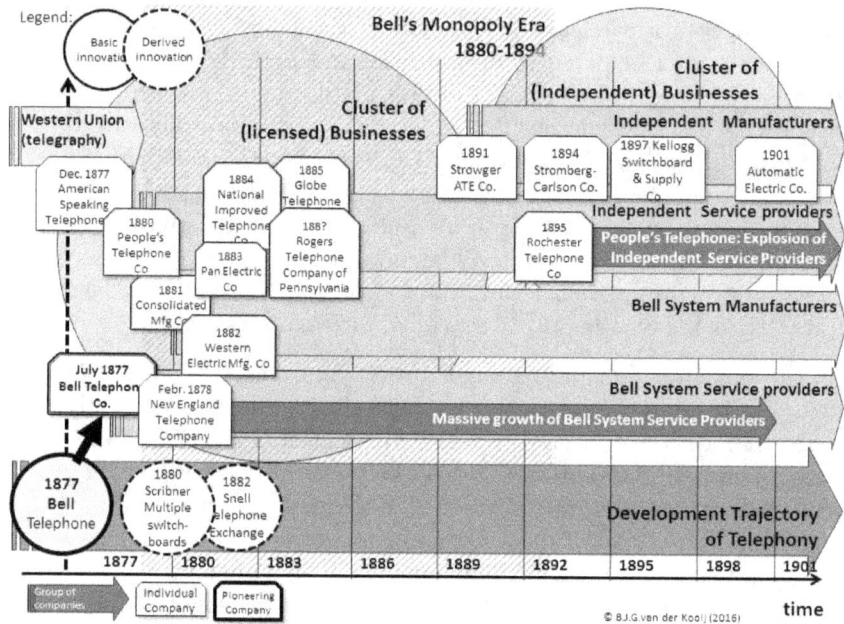

Figure 212: Cluster of businesses in US telephony: Bell system and independents.

Figure created by author.

Bell's telephone monopoly resulted in a myriad of service providers operating within the Bell system. After his patent protection ended, it saw the rise of a new industry of both telephone service providers and manufacturers of telephone equipment (Figure 212).

The other reason for the objection to Marconi's monopoly was that the US Navy did not want any British operating the vital communications on their naval ships. So the US Navy started cooperating with their own US-based industry. This resulted in the appearance of competing manufacturers that were supplying the American Navy test equipment based on Fessenden's invention and methods (Jolly, 1972, p. 96).

> In Britain and America in particular, his plans to provide a wireless service were openly by official obstructiveness, individual deviousness and very often open hostility.' ... the British Government was unwilling to grant him the commercial facilities he needed in England, the German government was set on displacing him as the principal supplier of wireless equipment throughout the world, and the

unknown laws of nature continued to wart his attempts to realize a commercial transatlantic service. (Jolly, 1972, p. 185)

From its start in 1897 the Marconi Company had been trying to develop its business concept. It started with the classic idea of simply selling the equipment and related services to interested parties, such as the government-dominated 'telegraph market' and the 'military market'. When that did not work, as the market for the new device was still immature, the idea was to combine the services to the equipment and copy the activities of the cabled 'telegraphic service providers'. That route being blocked by the private and state monopolies, they next developed the leasing concept as a business model serving the 'maritime market'. That proved a viable business model, and the Marconi system would soon monopolize wireless telegraphy. Part of the effort to develop the markets for wireless telegraphy was massive publicity. The technical development trajectory was constantly accompanied by Marconi inviting the press at every occasion: it was 'publicity, publicity, and publicity'.

Governmental Interests against Marconi

The preceding analysis illustrates the opposition the Marconi enterprise was facing. The Marconi Company had something in their possession that everybody wanted, and although Marconi had it well protected, he was attacked from all sides, like the attack from the British government that would eventually result in the *Wireless Telegraphy Act* of 1905 that claimed wireless communication belonged under the state's monopoly. In this act, it was laid down that the Post Office's permission had to be obtained to operate a wireless station. The Marconi Company was given a license for its shore stations for eight years. But as in 1909, the Post Office declared they wished to do the ship-to-shore communication themselves, so this was a short-lived permission. But British governmental opposition was not unique; Marconi faced a similar pattern everywhere.

The path of developing official recognition, interest and control followed for wireless telegraphy in England was repeated internationally during roughly the same period of half a dozen years at the beginning of the 1900s. Sectional interests opposed Marconi on is wider front as they had done in Britain. In this case the struggle was with Germany whose aggressive policy in support of her own wireless interests was in line with her general diplomatic attitude that is time. ... the German wireless system was based on Slaby's subsequent work, and the contention between the Telefunken Company and Marconi's Company intensified between 1900 and 1914 in step with the growing antagonism between Germany and England. (Jolly, 1972, pp. 139-140)

It was obvious that the implementation of Marconi's experimental work had resulted in an invention that was also important for other countries. Institutions like the military and imperialistic governments, traditionally opposed to commercial monopolies, considered wireless telegraphy too important to leave it under the control of a single monopolistic company, especially when one considers the fact that in 1914 the First World War, initiated by Germany, was about to begin. Already in the Russo-Japanese War of 1904, it showed how important wireless communication was:

> *Later it was generally accepted that the crushing defeat of the Russian Fleet was to a significant extent due to the superiority of the Japanese wireless equipment over that of the Russian Navy which - unlike the Russian army - was not using Marconi's apparatus.* (Jolly, 1972, p. 148).

The Russo-Japanese War (1904–1905): As Russia sought a warm-water port on the Pacific Ocean for their navy and for maritime trade, they came in conflict with the Japanese. Both nations were expanding their imperial ambitions. On February 8, 1904, the Russian naval base at Port Arthur, leased from the Chinese, was attacked by the Japanese in a surprise attack three hours before Japan's declaration of war was received by the Russian government. After a range of additional battles, the Russian were defeated, losing its entire Pacific fleet to Japanese minefields. The war crippled the Russian Empire and contributed highly to the discontent on the Russian home front that would result in the Russian Revolution. Both sides had the use of early wireless telegraphy. Marconi had supplied the Russian Army with equipment they never used. Telefunken also supplied equipment to the Russian Navy, and that proved a miserably failure. The Japanese Navy used a cloned version of Marconi's system that they had copied form the sets bought in 1902.

> *Early in the morning of May 27, 1905, the Japanese cruiser Shinano Manu, patrolling the East China Sea near the mouth of the Tsushima Strait between Korea and Japan, spotted the passing Russian fleet. The Shinano Maunu wireless operator signalled the news to Admiral Togo Heighachiro's flagship Mikasa, and the Japanese fleet was able to launch a surprise attack and sink or capture almost the entire Russian force. ... the wireless message that the Russian fleet had been sighted played a crucial role in Japanese victory. ... Wireless was now a key strategic, tactical, and military factor in what was soon to be known as electronic warfare.* (Raboy, 2016, p. 264)

Marconi's invention might have been of invaluable importance to the different states and their warfare, but for the Marconi Company, the military markets had not been a commercial success up to the mid-1900s.

Selling equipment was not too successful, as his designs were copied (Japan), and the business model of leasing was obviously not working in (British and American) naval applications.

For the Marconi Company, another option would be to take the wireless further at sea, focussing on the merchant marine. That would nearly take another decade, and it would require focusing on a specific segment of wireless communication: the long-distance ship-to-shore communications.

> *However, it was not until 1907 that his radio transmitting and receiving system had any commercial value. Marconi's plan was to set up a worldwide chain of radio stations on key shore sites transmitting long-wave signals. To accomplish is he had to set up subsidiary companies and establish his patent rights in various countries. Indeed by 1914, Marconi's companies had achieved market dominance in maritime radio communications. However, continuous wave systems were entering the market and had several advantages over Marconi's high-power spark system.* (Trainer, 2007, p. 360)

Organizational Development: Financing the Marconi Empire

As we have seen, Marconi's (and his nephew's) entrepreneurial aspirations had culminated in July 20, 1897, in the creation of the *Wireless Telegraph and Signal Co. Ltd.* As part of its business strategy, that first company had undertaken the creation of a manufacturing factory in Chelmsford in 1898, followed by a range of technical and commercial activities, as described before.

As a result of a lot of pioneering activities, between the year 1897 and 1908 the Marconi activities had expanded gradually. However, the company, in its different organizational forms, proved for the next decade to be a sinkhole of financial funds. A decade after it started, by 1908, the company had sunk £500,000 in experimental work,[310] no dividend had been paid to the holders of the ordinary or preference shares, and the stock market price of Marconi stock was 6s.3d. for stock that had cost £3 or £4 a decade before.[311] One hardly would call that a healthy high-tech enterprise, and it wasn't the star of the stock market that it could have been, considering the wireless mania in business that had appeared and influenced the stock market. It was quite a contrast with today's high-tech companies.

> Some eighty years later the venture Apple Computer, a high-tech company starting in personal computers and later moving into telecommunications with its iPhones, was organized. It was financed

---

[310] Equivalent to about $47 million in 2015 based on the historic standard of living. Source: http://www.measuringworth.com/ukcompare/
[311] Equivalent to about £29.47 resp. £301–401 (ibid).

with a $15,000 bank loan and some $1,750 private money on April 1, 1976,[312] when Steve Wozniak and Steve Jobs drew up a partnership agreement. In January 1977, Apple was valued at $5,309. Then early investors came on board in different investment rounds and share splits. Apple raised $518,000 in January 1978 and $704,000 in September 1978, creating the 'founders' shares'. This was followed by an investment round of $2.3 million in December 1980, just before the IPO. So, private investors had put up some $3.5 million by the end 1980.

Three years after the initial partnership agreement, by the end of December 1980, Apple Computer went public and sold 4.6 million shares at $22. The stock rose on the first trading day to $29. Jobs, the single largest stockholder with 7.5 million shares, was worth on paper some $217 million. Adjusted for splits since then, Apple's IPO price was $2.75 a share. It peaked in $14.82 in October 1987: for the founders it was from 'zero to hero' in a period of ten years.[313]

This example of the prosperous start of the Apple company shows how great a struggle it was for the early high-tech companies in the emerging field of telecommunication that were to be so valuable to society. For Marconi, the first ten years were anything but glorious.

Let's first go and look how that organizational development of the Marconi companies took place over time, followed by how the financial characteristics of the parent organization progressed over time.

Early Years of the Marconi Wireless Telegraph Company, Ltd[314]

The first start-up Marconi company, called the *Wireless Telegraph & Signal Company*, Ltd (in short: WTSC) was established, as we have seen before, on July 20, 1897. The first directors were James Fitzgerald Bannatyne, Henry Jameson Davis, William Woodcock Goodbody, John Mooney, Major Samuel Flood Page, Henry Spearman Saunders, and Guglielmo Marconi himself. The balance sheet showed that the company owned a capital of £125,281,[315] due to the 40,000 shares that were paid in full. The other 60,000 shares out of the 100,000 shares issued were given to

---

[312] Equivalent to about $58,700 (ibid). To get an idea about 2015-values, one could multiply the dollar by three.

[313] Data from different sources, among which are Linzmayer, O.W.: *Apple Confidential 2.0*.

[314] In the publications about Marconi, information on the financial aspects of the Marconi organizations is limited. We were given the opportunity to do some primary research in the Marconi Archives in the Bodleian Library. The data presented here originates from that source. Most of the financial data is given in actual pounds. To get an idea about present values, one could multiply the pound figures by 100.

[315] Equivalent to about $12,820,000 (ibid).

Guglielmo Marconi when he transferred his patent rights—granted on July 2, 1897—to the company. As one could expect, for the time being the company only had expenditures and no income. For example, the costs of patent application rose to £77,010.[316] (OX-3, 2017 Annual Statement 1897). This was a considerable amount that indicated the seriousness of the operation and the firm commitments of the financial backers.

In 1899 the WTSC was renamed and called the *Marconi Wireless Telegraph Company, Ltd* (in short: MWTC, also called British Marconi, Figure 213). It was this company that owned the patent rights. In addition, in the same year, another company—one could call it the operational company—was created to organize the commercial and manufacturing activities: the *Marconi International Marine Communication Company* (in short: MIMCC or Marine Marconi). For the shares of the Marconi International Marine Communication Company, £105,000 had been paid to the mother company MWTC. In that year, on September 30, 1900, the balance sheet for the MWTC showed additional revenue for the issue of 200,000 new ordinary shares. That increase in capital was much needed. The sales of equipment and the royalties from the start in 1897 till September 30, 1900, amounted to just a meagre total of £6,092. (OX-3, 2017 Annual Statement 1900).

By September 30, 1901, the annual sales and income for MWTC from royalties had risen to £11,642. On the balance sheet, the investment in patenting had grown to £78,164. To get new capital, new shares had been issued: 12,500 at two pounds each. This was the first year some lease

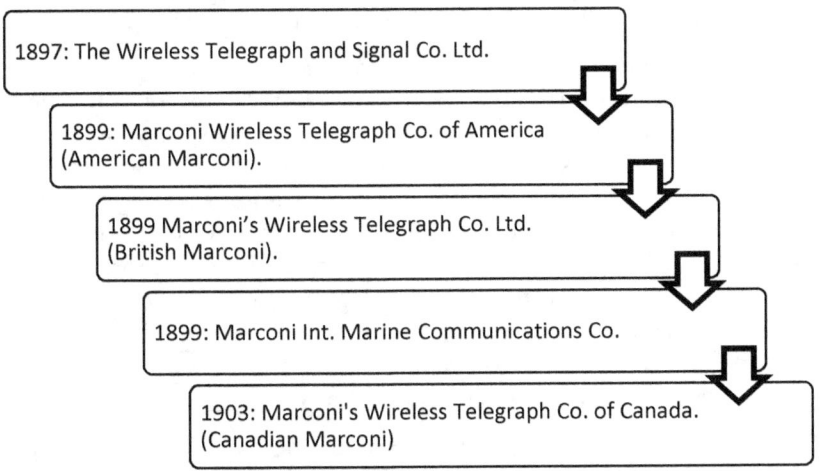

Figure 213: Early Marconi companies.

---

[316] Equivalent to about $7,880,000 (ibid).

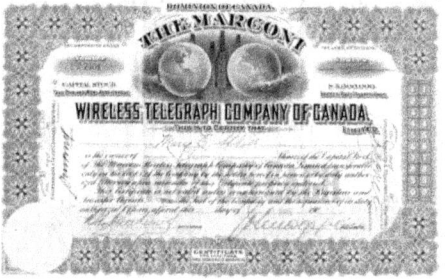

Figure 214: Stock certificate for MWTC of America (top, 1913) and Canada (bottom, 1905).

Source: http://scripophily.net/

revenues came from the sales of 32 wireless sets to the Admiralty and five sets to the War Office of the Naval Authorities in South Africa.

To organize the international expansion, on November 22, 1899, the *Marconi Wireless Telegraph Company of America* (MWTC of America) was created to cover American business. This was followed by the *Marconi Wireless Telegraph Company of Canada* (MWTC of Canada) in 1903 for the Canadian market (Figure 214). And, on the European continent, in 'Brussels (Belgium), the *Cie Francaise Maritime et Coloniale de Telegraphie san fill* was created to do business in France, Portugal, and Spain. It was fully owned by the MIMCC. It took some time, but in 1910 *La Compania Nacional de Telegrafia sin Hilos* was formed in Spain.

Balance Sheets for the MWTC (1903–1911): Associated Companies

The new international activities created business of their own—and, over time, revenues for the local shareholders—but also revenues for the parent company. However, it was not until March 30, 1903, that the report of the proceedings at the sixth general meeting of the MWTC noted that 'the income of the Company has exceeded its expenditure': it was the sum of £5,489. The positive effect on the revenues was largely due to revenues of long-distance communication. A year later the profit was reported to be £10,607. Business might have been improving, but the complicated problems with the British government in the form of the Post Office, and the support from the German government and German banks for the Slaby/Arco system, dominated the proceedings.

The annual report for the year ending September 30, 1904, mentioned the rise in ship installations: the Holland Amerika Line, the Navigazione Generale Italia, the White Star Line, etc. These ships were being served by 60 shore stations. As a result, gross profit had grown to £33,289, contributing to the total revenue of £34,119, not too much, as one observes that the total investment in patents had now risen to £80,589 on the balance sheet. But the value of the company had risen enormously, as the balance sheet showed the amount £1,437,000 as par value for the shares in associated companies like MWTC of America, MWTC of Canada, the MIMCC, and the newly formed Belgium company.

> The parent company MWTC now owned 35,650 fully paid shares in American Marconi, 660,000 fully paid shares in Canadian Marconi, 100,000 fully paid shares in the Marine Company and 68 bearer shares in the *Cie Francaise Maritime et Coloniale de Telegraphie san fill.* Total value on the balance sheet: £1,437,000[317] (OX-3, 2017 Annual Statement 1904).

In September 30, 1905, the gross profit of the parent company MWTC was £28,651, contributing to the total revenue of £39,319. For the first time, a dividend of 12% was considered (but not paid). The company had expanded its operations, and transatlantic communication had been established. Also, the British Navy had started communicating from its high power station in Cornwall over long distances. In the annual report the board quoted the ruling of Judge Townsend in the case against the De Forest Company. The 'cost of patents' in that year had been £13,243. New stations had also been erected all over the world, from China to Chili. The manufacturing was moved from the factory in Chelmsford to a new factory in Dalston: on the balance sheet, it was represented by the value of £41,221. By now, of the total shares available, some 251,127 shares had been issued, raising an additional capital of £251,127 (Table 7) (OX-3, 2017 Annual Statement 1905).

> At the London Stock Exchange the MWTC stock was dormant in 1905: no transactions in January, February, March, April, May, June, July, August and September. In October—the stock price at £1 3/8-1

Table 7: Extract from balance sheet of September 30, 1905.

To Capital		
Authorised	1905 value	2015 value*
500,000 Ordinary shares of 1 each	£ 500,000	
Issued		
251,127 Ordinary shares of 1 each, fully paid	£ 251,127	£ 24.32 M

Source: Marconi Archives, Bodleian Library. *Annual Report 1905.*
* http://www.measuringworth.com/uscompare/, M = Million

---

[317] Equivalent to about $139.6 million in 2015 based on the historic standard of living. Source: http://www.measuringworth.com/ukcompare/

½ and some business transacted at £1 $^{13/32}$. By 30 November the share price had fallen to £1 $^{1/4}$-1 $^{3/8}$ where it also finished on 31 December 1905 (GH)[318]. Hardly a revolutionary development for the stock of the MWTC in the five years of its existence.

Although confronted with the nationalization in Britain, by 1909 the financial tide was turning for the international Marconi Companies. As the MWTC largely participated in all these companies—the balance sheet of the year ending December 1908 shows a par value of £2,416,446 (Table 8)—it increased the value of the parent company considerably. Those participations had by 1909 already risen considerably to a total value (lower than the mentioned par value) of £291,032.[319] The board of directors of the MWTC reported:

> *A marked improvement in the business done by the Associated companies has also been experienced, several of which are now commencing to show profits, and with the ever-increasing appreciation of the Marconi system the Directors look*

Table 8: Extract from balance sheet of December 31, 1908.

By Patents and Shares in Associated Companies

	1908 value	2015 value*
200,278 fully paid shares of each £1 of the Marconi International Marine Communications Company Ltd.	£200,278	
80 Seven per cent. First Mortgage Debentures of £500 each of the Marconi International Marine Communications Company Ltd.	£50,000	
832,021 fully paid shares of each $5 each Series A of Cie Marconi de Telegrafia sin Hilos Rio de la Plata (Argentine Company).	£832,021	
88,250 Shares of each $5 each Series B of Cie Marconi de Telegrafia sin Hilos Rio de la Plata (Argentine Company).	£22,062	
34,224 fully paid shares of each $100 of the Marconi Wireless Telegraph Company of America Ltd.	£684,480	
614,855 fully paid shares of each $5 of the Marconi Wireless Telegraph Company of Canada Ltd.	£614,855	
273 fully paid shares of 1,000 Florins of Maatschappij voor Radiotelegraphie (Dutch Company)	£22,750	
Total par value	£2,416,446	£227.5 M

Source: Marconi Archives, Bodleian Library. *Annual Report 1908.*
* http://www.measuringworth.com/uscompare/, M = Million

---

[318] I owe much gratitude to Richard Crick for the many work he has done in searching the London Stock Exchange Archive in the Guildhall Library in London.
[319] Equivalent to about $27.4 million in 2015 based on the historic standard of living. Source: http://www.measuringworth.com/ukcompare/

with growing confidence to their large shareholdings in these companies. (OX-3, 2017 Annual Statement 1909: Report of Directors).

All in all it been a decade of operations before the Marconi-company started to show some real profits. Those early private investors from 1897 had to wait that long before their investment would to become profitable.

Stock Offerings for the MWTC (1904–1911)

As the profits from operations during the early years were too low to cover all the expenditures that had been made, often additional capital was needed. Over the years the capital of the company had increased as the result of several offerings of new stock. In 1904 the balance sheet of the Marconi Wireless Telegraph Company, Ltd. showed that 300,000 ordinary shares were available at a par value of £1. Of these shares, 221,127 fully paid had created a capital of £221,076.

In 1905 the offering of 200,000 new ordinary shares of £1 par value had raised the capital to £251,127 (Table 7). In 1908 another offering of 250,000 7% cumulative participating preference shares at £1 par value had added to the capital some £93,681. So, by 1911 the balance sheet showed a capital of £788,585 (Table 9). (OX-3, 2017 Annual Statements).

The profit and loss account for 1911 shows that the balance of contracts, sales, and trading had been £214,407. It shows that by 1911 the MWTC was slowly starting to become a profitable business. Orders in that year had increased to over £1,000,000, a considerable difference from the order position of £254,000 the year before. The shareholders were now entitled to their dividends, as the board declared at the annual meeting:

Table 9: Extract from balance sheet of December 31, 1911.

To Capital		
Authorised	1911 value	2015 value*
750,000 Ordinary shares of 1 each	£ 750,000	
250,000 Cum. Part. Pre. Shares of £1 each	£ 250,000	
	£1,000,000	£91.84 M
Issued		
508,266 Ordinary shares of 1 each, fully paid	£ 508,266	
241,734 Ordinary shares of 1 each, partially paid	£ 30,269	
250,000 Cum. Part. Pref. Shares of £1 each, fully paid	£ 250,000	
	£ 788,585	£72.43 M

Source: Marconi Archives, Bodleian Library. *Annual Report 1911.*
* http://www.measuringworth.com/uscompare/. M = Million

It is the intention of your Directors to recommend the payment of a final dividend of 10 per cent. For the year 1911, to which all classes of shares will be entitled. (OX-3, 2017 Annual Statement 1911: Report of Directors).

The preceding figures on the financial aspects of the Marconi Wireless Telegraph Company illustrate the financial side of the early development of Marconi's company. It shows the financial struggles of the company in its first decade of existence. It also shows that the different people who managed its development over time, although they were smart enough to understand the political/economical context in which they had to operate, were faced with an enormous business challenge. In our present-day vocabulary, we would say they were developing new products for new markets with a new technology, the most challenging of all innovation strategies.

So, it is time to step back in time and try to understand the driving force of technological development that brought the spark-based wireless technology step by step into maturity.

Marconi reaches Beyond the Horizon (1901–1905)

One issue of wireless communication is the distance that can be bridged by the electric waves. For a long time the scientific community occupied with wireless communication was restrained by the 'idea fixe' that electric waves only functioned in a straight line, just as visible light. For quite some time it was believed that the curvature of the earth would prohibit long-distance wireless telegraphy. Similar to electric light, it was thought that they would need an uninterrupted line of transmission between the transmitter and the receiver. That was a wrong assumption, as proven by Marconi..

Covering Longer Distances

Within the described business context the Marconi company was facing, in the meantime the experimental work on realizing long-distance transmission had continued. So, in January 1901 Marconi tested for longer distances over sea by experimenting on the south coast of England between the Isle of Wight and the Lizard in Cornwall (some 186 miles apart). In July, 1901, with an especially powerful apparatus, he was making intelligible transmissions for 225 miles, between Poldhu (Cornwall) and Crookhaven (Ireland) (Figure 182). So, maybe the problem of the earth curvature was not that dominant at all.

Figure 215: Marconi's circle of antennas at the Poldhu power station (1901).

Pictures show the antennas before and after the gale on September 18, 1901.

Source:
  http://www.oldradio.com/archives/jurassic/marconi.htm

Continuing on the path to increase the distance, not much later, on December 12, 1901, Marconi established another major event. That day a wireless signal consisting of the Morse code for the letter S was successfully transmitted across the Atlantic from a powerful 25-kilowatt transmitting station at Poldhu,

Cornwall, England, to Signal Hill, St John's, Newfoundland, Canada. The daring action of bridging the Atlantic distance of some 4,000 km created a furore in the press. And that was one of their objectives for all this activity, as America was a huge potential market. The publicity had also another result: after the restructuring in November 1899 the holding company was renamed the *Marconi Wireless Telegraph Company*, cashing in on the brand value of the Marconi name. His family was proud of their son, who was on his way to becoming a successful businessman.

Figure 216: Marconi kite experiment at Signal Hill, Newfoundland (1901).

A painting depicts Marconi's receiving aerial wire supported by a kite. Note the connecting wire leading from the aerial wire through the window to the receiving apparatus inside, and the ground wire from the window to the earth.

Source:
> http://www.newscotland1398.net/nfld1901/marconi-nfld.html

One could wonder what made Marconi motivated in his activities, as there must have been more than just technical curiosity that drove him.

> *Another reason is probably traceable to his daring and keen sense of the importance of capturing the imagination of the public at large that was always fascinated by this, to them, mysterious and intangible method of communicating through space. Although there was much to lose in case of failure this was more than compensated by the immense PR benefit that would be accrued by even a modest success and by the certainly interesting lessons learned even from a failure.* (F. P. Marconi, nd)

Marconi started to prepare for his experiments for the transatlantic crossing. However, that took some time, as they were faced with quite a few setbacks, some due to the atmospheric climate (Figure 215), and others due to the 'political climate'.

> *Marconi managed to convince his Board of Directors to invest in the powerful new stations required for the transatlantic gamble. These stations were built at Poldhu on the south-west coast of Cornwall, and at Cape Cod, Massachusetts. A renowned electrical engineer, Sir Ambrose Fleming, was consulted to design the*

stations. Oil fuelled engines driving large alternators replaced battery power supplies. The Poldhu station required a 25 kilowatt alternator, a then unheard of power for a wireless transmitter. ...

He then [after a gale blew down the antenna array in Poldhu] *decided on an experimental transmission from Poldhu to Newfoundland to prove that transatlantic radio communication was possible*

Figure 217: Earpiece used at Signal Hill experiments, Newfoundland (1901).

Source: Museum of History of Technology, Oxford. Press pack.
https://www.mhs.ox.ac.uk/marconi/presspack

and ease the concerns of his investors. Newfoundland was chosen to minimize the transatlantic distance because the destruction of his Poldhu antenna left him with less than ideal transmitting facilities. Nevertheless, he was attempting to bridge a distance of about 2100 miles when his best distance up to that time had been 225 miles.[320]

In the preparation before setting up the Atlantic connection experiment, Marconi travelled to New York by steamer. His arrival was a wild event, as he was met at the quayside and his hotel by dozens of reporters. He later commented on this event:

For some reason or the other it seemed to come as a shock to the newspapers that I spoke English fluently, in fact 'with quite a London accent' as one paper phrased it, and also that I appeared to be very young. (Jolly, 1972, p. 73)

He also was approached by a young graduate from Yale Sheffield Scientific School, who, having been working on his dissertation about Hertzian waves and looking for a way to fulfil his dreams, wrote Marconi a letter on September 22, 1899:

I have been exceedingly fascinated by the subject and hope for this opportunity of following work on that line. Knowing that you are about to conduct experiments for the US Government in the wireless telegraphy, I write you begging to be allowed to work under you. It may be that some assistants well versed in the theory of Hertz waves will be desired in that work, if so may I not be given the chance? ... It has been my greatest ambition since first working with electric

---

[320] Source: 'Marconi in Newfoundland: The 1901 Transatlantic Radio Experiment'.
http://www.newscotland1398.net/nfld1901/marconi-nfld.html (assessed March 2015)

waves to make a life work of that study. If you can, signor, aid me in fulfilling is desire, you will win the lasting indebtedness, as you have already the admiration of, Your obedient servant, Lee De Forest. (Jolly, 1972, p. 74)

The eloquent young author of this letter, being one year older than Marconi, was Lee de Forest (1873–1961), who would become known later for his work related to the invention of the vacuum tube, an invention that would change the world once again. But that did not happen as Marconi's employee. Even worse, the different companies he was involved in would later compete with Marconi's activities, and they would become entangled in costly patent litigation.

So, after arriving in New York, Marconi was to set up the facilities for his experiments, and that meant the search for a location.

The site chosen for the experiments was Signal Hill, a promontory near the mouth of the harbour that provided an unobstructed path to Cornwall and a marvellous view of the port of St. John's. His receiving apparatus, which was of the coherer variety, was installed in a room in a former hospital building near the present Cabot Memorial Tower. ... For the experiments on December 12, the aerial wire was supported by a kite, and Marconi abandoned the tuned receiver in favour of simply connecting the coherer detector and a sensitive earphone between the aerial and the buried metal plates which served as a ground connection. At 12:30 PM local time Marconi heard the signal, the three clicks of the letter "s". According to accounts of the history making event, he handed the earphone [Figure 217] to his assistant, Kemp, and asked "Do you hear anything, Mr. Kemp?" Kemp heard a few sequences of three dots before they faded back into the background noise. ... In the meantime, Marconi felt obliged to report his successes so far to his company and to the world. He notified the company by cable on the 14th, and on December 15, 1901 the New York Times announced: "St. John's, N.F.

Figure 218: Wireless room on board of the SS *Philadelphia* (1902).

http://www.marconicalling.co.uk/

Dec. 14th. Guglielmo Marconi announced tonight the most wonderful scientific development in modern times.''[321]

The Atlantic Ocean could now be crossed without using cables, and Marconi had proven it—not to everybody's satisfaction, though—with his wireless system. To prove once more that transatlantic communication was feasible, on February 22, 1902, Marconi boarded the SS *Philadelphia*, an American liner sailing to New York. The ship was equipped with antennas, and in the wireless room (Figure 218) daily transmissions with the Poldhu station were recorded. Using a syntonic receiving apparatus and a conventional coherer, messages were recorded on Morse inker tape and attested to by the ship's captain (Figure 219).

Step by step, Marconi gained the public's confidence in his wireless system, as was acknowledged later in 1903. On January 19, 1903, his wireless telegraph system was used to transmit a transatlantic message from President Roosevelt to King Edward VII of England:

In taking advantage of the wonderful triumph of scientific research and ingenuity which has been achieved in perfecting the system of wireless telegraphy. I extend on

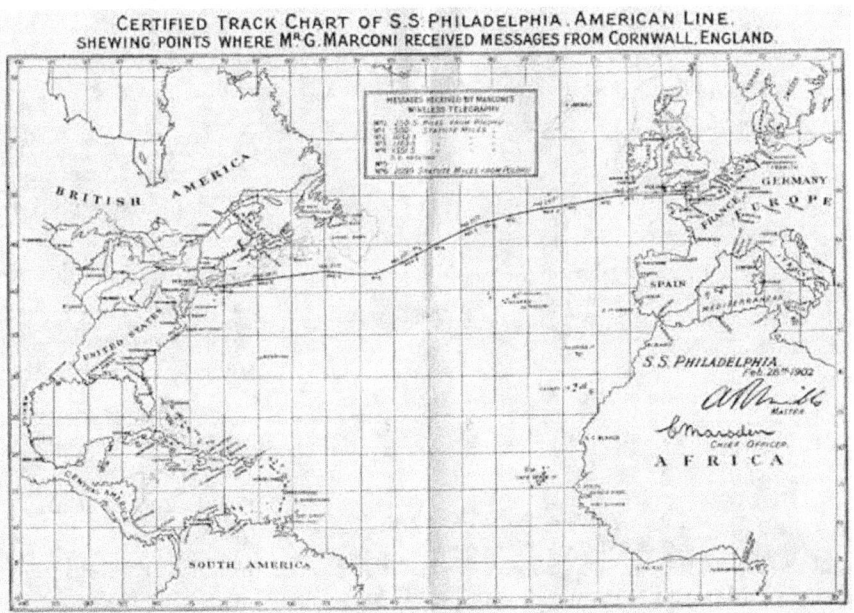

Figure 219: Attested track chart of Atlantic crossing experiments (1902).

Source: http://www.marconicalling.co.uk/museum/html/objects/ephemera/large_image/large_image-type_d__t04551.html

---

[321] Source: 'Marconi in Newfoundland: The 1901 Transatlantic Radio Experiment'. http://www.newscotland1398.net/nfld1901/marconi-nfld.html (assessed March 2015)

behalf of the American people the most cordial greetings and good wished to you and all the people of the British Empire.

Edward answered:

I thank you most sincerely for the kind message which I have just received from you, through Marconi's trans-Atlantic wireless telegraphy. I sincerely reciprocate in the name of the people of the British Empire the cordial greeting and friendly sentiment expressed by you on behalf of the American Nation, and I heartily wish you and your country every possible prosperity. (Fleming, 1919, p. 548)

Marconi had made his point: wireless long distance communication was feasible. The Earth curvature was no hindrance for electric waves.

Wireless Comes to America and Canada

Marconi had now technically proven that he could span increasingly long distance with his wireless telegraphy. He also illustrated the economic advantages to a journalist, who wrote about Marconi's achievements with the Atlantic experiment:

He [Marconi] *informed me* [the journalist] *that he would be able to build and equip stations on both side of the Atlantic for less than $150,000[322], the subsequent charge for maintenance being very small. A cable across the Atlantic costs between $3,000,000 and 4,000,000, and it is a constant source of expenditure for repairs.* (R. S. Baker, 1902, p. 298)

By now Marconi had demonstrated the technical and economic feasibility to the public as well as those in political power, not only to the British royalty and the American president, but also to those people in power in governments and military. It would help him in the many 'battles' that had to be

Figure 220: Marconi station at South Wellfleet, Cape Cod, Mass. (1903).

Source: http://earlyradiohistory.us/1903marc.htm

[322] Equivalent to $4,260,000 in 2015; calculation based on historic standard of living. Source: http://www.measuringworth.com/

fought with the vested interests of that time in the years to come. After his successful transatlantic transmissions, Marconi had been thinking about erecting a wireless station on American shores to compete with the undersea cables that crossed the Atlantic. He had spent months hunting for the right land to build his first US station along the coasts of New York, Rhode Island, Connecticut, and Massachusetts. He looked for a remote spot, but not too far from drinking water, labour, supplies, a rail line, and a hotel. He found the place on a cliff at South Wellfleet, Cape Cod, Massachusetts, where he erected a station in 1902 (Figure 220).

The station not only faced the rough climate, as in September the transmission masts were destroyed by storm, but also opposition from an unexpected side: the American cabled telegraph companies.

Opposition from American Cabled Telegraphy Companies

The kind words exchanged in 1903 between the president and king notwithstanding, there had been a different undercurrent in the preceding years. The general public may have enthusiastic about the prospects of the new way of communication, but not everybody else was enchanted. Already, before his demonstration of America's Cup in November 1899, upon his arrival in America, Dolbear's son had issued a statement on October 6, 1889, bragging:

> *Our patent issued on October 5, 1885 covers the entire grounds. It is for the art of wireless telegraphy. It does not matter what instruments of methods are used. Just as soon as anybody gets into the field of wireless telegraphy or telephony, they are poaching on our preserves.* (Jolly, 1972, p. 75).

Also, the solicitors of the *American Wireless Telegraph and Telephone Company* had already warned Marconi about infringing Dolbear's 1886 patent, a warning that was followed by a lawsuit on October 18, 1899: Marconi was accused of infringing on Dolbear's US Patent № 350,299 of 1886 (Figure 132), and damages of $100.000 were asked for.[323] However, the US Circuit Court ruled the complaint unfounded. It was the first sign of the opposition to follow.

Sometime later, around his 1901 experimental test of the Atlantic connection, Marconi was facing a new obstruction: the solicitors of the *Anglo-American Telegraph Company*, as they claimed to have the monopoly on communications in Newfoundland, threated a lawsuit when Marconi would not stop his experiments. That attack created a public outrage, with the Municipal Council of St. Johns passing a resolution and the newspaper

---

[323] Equivalent to $2,950,000 in 2015; calculation based on historic standard of living. Source: http://www.measuringworth.com/

publishing pro-Marconi opinions. Marconi got help from unexpected sides: Alexander Graham Bell offered the use of land he owned at Cape Breton in Nova Scotia (USA). The Canadian government invited him to discuss the matter, offered him a site at Glace Bay, and were willing to pay him a sum of £16.000[324] for assistance (Jolly, 1972, pp. 108-110).

So, sometime later Marconi moved to Canada and to the top of Table Head, a finger of cliffs pointing into Glace Bay on Cape Breton Island in the province of Nova Scotia. It was there on December 15, 1902, that the first transatlantic message was send. It was the beginning of transatlantic wireless communications from Canada, as in 1903 *Marconi's Wireless Telegraph Company of Canada* was organized.

> *The facts which were established by the end of 1902 were, that telegraphic messages could be sent 3,000 miles across the Atlantic Ocean by electromagnetic waves, at a speed and with a certainty which is not in any degree inferior to that effected by ordinary submarine cable telegraphy, when employing single transmission and hand sending.* (Fleming, 1919, p. 551)

For Marconi, this was a moment of victory. Marconi—his life by then already being a constant battle against the many vested interests of industry, government bureaucracy, military, and opportunistic scientists—was grateful for this expression of public support for the first time during his transatlantic experiments.

> *Indeed, at twenty-seven years of age, Marconi had overcome an unforgiving atmosphere and isolation of his father's attic in Pontecchio, the indifference of the Italian authorities, the scepticism of the scientific establishment, the conservatism of London's financial markets, the anxieties of his Irish investors, the scorn of the technical press, and the impatience of military bureaucrats in half a dozen countries.* (Raboy, 2016, p. 194)

It was also the beginning of a rivalry between Marconi and the telegraph cable companies that the legal impasse in Newfoundland had only delayed. Cable operators had already begun to protest less than two weeks after Marconi's Newfoundland experiment, when they observed that the value of their stock declined on the stock markets after the newspapers reported on Marconi's invention. The US businesses operating the long-distance cable system for telegraphy, like Western Union, were not too excited either. Marconi's plans were a direct threat to the *Atlantic Cable Company*, the *Commercial Cable Company*, and other parties that now carried telegraphs by underwater cables over (ie underwater) the Atlantic Ocean.

---

[324] Equivalent to $460,000 in 2015; calculation based on historic standard of living. Source: http://www.measuringworth.com/

This transmission over a distance of nearly 2,000 miles was a major achievement as even Marconi himself was unsure of the maximum range of is new technology. By the end of 1902, Marconi had established permanent and reliable wireless stations that Glace Bay in Nova Scotia, Canada and Cape Cod, US. By 1903, the Company had built a number of stations on shore and many merchant ships had been fitted with its wireless sets, which had to be rented from the company and were operated by Marconi personnel, who were allowed to communicate with operators using apparatus from rival companies during emergencies only. As a result of the growing maritime business the Marconi Company began to make a profit.[325]

Marconi had proven that transatlantic radio communications were possible in spite of the naysayers who claimed that radio waves, like light, would not bend over the horizon.

Looking backwards, it is amazing that the first transatlantic radio success was accomplished with such primitive equipment (no vacuum tubes or transistors!), and without our present knowledge of the propagation of radio waves at various wavelengths. ... Whatever the technical reasons for the success of the transatlantic experiment, it led directly to a commercial transatlantic radio service a few years later, and that in turn led to the worldwide wireless network that we take for granted today. (ibid)

Doing Business in the New World?

America, the land of opportunity, certainly had a capitalistic business climate with all its specific characteristics. On the one hand, there was the opposition from the cabled companies, who had their stake that they did not want to lose to a newcomer in the business. On the other hand, there were enough entrepreneurs in the United States willing to exploit a new opportunity.

In late 1899, Fred J. Cross contacted Guglielmo Marconi, and convinced the inventor to have the British firm partner with him in one of its first commercial efforts: an ambitious plan to build radiotelegraph stations on five of the Hawaiian Islands – then a U.S. territory – to provide inter-island communication. ... But after a promising start, operations were suspended.[326]

Contacts with American Navy—an obvious potential customer that had tested Marconi's equipment extensively—about purchasing a large number of installations were initiated. But that opportunity collapsed due to the

---

[325] Source: http://www.mhs.ox.ac.uk/marconi/collection/history.php (assessed March 2015)
[326] Source: 'Marconi Companies'. http://earlyradiohistory.us/sec007.htm#part040 (assessed March 2015)

business model of leasing the Marconi Company wanted to apply,[327] and it did not take long for the German competitor Telefunken to take over.

> *However, Marconi refused to sell its equipment outright, preferring to lease it, and the company also wanted to prohibit the Navy from communicating with stations using competitor's systems, except during emergencies. ... the Navy turned to other companies for its radio equipment purchases.* (ibid)

In the meantime the young Marconi—he was in his early 20s by that time—appealed to the American people and the media of that time. His business activities as well as his experiments and demonstrations were widely reported. For example, in 1899 the *Marconi Wireless Telegraph Company of America* (in short MWTCA) was created with the help of American financiers, and later—the by then already famous—Thomas Edison and Professor Pupin joined the board in 1903.[328]

> The MWTC of America was incorporated—by Guglielmo Marconi and the American Isaac L. Rice (president), the American financier August Belmont (treasurer), the American shipping magnate Clement A. Griscom, and the Irishman Robert Goodbody—with a nominal capital of $10,000,000.

STATEMENT FROM MARCONI

Says He Is Not Associated with Any American Company.

Owing to the number of companies advertising in America to sell stock in wireless telegraphy and telephone companies on large dividends are promised, G. Marconi, the inventor of wireless telegraphy, before sailing for Europe on the steamship Campania, requested his counsel, E. H. Morean of 34 Pine street, Manhattan, to cause the publication of the following letter:

"In view of the persistent representations of certain advertising parties, describing themselves as wireless telegraphy companies and claiming to be connected or suggesting that they are connected with the development of his system of wireless telegraphy in this country, and proceeding thereupon to appeal for subscriptions, Mr. Marconi feels that the time has come when it is only right and fair to the American public that he should make a personal statement on the subject. He, therefore, begs to state, and wishes it to be distinctly understood, that he is associated in no way whatever with any American company having for its object the development of wireless telegraphy in this country."

Figure 221: Statement published by Marconi (1901).

Source: http://earlyradiohistory.us/1901stat.htm

---

[327] The Marconi Company proposed to equip the vessels with their equipment for a fee of £100 per vessel.
[328] Source: *New York World*, May 28, 1903, p. 16. (assessed March 2015)

Such was the public interest in wireless transmission that—in a time where telephony was complementing telegraphy in a capitalistic America—this resulted in *wireless mania* and the *wireless bubble* on the stock market. And Marconi's name and reputation was often used by the bogus companies advertising to sell stock in wireless telegraph and telephone companies. Therefore, Marconi published a statement (Figure 221) in the *Brooklyn Eagle* newspaper on April 21, 1901, denying any relation with those companies.[329] Luckily, there were also serious business persons interested in Marconi's invention. It was the Boston banker E. Rollings Morse (one of the J. Pierpont Morgan group of financiers) who invested $6,000,000 in securing the patent rights for America, Cuba, Puerto Rico, the Danish West Indies, Alaska, the Philippines, the Hawaiian Islands, and all waters adjacent to these areas.[330]

International Demonstrations

As the maritime application of wireless telegraphy was seen by the Marconi Company as an important commercial field, the first success was from an unexpected client.

[It was during this time that] *some commercial encouragement had come the way of the company in the form of an order for a wireless installation aboard a transatlantic liner. Not, as might have been expected, from one of the British shipping firms but from a German line, Nordeutscher Lloyd of Bremen- a relatively unexpected source, seeing that extensive experimental work on wireless telegraphy was currently being carried out in Germany. However, there was the contract in black and white; an installation aboard the SS Kaiser Wilheldmder Gross and permission from the German government to fit wireless apparatus on the Borkum Riff lightship and at the Borkum Island lighthouse twenty miles or*

Figure 222: Marconi's lattice towers at Poldhu, Cornwall, carrying the antenna wires for the Marconi electric wave power station (1902).

Source: Salmon, A.L.: *The Cornwall Coast.*
www.gutenberg.org/files/26907/26907-h/26907-h.htm

---

[329] Source: http://earlyradiohistory.us/1901stat.htm (assessed March 2015)
[330] Equivalent to £581,800,000 in 2015; calculation based on historic standard of living. Source: http://www.measuringworth.com/

so distant. … On February 10 the apparatus was taken aboard the liner and, a few days later, in foul weather, other equipment was installed in the lightship and lighthouse. (W. J. Baker, 1970, pp. 57-58)

After Marconi had applied his system in 1902 during his experiments on the trip to the United States, it was the Maxwellian Silvanus Thompson who attacked him. Thompson stated that the transmitter at Poldhu (Ireland) had been designed by John Ambrose Fleming, the coherer located in St. John's (Canada) by Stephano Solari, and (again) that Oliver Lodge was the true inventor of wireless telegraphy (Hong, 2001, p. 102).

Long-distance Wireless Telegraphy: Space Telegraphy

Soon after returning in February 1902 to England Marconi concentrated on experimenting with long-distance wireless telegraphy (aka space telegraphy). In Poldhu, Cornwall, a permanent structure for carrying a large aerial was constructed. These lattice towers, similar to those erected in Cape Cod, Massachusetts, in America and Cape Breton in Nova Scotia, Canada, were 210 feet high (Figure 222).

Figure 223: Marconi's testing wireless telegraphy on the Italian warship *Carlo Alberto* (1902).

The antenna is noticeable between the masts.

Source:
  http://www.radiomarconi.com/marconi/pop

Traveling in February 1902 to Canada, he was active with receiving signals over increasing distance on board of the American liner SS *Philadelphia*. Readable messages were received up to 1,500 miles away. In July 1902, using the facilities of the Italian warship/royal yacht *Carlo Alberto*, he experimented on a voyage to Kronstadt on the Baltic Sea in 1902. There, he met with the Tsar and dined with the new Italian king Vittorio Emanuelle II (Fleming, 1919, pp. 546-547)

The wireless apparatus performed splendidly, receiving signals from Poldhu after nightfall when the cruiser lay anchor off Kronstadt, some 1,600 miles away. The meeting between the two potentates attracted world-wide interest, and the news of the long-distance wireless reception shared the headlines. (W. J. Baker, 1970, p. 77)

Still under attack from the British scientific community, Marconi published on the progress of his work in the lecture 'Progress of Electric Space Telegraphy' (G. Marconi, 1902). By that time the topic of 'space telegraphy' and 'electric waves' was the subject of attention of many scientists in England (eg Lodge and Muirhead) as well in Germany (eg Slaby and Von Arco) and America (eg Anderson, Fessenden, Stone).

That was the lasting conflict with the British scientific community. Soon he also ran into problems with the German government. However, that was based on a totally different topic: the commercial policies of the company.

> *The success the Marconi interests were having in their efforts to establish a global monopoly of radio communications began to cause considerable concern among some of the world powers and especially in Germany. In commenting upon the sentiment at the time, Barber reported: "The Germans are wild over the Marconi monopoly. Such a monopoly will be worse than the English submarine cable monopoly which all Europe is groaning under and I hope the Navy Department of the United States will not get caught in its meshes." Early in 1902 an incident occurred which caused the German Government to take official cognizance of the situation. Prince Henry of Prussia, brother of the German Kaiser, was returning to Germany, in the S.S. Deutschland, after a visit to the United States. Soon after sailing, he desired to send President Roosevelt a radio message thanking him for the numerous honors and courtesies which had been accorded him. The Deutschland transmitted this message to the Marconi station at Nantucket, but that station refused to accept it because the ship was fitted with Slaby-Arco radio equipment. The irate Prince brought the matter to the attention of his brother. Kaiser Wilhelm thereupon instructed his government to initiate action in an attempt to establish international control over radio communications.[331]*

Marconi had, as part of his monopolistic strategy, been forbidding 'intercommunication' between ships with Marconi equipment and ships using equipment from rival companies. At that time the competing equipment was supplied by Braun and Slaby. Henry's reaction, and the follow up by his brother Wilhelm II, created a 'Marconi-phobia' on the part of the Germans in the press. In London, the company moved into spin mode. Although apologies were made, Marconi insisted on his exclusivity rights. The whole affair triggered the German efforts to create their own wireless industry; not much later the Telefunken system became a serious competitor for Marconi (Figure 224). And it initiated the efforts to regulate wireless communication: the Berlin Conference of 1903.

---

[331] Howeth, L.S.: 'The Origins of Regulation'. Source: http://earlyradiohistory.us/1963hw07.htm#7footnote

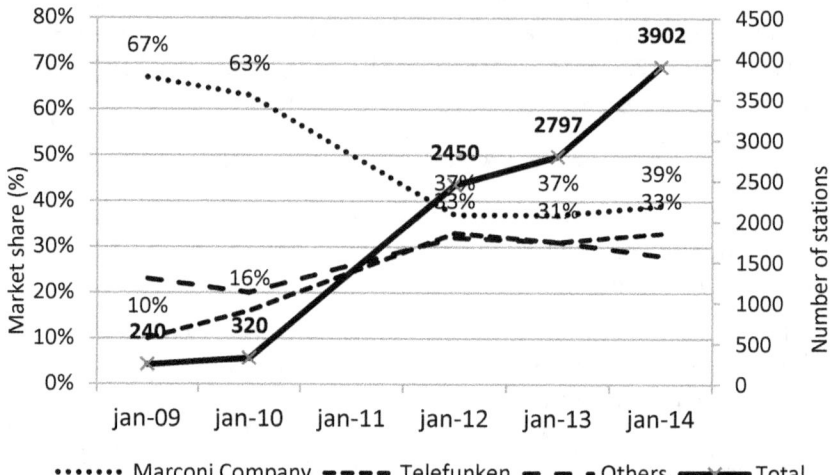

Figure 224: Market share of Marconi companies, Telefunken, and other companies in wireless telegraphy (1909–1914).

Source: Friedewald, M.: *The Beginnings of Radio Communication*, p.8.
https://works.bepress.com/michael_friedewald/71/

Although the conference was supposed to address a number of wide-ranging issues, the only real issue was the Marconi Company's refusal to communicate with other systems. All the countries at the conference, with the exception of Italy and Great Britain, favoured compelling Marconi to communicate with all ships because they opposed his 'de facto' monopolization of the airwaves. It was the beginning of the end of the Marconi monopoly (Raboy, 2016, pp. 197-199).

Commercial Successes for the British Marconi Company

By 1902 some commercial interest for Marconi's system materialized slowly. By that time the relationship with the Italian Navy had improved; he had visited the Italian royal yacht *Carlo Alberto* at Kronstadt in the Baltic Sea, where he was laurelled by the Russian tsar, and dined with the new Italian king Vittorio Emanuelle II. Even the first order from the Imperial Japanese Army showed his stature in the world. Also, in 1902 the contract with the Canadian government creating $80,000 in revenues was signed.[332] This agreement not only secured urgently needed cash, but placed Marconi,

---

[332] Equivalent to $2,270,000 in 2015; calculation based on historic standard of living. Source: http://www.measuringworth.com/

Canada, and the relatively remote region of Cape Breton in the forefront of world communication development. Nevertheless, financially, the company was in heavy weather.

> [At Cape Breton] *Not many inventors or commercial entrepreneurs had at their disposal the flagship of the Italian Navy, the resources of the Canadian Government, the spotlight of the* [newspaper] *Times of London, and the attention of an admiring local population. … At the same time, he had to deal with the company's chronic cash shortage. Remarkably, given the innovations Marconi was stewarding, the company's finances were in desperate straits. It was still in debt after taking out a £30,000 loan to finance Marconi's latest transatlantic adventure. …* (W. J. Baker, 1970 np)

He also managed to secure a contract with the Italian Navy on February 12, 1903: the Italian Parliament accorded £32,000 for the project.[333] Their coastal transmission station was to be built at Coltano (near Pisa) and would cost double the amount of earlier British or American stations. It was meant as an important strategic installation for the Italian colonial expansion in Africa (ie the Scramble for Africa). The contract also had another effect besides creating revenues for the company.

> *Even more important to Marconi than the agreement itself was the renewed recognition it brought him in Italy. Municipalities across the country, including Rome and his native Bologna, offered him keys to the city. He was hailed as Italy's greatest living genius, feted and coddled by the press, and, finally, completely vindicated in the eyes of his now ailing father Guiseppe.* (W. J. Baker, 1970 np)

On April 30, 1903, a crowd of thousand Romans cheered Marconi when he arrived at the train station. The king gave a dinner at the Quirinal Palace on May 4, 1903, where Marconi also met another guest: the German Kaiser Wilhelm. Soon the Kaiser brought up a sensitive subject—the refusal to transmit telegraphic messages from the equipment of other suppliers—with the remark, 'Signor Marconi, you must not think that I have any animosity against yourself, but I do object to the policy of your company.' To which Marconi replied: 'Your Imperial Majesty, I should be overwhelmed if I thought you had any personal animosity against me. However, it is I who decided the policy of my company' (Jacot, de Boinod, & Collier, 1935, p. 18) (Dunlap, 1941, p. 148).

On July 1903 the long-awaited order from the British Royal Navy also came through. It was a contract for a period of eleven years that had taken since 1901 to negotiate. All in all, the year 1903 was the year that the

---

[333] Equivalent to £3,099,000 in 2015; calculation based on historic standard of living. Source: http://www.measuringworth.com/

Marconi Company financially became healthier. For the first time, its revenues exceeded its expenditures.

New Manufacturing Facilities

All the equipment for the Marconi system had been manufactured in the Hall Street factory in Chelsmford (Figure 190). By 1905, when business was picking up, it was clear that the manufacturing facilities needed improvement. The works were therefore moved from Chelmsford to a large four-storey building at Dalston, North London (Figure 225).

However, this was not to last for a long time, as the Dalston works were eventually closed down in 1908 and the original factory at Hall Street in Chelmsford was reopened and refitted. In 1912 a new factory was opened in New Street in Chelmsford, built to replace the facilities in Hall Street (Figure 226). It was a large factory with a floor space of 6,500 m².

From conception to completion, the project took only 17 weeks. Built on part of the old Essex Cricket Ground, the new works were modern and equipped with the latest tools and laboratories. The changeover between Hall St. and New St.

Figure 225: The workshop in the Marconi works at Dalton (1906).
Source: http://www.marconicalling.co.uk/museum/html/objects/photographs/

Figure 226: Manufacturing facilities at New Street, Chelmsford (1912).
Source: http://www.marconicalling.co.uk/ museum/html/objects/photographs/

happened in just one weekend. The Works were ready for inspection in June by the suitably impressed International Radiotelegraphic Conference delegates on 22 June 1912.[334]

Regulation and Legislation

As early as around 1900 it was already obvious that wireless telegraphy was going to play an important role in society. As a chaotic situation existed, wireless telegraphy needed some regulations in order to be able to function properly in society. Also, a form of technical regulation was needed to prevent interference between stations, commercial dominance, competitive sabotage, amateur hacking, and technical interference of the electromagnetic waves.

Marconi might have been getting some considerable commercial successes, but his troubles were far from over, not only on his home turf, where the British government was unwilling to grant him the commercial facilities he needed in England, but also in Germany. There, the German government was anxious to displace the Maroni monopoly as the principle supplier of wireless equipment throughout the world. They had their own Slaby-Arco system supplied by the company Telefunken. Add to that the German imperial aspirations during the Scramble for Africa, where the German African colonies were totally dependent on communication through British wired telegraphic facilities. The Germans wanted to get rid of the British dominance and not only in the field of communications, which became clear not much later on.

> *Marconi's affairs, and the stakes at the international communications, were now at the centre of the most important geopolitical rivalries and it was all becoming increasingly indigestible.* (Raboy, 2016, p. 262)

The Maskelyne Affair

In the meantime, on the home front, Marconi's monopoly had become subject of criticism. During a demonstration at the Royal Institution on June 4, 1903, Marconi became the victim of a 'hacker'. This would become known as the 'Maskelyne affair', which happened when Ambrose Fleming gave a demonstration demonstrating that private messages of a confidential nature could be sent over great distances with his syntonized system.

> *Late one June afternoon in 1903 a hush fell across an expectant audience in the Royal Institution's celebrated lecture theatre in London. Before the crowd, the physicist John Ambrose Fleming was adjusting arcane apparatus as he prepared to demonstrate an emerging technological wonder: a long-range wireless*

---

[334] Source: http://www.marconicalling.co.uk/html/index.html (assessed June 2015)

communication system developed by his boss, the Italian radio pioneer Guglielmo Marconi. The aim was to showcase publicly for the first time that Morse code messages could be sent wirelessly over long distances. Around 300 miles away, Marconi was preparing to send a signal to London from a clifftop station in Poldhu, Cornwall, UK.

Yet before the demonstration could begin, the apparatus in the lecture theatre began to tap out a message. At first, it spelled out just one word repeated over and over. Then it changed into a facetious poem accusing Marconi of "diddling the public". Their demonstration had been hacked – and this was more than 100 years before the mischief playing out on the internet today. Who was the Royal Institution hacker? How did the cheeky messages get there? And why? …

Marconi would have been peeved, to say the least, but he did not respond directly to the insults in public. He had no truck with sceptics and naysayers: "I will not demonstrate to any man who throws doubt upon the system," he said at the time. Fleming, however, fired off a fuming letter to The Times of London. He dubbed the hack "scientific hooliganism", and "an outrage against the traditions of the Royal Institution". He asked the newspaper's readers to help him find the culprit. …

He didn't have to wait long. Four days later a gleeful letter confessing to the hack was printed by The Times. The writer justified his actions on the grounds of the security holes it revealed for the public good. Its author was Nevil Maskelyne, a moustachioed 39-year-old British music hall magician. Maskelyne came from an inventive family – his father came up with the coin-activated "spend-a-penny" locks in pay toilets. Maskelyne, however, was more interested in wireless technology, so taught himself the principles. … his ambitions were frustrated by Marconi's broad patents, leaving him embittered towards the Italian. (Marks, 2011)

Maskelyne would become the leading figure in the anti-Marconi faction.[335] Apart from Maskelyne's individual motives, however, behind him was another British party that had all reason to be critical of Marconi's effort: the wired telegraph industry that had invested heavily in the by the enormous telegraphic infrastructure spanning the world.

One of the big losers from Marconi's technology looked likely to be the wired telegraphy industry. Telegraphy companies owned expensive land and sea cable networks, and operated flotillas of ships with expert crews to lay and service their submarine cables. Marconi presented a wireless threat to their wired hegemony, and they were in no mood to roll over.

---

[335] Nevil Maskelyne was a self-educated electrician who became known after he used a wireless circuit to decharge gunpowder in 1899. He criticized Marconi's monopoly on wireless telegraphy and became the leading figure in the anti-Marconi faction.

The Eastern Telegraph Company ran the communications hub of the British Empire from the seaside hamlet of Porthcurno, west Cornwall, where its submarine cables led to Indonesia, India, Africa, South America and Australia. Following Marconi's feat of transatlantic wireless messaging on 12 December 1901, ETC hired Maskelyne to undertake extended spying operations.

Maskelyne built a 50-metre radio mast (the remnants of which still exist) on the cliffs west of Porthcurno to see if he could eavesdrop on messages the Marconi Company was beaming to vessels as part of its highly successful ship-to-shore messaging business. … Having established interception was possible, Maskelyne wanted to draw more attention to the technology's flaws, as well as showing interference could happen. So he staged his Royal Institution hack by setting up a simple transmitter and Morse key at his father's nearby West End music hall. (ibid)

The incident was followed by a large dispute in the technical press (eg the *Electrical Review, Morning Advertiser*), with Fleming siding with Marconi. Maskelyne's actions had a side effect, as they influenced the relationship between Fleming, the trusted scientific advisor who, being a widely recognized expert, often came to Marconi's defense, and Marconi himself.

The affair reflected, and accelerated, wireless technology's transition from a period in which public shows and sensations (such as the first transatlantic reception of SSS signals) had been essential to economic success to a period in which regulating frequencies and guaranteeing instrumental uniformity became serious issues. … The Maskelyne affair strikingly undermined his [Fleming's] *credibility, preventing him from serving as a trusted witness. Moreover, his contract with the Marconi Company, which terminated in December 1903, was not renewed. He had become useless to Marconi.* (Hong, 2001, p. 118)

The affair forced Marconi to moderate his love for public demonstrations. Fleming returned to his laboratory, where he would invent a device that would change wireless communication: the electronic valve. It had also become clear that the time for regulation of wireless telegraphy was approaching, but that would prove to become a hornets' nest of political manoeuvring.

International Conferences (1903, 1906)

In 1902 the Marconi policy not to transmit any messages coming from non-Marconi equipment—based on technical arguments but in reality just a monopoly issue—resulted in a situation that would have considerable consequences. In 1903 the German government called an international conference on the subject of wireless telegraphy. At this preliminary conference, held in Imperial Post Office in Berlin, representatives from Austria, France, Germany, Great Britain, Hungary, Italy, Russia, Spain, and the United States were present. The stakes were high, and certainly, the Germans had something to gain.

> *Britain's interests came to clash sharply with those expressed at the 1903 Berlin Wireless Telegraph Conference, where the German Arco-Slaby Company, arch-rival of the Marconi Company, strove to initiate international regulation of radio and enable all stations to communicate with all ships. This smacked of freeloading to the Marconi Company, which had the largest network of stations.* (Satia, 2010, p. 850)

At the conference, it was decided that coastal stations were to receive and transmit telegrams to and from ships at sea without distinction as to the wireless telegraphy system they used (article 1, section 2). Also, the technical specifications of the systems used should become available (article 1, section 3). Next, it was decided that the tariffs used for wireless telegraphy should be on a 'word basis' for both the ship charge and the charge for the coast station (article 1, section 4). In addition, priority was to be given to 'calls for help' transmitted by ships (article 5). For telegraph stations not open to the public, that is, the governmental stations used for diplomatic transmissions, the provisions did not apply (article 7). (Howeth, 1963)

Article 1, section 2, was intended to eliminate the possibility of a monopoly by one company, in this case the Marconi Wireless Telegraph Company of England, who by then was dominating the wireless telegraph market. Obviously, British interests were at stake, so the British delegation maintained a 'general reserve'. The Italian delegation, knowing that the Italian government had already entered into an agreement with Marconi, also objected to specific articles.[336]

> *In the end, both their obligations to the Marconi Company ensured that neither Britain nor Italy accepted the conference resolutions. From that point, protection of British pre-eminence in radio led to a jealous guarding of patent rights against foreign sale, a tactic inevitably resented and reciprocated.* (Satia, 2010, p. 850)

After a planned session in 1904 was postponed, in 1906 a second international radio conference was held in Berlin, Germany, to deal with issues left over from the 1903 Conference. Now a much greater range of countries participated (some 31 compared to the nine in 1903), some of whom were engaged in a military conflict (Japan and Russia). On the agenda was 'radiotelegraphy'. Again the issue was the exchange of radio telegrams regardless of the system used (article 3). The American

---

[336] The British and Italian governmental contracts with Marconi were based upon the exclusive use of their equipment. Those contracts also prohibited interchange of messages with stations equipped with other than Marconi apparatus. This way, Marconi protected his patent rights, as other companies were infringing its patents.

delegation, fed up with Marconi's influence and dominance, issued a declaration that asked for voting on article 3.

> This declaration brought forth a flood of polemics that rocked the chamber, but resulted in article 3 being immediately floored for final debate. The British delegation exerted its strongest effort to defeat it, but, with the vigorous support of the United States and Germany, it was adopted without alteration. Since it was impossible for the British to accept the provision without violating the conditions of their Government's contract with Marconi, its delegation intimated that it could not remain at the Conference under such conditions. The Conference then agreed that an easement in the final protocol, in the form of an article, might be inserted for the sole purpose of permitting such nations as were involved with the Marconi interests to adhere to their contracts. (Howeth, 1963, pp. 117-132)

Finally, parties came to an agreement in which both ship-to-shore communication and ship-to-ship communication should not be hindered by commercial interests of rivalling telegraph systems. And those commercial interests were massive, binding governments to contracts with their 'national industry' (such as Marconi in Britain). Wireless technology had become a political instrument.

> Having gone down in defeat as antagonists of article 3, the British delegation had submitted a proposal to the Committee on Regulations which provided that a surtax might be charged when stations employing one system were forced to receive and transmit messages emanating from a rival system. This proposal was entirely out of accord with the broad and liberal spirit which animated the Conference. Although the British proposal was presented most plausibly and with marked ability, it was plainly intended to make parties to the Conference generally assist in the payment of such royalties or indemnities as might be due the Marconi Co. by Great Britain. The proposition was being debated at the same time as the compulsory ship-to-ship one, but came to a final vote first. The American delegation fought it vigorously and exposed its character so effectively that the decision went overwhelmingly against the British position. As a result of this defeat the British delegation lost much of its following, and this was instrumental in the American victory on the ship-to-ship proposal. (Howeth, 1963, pp. 117-132)

The result was a comprehensive agreement, the *International Wireless Telegraph Treaty*. Great Britain reserved the right to organize a separate system of shore stations in fulfilling the requirement for compulsory communications between all coast and ship stations. Now it was to the separate states to ratify the convention. The treaty marked the end of Marconi's monopoly on seagoing ship-to-shore and ship-to-ship communications.

National Regulation Starts

On August 15, 1904, a British Act titled *An Act to Provide for the Regulation of Wireless Telegraphy* passed both houses of Parliament and received the royal consent. In this Act, the monopoly of the Post Office was re-established, as wireless communication was considered to be similar to wired communication. Using the radio spectrum, wireless receivers and transmitters had to be licensed from the Post Office, and licenses were given for commercial purposes or ship-to-shore communications. In 1908 the General Post Office built its first coastal station for wireless telegraphy. In 1909 they acquired—nationalized—most of Marconi's wireless coastal stations for £ 15,000.[337] Those stations by that time had complied with the requirement of compulsory communication as a result of the International Wireless Telegraph Treaty (Fleming, 1919, p. 635).

But before that was going to happen there was more at stake than just regulation of some technical issues (eg intercommunication). Now the regulation could cripple Marconi's activities that were based on his fundamental patent rights issued to him worldwide as the inventor of the wireless system. But there was more at stake on a national level, as was expressed by Marconi management:

> *The Berlin proposals were objectionable from the standpoint of British imperial and commercial interests ("represented in this case by our Company"). ... In other words, Marconi's interests and British interests were the same. (Raboy, 2016, p. 272)*

Now a new argument entered the discussion: the patriotic cause.

> *The crux of the matter was this: the Post Office had achieved its goal of bringing wireless under its wing; the Admiralty was eager to establish the system throughout the British Empire; but the government, as a matter of high policy, wanted to sign the Berlin protocols and could not do so without the company's acquiescence. (Raboy, 2016, p. 272)*

On the other side of the Atlantic Ocean, the parliamentary acceptance was less smooth, as American commercial interests disliked too much governmental influence. There, in the society of the New World, words like 'free enterprise' and 'monopoly' had a different status. In this case, that combination was challenged, as the treaty affected the interest of the Marconi Wireless Telegraph Co. of America. Not too amazingly, the American management declared that the ratification of the treaty:

---

[337] Equivalent to £1,410,000 in 2015; calculation based on historic standard of living. Source: http://www.measuringworth.com/

...would most seriously and injuriously affect the business and profits of the Marconi interests and practically destroy the advantages which the inventor and his assignees expected to receive and were entitled to obtain from the priority of their inventions and from the establishment of their system. (Howeth, 1963, pp. 117-132)

American Marconi was defending their patent rights, and they demanded compensation for their commercial losses. However, there were other interests supporting the ratification, such as the United Wireless Company and the American Navy, who embraced further innovation, not to be hindered by patent issues:

The opening of the field to any and all users, and thus fostering competition by promoting its use and increasing the demand for it, was clearly calculated to stimulate invention and encourage development and improvement. (Howeth, 1963, pp. 117-132).

Reflecting the political climate in the US, the American Committee decided not to make a decision on the acceptance of the treaty:

It is proposed to let the wireless Treaty remain in the committee indefinitely while the Navy Department watches the behavior of the wireless companies. If they refuse to transmit distress signals or show reasonable cooperation the treaty will be taken from its pigeon hole and ratified. The passage of the treaty would place all companies under international supervision. (Howeth, 1963, pp. 117-132)

In 1908 three bills were proposed trying to regulate wireless telegraphy. However, after long debates Congress adjourned without voting on any of them. This type of forestalling of legislation by commercial interests illustrates the dynamics of a new technology in its infancy. Governmental efforts to regulate 'radio' were influenced by the commercial interests of both those who'd developed the systems and those who used the systems. It would need a war to solve this stalemate.

A Fury of Patenting

After Marconi's early pioneering period and the experiments for the transatlantic crossing, the massive publicity in professional journals (such as *The Electrician*), but also general journals (like *Scientific American* and *Nature*) and the newspapers oriented at the general public (such as the *New York Times* and the *New York Tribune*), had created a massive awareness of wireless telegraph. In America, it created the *wireless mania*, both in the business community and the scientific and engineering world. To illustrate this massive interest of the many scientists and engineers who became involved in the development of 'radio communication', we mention just the most relevant, as they appeared in later patent litigation.

Further Improvement of the Wireless System: Mechanical Solutions

As described before, spark-gap emission is by definition a crude way to generate electric waves: they are damping, they have a broad frequency spectrum, and they are hard to create in a stable way. The world, waiting for an alternative, was ready for the next step in the development of wireless communication, a step that would be made when the switch from direct current (DC) to alternating current (AC) was made. At first the attention was focused on mechanical systems creating alternating current of a single frequency (eg 50 Hz, 400 Hz, etc). That current then was used to create a spark. The objective was to create Hertzian waves of a higher frequency, let's say 15 kHz. This new approach of using a 'continuous wave' instead of the 'on-off' method of bursts of waves that was used before would later prove to be fundamental.

One of the first to pioneer a new approach was the Canadian *Reginald Fessenden* (1866–1932) who, after moving from Canada to New York, started to work for Thomas Edison as an assistant tester for underground electrical mains.[338] He soon proved his intelligence and became a junior technician in Edison's laboratory. When Edison, running into financial problems, had to let go of most his laboratory employees, Fessenden took other jobs and ended up as professor at the electrical department of Purdue University. After helping George Westinghouse install the lighting at the Columbian Exposition in Chicago (1893),[339] he became professor at the University of Pittsburgh, where he started experimenting with wireless communication in 1898.

In 1900 he left the university and started working for the US Weather Bureau, and he experimented with a network of coastal radio stations to transmit weather information, thus avoiding the need to use existing telegraph lines. His early experiments with rotary spark transmitters resulted in the transmission over a distance of 1.6 km on December 23, 1900. Next he experimented with higher-frequency 'alternator' transmitters.[340] They produced low-frequency 'pure sine waves' (up to 10 kHz) as an alternative carrier of communications; the system of continuous wave was born.

On December 23, 1900 Fessenden said into his microphone, "One, two, three, four. Is It snowing where you are Mr. Thiessen? If so telegraph back and let me know." Thiessen replied by telegraph in Morse code that it was indeed snowing.

---

[338] See: B.J.G. van der Kooij, *The Invention of the Electric Light* (2015), pp. 164–166
[339] See: B.J.G. van der Kooij, *The Invention of the Electro-motive Engine* (2015). pp. 212–214.
[340] In its essence, an alternator is a dynamo creating an AC current. The frequency of the AC current is dependent on the rotation speed, the gear box, and the arrangement of the coils.

In great excitement Fessenden wrote at his desk, "This afternoon here at Cobb Island, intelligible speech by electromagnetic waves has for the first time in World's History been transmitted." This was almost a year before Marconi's transmission in Morse code from England to Signal Hill in Newfoundland, on December 12, 1901.[341]

In his US Patents № 706,735 through № 706,747, all granted on August 12, 1902, he protected his concept (Table 10). Over time Fessenden not only worked on transmitting electric waves, but he also conducted many experiments on detecting electric waves, for example, creating an electrolytic detector (his so-called 'Barretter', resulting later in US Patents № 727,331 of May 5, 1903 and № 793,684 of December 1904).[342] This was followed by a thermal detector: the hot-wire barrater (Sarkar & all, 2006, pp. 369-371)

Table 10: Patents granted to Reginald Fessenden on August 12, 1902.

Patent №	Granted	Description
US 706,735	August 12, 1902	Wireless telegraphy (filed December 15, 1899).
US 706,736	August 12, 1902	Apparatus for wireless telegraphy (filed May 17, 1900).
US 706,737	August 12, 1902	Wireless telegraphy (filed May 29, 1901).
US 706,738	August 12, 1902	Wireless telegraphy (filed May 29, 1901).
US 706,739	August 12, 1902	Conductor for wireless telegraphy (May 29, 1901).
US 706,740	August 12, 1902	Wireless signalling (filed September 28, 1901).
US 706,741	August 12, 1902	Apparatus for wireless telegraphy (filed Nov. 5, 1901)
US 706,742	August 12, 1902	Wireless signalling (filed June 6, 1902).
US 706,743	August 12, 1902	Wireless signalling (filed June 26, 1902)
US 706,744	August 12, 1902	Current-actuated wave-responsive device (filed July 1, 1902).
US 706,745	August 12, 1902	Signaling by electromagnetic waves (filed July 1, 1902).
US 706,746	August 12, 1902	Signalling by electromagnetic waves (filed July 1, 1902)
US 706,747	August 12, 1902	Apparatus for signalling by electromagnetic waves (filed July 22, 1902).

Source: USPTO.

---

[341] An Unsung Hero: Reginald Fessenden, the Canadian Inventor of Radio Telephony. Source: http://www.ewh.ieee.org/reg/7/millennium/ radio/ radio_unsung.html (assessed June 2015)

[342] Fessenden would later contribute to solving the basic problem of damping in spark-generated electric waves. He also developed the heterodyne principle, used to combine two frequencies. He was granted US Patent 706,740, filed on September 28, 1901, and granted August 12, 1902.

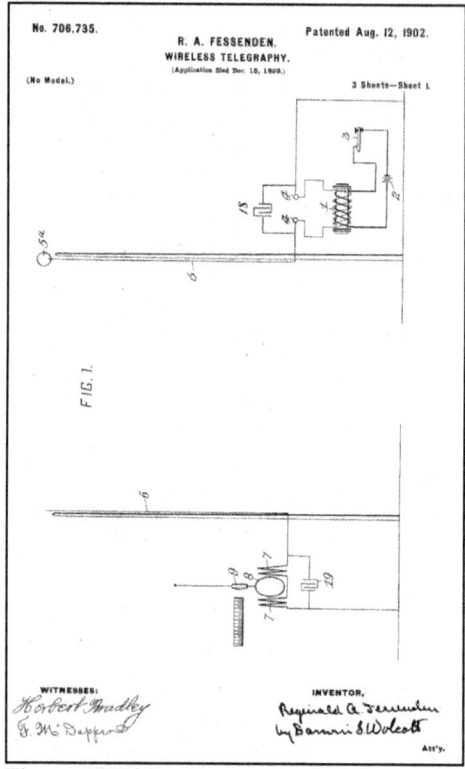

Figure 227: The Fessenden US Patent № 706,735 (1902).

Source: USPTO.

After leaving the US Weather Bureau as the result of a conflict about his rights to the inventions he had made, with the financial support of two Pittsburgh millionaires, Hay Walker and Thomas H. Given, he started in November 1902 the *National Electric Signalling Company* (NESCO). Fessenden contributed his patents, the millionaires the funding (for $330,000, they obtained a 55% share in the company). The company started trying to create business, building some experimental wireless stations with 400-foot antenna towers at Brant Rock, Massachusetts. As a result of their excellent performance, three more stations were built in New York, Philadelphia, and Washington. However, the company, trying to compete with Marconi, was not too successful, and the relationship between Fessenden and his financial backers soured.

> *Believing that he had already made major compromises to suit his backers, compromises that interfered with his experimentation and that required him to fill too many roles at once, Fessenden became increasingly uncompromising and abrasive. He came to see every negotiation over every detail as a battle over preserving the autonomy and discretion he had left. His backers, who by 1905 had already invested half a million dollars in Fessenden's visions, stoked the embers of their own resentment, which Fessenden fanned with each new demand. The increased tensions within the company, which were exacerbated by external events such as the panic, left both sides feeling beleaguered and frustrated. Fessenden, Given, and Walker had never been able to agree on and pursue long-term business strategies, and the erosion of their superficial alliance during these years precluded the discovery of a remedy for the situation. They continued to pursue short-term projects that were sustained only through the first intoxicating*

flush of enthusiasm. When the endurance and determination necessary to sustain a strategy over years rather than months was not summoned, one short-term plan replaced another. A productive alliance can provide the sustenance a company needs, but such an alliance did not exist at NESCO, and this deficiency had major repercussions not only on the company, but also on how and by whom radio would be developed. (S. J. Douglas, 1989, pp. 151-152)

Without informing his partners, he created the *Fessenden Wireless Company of Canada* (1906) for long-distance transmission between Canada and Scotland. He soon was confronted with technical and climate problems, as a gale broke down his transmission tower on December 6, 1906. The relationship between the partners took a turn for the worse, and Fessenden was dismissed in 1911. A decade after its start, the Nesco Company went bankrupt, to re-emerge in 1917 as the Int. Radio Telegraph Company.

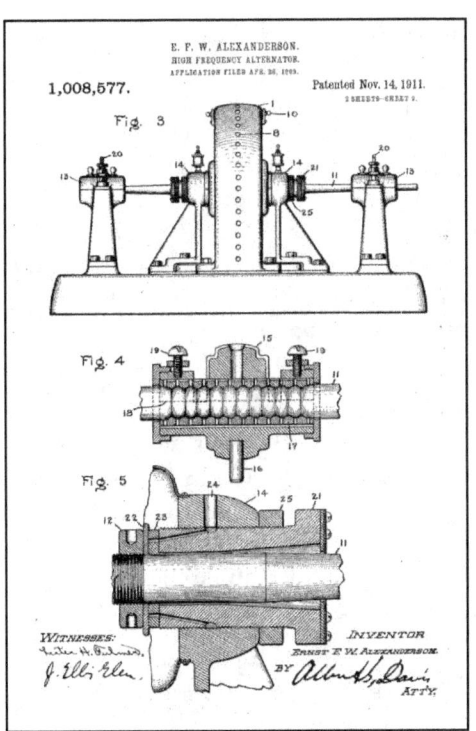

Figure 228: US Patent № 1,008,577 for the Anderson alternator (1911).

Source: USPTO.

Fessenden's scientific activities, however, were more successful than his business activities. They covered fields such as the incandescent lamp (US Patent № 452,494) and wireless telegraphy (US Patents № 706,735–706,747) granted on August 12, 1902 (Table 10). Fessenden's work, which would later continue with the development of 'radio' (the broadcasting of human speech and sound), was an early contribution to an alternative way of wireless communication that would replace the 'whiplash' effect of spark-based wireless telegraphy. Fessenden eventually became the holder of more than 500 patents. His US Patent № 706,735, with a priority date of December 15, 1899, would later cause problems for Marconi (Figure 227).

In 1902 the Swedish-American engineer *Ernst Alexanderson* (1878–1975) immigrated to America and started working for General Electric. That company had received an order from Reginald Fessenden to build an alternator able that could produce a high-frequency wave. So, in 1904 he designed, continuing on Fessenden's ideas, an alternator generating continuous waves in the frequency range of 50–100 kHz: the longwave alternator. In 1906 the machine was installed in Fessenden's radio station in Brant Rock, Massachusetts. By fall its output had been improved to 500 watts and 75 kHz. On Christmas Eve, 1906, Fessenden made an experimental broadcast of Christmas music, including him playing the violin, which was heard by Navy ships and shore stations down the East Coast as far as Arlington. He was granted US Patent № 1,008,577 on November 14, 1911, for his high-frequency 'Anderson alternator' (Figure 228). Anderson received in total some 322 patents for his work, and his 'Alexanderson alternator' would be widely employed in high-power/low-frequency wireless stations.

Improving on his system, he developed 50 kW machines that were massive in size and weight and were only suitable for land use. Next to maritime communication, they were used for commercial wireless telegraphy to transmit over intercontinental distances. Some were placed in Marconi's transoceanic station in New Brunswick, N.J. As the First World War (1914–1918) had interrupted the development of wire-based telegraphy and the military needed transoceanic communications with the military operations in France, in 1918 the installation was commandeered for official transoceanic service by the US Navy.

These were some of the major contributions along the 'electro-mechanical' trajectory of the electric alternators. In 1908 the Westinghouse engineer, professor *Rudolph Goldschmidt* (1876–1950) of German origin, devised an intricate method to enable an alternator to generate high-frequency waves without requiring excessive rotation speeds. His contributions were used in high-power longwave telegraphic transmission stations between the US and Germany, inaugurated on June 19, 1914.

At the same time that Alex Anderson was developing his mechanical wave generator, the Danish electrical engineer Valdemar Poulsen (1869–1942) followed another mechanical trajectory to create sparks of a more single frequency. He created the 'arc convector' that produced an undamped wave as he converted the direct current into an alternating current of a quite high frequency (2–20kHz). The arc was created within a glass vessel that contained hydrogen between a revolving carbon cathode and a copper anode. He combined it with a tuned circuit, added a traverse magnetic field, and the result was a continuous wave (and its harmonics).

He applied for a patent on June 19, 1903, and was granted US Patent № 789,449 on May 9, 1905, for a 'Method of Producing Alternating Currents with a High Number of Vibrations'. After further development in Britain and Germany, the system became used in America. Poulsen's patents were acquired by the Stanford-educated Cyrill F. Elwell, and the Poulsen Wireless Telephone & Telegraph Company was created in 1909. In 1910 the company was renamed the Federal Telegraph Company, and the capitalist Beach Thompson came on board.

This company would employ a remarkable man. Just after the company was formed, Elwell hired inventor Lee de Forest—fleeing from the East Coast to escape his problems there—to develop a practical receiver for the Poulsen wireless system, which he did using Fleming's valves and his own development of the Audion vacuum tube.

Poulsen's technology made wireless transmitters possible that could create high-power continuous waves (although still using a lot of bandwidth). It was used up till the First World War, when many battleships were equipped with Poulsen transmitters. However, after the development of the emerging new 'electronic' technologies—with the electronic devices like vacuum tubes developed by Fleming and de Forest—these technologies would make this development trajectory of spark-based telegraphy obsolete by the end of the war.

Improvement of the Wireless System: The Tuning Solutions

The attempt to create a continuous wave by mechanical means was soon complemented by efforts in another trajectory: the technical trajectory of electric resonance in which a tuning circuit created a continuous wave. For the time being, it proved more fruitful.

By that time Fessenden was not the only American interested in wireless telegraphy. The American mathematician *John Stone Stone* (1869–1943), inventor by profession, with many telephone patents on his name, also became interested. Building up experience while working in the experimental department of the R&D laboratory of the American Bell Company from 1890, he became acquainted with high-frequency transmissions and their problems in telephony. He would be the one who made the link between resonant circuits and the problems of wireless telegraphy, creating *selective wireless telegraph*.

After leaving Bell Company and becoming an independent consulting engineer, he started the *Stone Wireless Telegraphy Syndicate* in 1900, focussing on wireless problems—such as the interference problem—in the expectation of creating a commercial system. Continuing his tuning efforts with loose coupling to create 'selective' electric signalling, he filed from 1900 to 1902 several patent applications. In December 1902 he was granted a range of patents for 'electric signalling' (Figure 229). In US Patent № 714,756—applied for on February 8, 1900, and granted on December 2, 1902, and the antagonist of Marconi's British 7,777 patent—he patented a method of selective electric signalling. This patent would also become part of the patent war with Marconi.

Stone continued on this trajectory of resonant systems. From 1902 to 1905 he was granted numerous other patents (eg № 717,509 through 717,515; № 767,970 through 768,005; 802,418/432) for 'signal waves created by resonant electric circuits' and 'space telegraphy'. In total, John Stone Stone was issued about 120 patents in the United States (Table 11), and a similar number in other countries, covering telegraph and telephone devices and radio technology. In 1902 he became the chief engineer and

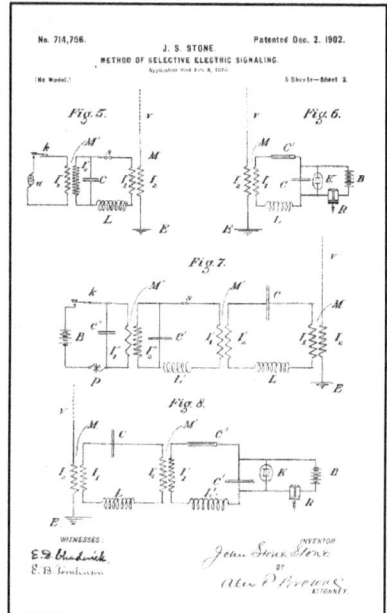

Figure 229: The Stone US Patent № 714,756 (1902).

Source: USPTO.

Table 11: Some of the patents granted to John Stone Stone, Dec. 1902.

Patent №	Granted	Description
US 714,756	Dec. 2, 1902	Method of selective electric signalling (filed February 8, 1900).
US 714.831	Dec. 2, 1902	Apparatus for selective electric signalling (filed January 23, 1901).
US 714.832	Dec. 2, 1902	Apparatus for amplifying electromagnetic signal waves (filed January 23, 1901).
US 714.833	Dec 2. 1902	Apparatus for amplifying electromagnetic signal waves (filed January 23, 1901).
US 714.834	Dec. 2, 1902	Apparatus for selective electric signaling (filed January 23, 1901).
US 716,139	Dec. 16, 1902	Apparatus for simultaneously transmitting and receiving space-telegraph signals (filed January 23, 1901).
US 716,134	Dec. 16, 1902	Method of determining the direction of space-telegraph signals (filed January 23, 1901).

Source: USPTO.

president of the *Stone Telegraph and Telephone Co*, which manufactured and leased wireless telegraph apparatus. However, he was not successful in business, and in 1908 he suspended its operations.

Numerous other inventors were also attracted by the phenomenon of electric waves and the challenges of wireless telegraphy. In America, *Harry Shoemaker* (1879–1932), patented over a hundred inventions, with some forty on wireless communication (Table 12). On February 16, 1903, he filed for a patent that was granted as US Patent № 824,676 on June 26, 1906. He designed a spark-based system with a coherer: the Shoemaker system. Testifying in April 1895, he claimed to have this done at the age of sixteen, but in the Marconi-de Forest litigation, the court would not accept his statement, bluntly declaring, 'His testimony is so utterly unsupported and insufficient and improbable that it will not be discussed.' From 1899 on he

Table 12: Overview of syntony patents granted in Great Britain, Germany, and the US from 1897–1905.

Patent №	Granted	Patentee	Description
GB 11,575	May 10, 1897	Lodge	Tuned circuit
US 609,154	Feb 1, 1898	Lodge	Electric telegraphy: syntonic circuits (filed February 1, 1898).
Ge 11,578	Oct. 14, 1898	Braun	n.a.
US 750,429	Jan 24, 1904	Braun	Wireless electric transmission of signals over surfaces (filed February 6,1899).
US 750,496	Jan 26, 1904	Arco	Spark telegraphy (filed (April 9, 1901).
US 763,345	June 21, 1904	Braun	Means for tuning and adjusting electrical circuits (filed August 5, 1901).
Ge 130,723	Oct 16, 1900	Slaby/Arco	n.a.
US 785,276	Mach 21, 1905	Slaby/Arco	System of wireless telegraphy with tuned microphone receivers (filed Sep. 27, 1901).
US 776,359	Nov 29, 1904	Slaby	Indicator for electric oscillations (filed July 26, 1904).
GB 12,325 /12,326	June, 1898	Marconi	Improvements in apparatus employed in wireless telegraphy: antenna design with cylinders and jiggers.
GB 6.982	April 1, 1899	Marconi	Improvements in transmitting electrical impulses and signals, and in apparatus (oscillation transformer).
GB 25,185 GB 25,186	December 19, 1899	Marconi	Improvements in transmitting electrical impulses and signals, and in apparatus (division of secondary circuit).
GB 7,777	April, 13, 1901	Marconi	Marconi's second important patent for wireless telegraphy (syntony patent) (filed April, 26, 1900).
US 676,332	June 11, 1901	Marconi	Apparatus for wireless telegraphy: cylindrical antennas (filed Feb. 23, 1901).

Source: USPTO. n.a.: Not available.

worked for a range of American wireless telegraphy companies, including the *American Wireless Telephone and Telegraph Company*, the *American DeForest Wireless Telegraph Company*, the *United Wireless Telegraph Company*, and the *Marconi Wireless Telegraph Company of America*, as we will see further on.

The Wireless Mania in Engineering

In just a couple of years wireless telegraphy had become a hot topic among engineers. This came after the initial—more scientific—interest in the phenomenon of the Hertzian waves that had occupied so many before Marconi patented his invention in 1896 (Figure 151). Now the question was no longer whether or not wireless communication was feasible; that had already been proven by Marconi. Now it was about the improvement of the system of wireless transmission by eliminating the unavoidable 'childhood defects', trying to create a better system of wireless communication.

By 1900 there was a range of thinkers and tinkerers in several countries who focussed on applying resonant circuits to the development of the syntonic system that created a continuous wave. Their work resulted in a number of patents in different countries (Table 12). In England, as Marconi was developing his syntonic system, Oliver Lodge and Muirhead worked on

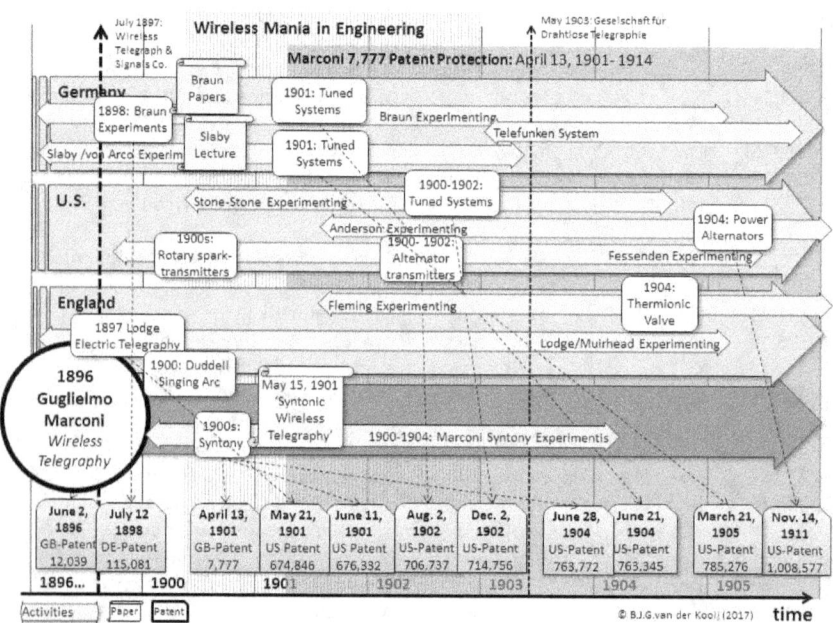

Figure 230: The wireless mania in engineering.

Figure created by author

their own system. And Fleming, scientific advisor to Marconi by now, was busy trying to improve the receiver. It was William Duddell's work on eliminating the noise from arc lamps that opened the way to use LC-resonant systems. Also, in Germany, both Braun and Salby/von Arco were working on tuned systems. And in the United States, a range of engineering work was done by the previously mentioned engineers Fessenden, Anderson, Shoemaker, Stone, and many others (Figure 230).

Marconi Marries into Aristocracy

The young, rich, and famous Italian Guglielmo Marconi—the Hero of Wireless—was to many women an attractive bachelor, not only due to the publicity he had received and which heralded him, but also through his good looks and social behaviour. And certainly the wealth he had already accumulated was not a handicap either. A wax likeness of him was displayed at Madame Tussauds in 1902, where it was the most popular exhibition. He also—being an 'auto aficionado'—was one of the first people in London who owned a motor car: a Napier Roadster, the car to own in the Britain of those days. He lived the life of an Edwardian bachelor, staying in the hotels and gentlemen's clubs of central London, having lunches most days in the Savoy, his favourite hotel. And he was constant on the move, travelling the European continent as well as the Atlantic Ocean.

Obviously, Marconi had a few girlfriends, some of them more serious than others. One of those serious relationships started during a boat trip from New York to Liverpool on the steamship *St. Paul* in 1900, when he met Josephine Holman. And he fell thoroughly in love, although she stayed in America and he travelled all around. They corresponded and met occasionally, but after some years that long-distance engagement was broken off in 1902.

> *Yet, for the two most intense years of Marconi's life, leading up to the success that sealed his reputation, Josephine Holman was his friend, lover, confidante, and occasional distraction: she was by far the person closets to him, possible closer than anyone had ever been.* (Raboy, 2016, p. 189)

Next, also on one of his experimental boat trips, now on the steamship *Luciana*, he met the 17-year-old Inez Milholland, already a strong-willed and gracious woman who became one of the most visible stars of the American women's movement in the 1910s. Marconi longed for family life and proposed marriage; the precocious Inez accepted. However, it was a short-lived engagement, as the relationship lasted only a year. There would be many women like Inez in Marconi's life, but they were typically not the ones he married (Raboy, 2016, pp. 235-236).

When Marconi was in Italy in 1904 to receive an honorary degree from the University of Bologna, his father Guiseppe died unexpectedly on March 26, 1904. Marconi went to Bologna and buried his father with the family. To his surprise, his father had left him the Villa Griffone. Marconi returned to England, and exhausted by the death of his father, the endless struggles with the Post Office, the US patent challenges, the continuing German diplomatic hostility, and his company's failure to take off financially, he headed to the tranquillity of the Haven Hotel in July, 1904, to recuperate (Raboy, 2016, p. 237).

Not far from the Haven Hotel on the Isle of Wight where Marconi had his fist wireless station was a privately owned island called Brownsea Island. There, the owners—the rich Dutch tobacco traders of the van Raalte family—hosted and entertained lavishly in their country retreat. On such an occasion, he met in July 1904 the daughter of one of the oldest families of the Irish peerage, the O'Briens. The O'Briens were direct descendants of the fabled eleventh-century High King of Ireland Brian Boru (941–1014), and they lived in a castle on their Irish estate of some 1500 acres: Dromoland in County Clare. However, like for so much of the Irish aristocracy, their wealth began to dwindle after the *Land Acts* of 1880 stripped them of their main income: land and tenants. And maintaining the estate and the London residence was quite a challenge.

Soon Marconi fell in love with the 22-year-old Beatrice O'Brien, daughter of Edward Donough O'Brien, 14th Baron Inchiquin, and Beatrice White of royal lineage, and the tenth of 14 children. However, when Marconi proposed marriage to her, she refused, leaving him in turmoil. Next, Marconi, as he was quite occupied with his company, experiments, demonstrations, and the related public appearances, left England for a high-profile trip to the Balkans for the inauguration of the first connection between Bari (Italy) and Antiveri (Montenegro), bridging the Adriatic Sea. It was during that trip that he contracted a case of malaria that would plague him for the rest of his life.

On his return to England, before leaving for New York on business on September 9, he met with the O'Brien family on the Dromoland estate. When he returned to England in December 1904—as arranged by Florence van Raalte—he met Beatrice at Brownsea again. After he proposed anew, Beatrice, who by that time had grown 'quite fond of him', accepted. However, her family was not too happy with marriage to a foreigner and asked her to return the engagement ring. Beatrice refused (Raboy, 2016, pp. 238-240).

Beatrice sister Lilah believed that Bea's main reason for marrying Marconi was to be a support to her family, in view of its shaky and unpredictable finances.

Figure 231: Marconi's marriage ceremony (1905).

Source:
http://www.marconicalling.co.uk/museum/html/o bjects/ephemera/large_image/large_image-type_d__t01872.html

From his viewpoint, he was finally making it into high society, while also hoping to introduce some playfulness to his life; he gave his bride two wedding gifts, a diamond tiara and a bicycle. (Raboy, 2016, p. 241)

Whatever their motives, Marconi and Beatrice O'Brien were married in London, in the fashionable George's Church (Figure 231), on March 16, 1905, amidst a cyclone of publicity and a gawking crowd. From the Italian family, only his brother Alfonso was present. In the tradition of the Edwardian upper class, she received, along with a diamond ring and a bicycle, £11,500 in stock and securities in her own trust, giving her an own income for life.[343] In addition, she received 12,000 shares in Marconi's Wireless Telegraph Company (Raboy, 2016, pp. 244-245). Their honeymoon in April 1905 to North America on the Cunard liner *Campania* turned out to be a working honeymoon—just as had been the case with Alexander Graham Bell's honeymoon to Europe some three decades before in 1877.

> *At Glace Bay, Marconi was immediately engrossed in experimental work, leaving his new bride to entertain herself in utterly foreign surroundings. The experience left her with some sharp observations about her husband; one of the moments she recalled was when some bad news arrived about one of the company's overseas stations and he reacted by playing on the piano with one finger until a solution was found.* (Raboy, 2016, p. 273)

When his first daughter died four weeks after she was born on February 4, 1906, Marconi had a nervous breakdown that left him ill for months. He had clearly suffered a burnout. Beatrice nursed him, and by May 1906, Marconi was back at work at Haven and Poldhu. For the next 25 years Beatrice would be Marconi's confidante and epistolary foil (Raboy, 2016, pp. 274-275).

---

[343] Equivalent to £1,103,000 in 2015; calculation based on historic standard of living. Source: http://www.measuringworth.com/

Overview of the Pioneering Years (1897–1905)

In the preceding analysis, we have observed in detail what happened from the invention of spark-based wireless telegraphy by Guglielmo Marconi up to his marriage. One could say that the period that started in the mid-1890s—when Guglielmo Marconi, with the help of his brother Alfonso, experimented in the garden of Villa Griffone in Italy—till the moment he married in England in 1905 were the pioneering years of Marconi. In a just a decade the simple ideas of a young adolescent (he was just in his 20s) had become transformed into a system of wireless communication that had conquered the world. In the process the young man himself had become famous and rich, the guest of monarchs and nobility. He was heralded by the international press and adored by the aristocracy. However, he was countered by their institutions, who claimed his invention as their own, and despised by his fellow entrepreneurs, who wanted to jump on the bandwagon.

A Wireless Tsunami

From a technological point of view, it was a hectic, chaotic, and confusing period in which experimentation with the electric spark—in general known as a phenomenon of the 'nature of lighting'—dominated the specific scientific agenda. The experimentation of many people resulted in several components for creating electric waves with sparks. In addition, they created components for detecting those electric waves at an increasing distance. These experiments were performed by electricity scientists and electrical engineers all over the world, from Russia to America and... in England, where a young Italian—called Guglielmo Marconi, 22 years old—travelling with his mother had set foot on land in 1896.

Coming from a country that had undergone massive social transition during the Italian Revolution, this more or less self-educated young man was entering the quite traditional British society. It was an enterprising, obviously militant society and seafaring nation, that had been shaped over centuries into the British Empire that was 'ruling the waves' with its powerful Royal Navy. But it had also grown into a democratic parliament-based society with a strong parliamentary-based governmental structure. The executive powers lay with the Cabinet, supported by different departments of state, from the Home Office, the Foreign Office, and the Colonial Office to the War Office (governing the British Army) and the Office of the Admiralty and Marine Affairs (aka the 'Admiralty', governing the British Navy). Over centuries, from an absolute monarchy with absolute, 'divine' powers, the country had grown into a constitutional monarchy fond of traditions and an elite-based societal structure.

One of these traditions was the state monopoly on postal communication originating from the Royal Mail monopoly in the seventeenth century. The department of the General Post Office was faced—and coped—with new communication systems like telegraphy and telephony. The Post Office was the office holder for the state monopoly on communication that would play such an important role in young Marconi's life. But there were also other institutions dating from way back, such as the scientific institutions like the Royal Society of London (1660s) and the military institutions like the Royal Navy, each with their long history of traditions. In those institutions, nobody was really prepared for the 'wireless tsunami' that, with its electric waves, would not only engulf Britain's institutions, but also the institutions of other nations like America, Italy, and Germany.

Seen from a personal point of view, for Marconi, the period after his arrival was a hectic time. All initial affairs, from technical demonstrations to getting patent protection, were coming together in a compact period of time. Seen from a contextual point of view, the decade after his arrival in England was also a chaotic period. By the end of the nineteenth century, there were the controversies between the 'theorists' (eg the Maxwellians) and the 'practitioners' (eg Preece) in the middle of which Marconi arrived. Borrowing from both sides, from the thinkers their constructs (eg Hertzian waves) and from the tinkerers their constructs (eg the coherer, the coil), his archetype was the working system that realized the much-needed function of wireless communication. The response was traditional: for some, it was the 'not invented here' syndrome; for others, it was a long-awaited breakthrough. Overall, in the scientific community, there was the institutional reluctance and denial followed by the slow awaking to the importance of Marconi's invention.[344]

After extensive investigation (including Marconi's experimentation and public demonstrations), institutions like the monopolist Post Office and the Admiralty—and even other nations like the antagonist German Empire and capitalist America—became more eager to implement wireless communication. The Post Office claimed its ownership because wireless telegraphy fitted within their monopoly. The British Navy claimed its ownership because it was important for their ship-to-ship and ship-to-shore communication and because it solved the incommunicado problem. For the Colonial Office, India Office and the Foreign Office administrating the colonies scattered over the word, it was important because wireless

---

[344] Here we touch on the age-old phenomenon of 'resistance to change' that is part of the individual and collective behaviour. It results in a behaviour of the organism that tries to maintain its equilibrium.

communication facilitated invulnerability in communication with the distant part of the British Empire.[345] In short, for the British, who were already 'ruling the imperial waves', this new method of communication without wires meant that Britain could be 'ruling the wireless waves' independently into the next century.

The early waves of the wireless tsunami might have been reaching British shores in 1896,[346] but it would not be long before the mass of it flooded Europe and America by the first decade of the twentieth century. Remarkably, it was a just in the one decade from 1897 to 1906 that wireless telegraphy conquered the world. By then Marconi—born in 1874—was in his early thirties. In that decade he had grown from a tinkering adolescent to a (at least on stock paper) wealthy entrepreneur-inventor, and into a married Nobel laureate (1909).

His entry into the game started in 1897, when Marconi, after having secured his patent position, to realize the further development of his concept for wireless communication, chose the vehicle of a private limited company—not too strange a choice considering the entrepreneurial context of merchant Britain in those days. He needed this organizational vehicle, as experimenting, prototyping and demonstrating his 'wireless machine' was expensive. Financing further technical development was not the only issue; opening up the diverse markets (by travelling to and demonstrating at events like the British Cowes Race and the American Atlantic Cup) was a costly affair. As always, with novelty, skeptical potential first-time users were hard to convince, delayed their decisions, and were slow in ordering. This reluctance is easy to understand, as early equipment tends to be not fully developed and plagued by technical imperfections.

That being the case, even worse, the newly formed company to exploit the invention of Marconi's wireless system was faced with a hostile business environment dominated by the previously mentioned conservative governmental institutions. For one thing, Britain was by now used to state monopoly for (cabled) telegraphy, and that created the first institutional constraints for entrepreneurial activities.

---

[345] The All Red Line network of the cabled telegraphy infrastructure was, by its nature, though designed to run over British-controlled territory, vulnerable not only to environmental conditions, but also to human intrusion and sabotage.
[346] A tsunami is a seismic sea wave caused by the displacement of a large body of water. This displacement of a water mass can be generated by sub-ocean volcanic explosions, earthquakes, etc.

The Act of Invention: From Idea to Concept and Artefact

Much is written about the heroes of discovery and invention,[347] as well as about their magic moment of insight, when, in the mind of the inventor, the magic *idea* sparked: the flash of genius. On closer look, the birth of novelty is much more complicated: it looks more like a process where that original idea is just a beginning. Take, for example, Marconi's case and what could be called his Act of Invention.[348]

It was in Italy that the young Marconi, excited by the possibilities of electricity and after being confronted with Hertzian waves in 1894, had conceived his *idea* for an apparatus for distant action, such as the 'wireless' ringing of a bell. This was a simple idea that had grew into a *concept*: how to realize a system using 'electric waves' for action at a distance. Thus, he had, over the period of some two years of thinking and tinkering in an experimental way, converted this concept into his *archetype* of a rudimentary system of wireless communication: the transmitter and receiver combination. Having realized that, he then pondered on what to do with his invention. With his background of a business-minded family (both in Italy as well in Ireland), the choice to go and commercialize his invention one way or another was not too surprising.

After Marconi's arrival in England in early 1896 the further development of his archetype for a wireless transmission system went quickly. From the first *prototypes* of the components that were the subject of his patent claims, as well as the claim for the overall system, much effort was put into both the technical development as well as the market development. The technical experimenting went hand in hand with the public demonstrations, which focused on interest groups as perceived by Marconi: the British aristocracy and the governmental decision makers within the Royal Navy, Royal Army, and the Post Office. In addition, the widely published demonstrations created awareness among society's decision-making elite; Marconi's trump card was the excitement of novelty.

Then came the period of production. From the early prototypes now a *commercial product* had to be made. First, it was an impromptu setup that created the quite crude wireless stations and wireless equipment; a lot of this work was done by subcontractors. But over time, as the product design stabilized, larger volumes were needed, and standardization and interchangeability of parts became a necessity. The products needed to be

---

[347] For example: MacLeod, C.: *Heroes of Invention. Technology, Liberalism and British Identity, 1750–1914* (2007). Cambridge Studies in Economic History—Second Series.

[348] We are confronted with semantics, the meaning of words. Here, we use the word 'invention' because at that time that was the word used for our phenomenon. Today we would call it 'innovation'.

able to endure 'real life' circumstances, as they were going to be used by non-technical operators. These products also needed additional services: from training the user to maintenance and repair of the equipment itself. In short, that total process of manufacturing needed to be organized. Just like product development and market development, organizational development was needed. This whole process, from the early conception of the idea to the birth of the company, often random, improvisational, and disorderly in nature, in its totality constituted Marconi's *Act of Invention*.

Looking in more detail at Marconi's activities, one can observe the following groups (Figure 232):

Concept development: It started with the moment that the young boy Marconi saw in his mind the possibilities the Hertzian waves could offer for the transmission of signals without wires—even in its most rudimentary form of ringing a bell at a short distance. Next, from the moment of his Orepa concept up to the moment he created his archetype able to transmit over the Celestine Hill, he was focused on creating a new method of transmitting signals over distance without wires. He had a vision and was anxious to realize it, oblivious to the problems that were lying ahead.

Technical improvement: The archetype of wireless communication that Marconi had created was just in a rudimentary form. It needed refinement and improvement on many aspects. After the concept was protected by a patent application in 1896, he tackled the dominant attribute of the system: increasing the distances that could be bridged by the signals. Soon he addressed the next problems of interference by tuning his circuit and obtaining his 'syntony patent' in 1900. And thirdly, he improved upon the sensitivity of the receiving detector: his 'Maggie'.

Public demonstration: Although communication over distance with electricity (aka telegraphy) had found solid footing in society by the end of the nineteenth century, *wireless* communication was not even thought of outside the technical community. Many people of influence (eg aristocracy, governmental decision makers, even the general public) had to be made aware of its possibilities. Marconi succeeded better than anyone in informing that public about his invention and its possibilities in a time that the British Empire was ready for his concept of wireless communication.

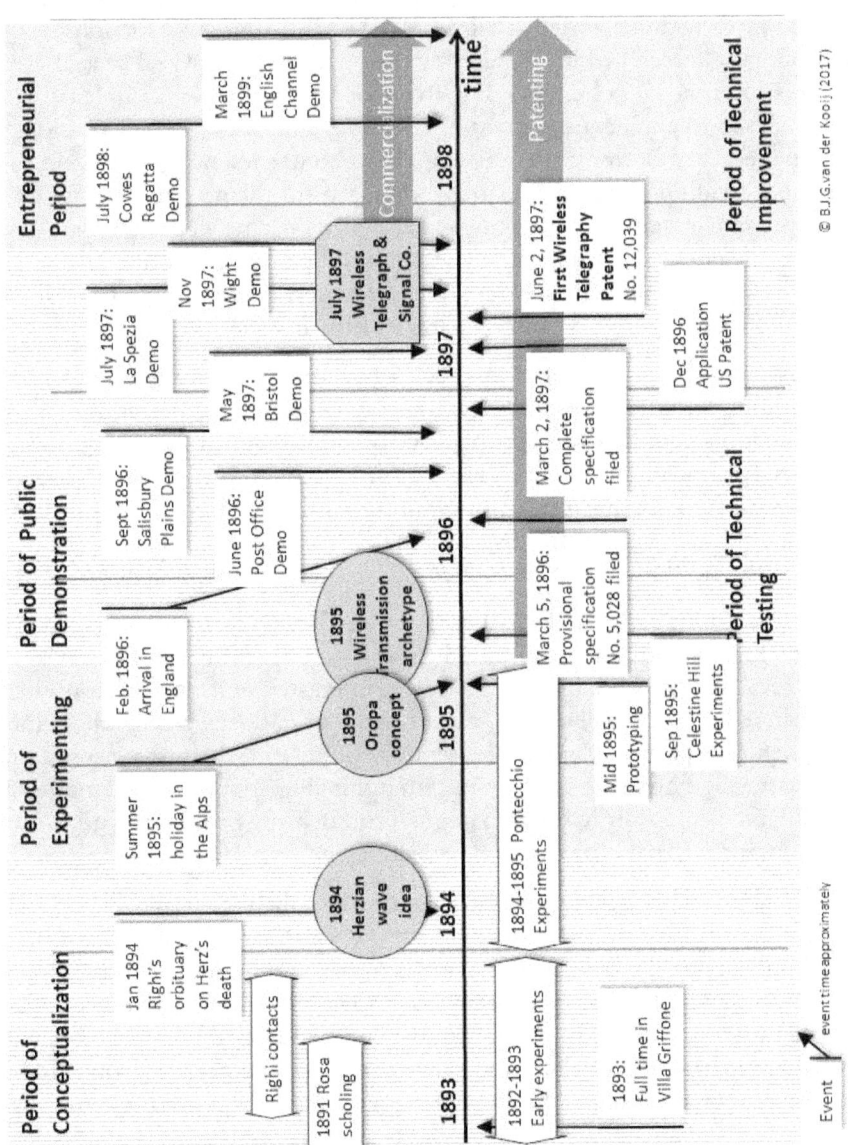

Figure 232: Act of Invention: the timeline of the pioneering years of Marconi's technical experimenting.

Figure created by author.

The Act of Business: From Artefact to Company

One could wonder if the young inventor-entrepreneur, excited with the invention he had developed and impressed by the initial and friendly reception he'd received in England by the Post Office, did realize at that moment in time what he was going to encounter in the next couple of years. Practical engineering might be his forte, and marketing his passion, but in creating a business, he was a novice who needed help. Such help for business creation was available in London, but for the rest, the young entrepreneur-inventor was on his own. Soon it became clear that his embryonic entrepreneurial activity was not to be born in a fertile business environment. By early 1900 Marconi had reached a stage in his young life in which he was opposed by several parts of British society.

Scientific community: First, there was the British scientific community with or without commercial aspirations and connections. They rejected Marconi's patent claims for the novelty of his concept by stating that they (eg Lodge) had already, at an earlier moment, proven the existence of a form of wireless communication (the priority discussion). For the scientific establishment, this was a matter of academic honor in the British tradition, not to be spoiled by a tinkering foreign upstart.

Post Office: In the second place, there were the British institutions working on wireless communication (ie the monopolist institution of the Post Office) or that were becoming slightly interested in its application (ie the Admiralty). Originally, there was a climate of curiosity and cooperation when they perceived young Marconi as fitting into their own aspirations. However, when that smart guy started thinking about his own interests, obtained an absurdly broad patent,[349] and became protective and commercial, that climate soon had changed.

The military and government: It was at this time that the early signs of decline of the British Empire were starting to show. Europe was under the influences of the day: imperial conflicts were simmering, and the madness of the times was starting to show. The emerging empires of that time—Russia, France, Germany, and budding Italy—were touching their horns in imperial conflicts that arose in abundance. European war was looming on the horizon. After a period of reluctance for the British

---

[349] The scope of a patent is formulated in its claims. Marconi claimed, for example, in claims 15/16 of US Patent № 763,772: 'Claim 15: A receiver consisting of a sensitive tube or other imperfect contact inserted in a circuit, one end of the sensitive tube or other imperfect contact being put to earth whilst the other end is connected to an insulated conductor. Claim 16: The combination of a transmitter having one end of its sparking appliance or poles connected to earth, and the other to an insulated conductor, with a receiver as described in claim 15.'

military community and the powers at the Colonial Office it was obvious that wireless communication was going to be important for ruling the British Empire, so important that it could not be left to the whims of a foreigner.

As a result, the young Marconi saw himself confronted by the British context of his time,[350] one that certainly made it difficult for him to execute his business activities.

> *One has to realize that wireless telegraphy was born at a critical moment in the development of the British warfare state, when colonial and industrial rivalries kept a diplomatically isolated Britain at the brink of conflict. Events like the Fashoda Incident of 1898[351] fed a sense of imminent European war, as did the Great Game with Russia[352]. In the shadow of a global arms race and a growing conviction that new technologies conferred military and imperial advantages to whomever was first in the field, the turn-of-the-century British state invested more deeply in scientific research, and scientists, in turn, relied increasingly on state support. In this time of science and technology for and by the nation, Marconi was an interloper; despite his mother's British ancestry, he was a foreigner, and worse, a tinkerer.* (Satia, 2010, p. 831)

A Time for Business Development

In that context, Marconi—and those close around him—was facing the challenge of developing a business that would bring his concept of wireless telegraphy from infancy to maturity. For Marconi, it started with the creation of an organizational infrastructure: the *Wireless Telegraph & Signals Company* (Figure 233). It was followed by a range of activities over the next years, his Act of Business, that consisted of a range of inherently connected activities that was related to that creation:

Organizational Activity: Marconi needed an organizational carrier to further the progress of his concept for wireless telegraphy. He could have intensified his cooperation with the Post Office (eg selling his patent to the state), but he chose for the legal structure of a private limited company: the joint-stock company. This vehicle was designed for financing, as it could raise the funds that were needed to transfer the

---

[350] See: B.J.G. van der Kooij, *Context for Innovation: British (R)evolutions in Perspective* (2016). pp. 521–522.

[351] The Fashoda Incident was the climax of imperial territorial disputes between Britain and France in Eastern Africa, occurring in 1898. It was the last crisis between the two that involved a threat of war (until 1940) and opened the way for closer relations in the Entente Cordiale of 1904.

[352] The Great Game (1830–1895) is a term used to describe a political and diplomatic confrontation that existed for most of the nineteenth century between Britain and Russia over Afghanistan and neighbouring territories in Central and Southern Asia.

infantile prototypes into reliable products and systems. He chose this vehicle not because it was the best way to realize a dream, but just because there was no other way for a budding entrepreneur.

Patenting activities: Creating something new is one thing, but protecting one's invention—and the related intellectual rights—is quite something else. Marconi soon enough understood, advised by his financial backers, the importance of patents. He hired the best patent lawyers and spent much time—and money—on the subject. It gained him his three wireless patents (GB Patents № 12,039, № 7,777, and № 10,245) that secured his monopoly for the decades to come. This monopoly was based on patents that had to be defended in court though.

Technical activities: A technical system, born in the protected area of a laboratory, is in a way primitive. Like a newborn child, it has to be nurtured, protected, and educated. In the case of a product, there is a similarity: the work done after its conception and birth being called engineering. Engineering—improving and even enhancing the technical properties and capabilities—requires quite some efforts and can be a costly affair. It was not only the engineering that was important; testing the equipment was also needed. So, Marconi started organizing the testing activities, often similar costly affairs that involved large constructions in faraway areas. In that respect, it is remarkable how he was able to get the support from royalty (eg the use of the Italian cruiser *Carlo Alberto*, made available by the Italian King) and others (such as the cooperation of the steamship SS *Philadelphia*). One could guess it was because those cooperators understood the importance of his experiments—or that they wanted to share in the glory.

Promotional Activities: Britain might have been the 'Workshop of the World' by the end of the nineteenth century after undergoing the Industrial Revolution; it was a society used to mechanical thinking in the Age of Mechanization. The world of invisible electricity was different. The electric waves, crossing at lightning speed the unseen ether, required a different mindset, one open to change and novelty. In a world full of conservative views, Marconi quite well understood how important it was to create public awareness and even public support. Just as Edison demonstrated his telephone to the decision-making British aristocratic elite some decades before, Marconi undertook similar activities with his invention (from his demonstrations at the Cowes Race and America's Cup to meeting with the tsar of Russia and the King of Italy[353]). And

---

[353] Marconi was often decorated, such as by the tsar of Russia with the Order of St. Anne. The king of Italy created him commander of the Order of St. Maurice and St. Lazarus, and awarded him the Grand Cross of the Order of the Crown of Italy in 1902. Marconi also

from the sideline watching and reporting on his endeavors, the press was his ally, heralding the progress of his experiments.

In addition to all these activities, the company itself also asked for his undivided attention on other matters, for example:

Commercial activities: After getting the company financed, creating revenues was a main point of attention. And just as with the development of the telegraph and telephone, revenues could come from the sales of products and services, and from the license fees for the use of the patents. Finding the right business model, however, proved difficult. When the pure selling of wireless equipment more or less failed, Marconi changed the marketing strategy and created the lease concept, supplying products as well as services. The maritime market became the first focus of attention and soon delivered his first customers (Lloyd's insurance company). Soon this was followed by the military market. With the first customers came the Italian Navy order, followed by the orders from the British and Japanese navies.

International expansion activities: Just from the frontier-less nature of the new technology, it was obvious at an early stage that the business potential for wireless telegraphy was large and international in nature. So Marconi created international subsidiaries to cover regional markets. After the *Marconi Wireless Telegraph Co. of America* (1899), the organizational structure was expanded with the *Marconi International Marine Communications Co.* (1899) and the Marconi *Wireless Telegraphy Company of Canada* (1902). Many more companies were to follow.

Organizational activity: In the industrial tradition of the emerging electro-mechanical industries, Marconi also organized his own production facilities in Chelmsford and Dalton (London region). There, the structured manufacturing of the wireless equipment took place, and the departments were organized along the production process: the workshop, paint shop, the assembly.

This overview shows the two dimensions of the inventor-entrepreneur: one to deal with technical affairs and the other to deal with business affairs. Both had their own complications.

---

received the Freedom of the City of Rome (1903) and was created Chevalier of the Civil Order of Savoy in 1905.

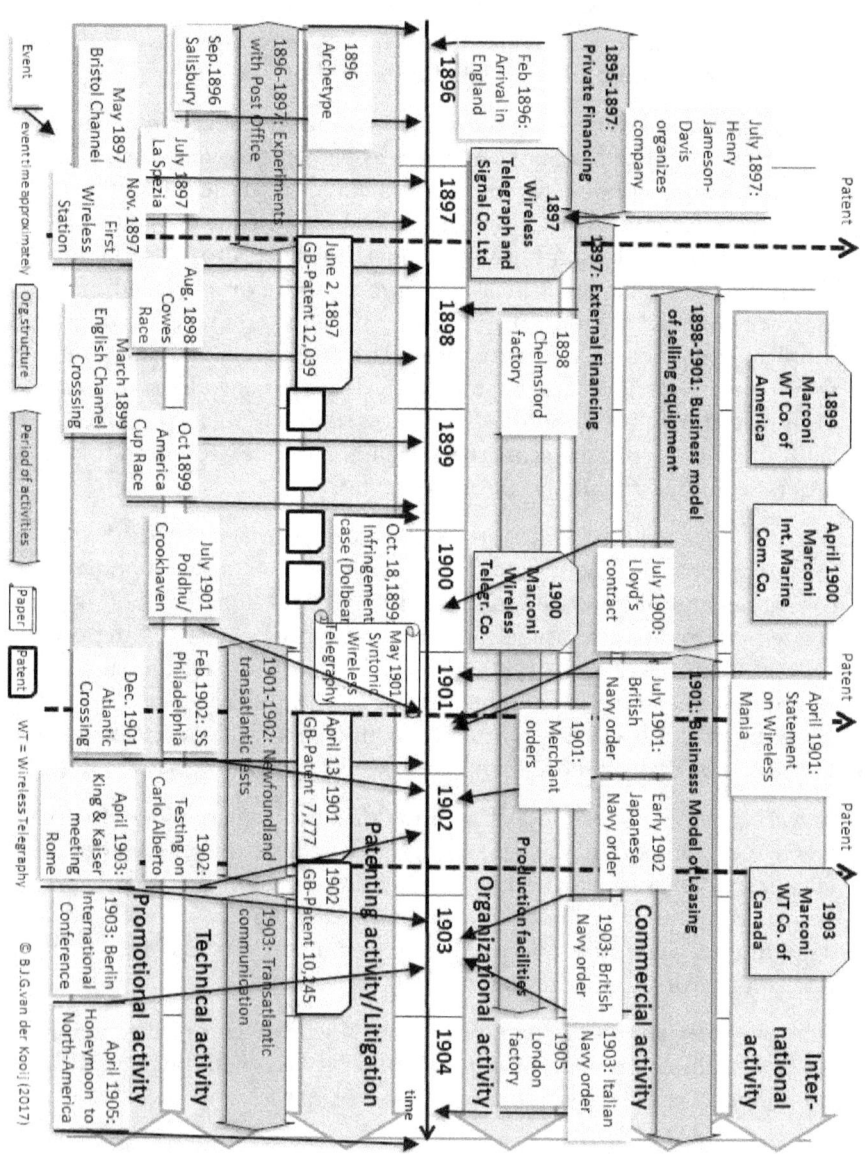

Figure 233: Act of Business: the timeline of the pioneering years of Marconi's business development.

Figure created by author.

The Marconi Monopoly Era (1905–1914)

In the first decade of the twentieth century, the Marconi Company grew in size and, inevitably, in complexity. The preceding pioneering years had been characterized by developing the technical aspects of the Marconi system: 'sales' was done by engineers, installation of the stations was often a matter of improvisation, and publicity was handled by Marconi himself. It was dominantly a British affair; from the Salisbury Plains up to the experimenting at the Haven Hotel on the Isle of Wight.

That changed after the success of the transatlantic crossing in 1901 and the start of the American and Canadian Marconi companies. Marconi, helped by his board members, expanded the wireless shore-to-ship activities worldwide. By 1903 the Marconi companies had forty-five coastal stations covering the entire globe. That was the beginning of growth, as by 1912 there were 13 overseas Marconi companies: along with those in the United States and Canada, there were companies in France, Argentina, Russia, Italy, and Spain. He had headquarters in Brussel, Paris, Rome, Madrid, and Buenos Aires. A network of dozens of stations covered all continents, from North and South America to Europe, Asia, Africa, and Australia. Based on his patent-position—with the three major wireless patents—he had created an *international wireless empire* with a wireless monopoly in all these regions.

> *…by being the first in the market, the Italian-born entrepreneur Guglielmo Marconi was able to create a de facto standard for wireless telegraphy. By refusing interconnection and by enforcing its patent rights, Marconi's British-based company,* [called the] *Marconi's Wireless Telegraph Company, was able to impose its standard worldwide, despite attempts by the British government to subvert it.* (Hills, 2002, p. 93)

By the mid-1910s the importance of wireless communication was no longer a question: the Golden Age of spark-based transmission had started. Both the military and maritime communities had understood its potential and importance to their activities. Wireless communication was even such an important issue that governments—such as the German and American governments—wanted more and more control over their own systems instead of being dependent on the monopoly of one country, or even the monopoly of one company. And that would have its effects on the Marconi monopoly. By the middle of the first decade of the twentieth century—let's say, 1905—navies (British, German, American) increasingly regarded wireless technology as a strategic weapon, and these strategic considerations vied with other commercial and regulatory priorities. This resulted in a change in the development of the Marconi monopoly. As we will see, many factors contributed to that change.

435

One factor could have been the fact that in Marconi's personal life, his circumstances certainly were influenced by his marriage, the death of his first-born child, and his physical 'burnout' in 1906. Another range of factors was created by the international context in which the Marconi companies found themselves entangled in a web of geopolitical affairs that soon dominated the application of wireless communication. One of them was the German-initiated effort to regulate wireless telegraphy and, at the same time, curb the hold the Marconi monopoly had on the international markets. The other was taking place in America: ending Marconi's Monopoly in the US.

> *Marconi's problems in America stemmed not so much from a lack of publicity or any scientific concerns as they did from economic nationalism. Marconi was trying to create a profitable enterprise, and this was what drove him to try and exclude competitors. The policy of leasing equipment and barring communication with rival equipment was very similar to that operated by the Bell companies for the telephone. By excluding competitors he effectively tried to create Marconi equipment as the industry standard.* (Nairn, 2002, p. 266)

In the next section of the book, we will try and address these consequences of that change of the geopolitical context.

The Priority Question and Patent Wars

It took a while, but after the pioneering years things changed for Marconi and his entrepreneurial activities. When a new invention like Marconi's system of wireless telegraphy makes its appearance—just like with Samuel Morse's invention of the telegraph and Alexander Graham Bell's invention of the telephone—and dominant players in society (eg, the military) slowly begins to recognizing its importance, that invention is going to be scrutinized. Sometimes these players reacted positively, such as the British Post Office initially recognizing an opportunity to add to their own investigations (into a dead-end technology). Sometimes they reacted more reluctantly, such as was the case with British Admiralty, who wanted proven performance.

Another type of reaction can be found within the scientific and engineering circles, where the priority debate emerged early on: the famous 'who was the inventor of...' discussion described later on, where different scientists claimed to have invented the new thing. Sometimes that claiming was about the honor; sometimes the scientists also had entrepreneurial aspirations and commercial interests (eg the Lodge-Muirhead Syndicate). All in all, the priority discussions created an atmosphere around the invention that went beyond the mere technicalities of the invention itself.

Along with those challenges, at a later moment, when it becomes clear to business circles that the invention has economic potential, the invention is challenged again. At least, that is the case when the invention enjoys patent protection that forms a (financial) restriction for others to use the invention at no cost for their own particular advantage.[354]

> *Gaining control over a key investment, such as a patent, was a crucial business strategy for securing dominance over—or even monopoly in—a particular field. And challenging historical claims of a patentee was an extremely demanding business involving recourse to the rituals and expertise of litigation to resolve alternate histories of who had been the first to make an invention and thus had the exclusive rights over it.* (Arapostathis & Gooday, 2013, p. 3)

Scientists Claiming Priority

The priority battle went in several stages. First came the scientists who opposed Marconi. This was the result of the wireless mania among scientists (Figure 151) that had started after Heinrich Hertz had worked on 'Hertzian Waves':

> *Hertz's discovery was the basis of Marconi's 1896 patent: between them a dizzying array of experiments took place in Russia, Germany, England, and France, as well as the United States and as far afield as India. Some of the period's most accomplished physicists, such as the future Noble laureate Ernest Rutherford immediately began working to improve the apparatus Hertz had used. Many of these distinguished scientists would soon be rejecting Marconi's claims of originality as spurious or invented.* (Raboy, 2016, pp. 52-53)

Among those scientists (eg Bose, Popov) we find the British 'Maxwellians' like Oliver Lodge, who—as we have described before—had demonstrated that electric waves could be used for short-distance signalling and lightning control (a major issue at that time). He even had developed—and duly patented—a receiver that would protect against the devastating lightning streaks: the 'coherer' based on Branly's concept. This was his lightning conductor, GB Patent № 5,844, 1889 for 'Protecting Electrical Instrument from Lightning'. As may be clear from the title, the patent indicated its application: lightning control. He also published widely on the lightning phenomenon, such as *Lightning Conductors and Lightning Guards: A*

---

[354] Patents are related to intellectual property and give the owner(s) an exclusive right for a limited period. One finds its base in constitutional documents such as: 'To promote the Progress of Science and useful Arts, by securing for limited Times to Authors and Inventors the exclusive Right to their respective Writings and Discoveries' (Copyright Clause: Article I, Section 8, Clause 8 of the United States Constitution). It is a right that can be licenced for a fee to other parties. In Great Britain, the copyright and patents are covered in Acts of the Parliament (eg the *Statute of Anne*, 1710), as the UK does not have one specific constitutional document but has an uncodified constitution.

Treatise on the Protection of Buildings, of Telegraph Instruments and Submarine Cables, and of Electrical Installations Generally, from Damage by Atmospheric Discharges (1892).

Sometime later Lodge had worked on tuned electrical circuits, 'syntony', as it was called. In the receiver circuit, he had applied his 'coherer' as a detector of the presence of electric waves. Already in 1898 in the United States he had patented his 'processes and combination of apparatus' to receive Hertzian waves under 'Electric Telegraphy'. After his patent application was applied for in America on December 20, 1897, he was granted US Patent № 674,846 on May 21, 1901. In the patent, he claimed again the 'coherer' (claim 4) without any specific motivation for its use in detecting electric waves, and he claimed its application (claim 5): 'In a receiver for Hertzian-wave signalling systems, the combination of the following instrumentalities: a coherer, a base or support upon which it is mounted, mounted in proximity to the coherer, and adapted to agitate its elements.'

Although this patent application was filed after Marconi's, it referred to the earlier patented coherer, and thus this patent was to become the basis for the claim of Lodge being the inventor of wireless telegraphy, as expressed by Silvanus Thompson in defence of Lodge:

> *The authorities of the United States Patent Office, under the provisions of their excellent patent law, don not issue patents—as the British Patent Office does— to the first fool who may apply for a patent to manufacture gold out of chopped straw. They search diligently, and require proof as to the actual first inventions. Lodge's sole right in the United Sates to use the coherer, automatically tapped, to relay wireless messages upon a telegraph instrument rest on no assertion of mine, … Neither by Signor Marconi's system nor by Professor Slaby's system can a wireless message be received in the States without infringing that patent.*
> (Arapostathis & Gooday, 2013, p. 165)

After giving it some consideration, Marconi responded to his rhetoric and made the choice to defend his patent rights.[355] The first opportunity to do so came in America. There, the Marconi Wireless Telegraph Company of America went to court in 1902 in the case of *Marconi Wireless Telegraph*

---

[355] With regard to patent rights, there are two courses of actions. One was the path of disputing the priority during process of application for a patent. The legal process is then within the domain of the patent-issuing organization: the Patent Office. The other was the path of infringement after a patent is granted. In that case, the owner of the patent defending his patent right is accusing that some party infringed on his exclusive right established by that patent. The legal course of action is to start a lawsuit by the owner of the patent.

Company of America vs DeForest Wireless Telegraph Company. It would be the start of a range of patent battles the Marconi companies undertook to defend their monopoly.

Marconi Claiming Priority: Early Priority Battles

Marconi was during his lifetime confronted with the 'priority' question, whether from the side of the scientific community, from the governmental institutions like the British Navy trying to evade Marconi's patent, or from those who had a commercial interest in wireless telegraphy. From the moment he was granted his first GB Patent № 12,039 on June 2, 1896, to 1903, just some six years later, Marconi had more than 130 patents in just about every country in which patenting was possible. And that patenting had consequences, as the Marconi board, later in time, vigorously defended his patent rights and attacked any attempt to build on it.

Patents were—by the very nature of their existence—an exclusive right given to the inventor. They were the cornerstone of a business strategy and were needed to defend the inventor's rights to his inventive work. As Marconi had transferred his personal right to the patents to the first Marconi Company during the commercialization of the Marconi system, the patents were defended by the Marconi companies. His invention was such a basic one that many of the other commercial, scientific, and governmental parties in wireless communication soon would be infringing on his rights, and that meant that to uphold his priority right, the Marconi companies had to go to court, a costly affair. This resulted in a range of legal patent battles in which the opponents basically tried to attack Marconi's priority right by claiming 'that somebody else had already invented his invention before him'. Or any other trick available in the book of lawyers.

> Lawyers were always trying to confuse the issue at hand in a priority suit and thus easy in concluding the invalidity of the patent. With the 7,777 patent, for example, a big issue was the 'tuning' of the apparatus. Thus, in court, the lawyer of the Massie Wireless Company stated that as they were selling the apparatus to the shipping companies in an 'untuned' version, they were not infringing on the patent. However, in a letter, Fleming responded: 'The difficulty which may arise is that "tuning" is an operation and not a thing' (OX-2, 2017, Letter by Fleming, April 26, 1910).

This illustrates the complexity of the issue at hand when legal affairs and technical affairs mingle. So, to obtain support in their legal affairs, the Marconi Company hired the services of one of London's top patent lawyers: John Fletcher Moulton.

Such expertise did not come cheaply, and although no detailed records of the payments to Graham and Moulton have survived, the fees must have been high. … Costly as Moulton certainly was, Marconi and the Marconi Wireless Telegraph Co. retained him as counsel, to work (along with a highly qualified team of patent experts) not only on the preparation of subsequent patents but also on the assessment of rival patents and the fashioning of the company's patent policy. (Guagnini, 2009, p. 360)

The First Wave of Patent Infringement Litigation[356]

From about 1903 until 1943, there were a large number of patent lawsuits initiated in the courts of the United States and Great Britain about Marconi's invention. There was a reason for this: the growing importance of wireless communication to business and society. And as the Marconi monopoly on wireless telegraphy dominated the scene, conflicts arose, and law suits were inevitable.

Antagonism toward the various Marconi enterprises around the globe was growing from a number of directions. In May 1903, the German government challenged what is construed as the Marconi Company's monopolistic refusal to communicate with rival systems… Among the British coastal stations for ship-to-shore communication, the cozy exclusivity formally accorded to Marconi by the Postal Office was gradually eroded in the wake of transparency required by the United Kingdom's 1904 Wireless Act. (Arapostathis & Gooday, 2013, pp. 167-168)

The result was, along with the public and political debate and governmental reaction, a first wave of priority battles, as it became clear that the Marconi patents were seriously threatening the cabled telegraph industry. After Marconi had secured in 1901 and 1902 his transatlantic transmission at a fraction of the cost of cabled telegraphy, the increased opposition came from the American side. It was the American scientist-inventor-entrepreneurs Dolbear, Shoemaker, Fessenden, and Lee de Forest who were opposing him forcefully.[357] In addition, the wireless mania in business that raged in America has to be taken into account, as many fraudulent entrepreneurs tried to jump on the (financial) bandwagon. More about that later on.

---

[356] Much on this subject is based on the material found in the Marconi Archive. I am grateful for the help of several of the staff of the Bodleian Library in giving me access to the archive.
[357] As we have observed in the case studies about the telegraph and telephone, in America, the people who studied the electrical phenomena and its application were often inclined to work together with enterprise, either in the form of selling their patent rights to others, who used them in litigation, or in cooperating in creating a company founded by themselves and financed by external parties with an interest in telegraphy. This was because capitalistic America supplied a totally different business context than Britain.

Marconi's original British Patent № 12.039, issued in 1897, was replicated on behalf of the company and its subsidiaries in many countries (many of them related to the British Empire) around the world. These broad patents—obtained at a considerable cost—covered such a wide range of applications of the invention that they de facto created a monopoly: ie the Marconi monopoly that was soon to be challenged internationally.

> *Litigation was never far from the surface. As Marconi sought to consolidate his company's global position, patent litigation became a standard piece in his repertoire. ... Marconi vigorously attacked any attempt by others to build on it. Marconi had a small army of legal and technical experts, including James Ambrose Fleming, who was on permanent retainer as a scientific advisor providing advice and testimony as to the primacy of Marconi's inventions, and keeping potential rivals at bay.* (Raboy, 2016, p. 248)

In America, it started early, in 1900, when Marconi was legally attacked for infringing on a previously issued patent.

Lyman C. Learned vs Guglielmo Marconi (1900)[358]: Before the US City Court Southern District of New York, Lyman C. Learned, as assignee of A.E. Dolbear of Boston, claimed that Amerson E. Dolbear with Francis M. Holmes had already invented the wireless in 1882, long before Marconi. And that the Dolbear Telephone Company had been granted a patent in October 1886: US Patent № 350,299, which improved a 'mode of electric communication' (Figure 132). Learned was the sole owner of the patent since July 2, 1899. This case was about Marconi infringing on Dolbear's patent, and Learned asked for $100,000 damages.

The lawyers tried all the tricks in the book, eg that the patent "gave the holder the sole right of connecting a telegraph transmitter and receiver to the ground", and Marconi was thus infringing on that patent, as he did something similar. Marconi, in his reply, admitted that patents had been issued to Dolbear, but as there never had been a practical demonstration of the commercial value of the inventions, he denied having infringed on Dolbear's rights. Men of reputation, such as Dr. J. A. Flemming, Pender professor of electrical engineering at the University College, London, and Charles R. Cross, Thayer professor of physics at the Massachusetts Institute of Technology, were witnesses for Marconi.

---

[358] As it is not always clear when a case was started from the material available to us, these years indicating the beginning of the lawsuit are indicative.

They testified that his inventions were the practical adaptation to wireless telegraphy of various discoveries in the field of electrical science. The case was dismissed by Judge Coxe.[359]

By then Marconi certainly had noted the flurry of patenting in America, among which were the thirteen patents granted to Fessenden on August 12, 1902 (Table 10). It did not seem to worry him too much, as he was much more worried with the patents granted to Lee de Forest. This became another court case.

Marconi Wireless Telegraph Company of America vs De Forest Wireless Telegraph Company (1902), In August 1902 the Marconi Wireless Telegraph Company of America brought suit against the De Forest Wireless Telegraph Company on the Marconi reissue of US Patent № 11,913. For Marconi, the case against de Forest was important, as by that time the De Forest Company was the owner of a range of telegraph providers.[360] All the assets of these companies were owned by the American De Forest Wireless Telegraph Company, parent company of the De Forest system, holding the patent, manufacturing the parts, and selling the apparatus to sub-companies paying yearly royalties to De Forest. The case took three years in court, but then the judge ruled on partial infringement (claims 3 and 5). It was an important decision, however.

> *After three years of litigation, Judge William Kneeland Townsend of the US federal circuit court ruled in Marconi's favour… The Townsend decision was the most important and lasting patent ruling for Marconi and anointed him practically unassailable. It established Marconi's original Patent as the foundation of wireless communication.* (Raboy, 2016, p. 247)

This ruling, although not a complete victory, was very important for Marconi's patent protection on which his monopoly was founded.

> *As a result of this Townsend decision, the DeForest Wireless Telegraph Company was restrained of further use Marconi's invention. More important, the decision sent a message to the industry and to would-be trespassers on Marconi's turf. Townsend established Marconi's patent as 'fundamental' … This was the clearest legal, if not scientific, statement ever of Marconi's accomplishment, and the Townsend decision became the corner stone of Marconi's patent protection strategy.*

---

[359] Source: New-York Daily Tribune, March 23, 1901, page 7: Suit against Marconi dismissed. http://chroniclingamerica.loc.gov/lccn/sn83030214/1901-03-23/ed-1/seq-7/
[360] Companies such as Consolidated Wireless Telegraph & Telephone Company, American Wireless Telegraph & Telephone Company, New England Wireless Telegraph & Telephone Company, Federal Wireless Telegraph & Telephone Company, Atlantic Wireless Telegraph & Telephone Company, North-western Wireless Telegraph & Telephone Company, International Wireless Telegraph & Telephone Company.

Townsend recognized that Marconi brought something new to the practise of communication. The decision meant that Marconi's basic patent 12,309 of 1896 had established 'an entirely new art'. (Raboy, 2016, pp. 257-258)

Shoemaker/Fessenden vs Marconi (1903): By May 1903 Marconi was sued by Henry Shoemaker and Reginald Fessenden for patent interference. They were claiming precedence over Marconi's magnetic detector. It was Marconi's 'Maggie' patent (US Patent № 884,988, filed February 2, 1903, but granted on April 4, 1908) that was supposed to infringe on the priority of their patent application (in the United States). They claimed they had invented something equivalent to Marconi's Maggie in 1901, while Marconi had only patented his device in 1902. On December 22, 1903, the ruling of the US Patent Office declared that Marconi's US patent № 141,398, filed on February 2, 1903, "is adjudged to interfere with others". Shoemaker immediately filed for another interference. Fessenden, however, followed another tack and tried to combine forces with Marconi to buy out Shoemaker. But that did not happen for another decade. The litigation process took years, with Marconi testifying in 1904 and 1905 in New York (Raboy, 2016, pp. 255, 257).

On October 23, 1905, the Marconi team called a sensational witness; none other than Lee De Forest, who gave his occupation as 'wireless telegraphy expert." ... *The following day, October 24, 1905, Marconi's testimony began. Once again he outlined the trajectory of his work since 1896, all the designing, erecting, testing, and operating of experimental stations he had done; all the government he had advised as chief technical expert to the Marconi company—the US government was conspicuously absent from the list. ... Marconi's deposition continued till November 6. ... It took more than another year before the case was decided in Marconi's favour.* (Raboy, 2016, p. 256)

Although he had won this case, it took another infringement suit in March, 1906, against the De Forest Company, for Marconi to win the battle against Fessenden and De Forest after spending much money.

Fessenden and DeForest were the first innovators to seriously shake Marconi's technical and conceptual foundations, and they drove the development of wireless communication on a new course, which Marconi would soon be following as well. But they never again represented a threat to Marconi's dominance in the field, not in the United States, or anywhere else. (Raboy, 2016, p. 260)

These were the early cases of patent litigation. Although most of the early patent battles were won, and Marconi had gained victory, the patent wars just had started, as this was just one of the many infringement cases that were brought to court over the next decade. In total, nearly 30 cases were filed between 1902 and 1917.

The Second Wave of Patent Infringement Litigation

By 1906 the court had reached a decision in the previously mentioned case *Marconi Wireless Telegraph Company of America vs De Forest Wireless Telegraph Company*. The court decision was in Marconi's favour, so Marconi was not too worried about his priority rights. Moreover, Marconi had other things on his mind, as on February 4, 1906, his newly wed wife had given birth to a daughter. However, less than four weeks later, the child died. This affected Marconi's health, and he was out of commission for some three months.

Also, by the end of the first decade of the 1900s the international scene for wireless telegraphy had changed considerably. It started a second wave of litigation that would take off after the nationalization of Marconi's British operation in 1909 (when the financial compensation of £15,000 brought some relief to the dwindling corporate finances).

> *In 1910, with their new capital, the Directors of the Marconi organisation 'resolved that it was time there should be an end to the infringement of the company's patent". The company took suit against BRT&T in 1911 and against the Helsby Company in 1913.* (Burns, 2004, p. 359)

It was then that the board of Marconi decided again to maintain their priority rights on Marconi's patents.

> *Of the many lawsuits there were three in particular that shaped the early history of wireless communication in general and of American Marconi in particular. Marconi vs United Wireless (1912), Marconi vs de Forest (1916), and Marconi vs E.J. Simon (1918). The outcomes of these suits were considered to be great victories for Marconi at the time, but in the end it became apparent that the victories were largely pyrrhic.* (Wenaas, 2007, p. 6)

One could say that this second wave started in Britain, where the Marconi board decided to start legal actions against infringements.

Marconi vs British Radio Telegraph and Telephone Company (1910): This was the lawsuit against British Radio Telegraph and Telephone Company (BRT&T) at the High Court of Chancery in England that started on December 12, 1910. Marconi sued for infringement of his patents, especially the 7,777 patent granted in 1900. The judgement by Judge Parker given in 1911 was in favour of Marconi.[361]

---

[361] Source: https://academic.oup.com/rpc/article/28/10/181/1641904/MARCONI-V-BRITISH-RADIO-TELEGRAPH-AND-TELEPHONE (assessed June 2015). This article is under Crown Copyright (1988), a British relic from feudal times when the royalty controlled the printing press. Generally speaking, this copyright is for 50 (published material), or 125 years (unpublished material).

Marconi Wireless Telegraph Company vs Helsby Wireless Telegraph Company (1913). Marconi started a court case against the Helsby Company. Again it was for an infringement of the 7,777 patent. The company had installed wireless sets on board a ship of the London & North Western Railway Co. The court ruled in Marconi's favour in February 1914 (Burns, 2004, p. 360).

Marconi against Telefunken[362]: The Marconi board decided to fight Telefunken on the home front as well. In Britain, the German company Siemens—a dominant shareholder in Telefunken and manufacturing its equipment—was infringing on the 7,777 patent. However, Telefunken did not offer legal resistance and admitted infringement in November 1912. In the following January they applied for a license that was granted on a royalty basis of £10 per kilowatt per annum. This case ended with other arrangements between the companies that were not disclosed. They were related to interests in other part of the world. As Telefunken and Marconi both also operated in Australia and New Zealand, and with its Australian Wireless Company selling to the Australian government, the Marconi board decided to sue both the German-owned company and the Australian government. A solution was found here as well: a new to-be-formed Australian company in which Marconi would hold 50% of the shares. In addition, the Marconi and Telefunken interests had reached a secret worldwide agreement to refrain from instituting litigation against each other (W. J. Baker, 1970, pp. 134-135).

In addition, the American Marconi focused on American infringements. It announced in May 1911 that it would start legal proceedings concerning infringement of US Patent № 763.772. The lawsuits started by focusing on shipping companies that were using equipment made by Massie Wireless Telegraphy Company (aka the Massie system) and connecting to the coastal stations erected by Massie. This resulted in the case *Marconi vs Clyde Steam Ship Company* and the case *Marconi vs New England Navigation Company* (March 25, 1912). Next, American Marconi attacked the manufacturers themselves. This action of industry-wide lawsuits against Massie and other American wireless company owners, charging patent infringement, crippled that industry. Many companies, among which Massie Wireless, closed their wireless stations or sold them to the American Marconi.

Marconi versus United Wireless Telegraph Co. (1912): By 1910 the American Marconi and the United Wireless Telegraph Company were left as the major wireless communication companies in America. Marconi controlled the Atlantic shipping communications with a handful of

---

[362] When a patent is licensed for a fee, this means the patent owner gets compensation for his invention. Court cases for infringement are not applicable then.

shore stations and several hundred ships under contract. United controlled the majority of coastal shipping communications, with less than a hundred land stations and claiming 262 merchant ships under contract. United Wireless had designed and manufactured its own communication equipment for wireless transmission, using exactly the same technology as covered by Marconi's 7,777 patent (with its American equivalent of US Patent № 763,772). So, in July 1911 Marconi sued United Wireless for infringement and quickly won the court case. However, the United Wireless company, already in problems due to prosecution of key executives for stock fraud, defaulted and filed for bankruptcy. British Marconi purchased United Wireless's assets—then some 500 ship stations and 70 shore stations—on behalf of American Marconi for a payment of $700,000 and turned them over to American Marconi (Wenaas, 2007, pp. 6-8).

Marconi Wireless Telegraph Company of America vs National Electric Signalling Company (1914): In 1902 Fessenden had joined two Pittsburgh financiers in organizing the National Electric Signalling Company to manufacture his inventions, which they intended to sell to customers such as the US Navy or shipping companies whose far-flung operations would benefit from wireless telegraph communication. The American Marconi company, however, considered their activities to be an infringement on Marconi's patents. The case was about the status of Marconi's patents in the United States. It would have broad consequences, as the US government had exploited, among others, their wireless station at Arlington, using copied equipment. The patents at issue were № 11,913 (July 13, 1897), № 609,154 (August 16, 1898)—Lodge's patent acquired by the Marconi Company—and № 763,772 (June 28, 1904), the equivalent of the 7,777 patent. The historic court decision on March 17, 1914, about the validity of these three patents was in favour of Marconi.

> *"Accordingly I find," said Judge Veeder, "that the evidence establishes Marconi's claim that he was the first to discover and use any practical means for effective telegraphic transmission and intelligible reception of signals produced by artificially formed Hertz oscillation."* ... *Thus the famous four circuit tuning patent* [the 7,777 patent] *established Marconi as the master of wireless.* (Dunlap, 1941, pp. 230-231, 234)

By 1914 the strength of Marconi's patents had been duly tested in courts of law or admitted without litigation by competing companies. But in Europe, the war had started, and the company's principal competitor, Telefunken, was now an enemy. By 1914 Marconi started an infringement case against Telefunken's subsidiary Atlantic Communication Company, which operated the Sayville station near New York. Of the 706 coast stations and 4,846 ship installations which existed at the end of March 1915, some 225 coast and 1894 ship stations were those of the Marconi International Maritime Communication Company (Burns, 2004, p. 361).

During the period of the war, the patent litigation continued, but now the focus was on America, as they were advised to strengthen their patent position. This resulted in quite a few court cases:

Marconi Wireless Telegraph Co. v. de Forest Radio Telephone & Telegraph Co.
(1914): On November 1914 Marconi filed suit against De Forest Radio Telephone & Telegraph Co. for infringement on Fleming's two-element valve (US Patent № 803,684). De Forest countered by claiming infringement on his Audion patent. After demonstration of the similarity between Fleming's valve and De Forest's Audion, the court ruled in favor of Marconi, dismissing De Forest's counterclaim.

Tesla against Marconi (1915): In 1915 the Marconi Company of America claimed 'ownership of all basic patent rights in the transmission of wireless messages'. At issue was the priority of Tesla's work on wireless for which he had filed patent applications in 1897 (granted in 1900) over Marconi's patent application of November 10, 1900, in America. The Patent Office had concluded in 1903 that many of Marconi's claims were not patentable over Tesla's patent numbers № 645,576 and № 649,621. However, in 1904, the US Patent Office reversed its decision and awarded Marconi the patent for radio. Tesla began his fight to reacquire the radio patent.

> He [Tesla] *sued the Marconi Company for infringement in 1915, but was in no financial condition to litigate a case against a major corporation. It wasn't until 1943—a few months after Tesla's death—that the U.S. Supreme Court upheld Tesla's radio patent number 645,576. The Court had a selfish reason for doing so. The Marconi Company was suing the United States Government for use of its patents in World War I. The Court simply avoided the action by restoring the priority of Tesla's patent over Marconi.*[363]

Marconi vs E.J. Simon (1918): In early 1915 the US Navy was in the process of buying wireless sets from American origin, and E.J. Simon offered equipment built using the technology Marconi had covered in his 7,777 patent. Marconi went to court, but the motion for injunction was denied. After appeals by Marconi the case came before the US Supreme Court, where it was reversed in a stunning victory for Marconi. The victory was short-lived, however, as Congress passed a law that absolved manufacturers from patent infringement liability when dealing with the

---

[363] Source: 'Tesla, Life and Legacy'. http://www.pbs.org/tesla/ll/ll_whoradio.html

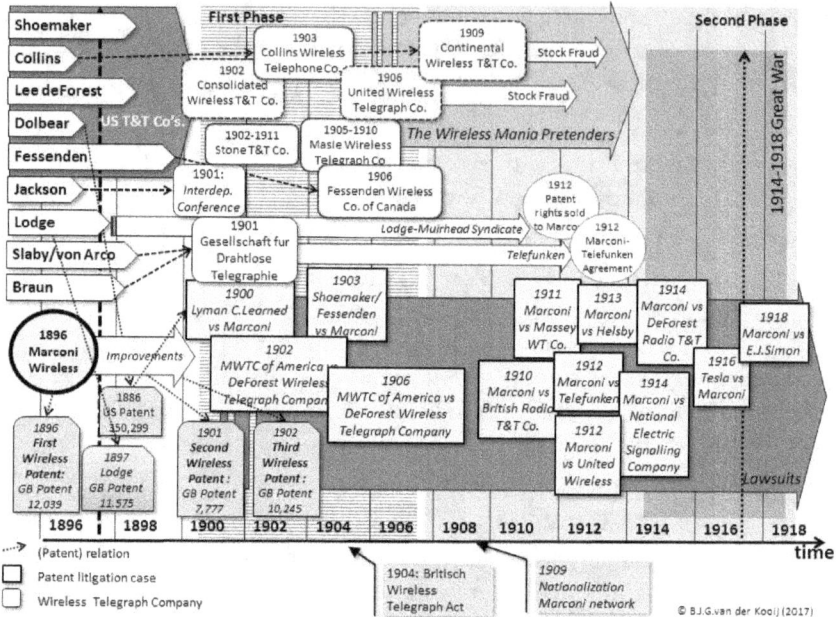

Figure 234: Overview of the Marconi patent litigation.

Figure created by author

government. As a result of this governmental intervention Marconi could not sue the competing manufacturers anymore.

By then, little by little, Marconi's patent empire was voided, as the patents started to expire after 14 years. But Marconi had fought his priority wars and patent battles (Figure 234). It had started with the British opposition from scientific circles (Lodge etc), followed by the struggle to protect his patent position against British wireless companies. Then he went to America, where his list of patent fights included almost all of the inventors and pioneers of radio communications. People like de Forest, Shoemaker, Fessenden, and many others were in an almost constant fight with Marconi and his company in the time of the wireless mania.

Nationalization and Monopoly Wars

Back to the early 1900s, when things started and Marconi had been busy with the technical testing and business development of the budding marine market. In that first decade of the twentieth century further experimenting with ship-to-ship communication (aka marine radio communication) and ship-to-land communication resulted in increasing distances being covered by wireless telegraphy. One potentially interesting client was to be found in the other institutions next to the Post Office that had a quite complex relationship with the Marconi Company: the military, especially the British and American navies. Both had by that time developed quite an interest now that wireless communication had begun to mature.

> *From the beginning, the military applications of radio technology were seen as immense; the US Navy asked Marconi for demonstration of his devices a few months after he introduced it to the public. The military's concern was prescient; wireless technology was used in the Boer War (1899–1902) and the Russo-Japanese War (1904). Wireless technology was seen as so important to international relations that Germany hosted International Wireless Conferences in 1903 and 1906 and the Institute of International Law in Belgium crafted guidelines to control wartime wireless use. In the United States, as the Army, Navy, journalists, the Weather Bureau, and other agencies began to compete for control of the airwaves. President Theodore Roosevelt formed an Interdepartmental Board of Wireless Telegraphy to handle problems arising from competition for the new technology. No regulations were actually created in the United States until a collision between two ocean liners in 1909 resulted in the Wireless Ship Act of 1910.*[364]

By 1905 the experimental period of time of the wireless communication was phasing out, and—next to the increasing marine communication operations—commercial long-distance telegraphy came on stage. In October 1907 communication for commercial purposes across the Atlantic Ocean between England and Canada started. This had already been predicted by Marconi in an interview in 1903:

> *Marconi says that there is no limit to the distance of wireless communication. It is only a question of increasing the power of the apparatus. He counts on sending messages across the Pacific Ocean from San Francisco to Japan in due time. England still refuses to allow Marconi to transmit messages overland, and until that is done public business by wireless telegraphy will be delayed. Within a short time stations will be erected at Seattle and San Francisco to open communication between these ports and vessels at sea. The State of Washington and Alaska are now being connected; and in a short time the whole of the Pacific Coast will be lined with stations for communication with shipping.*[365]

The Marconi companies—almost certainly as a consequence of its patent position as the owner of the emerging wireless technology—dominated the American marine market: the Marconi monopoly. In the coming decade a number of factors would influence the survival of the

---

[364] Source: http://www.enotes.com/research-starters/social-impacts-wireless-communication (assessed June 2015)
[365] Source: Perry, L.: 'Commercial Wireless Telegraphy'. *The World's Work*, March, 1903, pages 3194–3201. http://earlyradiohistory.us/1903marc.htm (assessed June 2015)

Marconi monopoly, many of them on a scale out of the sphere of influence or the control of Marconi. While, for the moment, the prospects for the American business looked good, the first clouds of the difficult times to come were appearing on the horizon.

Marconi Shares Noble Price in Physics

As we have seen, Marconi's contribution to wireless communication had been strongly disputed in scientific circles. However, that had changed, as for a number of years (1901, 1902, 1903 and 1908) Marconi had been nominated for the Noble Prize. This was a prestigious award for achievements in science—still issued by the Nobel Committee from the Royal Swedish Academy of Sciences in present days—and handed over by the king of Sweden. Then the Nobel Prize in Physics in 1909 was awarded jointly to Guglielmo Marconi and Karl Ferdinand Braun 'in recognition of their contributions to the development of wireless telegraphy'. Marconi concluded in his Nobel Lecture, held on December 11, 1909, the following:

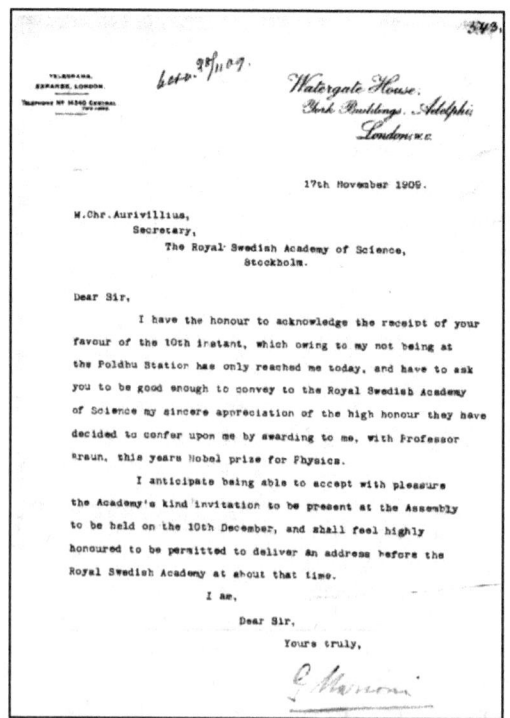

Figure 235: Marconi's letter of thanks for the Nobel Prize.

Source: http://ieeexplore.ieee.org/document/5525630/ citations

However great may be the importance of wireless telegraphy to ships and shipping, I believe it is destined to an equal position of importance in furnishing efficient and economical communication between distant parts of the world and in connecting European countries with their colonies and with America. ... Whatever may be its present shortcomings and defects, there can be doubt that wireless telegraphy—even over great distances—has come to stay, and it will not only stay, but continue to advance. (G. Marconi, 1909, p. 221)

One would expect the Nobel Prize for Physics to be given to a physicist, but now it had been awarded, remarkably enough, to a non-physicist. The reason was described as:

450

The crucial part of the contribution by Marconi was the following: "Marconi is indisputably the creator of the wireless telegraphy. His merit is not any epoch making physical discovery, but rather it is in putting it together and working it out into a practically useful system... Further arguments in the report emphasized the tremendous practical importance of the creation of wireless telegraph.[366]

This honour of the Noble Prize bestowed on Marconi might have been recognition of the technical work he had done; for his business endeavours, there was less admiration, especially considering the fact that, as described before, wireless communication by now had become an international affair in which many more players each had their own stakes. And Marconi, trying to maintain his monopolistic position, was faced with opposition from different sides.

One of those opposing factors was the international interest in regulation and further legislation of wireless communication that started with the 1903 conference in Berlin. After a range of conferences, an international agreement was signed at the International Radio Telegraph Convention in Berlin, Germany, on November 3, 1906.[367] This set the course for an international regulatory regime, first for wireless, then broadcasting, and ultimately all telecommunication. As a conference did not have legislative powers, it was left to the participating countries to establish national legislation, and that proved quite complicated, as national interests were quite diverse. As was the situation in Britain and America.

Nationalization by the Post Office

Britain had been following since the early 1900s a road towards a state monopoly in wireless communication. Ultimately, on September 30, 1909, the British Post Office took over all the British activities of the Marconi International Maritime Communications Corporation and Lloyd's coastal maritime communication stations. For this nationalization, the company was compensated £15,000,[368] a sum that was quite welcome because the British Marconi Company was at that moment without liquid capital. The postmaster general (Mr. Sydney Buxton), in the House of Commons, quite politically referring to the sacrosanct 'public interest', declared about the 'arrangements':

---

[366] Source: Marconi's Noble Prize. http://ieeexplore.ieee.org/document/5525630/
[367] Germany, the United States of America, Argentina, Austria, Hungary, Belgium, Brazil, Bulgaria, Chile, Denmark, Spain, France, Great Britain, Greece, Italy, Japan, Mexico, Monaco, Norway, The Netherlands, Persia, Portugal, Romania, Russia, Sweden, Turkey, and Uruguay.
[368] Equivalent to £1,410,000 in 2015; calculation based on historic standard of living. Source: http://www.measuringworth.com/

I am glad to say that arrangements have been completed with the Marconi Company for the transfer to the Post Office of all their coast stations for communication with ships, including all plant, machinery, buildings, land and leases, etc., and for the surrender of the rights which they enjoy under their agreement with the Post Office of August, 1904, for licences or facilities in respect of coast stations intended for such communication. In addition the Post Office secures the right of using, free of royalty, the existing Marconi patents, and any future patents or improvements, for a term of fourteen years. ...

Arrangements have also been made with Lloyd's for the transfer to the Post Office of their wireless stations for communication with ships, and for the surrender of all claims to licences for such communication. In return Lloyd's will receive the plant value of their stations, and will have transmitted to them (with due regard to the secrecy of private telegrams) information received at the Post Office stations in regard to the position and movements of ships, and other maritime intelligence. Lloyd's and the Marconi Company have mutually arranged to cancel an agreement between themselves which was made in 1901, and which has proved a source of dispute, and, therefore, an obstacle to the development of wireless telegraphy. I am satisfied that it is to the public interest, both from a commercial and a strategical point of view, that the coast stations used for communication with ships should be in the hands of the Government, and should be worked as part and parcel of the general telegraphic system of the country.[369]

Finally, the British State had succeeded, in the name of the 'public interest', at acquiring control over wireless communication, just as it had done before with cabled telegraphy and telephony. It had taken a decade, starting from 1899 up to 1909, to realize this British state monopoly on wireless. The contract of 1904 between the British Marconi and the Post Office with an eight-year license for the company's shore-to-ship communications, and the fifteen-year license for long-distance communication had been cut short in 1909. It was a bittersweet year, as in the same year Guglielmo received his Noble Prize. However, that honour went to Guglielmo Marconi personally; the British Marconi Company was the big loser, just as it had been the loser of the Berlin Convention in 1906 that had initiated regulation of wireless communication all over the world.

---

[369] Source: http://hansard.millbanksystems.com/commons/1909/sep/30/marconi-company-and-post-office (assessed June 2015)

The Beginning of the End of the American Marconi Monopoly

In America, Marconi was faced with a different situation. By the end of 1909 the Marconi Wireless Telegraph Company had equipped more than 300 ships with apparatus for wireless communication with each other and with coast stations. The British Navy had also adapted Marconi's system and equipped its vessels. It was obvious that wireless telegraphy was an important means of communication, during naval wartime operations as well as in cases of maritime distress.

Although wireless had existed in America for a decade, by this time the US Congress had not ratified either of the international treaties of 1903 or 1906, and the commercial companies—among which was the American Marconi—had successful lobbied against any type of regulation whatsoever. By 1912 numerous bills had been introduced in Congress to diminish pollution of the ether, as radio interference was called, though none were passed. The only new regulation that had passed Congress was Public Law 262 (aka the *Radio Ship Act* of 1910), and that was—as a consequence of the Titanic disaster—amended by Public Law 238 (1912). It was about safety at sea, and that did not create a commercial problem at all for the Marconi Company. On the contrary, it expanded the market, as now the installation of wireless became mandatory on all American ships carrying more than 50 passengers and crew on routes longer than 200 miles.

By the 1910s the Marconi Wireless Telegraph Company of America and the US Navy got into problems. As a result of intense patent litigation, by 1912 the American Marconi had achieved its goal of creating a wireless communication monopoly in America. Only the American United Wireless Company was a competitor. The US Navy, confronted by the dominance of the Marconi monopoly was annoyed by American Marconi's resistance to state regulation, their aggressive patent stance against their American suppliers of wireless equipment, and their treatment of the issue of 'intercommunication' between stations using equipment from different manufacturers. Additionally, as the British Marconi had a large interest in the American Marconi, the American government by that time had concluded that it was not in its best interest to have such a monopoly under foreign control—even more so in 1914, when the Great War in Europe started and America maintained a position of neutrality.

The consequence of this neutrality was a censorship on wireless communication, to be enforced by the Secretary of the Navy. Then, on September 2, 1914, an incident occurred with a wireless message sent from the British cruiser *Suffolk* to a private individual in New York via the

Siasconset Station on Nantucket Island.[370] The Navy intercepted this message—just ordering supplies and newspapers—and threatened to close the station. American Marconi opposed their authority forcefully and went to court. Not waiting for the court decision, the Navy ordered the station to be closed on September 25, 1914.

> *The stage was set for a historic—if not physical—confrontation between Marconi and its employees on the one hand and the Navy on the other. Fortunately for all concerned, American Marconi reconsidered its position (and closed the station themselves).* (Wenaas, 2007, p. 19)

On January 17, 1915, the station reopened under Navy supervision, which lasted for the duration of the war. This was the first confrontation where the American Marconi challenged the American government. It was a challenge the company could not win, especially with the war situation in the background. This was proved by 1916, when new legislation was proposed in the form of the *Alexander Bill*.[371] This bill sought to eliminate commercial interest in ship-to-shore communication: "In essence the Bill proposed that a government monopoly be formed to take over the wireless communication for both government and commercial purposes" (Wenaas, 2007, p. 20).

However, before the hearings on the Alexander Bill were concluded, America's neutrality ended as it entered World War I. Then, by executive order, the government seized control of all US wireless stations useful for naval and commercial operations on April 7, 1917, closing all other commercial and amateur stations. It was now a real war, and the American Marconi was to lose his wireless battle for America as the Navy now controlled a de facto state monopoly. It started with the acquisition— without Congressional approval, as it was war—of the Marconi patents and the shore stations of the Federal Telegraph Company. As a result, in 1918 the American Marconi sold its radio installations on over 300 vessels and 45 shore stations for $1.45 million.[372] It was left only with four stations for long-distance transatlantic communication using high-frequency Anderson alternators. An attempt by the American Marconi to acquire the sole rights for this alternator from General Electric also failed. The Navy instigated the

---

[370] The *Nantucket* lightship was the first point of contact for ocean liners bound for New York City. The wireless station was erected in 1901, and the Marconi Company's 'Sconset station became one of the most important in America.

[371] The bill, sponsored by Congressman Joshua W. Alexander, was passed to create the *Merchant Act*. This Act was designed to secure an American merchant marine owned by 'citizens of the United States'. Its instrument was the US Shipping Board which could create corporations owning merchant vessels.

[372] Equivalent to $ 22.8 million in 2015; calculation based on historic standard of living. Source: http://www.measuringworth.com/

creation of a new company, and the British Marconi interest was to be phased out. After the discussions on the Alexander Bill were revived and the Navy monopoly formally established, it was not long before the national interest had thoroughly eliminated the foreign-owned Marconi monopoly.

> On September 5, 1919, after several months of intense negotiations and diplomatic discussions, British Marconi agreed to sell the 364,826 shares they held in the American Marconi to GE [General Electric] for millions of dollars in cash and other concessions. (Wenaas, 2007, p. 23)

As planned by the Navy, in that way, General Electric was to create a new privately owned company that was uniquely identified with the US national interest. It was called the *Radio Corporation of America*,[373] and on October 17, 1919, American Marconi entered into a preliminary agreement with General Electric for such a merger.

> The deal was consecrated by an agreement signed on November 21, 1919, under which RCA took over British Marconi's interest in its American subsidiary, including the rights to use Marconi's patents in the United States. The agreement specified that each party would continue to improve its respective patents as well as develop new inventions "in order to establish world-wide public commercial wireless telegraphic and telephonic services by means of radio devices. Overnight the new company became a giant. ... Foreign domination of American radio was no longer an issue. (Raboy, 2016, p. 444)

At a special meeting on April 6, 1920, the shareholders of the American Marconi voted to dissolve the company. And all that was the result of the US Navy continuously trying to get control over wireless telegraphy. This had a range of consequences, some of which were quite unexpected.

> The creation of RCA did not happen overnight. For a number of years, the U.S. Navy had tried to extend its dominion over the airwaves and voiced its increasing frustration at the absence of a dominant American Company. ... The formation of RCA also paved the way for resolution of the main barriers to the development of the radio. First, there was the longstanding conflict between Fleming's diode and De Forest's Audion (triode). Now that RCA held the American rights to the diode, it was in a positon to negotiate with the licensee of De Forest's patents. ... However, the one thing that had not changed was the attachment to the notion that the commercial application of wireless technology revolved around point-to-point communication. It was not to be long before the huge opportunity that had been ignored and lefts in the hands of a minority of enthusiastic amateurs would explode on the market. (Nairn, 2002, pp. 279-281)

---

[373] RCA was a joint venture of the power players in the electric business at that time: General Electric, Westinghouse, and AT&T.

Radio broadcasting, due to the initial efforts of radio amateurs, would soon become an important activity for RCA. This development was comparable to that of the personal computer later in time, when the large mini-computer companies left the new personal computers to the entrepreneurial computer amateurs.

Figure 236: Stock Certificate for the Radio Corporation of America (1924).

Source: Weenas, 2007

Back to the 1920s, where one can conclude that the prelude for the creation of the RCA was not fitting the traditional American aversion of state influence. The nationalistic feelings that had developed during wartime, however, had prevailed in this pragmatic solution.

> *The Navy was* [after the war] *rebuked by congress for attempting to create a government monopoly and for misappropriating funds by purchasing radio stations during the war without congressional consent, was criticized for poor management of numerous radio stations they operated during the war, was forced to return numerous radio stations to private industry after the War, incurred millions of dollars in damages for patent infringements, and never again had a significant influence in the development or regulation of radio and other communication technologies thereafter.* (Wenaas, 2007, p. 25)

The Marconi monopoly in America had lasted a decade, and his American Company had lasted two decades. A little more than a century after the *War of 1812* (1812–1815), America had gotten rid of British dominance again. The Marconi wireless monopoly became under the control of a government-instigated private company. But one never knew with those Brits. RCA stock certificates (Figure 236) bore the following restrictions even far into the 1980s:

> *The voting rights represented by this certificate shall be transferable to loyal citizens of the United States and/ or corporations formed under the laws of the United States…free from foreign control or domination and not in any foreign interest.* (Wenaas, 2007, p. 25)

The end of the Marconi monopoly was the beginning of the 'Age of the Radio' after the Great War, with its radio boom that started to happen in the 1920s.

The Disaster of the Titanic

Back to the 1910s, when majestic steamers were crossing the Atlantic Ocean, connecting the Old World with the New World. These steamers had been incommunicado before the Era of Wireless, but now increasingly used Marconi's wireless equipment. On April 10, 1912, the brand-new RMS *Titanic* of the White Star Line left the harbor of Southampton for her maiden voyage to New York. On board were 1,224 people, among whom were some of the wealthiest people in the world,[374] hundreds of immigrants, and two 'marconists': Jack Philips and Harold Bride. These two men were employed by the *Marconi International Marine Communications Company* to operate the powerful state-of-the-art telegraphic equipment and transmit the 'marconigrams' for the passengers.

Disaster struck when the *Titanic* hit an iceberg in the early morning of April 15, 1912 (Figure 237), this despite many wirelessly transmitted warnings of drifting icebergs from fellow ships. Luckily, the *Titanic*

Figure 237: The *Titanic* sinking after the collision with an iceberg.
Source: Wikimedia Commons. Artist: Willy Stöwer.

---

[374] The *Titanic* was funded and built by the banker J.P. Morgan, who was to sail on the maiden voyage in his own private cabin, but he cancelled at the last minute. In April 1912 Marconi and his wife also booked passage for the transatlantic voyage on the RMS *Titanic* but their plans fell through. Marconi's wife remained at home, cancelling her trip altogether, and Marconi ended up sailing three days earlier on the *Lusitania*.

happened to be within contact range of twelve other vessels. The short transmissions sent by those ships' wireless operators, staccato bursts of information and emotion, tell the story of *Titanic's* fate that night: the confusion, the chaos, the panic, the fear.

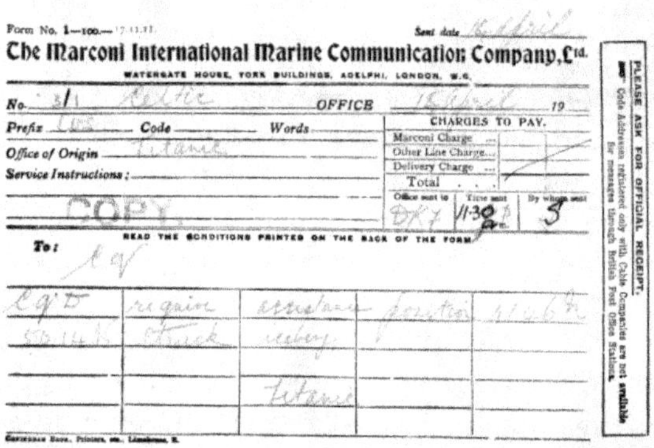

Figure 238: Telegraphic message sent from the HMS *Titanic* as received by the HMS *Celtic*.

Source: Presspack Museum of the History of Science, Oxford.
https://www.mhs.ox.ac.uk/marconi/presspack/

It started with the first transmission: 2.17 hr a.m., 15 April 1912, R.M.S. *Titanic* to Any Ship: 'CQD CQD SOS Titanic Position 41.44 N 50.24 W.[375] Require immediate assistance. Come at once. We struck an iceberg. Sinking.' Soon the ships in the neighbourhood responded and changed their course. The final message from the *Titanic* read around 2.15 hr: 'SOS SOS CQD CQD *Titanic*. We are sinking fast. Passengers are being put into boats. *Titanic*.' At 2.20 hr the ship broke apart and foundered, with well over one thousand people aboard. Around 4 hr a.m., the RMS *Carpathia* of the Cunar Line arrived on the scene in response to *Titanic's* earlier distress calls. In the end the *Carpathia* sailed to New York with 705 survivors aboard, while 1,500 people had lost their lives in the ice-cold seas.

When the RMS *Carpathia* docked in New York three days later, Guglielmo Marconi was among the 40,000 people greeting the survivors. He went aboard with a reporter from the *New York Times* to talk with Bride, the surviving operator. On June 18, 1912, Marconi gave evidence to the court of inquiry into the loss of the *Titanic* regarding the marine telegraphy's functions and the procedures for emergencies at sea. Britain's postmaster general summed up, referring to the *Titanic* disaster, 'Those who have been saved, have been saved through one man, Mr. Marconi...and his marvellous invention.'

The press coverage of the *Titanic* disaster resulted in an increasing interest in the American Marconi Company by American investors. At the same time as the disaster the American Marconi Company underwent a

---

[375] CQD was one of the first distress signals for ships. It is equivalent to SOS.

share restructuring, and the company's capital was raised to $10 million. It had a dramatic effect on the stock market.

> *The day after the shareholders meeting, April 19, 1912, a 'working agreement' to tie Western Union's US land stations to Marconi's transatlantic connection was announced. Plans revealed simultaneously in London to link the Pacific Islands, the US west coast, and Asia meant that, for the first time it would be possible to send a wireless dispatch entirely around the world. The American Marconi stock skyrocketed to $350, prompting the New York times to describe the rise as "the wildest boom which the curb market has ever witnessed." ... Twenty-four hours later the boom was over and the stock sunk to $150. Wall street had no explanation; Marconi himself just shrugged.* (Raboy, 2016, pp. 355-356)

The maritime disaster was—in commercial terms—a victory for Marconi, as it proved how important maritime wireless communication was for safety at sea. It boosted the sales of the Marconi Companies worldwide. But there was more to come, developments with a more long-term perspective.

Britain's Imperial Network Scheme

By the end of the nineteenth century, a worldwide network of cabled telegraph systems was in the making: the 'All Red Line' linking the countries of the British Empire (Figure 168). The highly redundant network—for security reasons, many additional lines were created—was completed in 1911. However, it functioned poorly and was overhauled by the new technology of wireless communication. But that took a while.

Just like in the days of the submarine cables, now the wireless connections were to span the globe. Several 'empires'—the German Empire with its African colonies, the French Empire with its African and Far East colonies, and the British Empire with colonies all over the world—all had plans to create long-distance wireless telegraphy networks they could use for governmental, military, and business use. And there was also interest from one small country with its own East India colonies (aka Indonesia) at the other side of the ocean: the Netherlands.[376] From the Chappe optical telegraph system covering Europe in the early nineteenth century, it would be the wireless telegraph system of the early twentieth century that was spanning the globe.

---

[376] It would take the Netherlands till 1918 before they could realize a longwave wireless connection with Telefunken equipment between Kootwijk (Netherlands) and Bandoeng (Indonesia). The buildings were inspired by the constructions at Nauen, Germany.

Stimulated by the popularity of wireless telegraphy after the publicity around the Crippen murder[377] and major disasters at sea—eg the sinking of the Titanic—that had led to an outcry for 'wireless telegraphy', in 1911 the Marconi company proposed a schema of 18 stations connecting Great Britain with its empire. As the state had its monopoly, Parliament decided to allow for contracting the construction of a commercial 'wireless company'. The only real candidate was the Marconi system. It took some political debate, partly due to the 1912 *Marconi Scandal* in which highly placed members of the governing Liberal Party and the brothers of Marconi's director Isaacs, with inside knowledge supplied by Isaacs, were accused of trading in Marconi's shares. The press smelled blood and wondered if the Marconi shareholders 'were the right sort of people to be entrusted with an all-British scheme of wireless telegraphy'. Another paper described the Marconi contract as 'one of the most disgraceful jobs to which a great public department like the Post Office has ever been made a party'. The Marconi Scandal was even compared to the South Sea Bubble.[378] (Raboy, 2016, pp. 366-367)

The anti-Marconi crusade by some members of the press resulted in a political affair of some magnitude that did not end before the beginning of the Great War in 1914:

> *The fundamental question in the subsequent hullabaloo* [of the Marconi Scandal] *was the propriety of these transactions, and the principal characters in the drama were the three Liberal Ministers, although there were many complexities in the story and very many other people became involved.* (Jolly, 1972, p. 195)

Despite these political issues, the financial secretary to the Treasury, Mr. Masterman, stated in a session of Parliament on August 8, 1913:

> *The necessity for the immediate construction of a long-distance Imperial wireless station was urged upon the Treasury many months—almost years—ago. The plea came not only that it was something that was desirable, but that it was something that was essential. ... The question therefore became not whether immediate action was desirable, but what sort of immediate action should be undertaken. ... There, therefore, came before the Lords of the Treasury three*

---

[377] In July 1910, the infamous Dr. Crippen murdered his wife in England and then fled to Canada with his new lover. The pair appeared suspicious to the captain of their ship, who then sent a wireless message to Scotland Yard, and hence they were arrested when they arrived in Canada. This was the first recorded instance of wireless telegraphy being used to aid police work.

[378] The South Sea Bubble was a stock-fraud scheme in the early 1700s. In Great Britain, a considerable number of people had been ruined by the share collapse, and the national economy greatly reduced as a result.

possible ways in which is work could be, or might be, immediately undertaken. The first was that it should be constructed either by one of the Departments of the Government or by some new Department created for the purpose by the Government. The second was that open tenders should be invited, and the most satisfactory accepted. The third was that if open tenders were impossible, that the work of construction should be given to any one of the wireless telegraphic companies.[379]

It was clear that government, neither Parliament, the Treasury, the Post Office, nor the military, could not cope with organizing the new system by themselves. They needed a private company, and the only candidate for such a project was the Marconi Wireless Telegraph Company. So, a—much criticized—modified contract was ratified by Parliament on August 8, 1913, for six major installations located in England, Egypt, East Africa, South Africa, India, and Singapore. The perspective of this contract brought the company for the first time into excellent financial condition. By April 1912, an eight-month period, its stock had more than quadrupled in value (Beauchamp, 2001, pp. 232-233).

Next, the outbreak of the Great War led to the suspension of the contract. After the war the Marconi company initiated in March 1919 legal action, claiming £7,182,000 in damages for breach of their 1912 contract.[380] Parliament acted and decided that the first two stations were to be realized (and they became operational in 1922). But that was all, as it was not before 1924 that Parliament finally approved the by then already adapted proposal for a system of longwave transmission.

> Although in 1920 the *Norman Report* of the Imperial Wireless Telegraphy Committee gave advice for a government-controlled and operated network, nothing had happened. By 1922 Marconi had installed stations for other nations, linking North and South America, as well as China and Japan. It took another report, the *Donald Report* from the Empire Wireless Committee with the same conclusion, for eight high-powered longwave stations to be used.

The New Beam System

Then, in renewed effort to connect the scattered colonies of the British Empire by wireless telegraphy, in 1924 the Marconi Company offered a new—now a shortwave based system instead of a longwave solution—

---

[379] Parliamentary Records: 'New Marconi Agreement'. Source: http://hansard. millbanksystems.com/commons/1913/aug/08/new-marconi-agreement (accessed March 2015)

[380] Equivalent to £297.7 million in 2015; calculation based on historic standard of living. Source: http://www.measuringworth.com/

Imperial Wireless Scheme to replace its old longwave proposals. The Navy and the Army responded with restraint. Nevertheless, it was in 1924 that the postmaster general of the Post Office, after a debate in Parliament, was granted permission to proceed with the construction of a wireless telegraph station on the new Beam system. This was part of the *Worldwide Imperial Wireless Scheme* that was to connect Britain with Canada, South Africa, India, and Australia. By October 1926 the circuit to Canada was ready and operating, and this was soon followed by stations in South Africa, Australia, India, and South America (Beauchamp, 2001, pp. 236-238).

> *And so, fifteen years after it was first proposed, Britain finally had its 'Imperial Wireless Chain'. It was certainly not to the government's credit. The British government was pushed, reluctantly, into the age of long-distance radio by the force of technology and the unflagging vision of Guglielmo Marconi.* (Headrick, 1991 np)

The government needed Marconi, and Marconi needed this major client, but it was a marriage made in hell. The political sentiment still was against Marconi, as Lieut. Colonel Moore-Brabazon, parliamentary secretary, illustrated in the parliamentary debate on August 1, 1924:

> *One of the difficulties of the Postmaster-General in regard to wireless is that he has to negotiate with the Marconi Company, and anybody who has negotiated with that company must have a very long spoon indeed. ... We have it laid down that the policy in future in this country, with regard to Empire wireless, is that the stations should be controlled by the State. With that I cordially agree. But let us remember that the Marconi Company have manœuvred us into a very difficult position. They have control of the other end. In all the Colonies and throughout the world we are not free to transmit and receive messages from another Government or another Dominion station, but always that the other end we have to communicate with the Marconi Company.*[381]

In the end the originally foreseen 'imperial high-power longwave wireless telegraphy system' was not realized, as technological progress offered the better performing shortwave communication systems. Then the Post Office started offering its 'Imperial Beam Wireless Service', a highly profitable service that threatened the viability of the cable telegraphy companies. Many of these worldwide operating companies that by 1900 had created a near monopoly on international communications, were British.[382] Thus, to curb the competition between the old cabled telegraphy and the

---

[381] Parliamentary Records: 'Post Office Contracts [Marconi Wireless Telegraph Company, Limited]'. Source: http://hansard.millbanksystems.com/commons/1924/aug/01/post-office-contracts-marconi-wireless (accessed March 2015).
[382] Such as the Eastern Telegraph Company, the Eastern Extension, Australasia and China Telegraph Company, Western Telegraph Company.

new wireless telegraphy, by the *Imperial Telegraph Act* of 1929, one operating company was created: the Imperial and International Communications Ltd, that later became Cable & Wireless Limited (1934) and its holding company Cable & Wireless (Holding) Ltd.

Conflict with Government-supported Telefunken

Back again to the 1910s, where the Marconi Company was facing growing competition (Figure 224) from a range of companies. In America, it was, as we will see in more detail further on, the small venture companies stimulated by the Navy that made inroads on Marconi's market dominance. However, in German-speaking Europe, it was mainly competition from one big player: the government-backed German company *Telefunken* that would create serious competition, not only in Europa's markets, but also in the American and Australian market, offering equipment for ship-to-shore service providers.

Telefunken could profit from considerable governmental aid, both financial as well as diplomatic. In addition, it was backed by the resources of the Germans banks. The company created—in business terms— an insurmountable wall of competition that was more than eager to block the Maroni Company from getting a worldwide monopoly.

Despite the agreement at the 1906 Wireless Convention that stipulated communication between stations using different equipment, Marconi still was blocking messages from non-Marconi systems (except distress calls). It was a trump card in Marconi's hand of cards, as he had it backed by his patent position and technological dominance at that moment in time. However, this policy created an operational problem for the Telefunken-equipped vessels passing through British-dominated areas like the English Channel, and for Marconi-equipped vessels passing through Telefunken-dominated areas (eg. North Sea). But there were more than those pragmatic problems, as Telefunken also started to operate on markets Marconi had set his eyes on (eg south and east of Europe, the countries of the colonial empire). For Marconi, it was a question of how to breach this 'Telefunken Wall', as Marconi and Telefunken were the dominating parties in wireless communications.

In short, the position was a stalemate and the need for diplomacy was being recognized by both sides. The situation reached a crisis point for the Marconi Company when in 1910, the German Government declared German henceforth no foreign wireless apparatus would be permitted on board German vessels. (W. J. Baker, 1970, p. 132)

As Marconi had managed to amicably settle patent infringements both in Britain and Australia, the situation in Europe became an issue. Although Marconi managed to sell equipment to Turkey, Romania, Portugal, and its colonies, the situation in the German-speaking countries was different. There, the German government had created that embargo that forbade the use of Marconi equipment aboard German vessels. In due course, a commercial solution was found in the creation of a new company, the Deutsche Betriebsgesellshaft fur Drahtlose Telegraphie (D.E.B.E.G), in which Marconi held a minority position. As the Austrians followed the Germans, German-speaking Europe was lost to Marconi's monopoly by 1914.

But that was not the only loss, as the Great War would mark the fast-approaching end of the international operations of the Marconi empire. As Telefunken was for the German military becoming a key asset, it created wireless systems for the German Navy and Army, including a station at Nauen (Figure 239) in 1906 that could reach German submarines. And those submarines were a considerable threat in case America would also participate in the Great War.

In addition to developing German military infrastructure, in 1911, Telefunken built a continuous-wave-type transmission system across the Atlantic Ocean by constructing three 180-foot towers in Sayville, New York. This station connected with the Nauen Station near Berlin some four

Figure 239: The Nauen Station of the Telefunken company (1906).
Source: http://de.academic.ru/dic.nsf/dewiki/548080. Bundesarchive

thousand miles away. The Sayville station was owned by the Atlantic Communication Company, a wholly owned subsidiary of Telefunken.

With the war looming, that company came under fire from different sides. Marconi contended that the company was infringing on his patents and went to the US to testify in court. Professor Ferdinand Braun arrived in December 1914 also. He was never to leave again, as he was detained during the war and died in 1918. The military suspected that Sayville was being used by the German government, and did not renew its license in 1914. When the war broke out, the Sayville station came under fire from the American public. Some suspected that the Germans were using their American installation to send coded signals to aid them in their undersea campaign. In February 1915 the German embassy warned the American public not to travel on ships carrying the British flag or sailing British waters. On May 7, 1915, the *Lusitania* was torpedoed by a German U-boat, confirming the use of Sayville for military operations.

The station may also have been used to transmit the infamous Zimmerman Telegram plotting a German-aided Mexican attack against the United States in the event of war.

> *We intend to begin on the first of February unrestricted submarine warfare. We shall endeavor in spite of this to keep the United States of America neutral. In the event of this not succeeding, we make Mexico a proposal of alliance on the following basis: make war together, make peace together, generous financial support and an understanding on our part that Mexico is to reconquer the lost territory in Texas, New Mexico, and Arizona. The settlement in detail is left to you. You will inform the President of the above most secretly as soon as the outbreak of war with the United States of America is certain and add the suggestion that he should, on his own initiative, invite Japan to immediate adherence and at the same time mediate between Japan and ourselves. Please call the President's attention to the fact that the ruthless employment of our submarines now offers the prospect of compelling England in a few months to make peace.*
> *Signed, ZIMMERMANN'*

So, the Americans seized Sayville during the war and retained it afterwards as part of the war reparations.[383]

---

[383] Source: http://ethw.org/Telefunken (assessed June 2015)

Wireless Business: Services and Equipment

By 1905 the market for maritime wireless communications was growing. Ships passed their positions, course, and speed to the nearest coastal station (to be passed on to Lloyd's shipping services). Especially ships crossing the oceans needed wireless communication to stay in touch with the harbours and their owners. Also, the passengers of the large passenger liners—often businessmen—were offered telegraphic services for their private and business affairs. All in all, the maritime market for wireless telegraphic services was immense, as seafaring was immense. And that attracted a lot of entrepreneurs on both sides of the oceans.

To realize those wireless telegraphic services, the wireless equipment for use on ships and shore had to be manufactured, and relay stations with large antennas had to be built along the coasts. Although those stations—each being engineered to fit the local situation—could be expansive, in terms of industrial volume, they were not that important. It was a different situation for the much larger volume of transmission and receiving equipment, as that was produced by the equipment manufacturing industry.[384]

Some Early Manufacturing Companies

As we have seen, the Marconi organization dominated the market for maritime wireless services. It owned manufacturing plants in Britain to build its own equipment, such as its spark transmitters, coherer receivers (Figure 191), and its reliable decoder 'Maggie' (Figure 205). It was not the only manufacturing entity, as other companies had entered the wireless market, like the German company Telefunken, which had its equipment manufactured by its parent companies AEG and Siemens & Halske.

As the telegraph and telephone manufacturing industry was well developed by that time,[385] many of the new wireless companies depended on their skills and technology to engineer their own systems. For example, it was General Electric—already used to building electric generators and electric motors—who built the high-frequency Anderson-alternators. In addition, there were many other suppliers for the subsystems (eg the Morse receiver) that were being used: the instrument manufacturers and the fine-mechanical industry (eg the clockmakers). For many of those

---

[384] Often the same company providing the wireless services built its own equipment, or had it built to order by subcontractors from the electric or scientific instruments industry that had emerged. It was not until in the 1920s that the radio industry emerged.

[385] See: B.J.G. van der Kooij, *The Invention of the Communication Engine 'Telephone'* (2015), pp. 302–310.

subcontracting companies, their details are lost in the fog of history. However, about some, we know what they did.

British Manufacturers of Wireless Equipment

In Britain, the publicity around Marconi's tests and demonstrations had stimulated many entrepreneurs already active in constructing (scientific) instruments and telegraph/telephone equipment to look into the possibilities of wireless telegraphy.[386] Among them were James T. Armstrong and Axel Orling (of Swedish origin) from London. By 1902 they had developed the 'Amorlsystemet'; a telephone system created by a dead-end technology. And there was the obscure *Helsby Wireless Telegraph Company* of London, which produced its Cymophone portable telegraph apparatus for ships, yachts, and trawlers. The *Lodge-Muirhead Syndicate* also manufactured systems based on the Lodge and Muirhead patents in a factory at Elmer's End in Kent. Their system was installed on cable ships in 1903. However, their application for a license to operate shore stations was refused by the Post Office, so they were limited to the manufacture and sales of apparatus, limiting their opportunities. In 1905 their system was installed in ferries operating from Heysham to the Isle of Man. One can assume that their systems were built by *Muirhead & Company* of Elmer's End, Beckenham (est. 1894).

A Queenslander from Australia, Graham Balsillie (1885–1924) had headed off to England in 1903 to study electrical engineering. While there, he devised a magnetic detector and joined a company erecting wireless telegraph stations in England and Russia. He later worked in Germany, Siberia, and China before returning to England, where he formed the *British Radiotelegraph Company* to market his own 'Balsillie' system of radiotelegraphy. His system, also applied in Australia and made by the Australian Wireless Company, was judged in 1911 to be an infringement of the Marconi patent.

American Manufacturers of Wireless Equipment

Already a very early stage, many American manufacturers of telegraph and telephone equipment had become interested in developing and selling equipment for wireless telegraphy. Many were located on the East Coast of America. Among them was the *United States Electrical Supply Company*, located in New York City, which sold experimental wireless equipment as early as the race for the America's Cup of 1899.

---

[386] Undoubtedly, there may have been more entrepreneurial activities of firms manufacturing electrical equipment that resulted in the production of the equipment needed for wireless telegraphy. However, their details have disappeared in the fog of history.

As building the wireless stations (with their large antennas) required a lot of engineering, the parts were often obtained locally from subcontractors. But the equipment used in the stations was often manufactured by the company that supplied the wireless telegraphy service (like the Marconi Company, which

Figure 240: IP-76 Receiver built by Wireless Speciality Apparatus Co. (c. 1910).

Source: http://www.sparkmuseum.com/WIREREC.HTM

manufactured its own equipment in Britain). In addition, many independent manufacturers of wireless equipment emerged over time.

On the East Coast of America, such as in the Boston area, after the telephone boom an electromechanical industry supplying electric parts had developed, among which the *Wireless Speciality Apparatus Company* (1907), organized by colonel John Firth to sell individual components such as condensers or receivers, primarily to the US Navy (Figure 240).

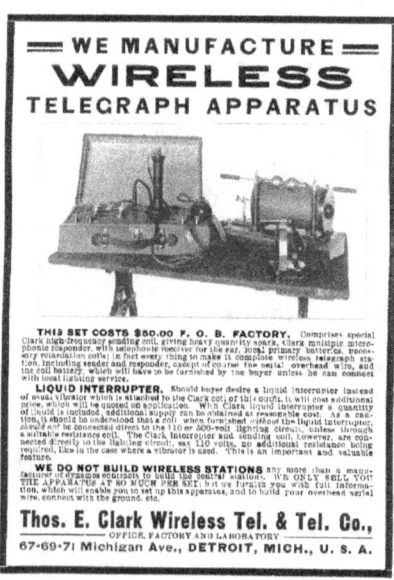

Figure 241: Advertisement for Clark wireless telegraph apparatus (1903).

Source: www.zianet.com/sparks/1903clarkad.jpg

Other inventor-entrepreneurs became 'service providers', using their own system, like the *Massie Wireless Telegraph Company* (1905), manufacturing the Massie wireless telegraph system. It used the patents granted to Walter Massie: US Patent № 787,780, granted on April 18, 1905, for a 'Wireless Telegraphic System'; US Patent № 800,119, granted on September 19, 1905, for a 'Coherer'; US Patent № 819,779, granted on May 8, 1906, for an 'Oscillaphone' (a receiver).

With the construction of some wireless stations (eg the Point Judith Station at Rhode Island) he expanded his business, but he ran into problems when Marconi started lawsuits against companies infringing on his patents, and the company closed down in 1910.

Other American inventors also developed their wireless systems and organized companies to market their inventions, such as Thomas E. Clark, who created in 1902 the *Clark Wireless Telegraph & Telephone Company* to exploit his US Patent № 805,412, which was filed on February 17, 1902 (and granted on November 21, 1905) (Figure 241). John Stone Stone, the owner of many patents, started, with the help of financial backers, the *Stone Telegraph & Telephone Company* in 1902, but he went bankrupt in 1911. Some of the inventors from the early years of the emerging American wireless industry became entangled in a Wild West scene on the stock market. They became part of stock frauds based on wireless telegraphy companies, headed by the gold mine and real estate promotor Dr. Gustav P. Gehring and by 'Honest Abe' Abraham White.

Service Providers for Wireless Telegraphy

The first company to be formed in America with the avowed purpose of making use of wireless telegraphy as a means of communication in a commercial service (aka 'service provider') was the *American Wireless Telegraph and Telephone Company*, which was organized in November 1899 in Philadelphia. Originally working on the patents issued to the engineer A.F. Collins, the owner of the company—Dr. Gustav P. Gehring—soon started working with Harry Shoemaker, holder of several wireless patents. In addition, the company also acquired the rights for US Patent № 350,299, which was issued on October 5, 1886, to Amos Dolbear. Based on this patent, they claimed loudly (but unsuccessfully in court) a monopoly for wireless communication in the US. The AWT&T Co. was the mother company that was to become the founder of other Gehring companies. Soon it had organized, just as companies had in the early days of the telegraph and telephone, regional subsidiaries.

> The parent company *American Wireless Telegraph and Telephone Company* had organized subsidiary companies as follows: *New England Wireless Telegraph and Telephone Company*; *Federal Wireless Telegraph and Telephone Co.*; *Atlantic Wireless Telegraph and Telephone Company*; and *North-western Wireless Telegraph and Telephone Company*. Not so much for technical reasons as well as financial considerations, these companies were later all merged into the *Consolidated Wireless Telegraph and Telephone Company*, which in turn was succeeded by the *International Wireless Telegraph and Telephone Co.*

About that time, the *American Marconi* and the *Canadian Marconi* companies were also formed, being licensees of the British Marconi, and in 1902 the first De Forest Company came into existence.

In January 1902 Lee de Forest met a promoter, Abraham White,[387] who would become de Forest's main sponsor for the next five years. White envisioned bold and expansive plans that enticed the inventor—however, he had quite dishonest business morals, and much of the new enterprise would be built on wild exaggeration and stock fraud. To back de Forest's efforts, White incorporated the *American De Forest Wireless Telegraph Company*, with himself as the company's president and Lee de Forest as the scientific director. The company claimed as its goal the development of 'worldwide wireless', but that proved to be too ambitious.

> *In the midst of the transatlantic experimentation, which he privately admitted was a complete failure, De Forest was summoned back to America because of patent problems. On arriving in New York in April 1906, De Forest learned that a warrant was out for his arrest, and the De Forest attorney advised the inventor to flee to Canada for a few months until White could raise the five-thousand-dollar bond. De Forest and White had been cited for contempt for continuing to market the electrolytic detector. The presiding judge also decided that De Forest and White should pay damages for using Fessenden's detector as their own during the Russo-Japanese War and the St. Louis exhibition. After extricating himself and De Forest from threatened incarceration, White demanded De Forest's resignation. White was furious over the patent suits and accused De Forest of misleading him about the rights to the receiver. He warned De Forest that failure to resign would result in White's rescission of the inventor's bond.* (S. J. Douglas, 1989, p. 168)

Their relationship seriously disturbed, in 1906 Lee de Forest resigned from the company, which was then reorganized by White as the *United Wireless Telegraph Company*. This incident did not withhold White from continuing with his plans and selling the stock of wireless companies, claiming by now he had control over the Marconi companies. But it was a fraudulent affair, as the press reported:

> *THE UNITED WIRELESS TELEGRAPH COMPANY has succeeded in the state of Washington in selling hundreds of thousands of dollars' worth of its stock, in reality worth less than $2 a share, at prices ranging from $20 to $40 a share. Probably for the first time in financial history a stock has steadily increased in price until it sold at more than four times its par value, yet it has never paid a dividend. Something like three-quarters of the company's*

---

[387] White's family name was Schwartz. He had made his fortune in real estate and on Wall Street.

$20,000,000 capital stock has been sold, and as it has gone for as much as four times its par value, it has brought in, probably $30,000,000. The total value of its equipment, stations, etc., according to the company's own figures, is about a million and a half. Its actual value may be much less.[388]

Next, in 1907 de Forest participated with his new partner James Dunlop Smith in organizing another company: the *Radio Telephone Company* to promote his wireless inventions, among which was the three-element grid vacuum tube called the Audion. The Radio Telephone Company was organized ostensibly to promote the use of wireless telephony for broadcasting music and speech. However, nearly almost its entire activity was in the distribution of stock. Also, the operations of the wireless installations proved unprofitable, and by 1910 the parent company and its subsidiary, the *North American Wireless Corporation*, went bankrupt. In 1914 it was reorganized, and in 1924 it became the *De Forest Radio Co.*

These are just a few of the many wireless companies that were created in the waves of the wireless mania that engulfed America in the early twentieth century. We will meet them again in the following section.

The 'Wireless Telegraph Hype' in the Making

At the end of the nineteen century and the beginning of the twentieth century, America was experiencing its Gilded Age, an era of rapid economic growth, industrialization, and the wealthy industrialists and financiers like the 'robber barons' in the railroad, telegraph, and telephone business. As a consequence of all those electricity-based inventions, everything around the magic word 'electricity' drew the attention of the public.[389] This was especially true for the remarkable invention of Edison's incandescent lamp and the subsequent invention of Marconi's 'wireless radio,[390] which stimulated the imagination of many. In the press, articles about the 'Wireless Age' appeared. For someone living in the twenty-first century, used to a smartphone glued to his ear, the following prediction may sound credible. However, at the time it was written, it sounded like pure fantasy:

Someday men and women will carry wireless telephones as today we carry a card case or camera. We shall switch ourselves on to the underground radiations

[388] 'The Great "United Wireless" Fraud'. *Seattle Daily Times*, June 16, 1910, pages 1–3. Source: http://earlyradiohistory.us/1910frd1.htm (assessed June 2015)

[389] The government was also interested in these developments, but they had more military applications as a topic of interest. For example: maritime communications and land communications to use these systems for spying (B. Lee, 2012).

[390] The word 'radio' is used here to mean spark-gap based transmission of signals over electromagnetic waves to be used as a telegraphy system. The use of radio for broadcasting was not until after the Great War.

through the medium of our walking sticks or boots, and then tune up our receiver to say tone No. 39,451, and tone No. 39,451 will go about his business undisturbed by other tones. For military purposes it soon will be no longer necessary to carry cumbrous coils of wire, which are always at the mercy of an enemy. The staff officer and the scout each will drive a wireless apparatus into the ground and await the magic touch of the sympathetic tone. Thanks to the Morse code, it will not even be necessary to await perfection in the conveyance of the human voice. A kindred apparatus will magnify the telephonic sound, and someday the mouse for which we shall set a telephonic trap, will be able to roar like a bull. A ship will proclaim her name loudly through the fog and Calais and Dover, in hazy weather, will announce themselves to approaching packets. Wireless torpedoes, probably, will provide the best solution of the difficulties of coast defense, and when a force of watchful and highly-expert electricians is sufficient to supply the torpedoes with guiding machines, how many expensive fortifications might not we do without?[391]

And it was this perspective that invited people who had some savings to invest, to buy stock in those exciting, new, and promising companies. Every new start-up company was interesting as long as it had the word 'electric' in its name. Articles described ventures such as the Chicago New York *Electric Air Line Railroad* as being 'the grandest opportunity the people have ever had to invest in a gigantic commercial undertaking of national reputation'; and the *Electric* Signagraph and Semaphore Company was 'an opportunity for another such financial whirlwind as was the Bell Telephone' (Fayant, 1907).

Not only the electric motor and the electric light, but also wireless telegraphy was by that time an object of fascination to engineers and businessman, as well as investors. Many of the inventors themselves became businessmen who sought to exploit their inventions covered by their patents (or they worked closely together with business partners). Soon several companies were created that manufactured products based on those patents, as one after another 'radio company' was founded.

Such as in 1902 the *National Electric Signaling Company* (NESCO) of professor Reginald Fessenden, and the earlier mentioned *American Wireless Telephone and Telegraph Company* of Dr. Gustav Gehring based on the patents of professor Dolbear (1901). Another company was the *Stone Telegraph & Telephone Company* (1902) based on the patents John Stone Stone of Cambridge, Massachusetts. Walter Massie obtained in 1905–1906 a range of patents and created the *Massie Wireless Telegraph Company*. And then there was the *De Forest's Wireless*

[391] Source: 'The Wireless Age'. *Los Angeles Times*, November 4, 1901, page 6. http://earlyradiohistory.us/1901age.htm (assessed June 2015)

Telegraph Company of America (1901). To name but a few of the many start-ups. (Scudder, 1926-1934)

From Engineering to Stock Fraud

And there was Archie Frederick Collins (1869–1952) experimenting with wireless telegraphy. He started his own company in 1903, but his inventor-entrepreneurship went downhill, and in December, 1909, the Collins Wireless Telephone Company was merged with three others—the Pacific Wireless Telegraph Company, the Clark Wireless Telegraph Company, and the Massie Wireless Telegraph Company—to form the *Continental Wireless Telephone and Telegraph Company*,[392] mainly a service provider. This is the story of how a prominent engineer became entangled in stock fraud.

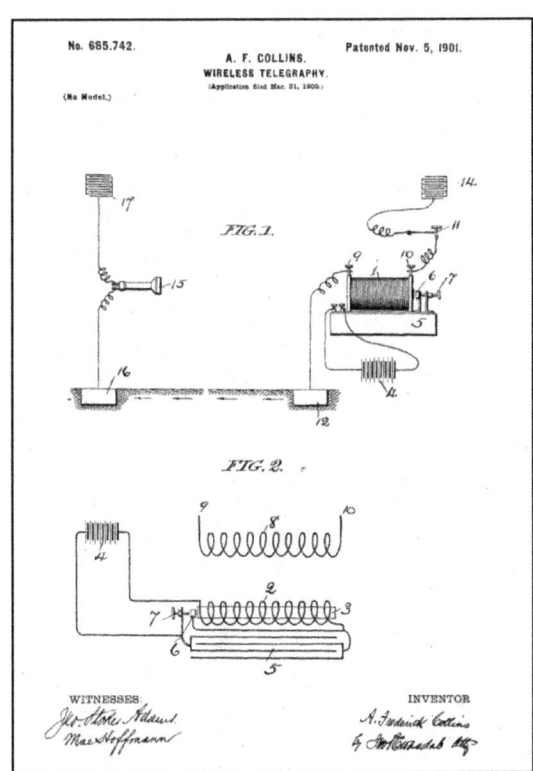

Figure 242: Collins US Patent № 685,742 (1901)

Source: USPTO.

Archie Frederick Collins had started as a prominent engineer. Based on that expertise, he wrote articles about wireless telephony for *Electrical World, Scientific American, Encyclopaedia Americana,* and other encyclopaedias. Over time he even became a prolific writer when, in addition to technical books (eg *The Radio Amateur's Hand Book*), he also wrote scientific adventure novels (eg *The Boy Scientist,* the Jack Heaton Series of a wireless operator).[393]

After finishing his education, he began his career working for the Thomson-Houston Electric Company in Chicago. By 1899 he was

---

[392] 'Merging the Wireless Companies'. *National Magazine*, June–July, 1910, pages 425–429. Source: http://earlyradiohistory.us/1910mrg.htm (assessed June 2015)
[393] Source: http://research.omicsgroup.org/index.php/Archie_Frederick_Collins

experimenting with electric waves, which resulted in a system for wireless telegraphy using an arc lamp. His first patent was granted in February 27, 1900: US Patent № 644,497 for 'Wireless Telegraphy'. On November 5, 1901, he obtained his next patent: US Patent № 685,742, also for 'Wireless Telegraphy' (Figure 242). By March 13, 1906, he was granted US Patent № 814,942 for 'Wireless Telephony'.

His system was based on inductivity using an electric arc. The transmitting coil carried current modulated by a microphone, which produced a magnetic field that varied with the speech of the speaker. The varying magnetic field produced an electric current in the receiving coil placed nearby, reproducing the speaker's voice in a telephone receiver. By its very principle, its distances were limited.

Already in May 1903 he created the *Collins Marine Wireless Telephone Co.*, a small workshop in Newark, NY, that did not have too much business. Its name was later to be changed to *Collins Wireless Telephone Co.* These names are interesting because they are about 'telephony' (and not telegraphy) and 'wireless'. Marconi's invention was about wireless telegraphy, as it could not transmit the spoken word. So, Collins was clearly ahead of the times of wireless telephony: a concept that would take nearly a century to develop into today's massive success of the 'mobile phone' (Figure 243).

FOR AUTOMOBILES

The Collins Wireless 'Phone Will Eliminate Many of the Troubles Experienced While Motoring at a Distance from a Garage.

MESSAGE FROM AN AUTOMOBILE.

Figure 243: Collins demonstrating his mobile phone (1909).

Source: http://www.sparkmuseum.com/collins.htm

Collins toured the United States, putting on demonstrations and selling stock in the Collins Wireless Telephone Co. The company made wild claims about his technology and was vocal in predicting the downfall of telegraph stocks such as Marconi. Usually two adjoining hotel rooms were rented for the demonstration, placing the coils on opposite sides of a wall. Celebrities and government officials would be invited to demonstrate the apparatus. These demonstrations were spectacular and resulted in appreciable stock sales. Unfortunately the money received was used by the company to cover the expenses of marketing its stock and to promote further speculation, not for building the assets of the company for the benefit of the stockholders. [394]

In 1909, Collins told the New York Times he had operated four separate wireless telephone links at the same time between Portland, Maine, and a nearby island, creating an impression that wireless telephony was on the verge of replacing wired telephone systems. The wild claims Collins made were not only verbal but also in appeared in print:

There is no doubt as to the position wireless telephony is to occupy. Its use on the ocean will be identical to the telephone on land, while its other fields of operation are practically without limitation. Governments will use it for their army posts and ships; islands in the ocean and harbors on the continents will employ it to speak with other parts of the world; every craft that sails the ocean must adopt it, isolated mining camps, rural districts and other places will be brought into touch with the civilized world in fact, wireless telephony will enter a field entirely its own, in addition to being an aggressive competitor of the present telegraph Graph and telephone system on land. It will enter a new, field by making it possible to telephone from automobiles to the garage when help is needed. There are thousands of automobiles in the United States, and, while touring the country in a powerful car is a delightful pastime, a breakdown several miles from a garage or other repair shop is not conducive to pleasure. Often some member of the party finds it his lot to walk to a house for supplies, while the rest of the party, patiently or otherwise, usually the latter, await his return. Mr. Collins proposes to eliminate is decidedly adverse feature of automobiling by employing the wireless telephone. Consequently every garage or shop will be equipped with the wireless telephone, as they are now with the tire pump and ignition plugs, and is latter day telephone will always be set up ready for use. Likewise, every auto will be provided with a portable wireless telephone. then in the event of the inevitable accident the 'phone can be taken out, set up ready for use and communication established with the nearest garage, and an auto with men and needful mechanism sent post haste to the scene to repair it. (Dubilier, 1908)

In December, 1909, Collins Wireless Telephone Company merged with *Continental Wireless Telegraph & Telephone Company*, with Abraham White as president and A. Frederick Collins as technical director. As this was a fraudulent operation, he later was charged with giving a fake demonstration of his wireless telephone on Oct. 14, 1909, at the Electrical Show in Madison Square Garden, New York, for the purpose of selling stock in the Collins Wireless Telephone Co. He was sentenced to three years in jail in Atlanta. After serving one year, he was released.

---

[394] Source: 'A. Frederick Collins... Tragic Genius'. http://www.sparkmuseum.com/ collins.htm (assessed June 2015)

Before his conviction he had been a respected engineer, considered an authority on wireless in general and a specialist in wireless telephony. This is just one of the many stories that can be told about the Wireless Mania that would create the Wireless Bubble[395].

The Wireless Mania

'Wireless radio'—to combine the 'wireless telegraphy' and the 'wireless telephony' under one label, aka 'space telegraphy' —was such an area in which the public was so highly interested that it became 'hyped'. Around 1900 the newspapers and magazines were full of advertisement relating the inventions of Mr. Guglielmo Marconi and his 'telegraph system without wire' (Figure 244).

The press spoke of the bright 'Future of Wireless Telegraphy' (*New York Times*, May 7, 1899). In *Harper's Weekly Magazine* of February 21, 1903, in an article called 'American Wireless Telegraphy', Lee de Forest and Reginald Fessenden were described as the two most prominent researchers in the United States during the first decade of the 1900s. The statements made by people like Abraham White were published widely:

> *Commercial wireless telegraphy, at a rate of one cent a word to the general public from Chicago to all principal points in the United States, will be an assured fact within ninety days, if the plans of the American De Forest Wireless Telegraph Company are carried out. Within sixty days it will be possible to flash messages from Chicago to steamers on the lakes, and to Detroit, Cleveland, Buffalo, New York, and the Atlantic seaboard.*

YOU HAVE
Only 4 Days
In which
TO BUY
Marconi Securities
AT $5.00

After June 20th, 1904,

Price Advances 20%

If you want to reap the full benefit of this, the best investment ever offered to the American public, send your application for not less than $100 nor more than $1,000 worth. Make all checks payable to

MUNROE & MUNROE,

Managers for the Marconi Underwriters.

25 Broad Street or Knickerbocker Trust Bldg., Fifth Ave. and 34th St.

Figure 244: Advertisements for stock introductions in the *New York Daily Tribune* by Munroe (1901).

Source: http://chroniclingamerica.loc.gov /lccn/sn83030214/1904-06-16/ed-1/seq-13/

---

[395] For more details: White, Th. H., *United States Early Radio History*. 'Early Radio Industry Development (1897–1914)'. Source: http://earlyradiohistory.us/sec006.htm (assessed June 2015)

Almost as soon, we will be in wireless communication with St. Louis, Omaha, Kansas City, and Fort Worth. A statement that these things would be accomplished was given out yesterday at the Chicago office of the company by Abraham White, president of the corporation, and Dr. Lee De Forest, whose inventions are claimed to have been made before those of Signor Marconi.[396]

Aside from all this glorification, there were also warnings, such as published in an article in *Electrical Age* in 1904:

The daily newspapers are still printing columns of advertisements relative to wireless telegraph systems, in which statements relative to the prospective earnings and profits of such systems are very alluringly set for; and some of the later statements imply that the success of certain wireless companies in Europe may be taken as a criterion of the success that will follow here. It is quite possible that the British Marconi Company is on a paying basis, owing to the large number of warships and mercantile vessels, light houses, etc., now equipped with that system. On this side of the Atlantic, however, the number of wireless equipments is much more limited, and what business there may be is divided among that least three different companies, with a corresponding division of earnings. Indeed, it is probably within bounds to assume that few, if any, of the American companies are that present realizing sufficient income from the commercial operation of their systems to meet even the expenses of the legal controversies over alleged infringements of patents which are now proceeding between that least three of those companies. One of these controversies concerns the use of an imperfect contact, such as the filings coherer or its equivalent, in wireless telegraphy. another is for infringement of the liquid barretter, or electrolytic detector, whichever the device may ultimately be held to be by the courts. (T.H. White, 1996).

Figure 245: Advertisements for stock introductions in the *New York Daily Tribune* by Clark (1901).

Source: http://chroniclingamerica.loc.gov/ lccn/sn83030214/1904-06-16/ed-1/ seq-13/

---

[396] Source: Fayant, F, 'Fools and Their Money: A Transcontinental Wireless Telegraph Dream'. http://earlyradiohistory.us/1907fool.htm (assessed June 2015)

Wireless Telegraph Bubble in the Making

Everything that was related to 'wireless' seemed to have become a hot item on the stock exchange in New York's Wall Street (Figure 245). It was an interest influenced by the huge profits of the first investors in the (wired) telegraph and telephone. Investors were lured with advertisements like:

> *"The most marvelous invention of the century." "Bell Telephone stock, when first offered, went begging that fifty cents a share, and those same shares to-day are worth $4,000." and "The career of the company controlling the basic patent in the richest field in America starts with a thousand times more flattering prospects than did the Bell Telephone."* (Fayant, 1907)

It's striking that in most of the advertisements (Figure 244, Figure 246), no dividends were mentioned. The perceived return on investment was not about dividend; it was about the increase in the value of the shares of these high-tech companies with their new and fascinating products.

Obviously, selling stock in new companies was serious business. It was L.E. Pike who in 1901 and 1902 spent around $150,000[397] on advertisement, quite a lot of money in those days. He promoted the purchase of shares in the *Federal Wireless Telephone and Telegraph Company*, a subsidiary of the *American Wireless Telephone and Telegraph Company* (Figure 246). That was done with advertisement texts like:

Figure 246: Advertisement in the *Washington Times* about the shares of the Federal Wireless Telegraph & Telephone Company (October 12, 1901).

Source:
  http://chroniclingamerica.loc.gov/lccn/sn87062245/1901-10-27/ed-1/seq-14/

---

[397] Equivalent to about $4.3 million in 2015 based on the historic standard of living. Source: http://www.measuringworth.com/uscompare/

The Whole World is Talking About Wireless Telegraphy. The inventor of this system, in an interview printed in the "New York Journal" August 20, states that he has no more doubt that he will in a short time be sending Wireless Messages across the Atlantic Ocean than he has that cars run up Broadway.

This recurring text was then followed by (Figure 246):

Have you read in the New York papers how ocean steamers 150 miles apart, communicated with each other through a dense fog. Do you suppose there will be a single steamer in 12 months' time without Wireless Telegraphy? Do you realize how many lives will be saved through is wonderful invention? It not only prevents accidents, but when a steamer is in distress she will be able to call for assistance. Do you realize that the Federal Company absolutely controls the great Gateway of Commerce on this side of the Atlantic and that the revenues from this source alone should be enormous? Don't Delay. The Opportunity is Yours. Will You Grasp it?

Pushing investors to buy stock did not cease, and promotions of stock introductions continued. In the May 1909 issue of *Wireless*, a promotional text issued by the New York Selling Agency stated:

"You should buy United Wireless now--without delay, because now is your opportunity", due to the fact that *"When the 'speculative' investors begin to fully understand and appreciate the wireless situation, United stock will undoubtedly be snapped up that whatever price is asked for it and will start bounding upward to quickly sell that big figures, the size of which would now seem impossible."* (T.H. White, 1996)

Basically, the trick with the shares was the issue of new shares during mergers and acquisitions, and the repeated offering of new shares. This was called 'watered stock' promotion, where an investment of $50 in a couple of years was watered down to a value of $0.85 (Table 13).

Table 13: Watering down the value of federal wireless stock (1902–1906)

Investment:	January, 1902
$50 (real money) bought $50 worth of Pike's Federal Wireless	
= $50 (certificate) watered Consolidated	February, 1902
= $10 (certificate) unwatered Consolidated	October, 1902
= $10 (certificate) International	February, 1903
= $10 (certificate) American De Forest	January, 1904
= $7.50 (company's money), subscription price of De Forest, St. Louis office	October, 1906
= $6 (company's money), subscription price of De Forest, New York office	October, 1906
= $0.85 (real money) cash market value.	

Source: Fayant, 1907.

The Bubble Bursts: The US Postal Department Strikes

It took a while, but then many investors realized they were being cheated, and calls for action started to appear—like the stockholders of the American De Forest Wireless companies, who applied for receivership. After years of complaints, in June 1910, inspectors from the United States Postal Department finally moved to shut down what was described as 'one of the most gigantic schemes to defraud investors that has ever been unearthed in this country'. Lee de Forest's name would be tainted in the process.

George H. Parker, fiscal agent for the United Wireless Telegraph Company for the territory west of the Mississippi River, was arrested on a federal warrant charging the use of the mail to defraud. Wilson, Bogart, and Tompkins of the United Wireless Company were also arrested. The *New York Times* of May 30, 1911, reported about their conviction: "A verdict of guilty on all four counts of the indictment charging misuse of the mails in a scheme to defraud was returned yesterday before Judge Martin in the United States Circuit Court." The defendants were given jail sentences of up to three years (T.H. White, 1996).

United Wireless Telegraph Co. wasn't the only company under scrutiny. On December 11, 1911, four officers of Continental Wireless Co. were also charged with fraud, among them was Collins, whom we met before:

> *In addition, A. Frederick Collins was charged with giving a fraudulent demonstration of his wireless telephone on Oct. 14, 1909 that the Electrical Show in Madison Square Garden, New York, for the purpose of selling stock in the Collins Wireless Telephone Co. It was developed that the trial that the four Collins officers had claimed in their prospectus that the Collins wireless telephone had been perfected to such an extent that in a community equipped with it, any two subscribers could talk to each other with total exclusion of all other subscribers, that the Collins wireless telephone would do away with all central exchanges. The necessity for wire lines, etc, that an automobile so equipped would be in constant touch with a garage so as not to be stranded in case of trouble[398], that because of the lower cost of the wireless telephone, with no wires needed, the telephone and telegraph systems would soon be put out of business and that the demand for the equipment would increase so rapidly that the stock price would quickly increase. ... Four officers were convicted on all five counts. Three were*

---

[398] It wasn't until the twenty-first century that this became available for automobiles, such as BMW with its Intelligent Emergency Call system that—after the airbag is activated—connects with the BMW call center to offer support, automatically transmitting data about the situation (GPS location, number of passengers, impact).

fined and sentenced on January 10. 1913 to prison terms of up to four years.
This was the end of the Continental Wireless Tel. & Tel. Co.[399]

Millions of dollars invested by the public were lost. Those guilty of the fraud were found guilty and got sentenced. The June 1912 issue of *Munsey's Magazine* concluded:

> *More men are now in prison or under indictment for selling stock in wireless*
> *telephone and telegraph companies than is the case with any other line of*
> *industrial promotions of which I have knowledge. Lodged in various jails and*
> *penitentiaries are C. C. Wilson and four associates, who exploited the United*
> *Wireless Company swindle, and one of the Munros, of the famous firm of Munro*
> *& Munro, who pretended to sell Marconi Wireless stocks, but did not deliver the*
> *shares. All the enterprises mentioned by the above correspondents have their quota*
> *of men under indictment on charges of using the mails for the purposes of fraud.*
> (T.H. White, 1996 Munsey's Magazine, June, 1912, page 424)

The collapse of Continental was mirrored by the downfall of other companies such as United Wireless and the De Forest companies. The era of bogus stock selling had come to an end. And the Great War was to begin.

> *With the elimination of three major fraudulent U.S. radio firms, the field was*
> *cleared for legitimate companies. And with its takeover of the United Wireless*
> *assets, the American branch of Marconi Wireless was now by far the largest*
> *radio company in the United States, a status it would hold until after World*
> *War One.*[400]

Mergers and Acquisitions in the US

Similar to the development of the wired telegraph and telephone systems, after the initial boom in start-ups came a period of consolidations in which the stronger companies merged with, or just acquired, other companies (Figure 247). One such example is the previously mentioned *Continental Wireless Telephone and Telegraph Company* that 'embraced' *Collins Wireless Telephone, Pacific Wireless Telegraph, Clark Wireless Telegraph and Telephone*, and *Massie Wireless Telegraph* in 1909. As described in their prospectus:

> *The Continental Company enters the field with not only the best there is in*
> *wireless telephony and telegraphy, but with the most experienced and efficient*
> *management that it is possible to concentrate in one enterprise.* (Dubilier, 1909)

---

[399] Source: 'A. Frederick Collins... Tragic Genius'. http://www.sparkmuseum.com/collins.htm
[400] Source: White, T.H., 'Early Radio Industry Development (1897–1914)'.
http://earlyradiohistory.us/sec006.htm#part050

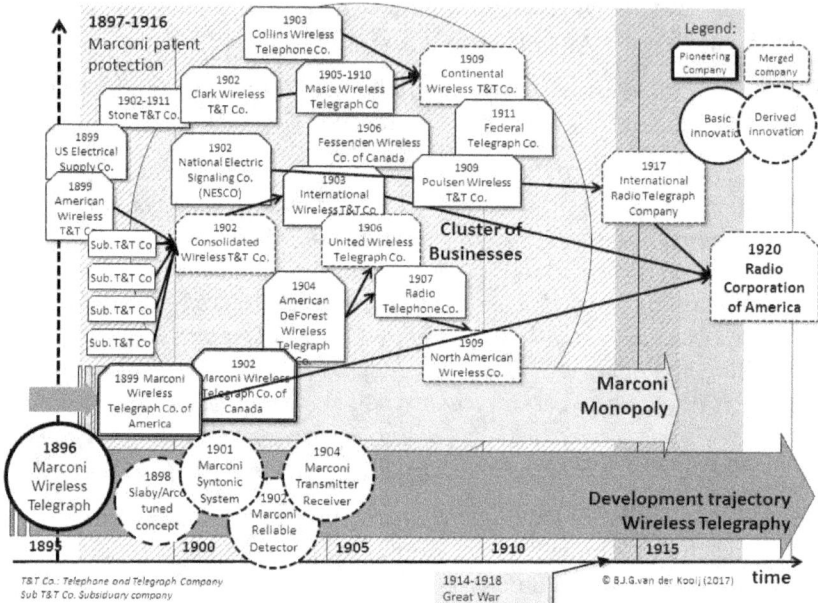

Figure 247: Mergers and acquisitions of US wireless telegraph and telephone companies.

Figure created by author.

There were also others who had understood the game, as there were numerous companies that had only patents and had no operational activities (such as the *American Wireless Telephone and Telegraph Company*, in short the *American Wireless T&T Co.*). The patent holder, in this case Professor Amos Emerson Dolbear (1837–1910), released his patent (US Patent № 350.299) and got a limited share of the start-up company. The rest was for the investors, headed by the before mentioned Dr. Gustav Gehring, former real estate and gold mine investor.

It was Gehring who had a much larger plan in mind. In addition to the parent company, which owned the patents, a series of daughter companies were formed on a regional basis. So, the American Wireless T&T was not an operating company. It owned the patents and divided the country from coast to coast into fields in which the sub-companies were to operate. In the case of the American Wireless T&T Co. those were: *New England T&T Co., Federal T&T Co., Northwestern T&T Co., T&T Atlantic Commercial Co., Central Western T&T Co., T&T Gulf and Pacific Co., Continental T&T Co.* Each was capitalized with $5 million, and the stocks of all those companies were introduced at the stock exchange. The design was great: the total offering of the American Wireless T&T Co. added to a sum of $ 55 million.[401] This was quite a bit of capital at a time in which the weekly salary of a laboratory assistant (like Lee de Forest in his early years) was $10 a week. And the profit went to the promoters:

> *It is claimed that the money derived from the sale of these shares was diverted to the pockets of the promoters of the subsidiary companies and was not used in the development of the wireless system. (WP-staff, 1903).*

To make it even more complex, shortly thereafter a new company was capitalized with $25,000,000[402]: the *Consolidated Wireless T&T Company*. This new company took over American Wireless T&T Co., New England T&T Co., Federal T&T Co., Atlantic T&T Co., and then Northwestern T&T Co. The Consolidated Wireless T&T Company, in turn, was then taken over by the *International Wireless T&T Company*, also capitalized with $25,000,000. This was all very opaque, but that was exactly the intention (Figure 248). In reality, there was not too much substance to the operations:

> *The company at the present time has no stations in operation. It has nine stations at various points on the Atlantic coast, but these have been closed during the winter. It has a laboratory at No. 827 Arch street, Philadelphia where its experts are employed in experimental work. The company owns 61 patents, the greater number of which have been granted to H. Shoemaker, the company's electrical expert and include a variety of applications of wireless not hereto made public, including a wireless "autotorpedo" and a wireless "auto-motor." The company claims priority of invention on many of the principles and instruments*

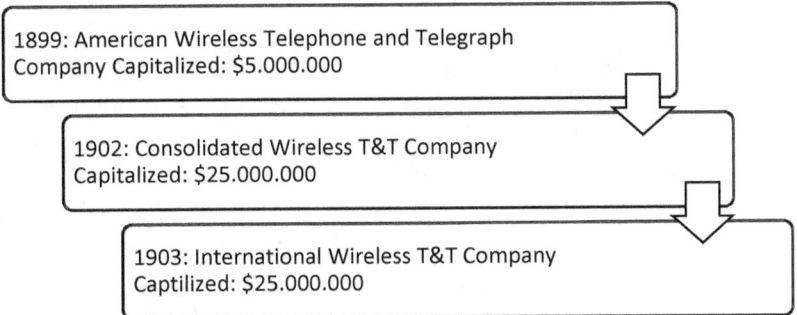

1899: American Wireless Telephone and Telegraph Company Capitalized: $5.000.000

1902: Consolidated Wireless T&T Company Capitalized: $25.000.000

1903: International Wireless T&T Company Captilized: $25.000.000

Figure 248: Successive companies based on the patent of Dolbear, also called the 'Gehring companies'.

---

[401] Equivalent to about $1,503 million in 2015 based on the historic standard of living. Source: http://www.measuringworth.com/uscompare/
[402] Equivalent to about $695 million in 2015 based on the historic standard of living. Source: http://www.measuringworth.com/uscompare/

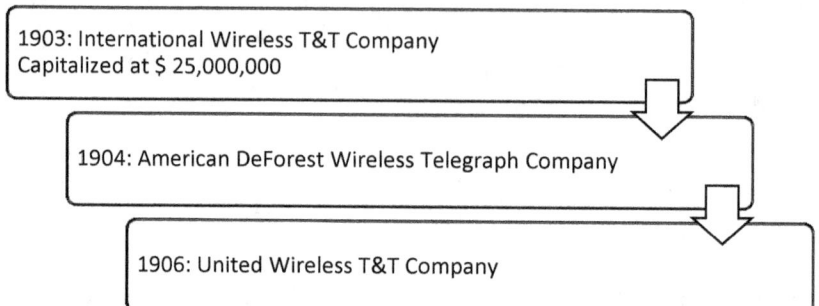

Figure 249: Companies after International Wireless Telegraph &Telephone Company.

which they say both Marconi and De Forest are now using. They also have a patent on the word "Aerogram" as a trademark which word they claim De Forest is using in contravention of their rights. In addition to the Shoemaker patents they have purchased Prof. A. E. Dolbear's patents on the "art of wireless transmission", granted in 1886, before the publication of Hertz's discoveries. (WP-staff, 1903)

In early 1904 the *International Wireless T&T Company* (the successor to the American Wireless T&T Co.) in turn was taken over by the *American DeForest Wireless & Telegraph Company*. The promotion of the shares went rough, inter alia, with an advertisement in the *New York Tribune* of 16 June 1904 (Figure 249) with the text: 'For every $100 invested, it will return thousands.'

At the end of 1906 the American DeForest Wireless T&T was reorganized as *United Wireless T&T Company*, and the tireless promotion now said: 'because the shares of the United Wireless Telegraph Company are the nest eggs of fortune' (Figure 249).

There is not enough stock to go around. Consider the matter carefully. You have the opportunity. Will you grasp it 'at the flood tide' (now) and ride on to the shore of plenty, high and dry above the adversities which often beset old age, to the land of our dreams, where weal is unbounded and every wish gratified, where comforts admit of enjoyment and weal admits of opportunities for yourself and those you love? Or will you hesitate and doubt, and let the chance go by, to remain in senile dependency upon the bounty of others? Think! It is for you to decide! Think well! Buy! Do it now! (Fayant, 1907).

By the time the Great War had started in Europe (1914), with the elimination of three major fraudulent US radio firms, the field was cleared for legitimate companies. And with its takeover of the United Wireless assets, the American branch of Marconi Wireless was now by far the largest radio company in the United States. But that was not a situation that was to last for a long time, as we have seen before.

> *By the start of World War I, British Marconi stood preeminent in the field. It controlled wireless communication throughout the British Empire and had taken over the assets of De Forest's United Wireless, thus controlling, through its subsidiary, the American Marconi Company—90% of all American ship-to-shore commercial communication.* (Sterling & Kittross, 2001, p. 39)

Later Developments in Wireless Technology and its Applications

As described extensively, the principle of the Hertzian waves used to communicate was based on the properties of electromagnetic waves, properties that could be used to bridge distances without wires. Just as the waves emitted from a source in the visible spectrum (Figure 138)—known as the rainbow colors of light—are detectable by the physical receiver of the human eye, the electromagnetic waves of another frequency band—known as radio waves—became detectable by a special 'wireless receiver'. And, similar to the eye, which can distinguish between the different colors—ie the different frequencies in the spectrum of visible light—the different frequencies of electromagnetic waves (short waves, long waves) were used for communication over distance. First in crude bursts of spark-generated emissions and then in single continuous waves of single frequencies. When it became possible to control the frequency spectrum that was used to connect transmitter and receiver, the different forms of communication developed (short-distance, long-distance communication). That control of frequencies offered a rich source of opportunities for further evolution of wireless systems.

Because wireless communication, like the other forms of cabled communication called telegraphy and telephony, fulfils a basic human need to communicate over distance, its application is also broad. As a consequence, the later technological developments after its conception and birth followed different trajectories (Figure 250):

Improvement point-to-point trajectory: The basic form of wireless equipment was *point-to-point communication*, such as its maritime use in ship-to-shore, ship-to-ship, and shore-to-ship communication. A basic need was fulfilled when the wireless telegraphy solved the incommunicado problem. This application was, as we have seen, constantly improved upon. It resulted in the gradual improvement of the spark-based wireless equipment: better-tuned transmitters, more reliable receivers, the (nearly) constant wave generation, and more sensitive detectors. Due to its mobile properties, it soon became clear that wireless could also be used in other forms such as air-to-ground and air-to-air communication. For that aircraft use, the equipment had to be redesigned to fit this particular application. The Great War highly stimulated the development of lighter equipment that could be used in the airplanes of the British Royal Flying Corps. And it resulted in a range of portable 'trench' sets for military use in the field.

Application spawning trajectories: Sometimes the experimenting with Hertzian waves led to unintended applications, such as 'wireless direction finding'. Now the Hertzian waves detected by a layout of several antennas could be used to pinpoint the physical location of a source of radiofrequency energy, for example, the location of a ship in distress, the location of a thunderstorm, or the bearings of a German Zeppelin or U-boat. Later in time another property of high-frequency electric waves—their reflection by objects—led to a completely new application that became known as 'radar'. Then reflection of the electric waves by objects was used to pinpoint the position of ships and aircraft, but that was not until well into the 1930s[403].

Technological spawning trajectories: Originating from the crude sparked-based devices was a different range of developments that created a new technology called 'electronics' to be used in wireless communication. More or less a spin-off from the incandescent lamp developed by Thomas Edison, it would create the new development trajectory of vacuum-bulb-based 'electronic devices'. It started with the early solid-state technologies—for which the seeds were sown when Ferdinand Braun, in the early 1870s, started experimenting with crystals. Application for wireless telegraphy resulted in the crystal diode rectifier (aka cat's whisker' diode). Then the development of vacuum tubes by Fleming (1904) and Lee de Forest (1907) would create a totally different range of wireless equipment. The technology of vacuum tubes had a course of its own: the 'Era of Electronics', which started after the Great War. And Marconi was going to be part of it.

Point-to-many application spawning trajectory: The use of electric waves started with point-to-point communication: a 'transmitter' sending the message and a 'receiver' receiving it. Originally, as other receivers also received the emitted signals, this created a problem of interference, hindering the regional use of different wireless transmitters at the same time. Tuning the systems solved that problem more or less. This property was quite useful, however, to creating *point-to-many communication*, especially when the continuous wave transmitters and valve receivers were developed. Now a tuned receiver (aka 'radio') could

---

[403] In Britain, during February 1935, Robert Watson-Watt of the government's Radio Research Laboratory conducted the first successful ground-to-air test involving radio detection with an RAF bomber aircraft flying through the beam of the BBC's shortwave transmitter (built by Marconi) at Daventry. In December 1935 the British government asked the Marconi Company to design and manufacture the transmitting 'curtain' antenna arrays for the first five 'Chain Home' radar stations covering the Thames Estuary and approaches to London. It served the country well when the Second World War broke out. Source: http://www.marconicalling.co.uk/

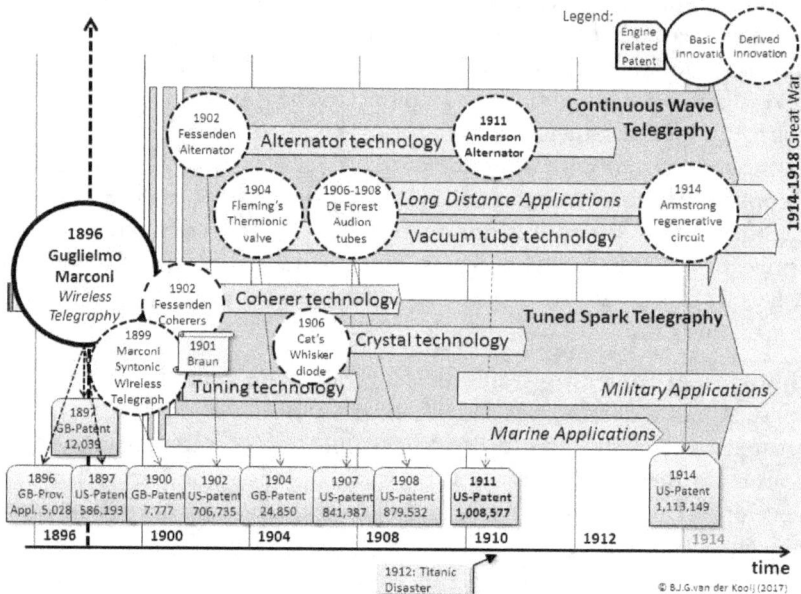

Figure 250: Technological and application trajectories for wireless technology (1900–1914).

Figure created by author.

receive the signal 'broadcasted' by the transmitter. Soon, just as had taken place with telegraphy and telephony, 'radio broadcasting' developed. It was the dawn of the 'Era of Radio Broadcasting'.

The exploration of these trajectories is outside the scope of this case study and might deserve some special attention.[404] However, we will try and have a peek at them to illustrate the contours of times to come after Marconi brought his system of wireless telegraphy to maturity.

Application Development

During the first decades of the twentieth century, the spark-based wireless equipment matured as the result of a continued range of improvements. Equipment that was adapted for specific applications was improved upon technically (more power, better antennas, and additional instruments like wave meters). Even equipment for emergency situations, like the lifeboat sets (Figure 251, bottom) and marine auto-alarm sets, were developed.

---

[404] See: B.J.G. van der Kooij, *The Invention of the Electronic Engines* (Provisional title, to be published).

Marine Application of Wireless Telegraphy

In the early twentieth century, next to long-distance (transatlantic) communication, the marine application of wireless telegraphy was becoming the mainstream activity of the Marconi companies. One reason for its growth was the safety issue at sea. Stimulated by accidents at sea—such as the collision of the ocean liner SS *Republic* with the SS *Florida* in thick fog in 1909—ship owners like the Cunard Line and the White Star Line increasingly closed contracts with the Marconi International Maritime Communications Company (MIMCC) for wireless services.

Thus, wireless telegraphy was regarded as an essential part of the equipment of large ocean-going passenger vessels. At the beginning of 1909 (after eight years of development) there were 125 ships of the mercantile marine fitted with Marconi apparatus. At the end of 1909 the number had risen to nearly 300. Soon it indeed proved its use when the *Titanic* stuck an iceberg in 1912.

Technical progress created the Marconi standard marine sets in a range of different models and powers.

The development of regular communication between an increasing number of moving stations has necessitated not only a carefully devised organisation, but a uniform method of working. This, in turn, has necessitated a practical standardisation of apparatus. At the same time, the demand for absolute reliability in the hands of ordinary operators has led to the evolution of a type of apparatus which is free from complications and is constructed to work continuously without derangement.[405]

Figure 251: Marconi Wireless 103 Radio Receiver used on board ships (1913, top), and Lifeboat Transmitter Type 241C (1919, bottom).

Source:
http://www.northlandantiqueradioclub.com/rdaze13/marconi103.jpg.
http://www.radiomuseum.org/r/marconi_wi_lifeboat_transmitter_type_241c.html

---

[405] Bradfield, W.W., 'Wireless Telegraphy for Marine Inter-Communication'. Source: http://marconigraph.com/titanic/electrician/elec_100610.html

The marine market was becoming the bread and butter application for Marconi equipment. But with the war looming more and more on the horizon, other applications followed suit.

Military Applications of Wireless Telegraphy

As soon as the military in both Britain and America understood the importance of wireless telegraphy for military use, they started developing their own sets (America) or procured adapted Marconi sets (Britain). These sets had to be adapted to conditions (in the airplanes and in the trenches) under which they were going to be used.

Around 1910 testing wireless equipment in aircraft started. Not only were aircrafts still in a state of infancy, but the wireless equipment was not suited for use in them. Altogether, the wireless apparatus—which could only transmit, mind you, not receive—weighed about 110 kilograms. The available wireless sets weren't just too heavy; their operation was too complicated for the pilot, who was already busy with other tasks. Until 1917 wireless communication was one-way, as no receiver was mounted in the aircraft and the ground station could not transmit to the pilot (Figure 252, left). But, the war changed that, and by its end the important role of wireless communication had resulted in the Royal Air Force having some 1,000 ground stations, 18,000 wireless operators, and sets in some 600 aircraft.

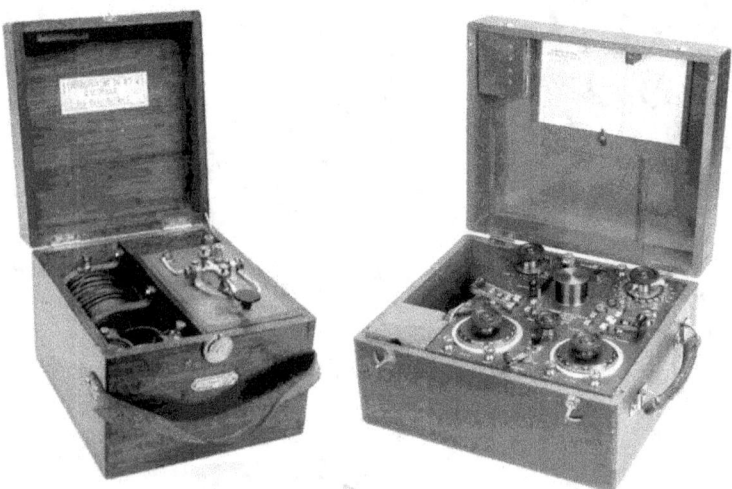

Figure 252: Marconi's portable aircraft spark transmitter (left) and ground set (1917, right) for military use.

Source: http://www.marconicalling.co.uk/, http://marconiheritage.org/ ww1-land.html, ww1-air.html

Figure 253: Marconi's field set on wagon (1911) for military use.

Source: http://marconiheritage.org/ ww1-land.html, ww1-air.html, http://www.marconicalling.co.uk/

By 1911 the Marconi Company had already provided the Army with a battlefield wireless station by designing two small 'portable sets' (Figure 252, left) as well by other sets mounted on a wagon for field use. The wagon had two compartments, one for the wireless equipment and one for the accumulators (Figure 253).

In 1912 the British Royal Engineer Signal Service was formed and made responsible for visual, telegraph, telephone, signal despatch and later wireless communications from HQ down to brigades and for artillery communications down to batteries. In addition, the techniques of communications intelligence (comint) such as message interception, cryptoanalysis, direction finding, jamming, and intelligence gathering developed rapidly.

The Rise of (Ham) Radio

In its essence, wireless transmitters and receivers were not that complicated to build. True, the wireless technology as a whole was in its infancy at the start of the twentieth century, but soon the curiosity and ingenuity of many creative people resulted in the construction of amateur-made devices: the 'ham radio' was born. Often these pioneering amateurs, already experimenting with private telegraph lines, were now starting to experiment with the newly discovered electric waves. Influenced by the publications in the press, stimulated by the success of Marconi, they would become professionally involved in the world of wireless communication, not only as wireless operators but also in manufacturing.

One could even say that along with the wireless manias in science, engineering, and business as described before, a wireless mania among the often young radio amateurs developed in America. By 1910 thousands of—often crudely built—wireless transmitters and receiver were clogging the ether, increasingly causing interference problems for the commercial operators of wireless telegraphy.

Many early amateurs were young, and most built their own spark transmitters and receivers (Figure 254). Using the scarce available information on how to build a wireless telegraph apparatus,[406] they constructed the transmitters and receivers, and bridged short distances. Soon small companies started to sell affordable radio equipment to experimenters and amateurs, including components like Morse keys, the Morse telegraph, and the Rhumkorff coils to complete systems. Among them, we find the *Electro Importing Company* of New York City, set up in 1904 by Hugo Gernsback, an 18-year-old immigrant from Luxembourg. Beginning in 1905, this company sold what may have been the first complete radio system—including both a simple transmitter and receiver— offered to hobbyists on a national scale, under the name of *Telimco Wireless Telegraph Outfits* (Figure 255). Gernsback later started three magazines with large amateur followings: *Modern Electrics* in 1908, The *Electrical Experimenter* in 1913, and *Radio Amateur News* in 1919.

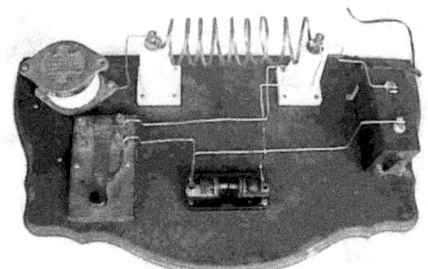

Figure 254: Homemade amateur radio spark transmitter.

Date unknown.

Source: http://w1tp.com/imperad.htm

It was not too long before thousands of radio amateurs were active, especially in the industrial northeast part of America, though they were soon followed by amateurs in California. They organized in amateur radio clubs, such as the Junior Wireless Club of New York City formed in 1909 and the Wireless Association of America with some 3,000 members. By 1912 their numbers had grown dramatically:

> *Although it is not generally known, there are to-day* [1912] *close to 400,000 wireless experimenters and amateurs in the United States alone. The Wireless Association of America to-day numbers 16,189 members, and there are now 122 wireless clubs and subsidiary wireless associations scattered from coast to coast.*[407]

---

[406] Such as the publication of A. Frederick Collins, 'How to Construct an Efficient Wireless Telegraph Apparatus at Small Cost'. *Scientific American Supplement*, February 15, 1902, pages 21,849–21,850. Source: http://earlyradiohistory.us/1902cons.htm

[407] Gernsback, H., '400,000 Wireless Amateurs. *New York Times*, March 29, 1912, page 12. Source: http://earlyradiohistory.us/1912gern.htm

WIRELESS TELEGRAPH

The "Telimco" Complete Outfit, comprising 1 inch Spark Coil, Strap Key, Sender, Sensitive Relay, Coherer, with Automatic Decoherer and Sounder, 4 Ex. Strong Dry Cells, all necessary wiring, including send and catch wires, with full instructions and diagrams, $8.50. Guaranteed to work up to one mile. Send for Illust. Pamphlet & 64-page catalogue.

ELECTRO IMPORTING CO., 32 Park Place, New York

Figure 255: Advertisement for wireless apparatus from Telimco (1905).

Source: http://earlyradiohistory.us/1905teli.htm

The amateur activities became so intense that the 'interference problem' of spark-based wireless telegraphy interfered with the commercial and governmental stations. This soon raised problems and initiated legislation. 'In the early days of radio, many US amateurs operated with skill and efficiency, but a few others did not, and in this unregulated era they were a nuisance to both commercial stations and fellow amateurs[408].'

In 1912 it resulted in the American *Act to Regulate Radio Communication* (the *Radio Act* of 1912), which made having a license a requirement for operating wireless equipment. But although the new regulations restricted amateur activities, it also forced them to become more disciplined and proficient. Soon the amateurs organized and created the *American Amateur Radio League* and the *National Amateur Wireless Association* with Guglielmo Marconi on its board. Following the start of World War I in Europe in August, 1914, US radio amateurs watched with special interest as to whether the United States would be drawn into the conflict, due to the fact that the 1912 *Radio Act* gave the president permission to shut down radio stations 'in time of war'. That did not happen, as America stayed neutral for the first two and a half years.

However, all amateur and commercial use of radio came to an abrupt halt on April 7, 1917, when, with the entrance of the United States into World War I, most private and commercial US radio stations were ordered by the president to either shut down or be taken over by the government. For the duration of the war it became illegal for private US citizens to even

---

[408] Source: http://earlyradiohistory.us/sec012.htm#part010

have an operational radio transmitter or receiver. Then, in 1917, Guglielmo Marconi, as president of the Association, proclaimed:

> *My word to the amateurs of America is: Begin at once some form of military training. Begin with essentials, and later take up the study of map reading and observation; it will help wonderfully in increasing wartime efficiency and will be invaluable to those subject to draft.*[409]

Also stimulated by the press coverage of Marconi's experiments, soon, in Britain, the same phenomenon of amateur radio took place. In January 1898, in the British hobby magazine *The Model Engineer and Amateur Electrician*, Leslie Miller published an article for experimenters called 'The New Wireless Telegraphy'. The first amateur radio station was created by M.J.D Dennis in London 1898. As amateur activity grew in Britain, regulation was also needed. That was the task of the British Post Office, which—acting in the 'public interest'—got more and more control over wireless telegraphy, thereby frustrating Marconi's plans (and others, like the Lodge-Muirhead Syndicate) for offering wireless telegraphic services in Britain. By 1904 the *Wireless Telegraph Act* made a license a requirement, and some 60 amateurs were licensed for experimental activities.

By 1910 the 'call signs' were introduced, and license holders got a three-letter identification they used to identify their station. In spring 1911 the *Wireless Club* in Derby, near London, was founded. This was followed by the *Wireless Society of London*, which was created in 1913 and had some 200 members by 1914. However, as the ban on amateur radio became active at the outbreak of the war in 1914, it silenced amateur activities during wartime. After the war of the trenches came to an end, British hams were encouraged to reapply for their licenses. Again, fitting in the British tradition, this was a controlled activity, all in the name of 'public interest'.[410]

> *Obtaining an amateur radio license in 1919 required proof of British citizenship, two written references and an agreement to observe the secret and confidential nature of messages. Licensing at the time was actually quite restrictive, as British hams were limited to a maximum of five contacts and such contacts required authorization by the government. Power was limited to ten watts. ... However, to receive the coveted "radiation licence," it was necessary to demonstrate to authorities the scientific value of one's experimentation or overall benefit to the general public.*[411]

---

[409] Marconi G., 'Send the Wireless Men Abroad Immediately'. *Wireless Age*, September, 1917, pp. 888–891. Source: http://earlyradiohistory.us/1917send.htm.

[410] Source: www.scribd.com/document/258114571/The-History-of-Amateur-Radio.

[411] Source: http://www.dxcoffee.com/eng/2013/radio-society-great-britain-100-years-dedication-amateur-radio/

However, whatever the specific national situation, the activities of all these amateur radio experimenters, the tinkerers of the Wireless Age, contributed highly to the development of a new application for the 'wireless': the radio. With their stations, experimenting in the broadcasting of music programs, the new concept of *point-to-many communication* was born.

The Invention of Radio Broadcasting

From the early times of (cabled) telegraphy, the development of electric speech had added (cabled) telephony to telecommunications. As a result, along with coded text messages in a *point-to-point system*, people could transmit the spoken word and sound from one person to the other. In addition to these point-to-point communication engines called 'telegraph' and 'telephone', the cabled infrastructure was used to transmit information from one point to multiple receivers: the *point-to-many communication*, also called *broadcasting*. Early examples were the printing telegraph to distribute news items to newspapers (eg the Reuters Telegram Company of 1851) and the stock market ticker (as developed by Thomas Edison in 1867). It is not too surprising that soon the cabled communication infrastructure also would be used to distribute the spoken word from one transmitter to many receivers: such as the Telefon Hirmondo broadcasting music and spoken word, which started in Budapest (1893).[412] By 1890, in France, the Théâtrophone system of Paris allowed the subscribers to listen to opera and theatre performances over the telephone lines. In Britain in the late 1890s, the Electrophone, using conventional telephone lines, was used to distribute live theatre, music hall shows, and Sunday church services.

But the arrival of wireless communication would change all those services. As we have seen, the cable-less transmission—also called wireless communication—was the result of the creative efforts of thinkers and tinkerers. Electricity, traveling with the speed of light over a copper cable as carrier of the information, was now replaced by the lightning-fast electromagnetic waves. Not too surprisingly, the early efforts for the application of a wireless system had focussed on the telegraphic services over land, replacing the existing telegraph system. But that replacement strategy proved not so easy, as the telegraph establishment in Britain and America defended their monopolies vigorously. Thus, instead of fighting an uphill battle, the search for application of wireless communication focussed on a terra incognita. And that was the application area that could not be served by cabled communication system: the *marine wireless communication* with its ship-to-shore and ship-to-ship communication. They built their spark-based wireless systems to transmit coded information from

---

[412] See: B.J.G. van der Kooij, *The Invention of the Communication Engine 'Telephone'* (2015).

lighthouse to ship and between ships themselves, solving the incommunicado problem. This application was massive, both in the civil Merchant Marine and the Navy.[413] It started small with a transmitter on the shore serving a local area, but soon was serving longer distances, crossing the Channel and even the Atlantic Ocean. Radiotelegraphy, within a decade, conquered the world.

Not too surprisingly, considering the success of Alexander Graham Bell's telephone, their attention soon also focussed on transmitting the spoken word by wireless means. Or, just as wireless telegraphy had become known as 'radiotelegraphy', they focussed on 'radiotelephony', 'radio' being the word to indicate the use of electromagnetic waves as the carrier of information. However, before the early forms of radiotelephony could emerge, something had to replace the crude method of spark-tuned transmission. The solution was to be found in the creation of the single-frequency electric waves. And that needed the invention of single-wave transmission and the hardware that went with it.

The use of point-to-many-communication as 'radio broadcasting', with a single radio station transmitting speech and music to multiple listeners, started with early experimentation in 1906 by people like Reginald Fessenden[414]:

> *At 9 p.m. on Christmas Eve, 1906, he beamed a "Christmas Concert" to ships of the United Fruit Company, broadcasting from 125-metre towers at Brant Rock, Massachusetts. Astonished crews heard Mr. Fessenden give a speech about the concert, followed by a phonograph recording of Handel's Largo, making Mr. Fessenden, in another first, the earliest of all disc jockeys. Amateur violinist Fessenden then scratched his way through the carol, O Holy Night, with his own vocal obbligato, which won him double bragging rights to radio's first live vocal and first live instrumental performances. The United Fruit Company ships heard the concert, but so did sister vessels all along the Atlantic coast. The program was repeated on New Year's Eve.*[415]

But it needed a technology other than the spark-generated electric waves: the real single-wave transmission that could be modulated. In addition, for audio broadcasts to be possible, electronic detection and amplification devices had to be incorporated, such as de Forest's triode vacuum tube. It would take into 1920 before valve technology matured to the point where radio broadcasting was becoming viable.

---

[413] The merchant marine consists of privately or publicly owned commercial ships of a nation. The navy consists of a nation's ships of war and their support.

[414] O'Neal, J.E., 'Fessenden, the Next Chapter'. Source: http://www.radioworld.com/columns-and-views/0004/fessenden--the-next-chapter/300426 . (accessed April 2017)

[415] One against the World. https://www.ieee.ca/millennium/radio/radio_about.html

Single-wave Transmission

The original wireless telegraph was—as we explained exhaustively before—spark-based, creating long and short sparks—with DC electricity from a battery—resulting in a stream of bursts of electric waves to transmit the Morse-coded transmission. It was a rather crude method that had many drawbacks, both technically and in terms of application. Despite early improvements that made the spark-based telegraphy quite usable, it had its limitations, as it used a lot of bandwidth, and that was something that was going to become scarce.

Over time the spark-based system was constantly improved upon, and after the application of condensators (C), inductive coils (L) and resistances (R), a more efficient single frequency wireless transmitter (aka the *arc oscillator* and *arc convertor*) could be created. These LCR oscillator circuits operated at their resonant frequency in a tuned circuit. Next, the creation of single-frequency waves was made possible by using the mechanical 'alternators'. The single wave, which could be switched on and off, created improved wireless communication.

That—in short—was the initial system of the spark-tuned wireless telegraph. But the signs of new technological improvements were already visible when Reginald Fessenden—looking at electric waves from a different background—created his *continuous wave transmitter* based on an electric dynamo (aka generator, alternator).

> He [Fessenden] *came to wireless telegraphy, not by way of spark gaps as Hertz and Marconi did, and not from curiosity about lightning conductors and resonant circuits as Lodge did, but from working with electric motors, transformers, and dynamos—which is to say, from power engineering.* (Aitken, 2014a, p. 50)

As we met him already and described his earlier contributions to wireless telegraphy, we now will focus on his contributions to 'radio'.

The Heterodyne Principle of (De)Modulation

So, Fessenden had built his high-frequency alternator to create a continuous wave of a single frequency. And the use of alternating current proved to be essential for the further progress of wireless transmission. In addition to that, he added something completely new to the transmission of electric waves: the heterodyne principle of mixing two frequencies of alternating currents.[416] It resulted in the new art of 'amplitude modulation' (Figure 143).

---

[416] The Greek word heterodyne means the mixing of two forces.

Fessenden's contribution was to apply it [the mixing of two frequencies] *at radio frequencies, and no one had done that before. He was perhaps fortunate to find judges who would uphold it is something that could be patented—who would acclaim it, indeed, as "invention of a high order ... a new contribution to the knowledge of the time." ...*

The heterodyne receiver was an invention that grew in importance as radio technology and the radio industry developed. At the time of discovery, despite Fessenden's enthusiasm, it had little impact. There were two reasons for this. First and obviously, there were no continuous wave transmitters on the air in 1900–1901 and a few in the next decade and a half. The period from 1900 to 1915 was, indeed, the golden age of the spark transmitter, ... The second reason [was that] *the heterodyne receiver was a brilliant conceptual breakthrough, but one made before the hardware was at hand to implement it properly. The situation was to change drastically in 1912, with the invention of the vacuum tube oscillator; but in 1901 that was neither available nor contemplated.* (Aitken, 2014a, pp. 58, 59)

But before that was going to happen, another trajectory of development was followed, one that would create the foundations for the later semiconductor technologies.

Crystal Detectors

One of the problems of the sparked telegraphy was the detection of the weak signal. First, the not-too-sensitive coherer was used for detecting the electric waves, but many efforts were undertaken to create better-performing devices, such as Fessenden's electrolytic detector (his so-called 'barretter') (Figure 256, top) and thermal detector (the hot-wire barretter), and Marconi's reliable receiver 'Maggie'.

Figure 256: Fessenden's electrolytic detector (1902, top) and cat's whisker crystal detector.

Source: Wikimedia Commons (top), http://www.demajo.net/museu m /page4.htm

After the experimenting with crystals (such as the semiconductor carborundum[417], SiC) by Ferdinand Braun in the 1870s, others created a new type of rectifier: the cat's whisker detector (Figure 256, bottom). Among them was G.W.

---

[417] Carborundum was originally used in lightning devices, as it had a voltage-dependent resistance.

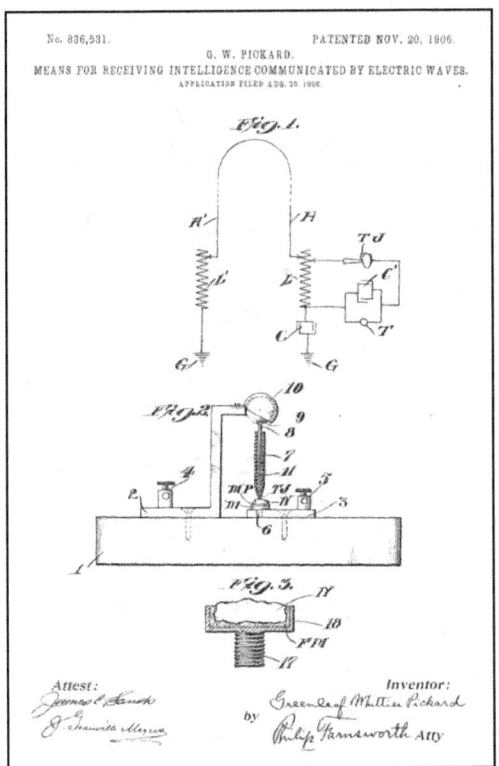

No. 836,531.   PATENTED NOV. 20, 1906.
G. W. PICKARD.
MEANS FOR RECEIVING INTELLIGENCE COMMUNICATED BY ELECTRIC WAVES.
APPLICATION FILED AUG. 30, 1906.

Figure 257: Pickard's Patent 836,531 (1906).

This figure shows the thermos junction TJ where the diode effect is realized.

Source: USPTO.

Pickard, who was granted US Patent № 836,531 on November 20, 1906 (Figure 257). It applied the diode effect; the current would flow only in one direction from a thin metallic wire (the 'cat's whisker') to the carborundum. Other experimenters created similar rectifiers, such as Henry Dunwoody, who was granted US Patent № 837,616 on December 4, 1906.

Interestingly, it was H.J. Round, an engineer consultant employed by Marconi in 1902, who performed a number of experiments on the crystal detector using a range of different materials. He also applied a direct current to them and noticed that some actually emitted light. H.J. Round reported this in the February 9, 1907, edition of *Electrical World*. This is the first known report of the effect of the light-emitting diode, but it would take until the 1960s before that trajectory took off.

However, the receivers built with the cat's whisker were not that successful, as it was very sensitive to the exact geometry and pressure of contact between wire and crystal. It needed constant adjustment. Soon the vacuum tube diode would realize the same effect in a much more reliable way, and the crystal decoder would fade away as other technologies came available that created more sensitive receivers that could convert the received AC signal to a DC signal. But the use of crystalline materials, aka semiconductors, heralded the semiconductor electronics to come.

Fleming's Thermionic Valve

It is interesting to see how a supporting technology of the making of glass vessels (as use to create incandescent lamps), influenced the development of wireless communication. Originally, electric waves were detected by a coherer, a product of lightning research. The coherer was a glass vessel which contained metallic scraps (eg carborundum) which changed in conductivity under the influence of a burst of electro-magnetic waves. The next big step in increasing the sensibility of the detector was another component encapsulated in a glass vessel: the thermionic valve that John Ambrose Fleming developed in 1904 (Figure 258).

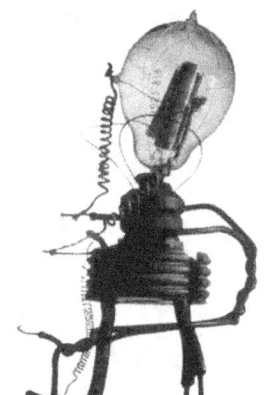

Figure 258: Fleming valve (1904).

Source: http://collectionsonline.nmsi.ac.uk/

John Ambrose Fleming (1849–1945), the eldest son of seven born to a Congregational minister, was educated in London. Working for a living at a stock-jobbing company in the London Stock Exchange, he studied in the evening. Continuing his job as a teacher, he managed to further his education. Having been corresponding with Professor James Clark Maxwell at the University of Cambridge, and having earned enough money, he entered Cambridge in 1877, at the age of 28, to study under Maxwell. After graduating and getting his doctorate he became professor of physics and mathematics at University College Nottingham (1882). Next, he took up a position as a consultant to the Edison Telephone Company. By 1885 he became professor of the newly formed electrical department at University College London, the premier college of London University, a chair he would hold for 41 years. By 1899 Fleming had become scientific advisor to Guglielmo Marconi.

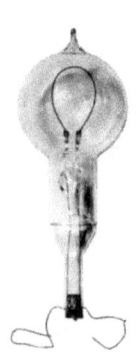

Figure 259: Edison's incandescent lightbulb (1879).

Source:
www.edison.rutgers.edu/company.htm

Fleming had previously become involved with wireless through Maxwell who had devised his electromagnetic theory that proved the existence of these [Hertzian] *waves. Now Fleming was involved in moving this further forward. As such he was aware of the limitations preventing further progress. He was acutely*

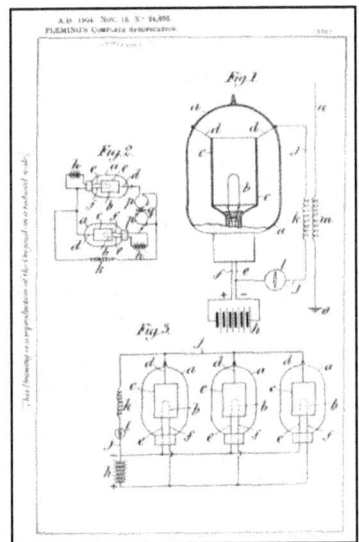

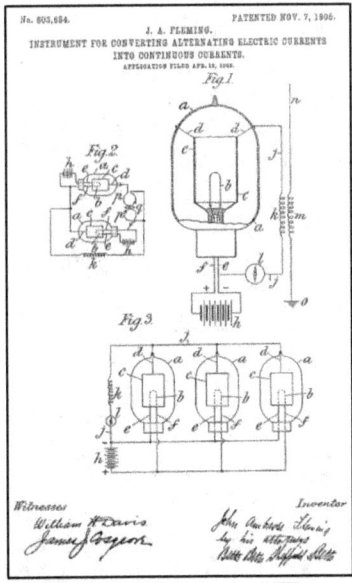

Figure 260: Fleming's UK Patent 24,850 (1904, top) and US Patent 803,684 (1905, bottom).

Source: USPTO (bottom).

aware of the insensitivity of the detectors. The coherer that was the most common way of detecting signals was not particularly sensitive, although it had been refined to a remarkable degree. It was now that Fleming recalled the Edison Effect he had seen many years earlier in the USA. Although he had experimented with it in 1883 and later in 1896 and had special lamps made he could find no use for it. Others had also seen that the effect and noted that it could be used to rectify alternating currents, but it was Fleming in a stroke of genius who realised that it could be used for detecting radio signals.[418]

His experimenting resulted in the 'thermionic valve', which originated from an earlier development in the days of another famous glass-based vessel: the incandescent lamp. Invented in 1879 by Thomas Edison (Figure 259) at the end of the nineteenth century, the further improvement of the incandescent lamp was in full steam[419].

During the initial development of the incandescent lamp Thomas Edison had observed that the inside of the bulb blackened unevenly. This was the result of a phenomenon that was soon to be called the 'Edison effect': the flow of electrically charged particles (called thermions in those days, today electrons) from the hot glowing carbon filament to the glass surface. Edison built several experimental lamp bulbs with an extra wire, metal plate, or foil inside the bulb that was separate from the filament and thus

---

[418] Source: Poole, I., 'Ambrose Fleming's Oscillation Valve'.
https://www.electronics-notes.com/articles/history/pioneers/ambrose-fleming-oscillation-valve.php
[419] See: B.J.G. van der Kooij, *The Invention of the Electric Light* (2015), pp. 135-152.

could serve as an electrode. And he observed that the electric current between filament (cathode) and metal plate (anode) would depend on the polarity of the battery. This effect of conductivity depending on the polarity would become the 'diode effect'. Edison was granted US Patent № 307,031 on November 15, 1883.

Thus, Fleming, by then no longer working as scientific advisor for the Marconi Wireless Telegraph Company,[420] remembered the Edison effect and discovered that it could be used for the rectification of an AC current into a DC current. His thermionic valve (Figure 258) was an 'instrument for converting alternating electric currents into continuous currents', a device that could be used to rectify the weak AC signal of the wireless receiver into a DC current: he had created the vacuum rectifier (aka 'diode'). As he described it later:

The next step of invention was made in 1904, when I placed around the filament of an electric lamp a cylinder of metal sealed inside the bulb, and found that a current of negative electricity could be sent from the filament to the cylinder but not in the opposite direction. This at once gave us a means for converting the feeble but rapid to-and-fro motions of electricity in an aerial wire, which are set up when electric waves from a distant transmitting station strike the aerial wires, into a current of electricity all in the same direction by including in the circuit such a lamp with a cylinder round the filament. I therefore called the instrument a valve because it acts, as regards electricity, as a valve in a pump acts for air or water. It is now called a thermionic valve. (Dylla & Corneliussen, 2005, p. 9).

Figure 261: Marconi's radiotelephone set using a vacuum tube (1914).

Source: Wikimedia Commons.

Fleming had what he described as a "sudden very happy thought" one day when he realised the possibilities. So in October 1904 he asked his assistant, G.B. Dyke to put the idea to the test, and it worked. Thus it was that on one day in November 1904,

---

[420] The Maskelyne Affair and Fleming's public reaction tarnished the cooperation between Marconi and Fleming. It also hurt Fleming's reputation and credibility as an independent scientist. When, in December 1903, his three-year contract as scientific advisor expired, Marconi did not renew it.

Fleming was seen "scudding down Gower Street" in London, oblivious to all around him on his way to patent the idea.[421]

With his invention he was able to detect high-frequency radio signals. He was granted GB Patent № 24,850 on November 16, 1904. In the US, Fleming applied for a patent on April 19, 1905, and was granted US Patent № 803,684 on November 7, 1905. His 35 claims(!) for both the wireless transmitter and receiver were based on

...the combination of a vacuous vessel, two conductors adjacent to but not touching each other in the vessel, means for heating one of the conductors, and a circuit outside the vessel connecting the two conductors. (text US patent).

On November 7, 1905, he also obtained US Patent № 804,190 for a 'cymometer': an instrument to make electrical measurements of the wavelength and frequencies employed in Hertzian waves.

Shortly after his discovery Fleming wrote to Marconi to tell him of his discovery. In the letter he mentioned that he had not mentioned the idea to anyone as he thought it might be very useful. Little did he know of its importance, although he entrusted the patent to the Marconi company. However it did not bring any money to them. Any returns from the invention made were used in fighting the legal battles that were to arise later. [422]

The thermionic valve also managed to re-establish the relationship between Fleming and Marconi, as in May 1905 he was rehired as scientific advisor to the Marconi Company. It was Marconi who started experimenting with the valve and transformed it into a sensitive detector for wireless telegraphy (Hong, 2001, p. 148).

As soon as it became clear to Marconi that the thermionic valve was improving the sensitivity of the receiver (Figure 142, middle), it was adapted into his systems (Figure 261). It was Henry Round, consulting engineer for the Marconi Company, who designed them. The tubes—handmade in small

Figure 262: Marconi's Round Valve, Type N (1914–1915).

Source: http://www.r-type.org/exhib/aac0042.htm

---

[421] Source: Poole, I., 'Ambrose Fleming's Oscillation Valve'.
https://www.electronics-notes.com/articles/history/pioneers/ambrose-fleming-oscillation-valve.php
[422] Source: Poole, I., 'Ambrose Fleming's Oscillation Valve'.
https://www.electronics-notes.com/articles/history/pioneers/ambrose-fleming-oscillation-valve.php

quantities—were either manufactured by the Edison Swan Electric Company (Ediswan) or by the Osram-Roberson Lamp Works of the British General Electric Company (GEC) (Figure 262). This was a propriety design for use in Marconi equipment, and no patents were applied for.

Around 1910 Captain Round began work to develop wireless telephone (voice) communication using triode valves. Marconi's had been using Fleming diodes for some years. Similar work was under way in Germany (of which Round was aware) and in the US. However, it is not clear what, if anything, Round knew of the American work. By 1912 he had not only designed a sea-going table-top transceiver giving good voice communication over ranges up to 50 km, plus the valves to go in them (2 types) but had successfully organised substantial production batches of these.[423]

The Fleming valve proved to be the start of a technological revolution based on vacuum tubes. The ramifications of the thermionic valve were myriad and far-reaching. It was a key component of radios for nearly three decades until it was replaced by the transistor, and was integral to the development of television, telephones, and even early computers. Fleming's discovery would lead to a range of infringement cases, among which was the one based on the invention Lee de Forest patented in 1908.

Lee de Forest's Audion

Figure 263: De Forest's Audion triode bulb (1908).

Source: Wikimedia Commons

It was the former PhD student writing a dissertation on Hertzian waves, Lee De Forest —'Reflection of Hertzian Waves from the Ends of Parallel Wires'—who created, after a tumultuous career as inventor-entrepreneur, a device that would turn the tables in wireless communication: it was called the Audion.

Lee de Forest (1873–1961), the son of a Congregational Church minister, was a descendent of Walloon Huguenots who'd fled Europe in the seventeenth century due to religious persecution. Lee, after finishing his education, was fascinated by Marconi's work on wireless telegraphy and even applied (unsuccessfully) for a job with Marconi. For years he worked on several devices related to wireless receivers. He became involved in entrepreneurial activities with small working

---

[423] Source: Vyse, B., Jessop, G., 'The Saga of the Marconi-Osram Valve: the Round Valves'. Sources: http://www.r-type.org/articles/art-014.htm (accessed April 2017).

companies and ultimately created, with the backing of Abraham White, the *American De Forest Wireless Telegraph Company* in 1902.

The company's most important early contract was the construction, in 1905 and 1906, of five high-powered radiotelegraph stations for the US Navy, located in Panama, Pensacola, and Key West, Florida, Guantanamo, Cuba, and Puerto Rico. It also installed shore stations along the Atlantic Coast and Great Lakes, and equipped shipboard stations. But the main focus was selling stock at increasingly inflated prices (during the wireless bubble described before), spurred by the construction of promotional inland stations. After conflict with the company's management, de Forest was dismissed, and he organized in 1906 the *Radio Telephone Company*. This company was one of the pioneers in promoting full-audio radio transmissions, in contrast to the Morse-code telegraphy that had dominated the airwaves to date. However, it was ahead of its time and failed (going bankrupt in 1911).

Trying to find a solution for the insensible wireless receivers available at that time, de Forest experimented with flame detectors (his later US Patent № 979,275). Then, like Fleming, he started to develop glass-tube devices using two electrodes. On January 15, 1907, he was granted US Patent № 841,387 for a 'Device for Amplifying Feeble Electrical Currents'. (Figure 264). This device acted as a diode (Figure 1 in patent № 841,387).

Further experimenting with a range of different configurations of a third electrode (Figure 2 in patent № 841,387)— placing it between the plate and the filament— resulted in a breakthrough.

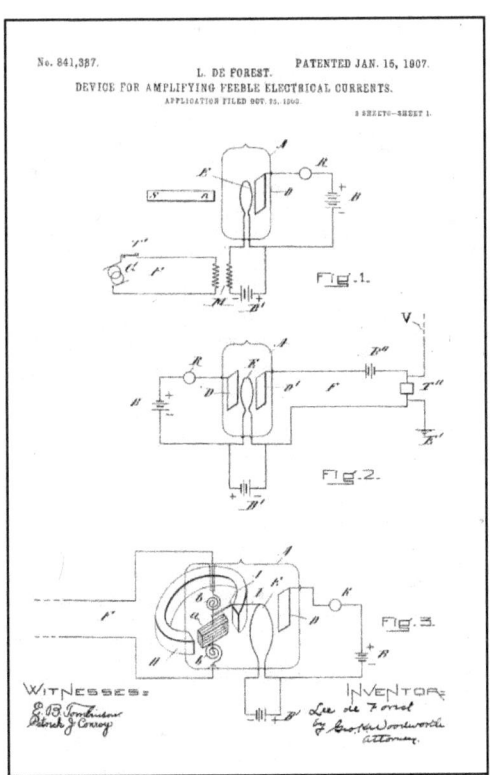

Figure 264: Lee de Forest's US Patent № 841,387 (1907).

Source: USPTO.

It was this third electrode (aka grid) that created a triode, and that would later show some unintended properties. On February 18, 1908, he was granted US Patent № 879,532 for the device (Figure 265). It seemed that he'd just added a third electrode to a device that had already been created, and patented by Ambrose Fleming: his thermionic valve. However, the device proved to have amplifying capabilities, and that caused it to become an *active* electronic component that would have a major impact on electronic technologies to come.

One indication of its huge impact was the great uproar it caused in the scientific community, as many saw it as an infringement on Fleming's invention. This was a view de Forest contradicted by playing down Fleming's work:

> *De Forest tried to undermine Fleming's earlier contribution as much as he could. He asserted that Fleming's work on unilateral conductivity* [the diode function] *had been anticipated by the German scientists Julius Elster and Hans Geitel. He argued that the use of Fleming's valve was confined to "quantitative measurements over short distance." Concluding that "the value of [Fleming's valve] as a wireless telegraph receiver is nil," he argued that the Audion was "tremendously more sensitive and available in practical wireless" than the valve.* (Hong, 2001, p. 149)

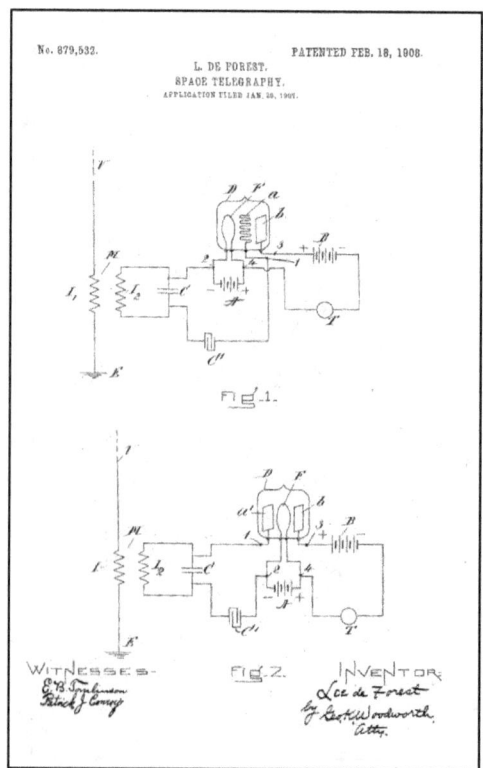

Figure 265: Lee de Forest's US Patent № 879,532 (1908).

Source: USPTO.

This started a lifelong animosity between de Forest and Fleming. After a polemic in professional publications, the controversy was brought to court. However, Fleming's own suit for patent infringement as to the Audion tube's thermionic technology failed. Even more, in 1943 the American Supreme Court, in the case *Marconi Wireless Telegraph Company of America vs US Government*, ruled that Fleming had made an improper disclaimer. Therefore, because US patent law holds that an invalid disclaimer automatically invalidates the patent to which it refers, Fleming's patent was declared to be invalid.[424]

De Forest had, in the beginning of 1907, started a new company called *DeForest Radio Telephone Company* to sell the grid Audion. Manufacturing was done by the McCandless Company that produced between 200 to 600 vacuum tubes a year for him. The devices sold for $5–8. All in all, it was not that much of a business success nor a technical success, as it did not create a more sensitive receiver at all (Hong, 2001, p. 181).

It took another creative mind, Edwin H. Armstrong, to discover that: 'The action of the Audion as a detector of radio frequency oscillations is very different from its action as a simple amplifier.'[425] This amplifying property of the Audion, when configured in a proper circuit with positive feedback, made it possible to use it both as a transmitter of a continuous wave (Figure 266, top) and as a receiver (Figure 266, bottom).

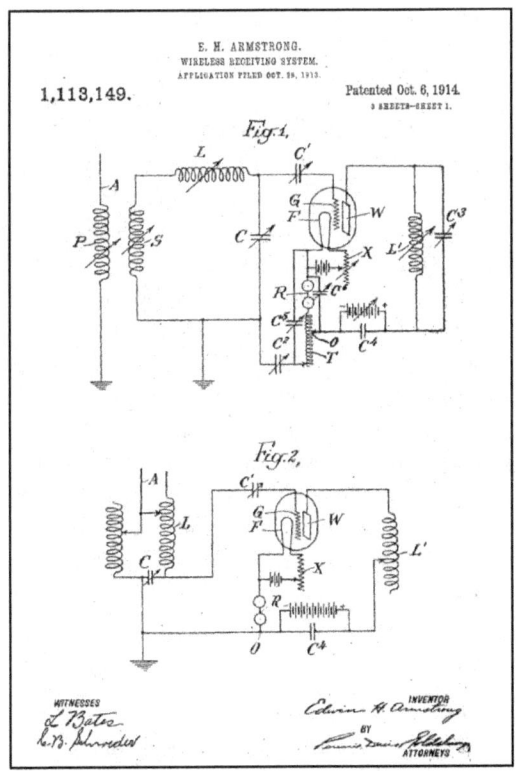

Figure 266: Armstrong's Patent № 1,113,149 (1914).

Source: USPTO.

[424] Source: 'Fleming, John Ambrose'. *Encyclopedia of World Biography*. Accessed 24 Feb. 2017. http://www.encyclopedia.com. Wunsch, A.D., 'Misreading the Supreme Court: A Puzzling Chapter in the History of Radio'. http://www.mercurians.org/1998_Fall/ misreading.htm
[425] Armstrong, E.A.. 'Some Recent Developments in the Audion Receiver: The Audion as Detector and Amplifier. Proceedings of the Institute of Radio Engineers', September, 1915, pp. 215–238: Source: http://earlyradiohistory.us/1915reg.htm

On October 6, 1914, US Patent № 1,113,149 was issued for his discovery of the regenerative circuit. Not too surprisingly, his work would also lead to a range of patent conflicts.

The invention of the triode vacuum tube made it possible to apply 'amplification' in circuits. Together with Armstrong's technique of mixing signals, now it became possible to modulate the wireless transmission frequency with another frequency. Amplitude modulation (AM) was born, and that modulation frequency could be music as well as the human voice (Figure 143).

> *The first acknowledged radio transmission of human voice was indeed made in 1915, but was done by the Western Electric Company, a subsidiary of AT&T, from the US Navy Station at Arlington, Virginia to the Eiffel Tower in Paris.* (Raboy, 2016, p. 383)

That was the birth of the wireless telephony that made the transmission of speech possible. The transmission of music would follow later. In Britain, for example, Marconi started experimenting at his Chelmsford factory. On June 15, 1920, the Australian soprano Dame Nellie Melba sang in a makeshift studio at the Chelmsford factory, using a microphone created with a telephone mouthpiece and wood from a cigar box. She opened her recital at 19:10 by singing 'Home Sweet Home' and, after other popular favourites and several encores, closed with the national anthem. Her voice, carried from an aerial with towering masts, was heard as far away as Iran and Newfoundland. Other Marconi stations were the receivers of the transmission. The devices we today call radios—the wireless receivers that could receive the AM signals carrying human speech and music—with their vacuum tubes, would not until sometime later become available to the general public. People could listen to the early broadcast of the BBC (created in 1922), but that is another story…

Prelude to the Great War (1914–1918)

By the 1910s the role of point-to-point wireless communication had changed. Marine wireless communication started to mature, both in equipment as well in application. Along with marine wireless communication, other applications started to emerge. Keeping up with the improvement of airplanes (the brothers Wright had improved their machine after its first powered flight in 1903), wireless telegraphy became airborne, and air-to-ground-communication developed. After additional testing and design of special equipment, military airplanes became 'wireless'-enabled with special wireless sets. These planes were to be used for reconnaissance and artillery spotting.

Other technological developments like the new devices created with the vacuum-tube technology, Flemings' tube and De Forest's Audion, started to influence wireless communication. The first experimentation with wireless telephony was done and was soon followed by experimental broadcasting. Clearly, wireless technology, though still in its infancy, fitting in the Spirit of the Time,[426] was on the move, although the dark clouds of military conflict were already visible on the horizon. Wireless would become the invisible weapon of war.

Then started the Great War, triggered by the assassination of Archduke Franz Ferdinand of Austria on July 28, 1914 (Figure 267). The war was a nationalistic conflict that exploded when entangled international alliances formed over the previous decades were invoked. Within weeks, fuelled by the Madness of the Time, the major powers were at war.[427] The conflict spread like a wildfire over Europe, as many countries had things left to settle from preceding times,[428] creating an Eastern Front and a Western Front. Being outnumbered on the Eastern Front, Russia urged its Triple Entente ally France to open up a second front in the west. In response, France mobilized, and consequently, Germany declared war on France.

Figure 267: The assassination of Archduke Franz Ferdinand of Austria (1914).

Source: Wikipedia Common. Artist: Achille Beltrame.

---

426 The topic of Spirit of the Time and the Madness of the Time have been described in the first part of this case study.

427 The Central Powers of Germany and Austria-Hungary were on one side, opposed by the Triple Entente of the Russian Empire, the French Third Republic, and the United Kingdom of Great Britain on the other side.

428 *French Revanchism* originated in the fact that France, being defeated in 1870, had lost the provinces of Alsace Lorraine to the newly formed German Empire (dominated by Prussia), and had to pay large war indemnities. By 1918 the French had their lost territory back and was receiving the war indemnities now being paid by the Germans.

By chance, in the same period of time, Marconi people were in Germany. After the legal battles against Telefunken that started in 1912, an agreement was reached to exchange patents, to accept each other's spheres of commercial influence, and to maintain personal contacts. The latter was realised when, on July 29, 1914, senior Marconi engineers were shown Telefunken's factories, research establishments, and its 200-kw high-frequency station at Nauen, near Berlin, as guests of the company. On July 30, the day after their visit to Telefunken, the Admiralty in London sent a wireless message to the Grand Fleet, then assembled for annual review at Spithead. Ships of the Royal Navy were to disperse to battle stations all over the world. On August 1, the Admiralty issued an order forbidding all but British ships to use wireless in British territorial waters. The following day, German troops entered France, and in Britain, the government took control of all wireless messages.[429]

After the Russians ordered general mobilization on July 30 the Germans responded on August 1 with their own mobilization. Britain entered the war on August 4, 1914, when First Lord of the Admiralty Winston Churchill sent a wireless message to the fleet to commence hostilities against Germany, and they did what had always been feared by the Germans: they cut the communication cables.

Figure 268: Trench warfare communication: (wireless) telegraphy.

Source: http://misscosta2.wixsite.com/yr9ww1/nature-of-warfare

Within hours, British ships cut German cables in the Atlantic Ocean and North Sea, leaving the four-thousand km wireless connection between Nauen[430] and Sayville, as the only German communication link to neutral United States.... Effectively, Nauen was German's only direct link to the rest of the world. (Raboy, 2016, p. 385)

---

[429] Source: 'WWI'. http://www.marconicalling.co.uk/ (assessed January 2016)

[430] Telefunken had erected in 1906 a high-powered long-distance wireless station in the city of Nauen, Brandenburg, Germany. By 1914 the installation had been upgraded with massive antennas.

By August 4, 1914, the German forces, circumventing the frontier fortifications between France and Germany, invaded Belgium to move for France from the north. After the *Battle at the Marne* (September 7–12, 1914) the confronting military forces became stuck in the trenches. The ensuing static trench warfare (Figure 268) became a conflict of barb wire and mines, death by gas poisoning, death by disease, and death by the improved weaponry. It cost millions of human lives,[431] as well as the lives of millions of animals. Animals like pigeons and dogs which were used for communication, and horses and mules[432] which were used for transportation.

After other nations like Italy and America joined the war in 1917, by November 1918 it ended with the *Armistice of Compiègne*. Germany was facing an internal revolution, the German Empire had collapsed, and the Kaiser had fled to the Netherlands,[433] and it took six months of negotiations to conclude the Treaty of Versailles. By that time the Russian government also had collapsed, the Russian February Revolution had started, and Tsar Nicolas I had abducted, ending the rule of the Romanov Dynasty. It took the Russian October Revolution to also end the lives of the former tsar and his family on July 17, 1918.

Marconi Becomes Involved Diplomatically

With most of Europe at war, for the time being Italy remained neutral, and Marconi's personal status was ambivalent. The importance of his company to Britain was in no such doubt, however. This would not only be the *First* World War, but also the first war involving wireless communication in every theatre, from beginning to end. Its application had a slow start, however.

> Again, in spite of the company's pre-war ground-to-air research, the first units of the Royal Flying Corps embarking for France possessed only one airborne spark transmitter and one ground-based receiver between them. Similarly, on landing in France, the soldiers of the British Expeditionary Force were supported by a mobile wireless unit consisting of a single lorry carrying one receiver and

---

[431] The Battle of the Somme in 1916 cost the British Army some 420,000 casualties. The French suffered another estimated 200,000 casualties, and the Germans an estimated 500,000. This was just one of the many battles fought (eg Battle of Verdun).
[432] Source: Shaw, M., 'Animals and War'. http://www.bl.uk/world-war-one/articles/animals-and-war#, http://www.mirror.co.uk/news/real-life-stories/9-million-unsung-heroes-ww1-3939895 (assessed January 2016)
[433] Being a relative of the Dutch Queen Wilhelmina, he took up residence in Doorn, where he had acquired a mansion. He started living there on May 15, 1920, having brought some five wagonloads of furniture with him. He died there on June 4, 1941.

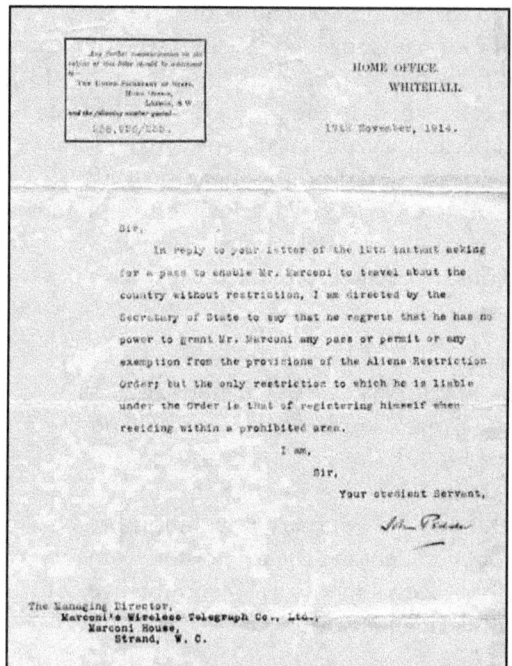

Figure 269: Letter from the Home Office regarding Marconi's status in England (1914).

Source: http://www.marconicalling.co.uk/
museum/html/objects/ephemera/large_image/large_ima
ge-type_d__t04719.html

transmitter set. A month later, as the Battle of the Marne began in September 1914, they possessed just ten sets and, like the other armed services, had some rapid catching-up to do.[434]

For Guglielmo Marconi, the outbreak of the war would become a dramatic event, as—again—the fact that he was Italian (he had kept his Italian citizenship all these years) would be held against him. His movements—being a foreigner—to travel in England were restricted, as the Home Office saw no reason to exempt him from the Aliens Registration Order (Figure 269).[435]

At Buckingham Palace in July 1914, King George V awarded Guglielmo Marconi an honorary knighthood. This was an act of grateful recognition but also, perhaps, a tacit apology for the gratuitous harm done to his name by innuendoes of financial scandal two years before. Later that month, at Spithead, Marconi and his wife Bea lunched on one of the Royal Navy battleships anchored for the annual review. Next morning the ships were gone, and within days, as a citizen of Italy, still neutral in spite of its 'triple alliance' with Austria and Germany, Marconi became an 'alien' and the subject of suspicion. For a while, down by the Solent at Eaglehurst,[436] his country house, he was treated by locals

---

[434] Source: 'The Marconi Company at War (2)', http://www.marconicalling.co.uk/
[435] The 1914 *Aliens Registration Act* was rushed through Parliament in a single day on the eve of the First World War. It allowed stricter controls than before, including the power to make aliens over the age of 16 register with the police.
[436] In 1912 Marconi had leased one of the most beautiful places on the Hampshire coast, on the shores of the Solent looking towards Cowes on the Isle of Wright (where much of his early demonstrations had taken place in 1897).

as an enemy spy. He nonetheless remained there with his family until, in the
words of his daughter Degna, 'sanity reasserted itself'.[437]

On August 14, 1914, Marconi left England to travel by train through
France to Italy. Although still neutral, the newly formed kingdom was
anxious about how the conflict was going to develop. The bad experiences
with the domination of the Austrian Empire from over half a century ago
had not been forgotten by the Italians. Well informed through his travels
and with transnational connections all over the world, Marconi became
involved and became an influential figure in diplomatic circles.

The tinkering boy who had left Italy in 1896 came back to serve his
country in 1914 in a diplomatic role.

For the rest of his life he would be more or less actively engaged in diplomacy,
often acting as a bridge, a go-between, or an informed advisor to governments and
embassies on military and political questions. (Raboy, 2016, p. 387)

Back in Britain, the British Post Office ordered the wireless stations to
be sealed up and the aerials taken down. The airwaves were being
monitored for illegal transmissions. The Marconi Company, due to the
nature of its operations, was put under governmental control. Soon it
became more involved than just as an equipment manufacturer. Marconi
wireless operators were taken off merchant vessels and transferred to the
Royal Navy, and some additional 2,000 new wireless operators were trained
to operate communication equipment in the trenches (Figure 268). On the
home front, wireless direction finders were used to develop a defense
system. This 'Ashmore system'[438]—named after Major General E.B.
Ashmore—was to protect London from daylight bombing raids by
Zeppelin airships and formations of Gotha heavy bombers.

By 1914, as the European cauldron was simmering with the coming war,
the era of the Marconi monopoly was on the brink of dissolution. Wireless
communications had become too important to be left to the whims of one
large dominating enterprise. It also ended for Guglielmo Marconi a time in
which he'd created his 'wireless empire', as now he became involved in the
turmoil of the Great War, which caused the British to not really trust the
foreigner and to restrict his movements in Britain. They also looked with a
suspicious eye at the dealings of his company.

Despite the Marconi company's involvements in the British war effort and
Marconi's personal enthusiasm on the Italian front, unwelcome questions were

---

[437] 'Guglielmo Marconi's War (1)'. http://www.marconicalling.co.uk/ (assessed January 2016)
[438] The author was the proud owner of a small Dutch venture company he called *Ashmore*
Software BV (1986–2003), which developed and marketed personal-computer-based software
for tax applications.

being raised about the company's loyalty… The flurry of bad feelings contributed to further souring Marconi's taste for the British. The government's attitude towards the company during the war "was such as to disgust me", he later wrote, and he several times expressed a desire to resign. It was only the Urging of Isaacs and some other directors that prevailed upon him to remain. (Raboy, 2016, p. 412)

In this turmoil, the Italians lauded him, making him a member of the Italian Senate and promoting Lieutenant Marconi to captain. During the summer of 1915 Marconi was commissioned as a lieutenant of engineers and put in charge of organising the Italian 'Army Wireless Service'. Then, as a lieutenant commander, he was asked to do a similar job for the Italian Navy, which was suffering from the congestion of longwave wireless traffic in the Mediterranean. After the Americans entered the war in April 1917, the Italian government sent Marconi on a number of goodwill diplomatic visits to the USA and Britain. By the end of the war Marconi was more and more to be found in Bologna.

By the beginning of 1918, Marconi's diplomatic career was in full swing taking up most of his time. He was still chairman of the board and technical advisor of the London-based Marconi company, president of the Banca Italiana do Sconto, and overseeing a talented bevy of research engineers, but by far most of his energy was now going into helping to position Italy favourably in anticipation of the end of the war. (Raboy, 2016, p. 419)

His diplomatic career was not to last that long, as it was in Rome, however, when listening on earphones early in the morning of 11 November 1918 that Marconi heard the end of hostilities announced by Marshal Foch in a wireless message transmitted from the Eiffel Tower. The war had ended…[439]

Marconi's Later Years and Death[440]

After the Great War had ended, Marconi returned to the love of his life: tinkering—now with a research group of capable technicians—with his wireless systems and promoting it to the powerful people of the world. Marconi discovered that short waves, reflected off the ionosphere, offered a much better communications method requiring substantially lower power and more compact antenna systems and radio sets. He (finally) created for the British government the Imperial Wireless Scheme with the shortwave Beam system. He also guided his company through a financial crisis in

---

[439] Source: 'World Tour'. http://www.marconicalling.co.uk/ (assessed January 2016)
[440] For more biographic information: Baker, W. J., *A History of the Marconi Company 1874–1965* (1970). Marconi, D., *My Father, Marconi* (2001). Raboy, M., *Marconi: The Man Who Networked the World* (2016).

1924. Having again helped steer the company back into profit, Marconi resigned as chairman in July 1927 and became 'technical adviser' to the board.

Wireless by that time had become a world affair, and the Marconi Company was again confronted by the great powers: the British government and the international cable companies, among which was the British Eastern Telegraph Company.

> By 1927, cable had lost half of its international business to shortwave wireless, and the cable companies were claiming they might be forced into liquidation. … The cable companies had been floating the idea of a merger for some time, but there was no obvious advantage to the Marconi's. The company's new leadership, however, had no ties to the old ways of doing things; from a shareholder perspective, in fact, there were a number of potential benefits. The company was still in a precarious financial position; it could not count automatically on government support as new uses and markets for wireless developed; there was a lingering antipathy towards it in some political quarters; and the Eastern's deep coffers made the outcome of a protracted intercorporate fraught and unpredictable. (Raboy, 2016, pp. 515-516)

Mergers and Acquisitions were in the air, times had changed now that the old pioneers had gone, and nations were looking to create their national industries in the field of communication. The British government organized in 1928 the Imperial Wireless and Cable Conference. It was to be the end of the Marconi Wireless Telegraph Company. A new company, the *Imperial and International Communications Ltd* (IIC) under the holding company *Cable & Wireless Limited* (C&W) was created and MWTC was merged with the Eastern and Associated Companies.

> In April 1929, with terms now ratified by the dominions and the British government, a new monopoly organisation was formed, Cables & Wireless Ltd, shortly afterwards renamed Cable & Wireless Ltd. In effect no sooner had Marconi achieved his ambition of a worldwide wireless network than he-and his company-had been robbed of the prize.[441]

There were more changes in the air for Guglielmo Marconi personally. His marriage with Beatrice was winding down to a divorce, and he was increasingly focussing on Italy, getting involved in politics and diplomacy in a state that was dominated by the new fascist government. Marconi became a servant of the regime (Raboy, 2016, pp. 549-576).

---

[441] Source: 'Cable & Wireless'. http://www.marconicalling.co.uk/ (assessed January 2016)

Figure 270: Marconi's Funeral Procession on the Via Indipendenza, Rome (1937).

Source: http://www.radiomarconi.com/marconi/marconi_mausoleo.html

In September 1933 Marconi and his new bride, Maria Cristina Bezzi-Scali—the daughter of Vatican nobility, whom he had married in 1927—travelled around the world on a world tour, and they were feted everywhere they arrived. In Europe, Adolf Hitler and Benito Mussolini were coming into power, and the Second World War was looming on the horizon. With his health deteriorating, after several heart attacks between September 1934 and March 1935, Marconi died on July 20, 1937, following his nephew Henry Jameson Davis, who had died on Christmas Day 1936, to the grave. In the prelude to another war, it was the end of an era, with the two pioneers of the Marconi empire gone.

Marconi was given a state funeral in Italy on July 21, 1937, in Rome. The funeral procession was over a mile long and was witnessed by over half a million people (Figure 270). At 6.00 hrs, Roman time, telegraph and radio stations in Italy, Britain, the United States, and Canada went silent. Internationally, the press heralded the contributions of the ruler of the electric waves. Even the Vatican daily, *L'Observatore Romano*, paid attention to his passing (Raboy, 2016, pp. 653-655).

This also ends our analysis of the emergence of wireless telegraphy from Guglielmo's experimenting in the 1890s to the worldwide use of wireless in the 1910s. It had been two decades in which a young foreigner had to fight numerous battles to go from a 'nobody' to a 'somebody', not only fighting the scientific, technical, and commercial battles, but also the battles against the institutions (scientific, governmental, and military) of Britain. And the battles when he was opposed by the national interests of other countries like America and Germany. However, together with the numerous capable and motivated people working with him, he created the Marconi monopoly, controlling a large part of worldwide wireless communications till the Great War. Operating on different playing fields, he changed from the young engineering tinkerer with entrepreneurial aspirations into a recognized and negotiating diplomat. In the process, he became a multimillionaire, owning Villa Griffone and stocks in his companies.[442]

It is time to return to our satellite view. In order to observe Marconi's contribution in the perspective of the Communication Revolution, we will, in short, recall the preceding phases in the Age of Communication before we evaluate Marconi's contribution itself.[443]

---

[442] At his death, the estate declared in England was some £48,000. His assets in Italy totalled $60,000, and Villa Griffone was valued at some $50,000, equivalent to respectively £2.793 million and $1.8 million in 2015; calculation based on historic standard of living. Source: http://www.measuringworth.com/.

[443] See: B.J.G. van der Kooij, *The Invention of the Communication Engine 'Telegraph'* (2015); B.J.G. van der Kooij, *The Invention of the Communication Engine 'Telephone'* (2016).

The Communication Revolution (1837–1914)

The Communication Revolution is the series of technical events that created *telecommunication*: the transmission of information over long distances in different forms (code, speech, music) and by different means (cables, wireless). It resulted in the Age of Communication, a social development that accompanied the distinct phases of technological innovation.

Phases in the Communication Revolution

The first phase of the Communication Revolution was the invention of the communication engine called 'telegraph' (Morse; Cook and Wheatstone) in the early nineteenth century. The second phase was the invention of the communication engine called 'telephone' (Bell) in the second half of the nineteenth century. And the third phase was the invention of the 'wireless' communication engine (Marconi), at the end of the nineteenth century. Due to the societal impacts of each of these developments, this period in time often is called the Age of Communication, with its Era of Telegraphic Communication, Era of Telephonic Communication, and Era of Wireless Communication (Figure 271).

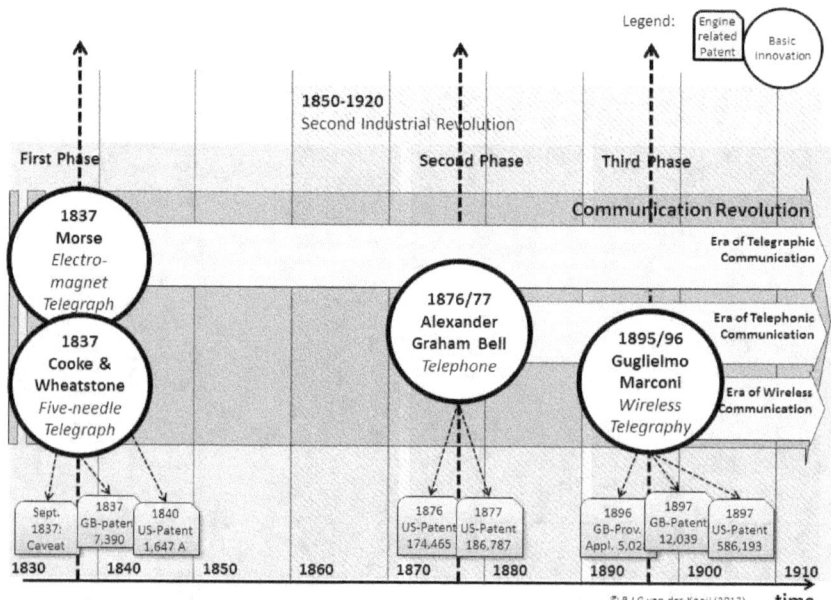

Figure 271: Overview of the phases of the Communication Revolution.
Figure created by author.

Phase 1: The Invention of Telegraphy

The invention of electric telegraphy occurred in a period of time when much of the Old World was in the aftermath of a period of considerable turmoil (see the context of the Age of Empires described before). It was one of the many inventions that originated from the general purpose technology of electricity that proved to be very pervasive in new application areas such as electric motors, the electric light, and electric communication. It was to become one of mechanisms that created the Second Industrial Revolution.

The electric telegraph was conceived in a world where a specific form of long-distance communication already existed. Chappe's optical telegraph (the semaphore) was well established in mainland Europe. The principle was simple: the positioning of semaphore panels (eg arms of a semaphore) transmitted a letter or sign. The new electric telegraph, however, operated on an entirely different technical principle than the optical telegraph and promised to be faster and more reliable. It was in no way a modification of the then existing visual telegraph technology, save perhaps the in use of codes to create a large corpus of relatable information from a relatively small base of possible signals, if that could be considered a technology. Nor was the electric telegraph really a modification of any other technology. Electric telegraphs used metal wires, galvanometers, needles, relays and voltaic piles (and, much later, generators and induction coils). The electric telegraph used existing technologies within its construction, but arguably it was not a modification of any of them. It was in its essence a new art (Bowman, 2005).[444]

> *The understanding that an electrical current through wire can affect magnets lead to a new group of prototype electrical telegraph systems, principally working through using the relayed current to move a magnetic needle or armature. This allowed for greater reliability in conveying messages, and greater simplicity in the design of electric telegraph systems; the pioneering patented systems being reliant upon knowledge of the principle. ... The invention of the electric telegraph was made possible by many scientific and technical developments in electricity in the years leading up to the 1930s and beyond. The earliest electrical experiments concerning signals over distance in the 1740s were considerably different from the broadly similarly intended experiments of the 1830s. Over that period of time*

---

[444] The word 'art' is here used as the work of an 'artisan' (craftsman), a meaning reflected in the expressions of 'state of the art', and 'prior art'. In the English Elizabethan Era rights were granted to create monopolies on specific artisan activities (ie new art): the Letters Patent. So patents could be given to new arts like the cutting of glass. A particular form of this later became the patent granting exclusive rights to an inventor. As the work of the artisan was based on his 'knowing how to make things' (ie technology), patents were related to material objects and how to make them (their specific technology).

new understandings of several of the processes involved in sending electricity through a wire: conduction, resistance, electromagnetism, current, charge and voltage to name a few, drastically altered the understanding of experimenters, scientists (including the recently minted Physicists) and the general public. Improved understanding of the nature of electricity allowed better inventions and innovations to be made within the field of electrical telegraphy, which sought to replace the existing proven but flawed system of long-distance communication. ...

Technological and scientific innovation, however, were not the sole mothers of invention. The experiments, trial-runs, funding (state and private), implementation and eventual institutionalization of electric telegraphy wouldn't have been possible without changes in public opinion, political consensus, or military thinking. Part of these changes were themselves brought about by the publication of scientific discoveries, but broader sociological and political changes in Europe and America certainly contributed towards the invention and development of the electric telegraph. (Bowman, 2005, pp. 8, 9)

The development of electric telegraphy found its core in a cluster of innovations to which many people contributed. Surprisingly, this took place on two continents, resulting in the contributions of Cook and Wheatstone in Britain, as well as Samuel Morse in America (Figure 272).

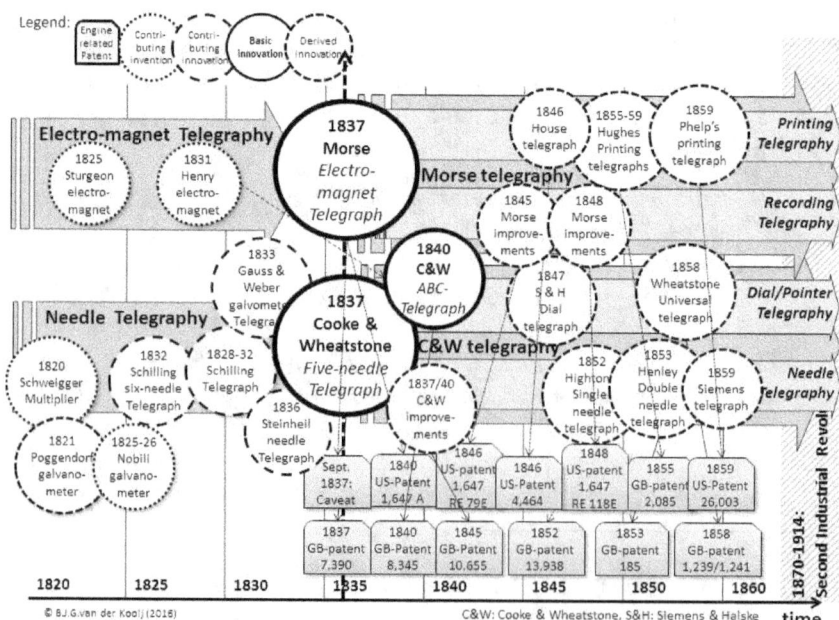

Figure 272: Phase one of the Communication Revolution: the cluster of innovations of telegraphy.

Figure created by author.

Phase 2: The Invention of Telephony

The invention of telephony was a logical expansion of electric telegraphy and built on its technology. However, it was realized thanks to the further evolution of electricity. From its conception by Volta in the early 1800s to the mid-1850s, electricity came in one flavor: the form called direct current, generated from cumbersome electrochemical batteries. By the 1860s another flavor had emerged: the alternating current. Now electricity generated by rotating electric dynamos came available in abundance. Where the telegraph was based on direct current (DC) from Voltaic batteries, electric speech was based on the rise of that alternating current (AC). The analogue signal of the human voice was transferred into the frequencies of the alternating currents, carried over the cabled infrastructure already in place. Technically speaking, it was the result of experiments to increase the capacity of the telegraph lines: the socalled harmonic telegraphy. However, with the development of sound telegraphy, all those efforts to transmit electric speech over copper lines coincided. And it was the contribution of a teacher of the deaf, Alexander Graham Bell, that would create the telephone (Figure 273). A very basic contribution, indeed, that was also considered to be a new art.

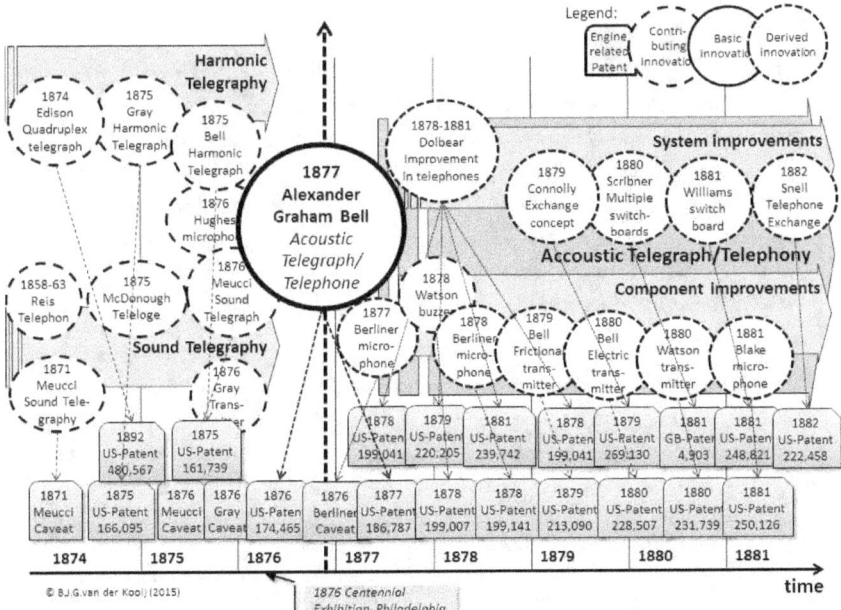

Figure 273: Phase two of the Communication Revolution: the cluster of innovations of telephony.

Figure created by author.

Thus it happened that when Bell invented the telephone, he surprised the world with a new idea. He had to make the thought as well as the thing. No Jules Verne or H. G. Wells had foreseen it. The author of the Arabian Nights fantasies had conceived of a flying carpet, but neither he nor anyone else had conceived of flying conversation. In all the literature of ancient days, there is not a line that will apply to the telephone, except possibly that expressive phrase in the Bible, "And there came a voice."

In these more privileged days, the telephone has come to be regarded as a commonplace fact of everyday life; and we are apt to forget that the wonder of it has become greater and not less; and that there are still honor and profit, plenty of both, to be won by the inventor and the scientist. The flood of electrical patents was never higher than now. There are literally more in a single month than the total number issued by the Patent Office up to 1859. The Bell System has three hundred experts who are paid to do nothing else but try out all new ideas and inventions; and before these words can pass into the printed book, new uses and new methods will have been discovered. There is therefore no immediate danger that the art of telephony will be less fascinating in the future than it has been in the past. (H.N. Casson, 1910, pp. 307-308)

Like telegraphy, telephony had a massive social impact. It influenced the way people did their business, it affected their social life, and it enhanced the newsgathering.

What we might call the telephonization of city life, for lack of a simpler word, has remarkably altered our manner of living from what it was in the days of Abraham Lincoln. It has enabled us to be more social and cooperative. It has literally abolished the isolation of separate families, and has made us members of one great family. It has become so truly an organ of the social body that by telephone we now enter into contracts, give evidence, try lawsuits, make speeches, propose marriage, confer degrees, appeal to voters, and do almost everything else that is a matter of speech. ...

Public officials, even in the United States, have been slow to change from the old-fashioned and more dignified use of written documents and uniformed messengers; but in the last ten years there has been a sweeping revolution in is respect. Government by telephone! is is a new idea that has already arrived in the more efficient departments of the Federal service. And as for the present Congress, that body has gone so far as to plan for a special system of its own, in both Houses, so that all official announcements may be heard by wire. ....

In news-gathering, too, much more an in railroading, the day of the telephone has arrived. The Boston Globe was the first paper to receive news by telephone. Later came the Washington Star, which had a wire strung to the Capitol, and thereby gained an hour over its competitors. To-day the evening papers receive most of their news over the wire a la Bell instead of a la Morse. This has resulted in a specialization of reporters --one man runs for the news and another man writes it. Some of the runners never come to the office. They receive their assignments by telephone, and their salaries by mail. There are even a few who are allowed to telephone their news directly to a swift linotype operator, who clicks it into type on his machine, without the scratch of a pencil. This, of course, is the ideal method of news-gathering, which is rarely possible. (H. N. Casson, 1910, p. 199)

Phase 3: The Invention of Wireless Telegraphy

And then came Marconi's contribution, as described in detail. Again it was a cluster of innovations (Figure 274) that was fuelled by contributions from scientific experimentation that had resulted in a range of components to create and receive electric waves. In addition, the new concept of point-to-point wireless telegraphy could build on earlier technologies that the engineering experience had accumulated during the earlier development of wired telegraphy and telephony.

Marconi's contribution would become the core of the third phase of the Communication Revolution: the time that saw the development of wireless telegraphy in particular, and the dawn of a spawning wireless

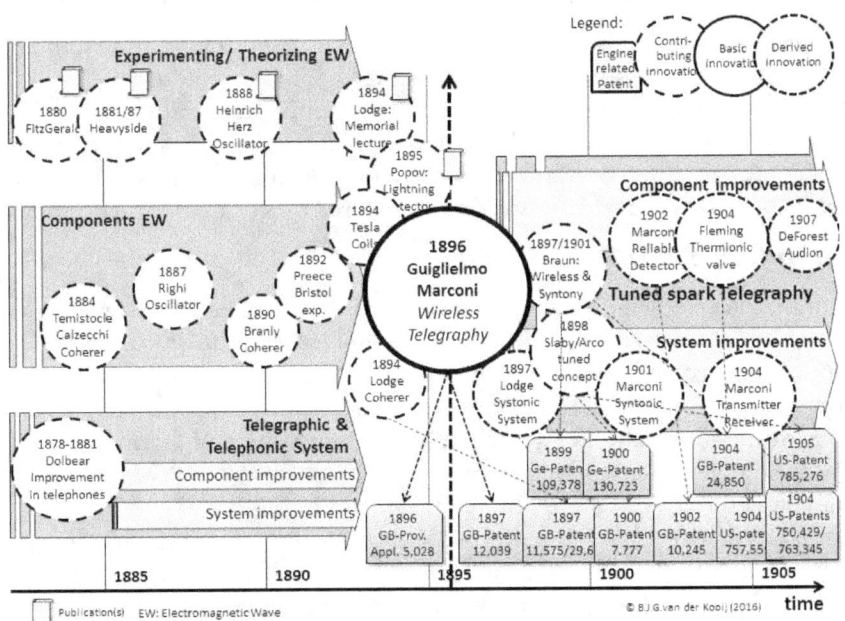

Figure 274: Phase three of the Communication Revolution: the cluster of innovations of wireless telegraphy.

Figure similar to preceding figure.

Figure created by author.

communication in general (aka radio broadcasting). It was part of a range of developments that changed how the world communicated.

Like telegraphy and telephony, wireless communication had—after some time—a massive social impact. It influenced the way people did their business, it affected their social lives, and it enhanced military warfare. But, that was not to happen until after the interruption of the Great War. In terms of business, Marconi's invention had not brought to him and his investors what the inventions of Morse's telegraph and Bell's telephone to their inventors and investors.

> *The overall conclusion has to be that without a complete monopoly, point to point radio operations* [as in marine wireless and long distance wireless] *were simply not a sustainable high-margin growth-business. The high growth business proved to be in broadcasting. Unfortunately, the principle of public ownership in the United Kingdom prevented Marconi from participating on one side of the Atlantic, and government regulation aided by RCA prohibited it on the other. Investors in Marconi therefore had the unfortunate outcome of being largely excluded from an industry the company helped to create.* (Nairn, 2002, p. 266)

Rise and Fall of an Empire: A Revolution in Just Two Decades

Over a range of decades starting in the 1840s with Morse's contribution, telecommunication matured in a chain of developments that had a slow start and that would have to wait till Bell's contribution in the 1870s. Then, as the GPT-Electricity matured, the development picked up speed, and just two decades later we see Marconi's contribution taking shape. A new phase started that would cover just two decades more.

The First Decade (1895–1905)

In the preceding analysis, we have seen how the wireless tsunami engulfed Western society at the end of the nineteenth century. The original eruption—Marconi's invention far away in Italy in the mid-1890s—was barely noticed, but that was not the case with its consequences. The big waves of the new technology reached Britain, America, and Germany, causing a range of disruptions. Italy may have been its birthplace, but Britain was the country where Marconi landed his invention, got the blessing of its Patent Office, and brought it to maturity. It was also the country that initially supported and later thoroughly frustrated his further activities as the Post Office exercised its state monopoly on communication. On the other side of the Atlantic Ocean capitalistic America was the country where the wireless mania created a business climate in which the new wireless technology played an important role. Just as for telegraphy and telephony, now 'wireless' was seen as the new business opportunity, resulting in the wireless mania. Finally, the emerging industrial Germany had its own agenda, as it wanted to include wireless communication, similar to Britain, in its imperial aspirations.

In that context, we see the rise of Marconi as inventor and entrepreneur. Based on the vision of Guglielmo—shared, supported, and enhanced by others employed by his companies—the Marconi Company developed the technical dominance of the *Marconi system*, a wireless communication system that was first applied in marine applications over relatively short distances, and then was applied in long-distance communications across the oceans. Within a decade the system was technically so dominant and well protected by a range of patents that it created a de facto, although not absolute, commercial monopoly.

That creation of the *Marconi monopoly* was not the work of one man alone, as many around Marconi contributed their own expertise. Along with the technical skills of the engineers and consultants came the managerial skills of those who were the executives, either commercially or organizationally. They were coping with the military, the politicians, and the legislators. In this group of people, the personal contributions of Guglielmo himself are remarkable. He was a man heralded by the press, received at the tables of the royalty and aristocracy, endlessly traveling Europe and America, highly motived to transform his vision into reality, a man who showed the guts and perseverance to oppose his different opponents: eg German royalty, British scientists, American military, as well as those in competing businesses. Marconi had something that everybody wanted, but each on his own terms—and that all before he was even 30 years old.

The Second Decade (1905–1915)

In the mid-1900s the socio-political 'environmental conditions' for the Marconi monopoly had changed. With the Berlin Preliminary Conference in 1903, a path of international legislation had started that led to the International Radiotelegraph Conference (Berlin, 1906), where an agreement was reached. It was an important agreement in which specific national laws regulating the use of the airwaves were to be created. In Britain, it paralleled the national legislation—eg the British Wireless *Telegraph Act* of 1904—that was aimed at nationalization of Marconi's activities. In America, it originally resulted in little action, and America saw no legislative action till the 1910s. But the approaching war would give rise to national economics stimulated by the Navy. The combination of the international regulatory efforts and the specific national aspirations would prove to be an insurmountable hurdle for Marconi.

It started in Britain, where the British national market for local telegraphic services became inaccessible for Marconi due to the state's monopoly. In the process, Marconi was attacked from all sides (the Post Office, the Navy, and the scientific community by men like Lodge) for wanting to create a national Marconi monopoly based on his GB patents. By 1909 all the Marconi operations in Britain were nationalized. On other hand, the international marine communications market had flourished and saw increasing business activity with companies like Lloyd's (during a large part of the 1901-contract) and the large shipping companies (eg North German Lloyd Company) that started adapting Marconi's wireless systems aboard their steamers. As maritime transportation was extensive and growing, marine communication was also a growth market. This became even more true when further legislation, after some attention-gathering maritime accidents in the period from 1910 to 1912 (eg *Titanic*), required wireless sets on board US vessels. On the other hand, access to that market became difficult for Marconi due to nationalistic aspirations. The German government declared in 1910 that henceforth, no foreign wireless apparatus would be permitted on board German vessels. All in all, notwithstanding the long process of international diplomatic wheeling and dealing, in the end, Germany's agenda to destroy Marconi's monopoly succeeded, and intercommunication between different wireless systems became obligatory in ship-to-ship communications.

In the field of long distance (land-based, longwave) wireless telegraphic services, originally primarily used by New York papers like the *New York Times*, after the mid-1900s the business picked up and became profitable. However, in America—and especially the US Navy—over time an institutional antipathy had developed as the result of Marconi's business model (the leasing of equipment and personal), its strong patent enforcements, and its intercommunication standpoint. So, the US Navy was stimulating local inventor-entrepreneurs to create wireless equipment and was at the same time buying Telefunken equipment. As a result, by the 1910s the American Marconi Company was facing growing competition from a range of companies. In that fast-growing market, Marconi's market share dropped considerably (Figure 224). Then came the Great War, and wireless transmissions were shut down (Britain) or restricted, and in a short time the military took it over (in America). Clearly, maintaining a monopoly was a battle Marconi could not win, as he ran out of patent protection. Also, the Great War, starting in 1914, intervened, and the national interests in 'wireless' were too important to leave to the responsibility of one person's wireless empire.

The Lost Battles

As we have seen, at the beginning of its business activities there were three major strategy items on the agenda of the Marconi Wireless Telegraph Company: patenting, internationalization, and manufacturing. These three items were the cornerstones of its business model, which, in a decade, would create the Marconi empire.

Obtaining patents was establishing the priority right, which gave exclusivity to its owner.[445] Of the many patents—over one hundred in the period from 1897 to 1914 (Figure 188)—applied for by Marconi, three patents created the cornerstones for his monopoly: starting with GB Patent № 12,039 in 1896, followed by GB Patent № 7,777 in 1900 and the GB Patent № 10,245 in 1902. They gave the Marconi companies a strong patent position that lasted into the time of the Great War. And that war would turn the tables anyhow.

As communication over distance was so fundamental—proven by the booming wired telegraph and telephone business—national geographic borders were hardly relevant, and this was true for the maritime communication and for the long-distance communication that developed. So, the business of wireless communication services was by nature international. The result was that Marconi was organized in a range of companies operating in a part of the world (eg the American Marconi) or a specific market (eg the British Marconi), a strategy doomed to fail, as national interests were too important.

The early technology for wireless equipment resembled the electric technologies for building equipment like motors and dynamos,[446] and the electricity distribution infrastructure. The development pace of the wireless technology was high, not only due to product improvements but also due to improvements in system performance. Exclusivity of that technology may have been given by the patents, but the development process (aka as R&D) and the manufacturing process had to be controlled. It gave the Marconi Company the first-entry advantage in the market.

---

[445] This was an exclusivity that was commercially interesting, as others using that technology to develop similar products and doing similar business would have to take a license (and pay for it). That exclusivity could create revenue, but it came at a cost for the owner too, as many legal activities were needed.

[446] Electric technologies are dominantly mechanical technologies. The bulk of the equipment is hardware. The 'software' is in the system design and the design of the infrastructure.

These three strategic issues were at the foundation the Marconi empire that resulted from the Marconi monopoly. And following that strategy over the two decades after its establishment, the Marconi companies were bound to be confronted by some of the great powers of the late nineteenth century: Britain, America, and Germany—and these powers were soon locked in battle during the Great War.

Thus, it seems fair to say that the Marconi companies were operating in a complicated geopolitical context with growing political tensions—tensions in which each of the dominant players had its own agenda in its own regions of influence. And in those different national theatres, the different 'business battles' developed.[447] To paint the picture in bright colours: the Marconi companies had to fight a battle with the British, a battle with the Americans, and a battle with the Germans. And they lost them all.

The Battle with the British: As described, in Britain, the British Marconi Company had the state as its great adversary. Britain claimed—in the interest of the public—communication as a historically grown state monopoly. But the state's institutions—neither the Post Office nor the military—could adequately develop the equipment for wireless communication. This was due not only to their very nature, but also because Marconi held the dominating patents. The Post Office was quite overloaded with other obligations and problems by that time, and military institutions, by their existence, have a different nature than commercial organizations; doing commercial business and developing a new technology is certainly not their dominant objective. This was even more true at the end of the nineteenth century because the British Empire was at its zenith of imperial power. So, as an unavoidable consequence, an ambivalent relationship developed between these British institutions and the British Marconi Company. Marconi needed the business of the Post Office and the military market. The institutions needed the equipment developed by the patent holder.

From the start, Marconi was already confronted with the British monopoly that prohibited him from offering wireless services on the British Islands (and the seas around it). So he had to circumvent this by adapting his business model: for the maritime market, he offered services in a leasing concept (putting personal and equipment on board ships). But providing wireless communication services in the homeland was impossible, and the British *Wireless Telegraph Act* of 1904 restricted

---

[447] In the world of business, the very existence of companies can be at stake in fierce competitive confrontations. Like in military operations, these 'wars' have casualties, as can be observed every day by companies going bankrupt.

him even more. In 1909 the battle with the British was lost when the Marconi's shore-to-ship operations were nationalized by the British government.

The Battle with the Americans: The former British colony of the US had historic ties and institutional hang-ups with Britain. In contrast with the state-dominated class-oriented British society, the egalitarian, capitalistic America rejected state dominance. So, the resistance Marconi encountered when starting to do business in the US was not from the state so much as from the monopolistic telegraph and telephone industries. He was threatening their business. Like the British military institutions, Marconi was also obstructed by the American military institutions. The US Navy was not too charmed by the Marconi business model of leasing, nor by its patent policy prohibiting new business entrances. So they looked for alternatives: at home from the emerging new wireless companies, and abroad from the German Telefunken. In the meantime American business went through its own capitalistic turbulences. Nevertheless, from the dust clouds of the wireless mania, boom and bust, the American Marconi Company had emerged in America by 1913 in a strong position, dominating the maritime market and long-distance communications (eg transatlantic).

But that changed when the US Navy, after America ended its neutrality in World War I in 1917, decided to shut down commercial and amateur wireless radio. That was the beginning of the end for the American Marconi. By 1920 the last Marconi operations had been sold by its shareholders to the government-instigated new Radio Corporation of America.

The Battle with the Germans: The German Empire had by the end of the nineteenth century a national telegraph and telephone infrastructure. Germany also had its own electric industry supplying the needed equipment, and it had prominent scientists developing the wireless technology (eg Braun, Slaby/Arco). So, a natural competition in the new emerging field of wireless telecommunications was to be expected. However, for the German government, wireless communication was not just a new technology offering new opportunities. For the imperialistic aspirations of the Prussian-dominated German government, it was a tool of the empire to become independent from the British dominance in communications. Hence, there was massive (financial and diplomatic) support for the newly created Telefunken, a company that could depend on technical capabilities of the parent companies Siemens & Halske and AEG, and draw from the rich German banks. It was a wireless company that also had internationalization (eg Australia) on its agenda.

In addition, the Germans initiated international regulation (1903, 1906) changing the scene for the Marconi monopoly. Marconi's refusal to participate in intercommunication between different wireless systems ultimately resulted in a confrontation. By 1908 Telefunken had its first transmission station ready in Nauen, near Berlin, and covered distances up to Tenerife, 3600 km away. By 1910 Marconi equipment was no longer allowed on board Germans vessels.

In a period of one decade a young man had grown from a 'zero' to a 'hero': heralded in the press as the Hero of Wireless, at home with European royalty, the masthead of a high-tech company spanning the world. In the next decade that hero was brought down by forces too big for one man. His legacy, the British Marconi Company would live into the twenty-first century, not as a monopolist, but just as a normal struggling company coping with the ebb and tide of times, floating on the currents of new technologies to come.

Brothers in Arms: Morse, Bell, and Marconi

At the end of the nineteenth century the young Guglielmo Marconi found himself in a similar situation to that of Samuel Finley Morse some sixty years earlier,[448] and to that of Alexander Graham Bell some two decades before. Morse, American born, had travelled Europe extensively. The young Bell, of Scottish origin, had left Britain with his parents for Canada and then America. He had immigrated to the New World of opportunity. The young Marconi was an Italian who'd accompanied his British mother back to England. He was entering the Old World, where Britain was at the peak of its imperial powers and considered to be the 'Workshop of the World'. But there was much more that the inventors had in common:

No formal education in electricity: None of these men were solidly educated in the emerging field of electricity, but they were fascinated enough by the phenomenon to acquire the knowledge, knowhow, and hands-on experience needed to create their first rudimentary archetypes.[449] Morse was a painter fascinated by the new phenomenon. Bell was a teacher of the deaf, and Marconi a young, self-educated man fascinated by electricity and electric waves.

---

[448] We excluded the contribution of Cook and Wheatstone as their needle telegraph proved to be a dead-end technology. Nevertheless, quite a few similarities could be found in their case: such as the use of technical assistance, where the businessman Cooke started working together with the electrician Wheatstone.

[449] Bell was quoted to have said: 'Had I known more about electricity, and less about sound, I would never had invented the telephone' (H.N. Casson, 1910, p. 27).

Driven by fascination: All three men had the spark of an 'idea' that they transformed, with a lot of experimentation and in their own workshops, into artefacts. Morse worked in the Vial's Speedwell Iron Works facilities of his financial backer, Bell in his workshop in the cellar of the Sanders' house, where he lodged, and Marconi in the attic of his parents' house, Villa Griffone. Their fascination led them on a path of discovery and experimentation.

Developing a system from an idea: Each inventor developed his original idea—similar to a mental image of something to become—into a 'concept of communication': the system of coded information by Morse, the system of electric speech by Bell, and the system of wireless transmission by Marconi. All three seemed to be quite aware of the magnitude of their vision. Bell feared others would steal his idea, and Marconi was quite afraid someone else might have done something similar at the same time he was busy experimenting.[450]

Simple basic concepts: The concepts on which their inventions were founded were simple. For Morse, it was a DC electromagnet commanded at a distance by a switch (ie the telegraph key). Using a marker, that action would be recorded at a distance on a moving strip of paper: the dots and dashes. For Bell, it was sound that initiated a variable resistance (in the microphone) creating an alternating current travelling through a wire that moved the coil (of a loudspeaker) creating the 'electric speech'. For Marconi, it was the burst of electric sparks creating electric waves that carried the dots and dashes to be detected by a receiver.

New Art: As simple as they may be, their concepts were 'new to the world'. Morse was working on the use of electric devices in distant writing that others before him—eg Weber and Gauss in Germany—had initially explored also but failed to commercialize seriously. He created a new system of *coded telegraphy* of hardware (the receiving apparatus), software (the Morse code), and the network of copper cables. Bell, also building on the insights, experience, experiments, and early apparatus of many others, created a system for electric speech: the *speaking telegraph*. It was a network of equipment (ie the telephone equipment) and infrastructure (ie the copper cables spanning the globe and the exchange stations interconnecting them). Marconi, in a similar way, using the contributions of many others, created a system for cable-less transmission: the *wireless telegraph*. It was a network of equipment (the Marconi sets on board

---

[450] Marconi was quoted as saying: 'My chief trouble was that the idea was so elementary, so simple in logic, that it seemed difficult to believe no one else had thought of putting it into practice' (Dunlap, 1941, p. 12).

ships) and infrastructure of (coastal) wireless stations around the globe transmitting those messages 'without wires'. In terms of novelty, as required by the patent-issuing institutions, they all had created a new art.

Technical assistance: The further technical development of their invention into a working system also shows quite a bit of similarity. Being a painter, Morse was assisted by Alfred Vail, a skilled mechanic. The teacher of the deaf, Bell, was assisted by the mechanically skilled Thomas A. Watson. Marconi got a lifetime assistant in George Kemp, the former technician of the Post Office. Vail assisted Morse in his first experimental transmission over two miles at the Speedwell Iron Works. Watson was there at the magic moment when Bell called for him over the experimental telephone line in March 1876: 'Watson, come here. I need you.' Kemps was there when Marconi asked him on May 13, 1897, over the wireless telegraph line in Morse code: 'Can you hear me?'

Entrepreneurial assistance: Both inventors had transformed an idea into a physical archetype. In that process, both had obtained the help of relatives and friends. Morse was financially supported by Vail's father, Judge Stephen Vail, owner of the Speedwell Iron Works, and his son Alfred Vail, building the instrument and paying for the patent, with whom he created a partnership. Bell was supported by Gardiner G. Hubbard (later his father-in-law) and (business)friends like Thomas Sanders coughing up early 'angel financing',[451] in the process of creating in July 9, 1877, the *Bell Telephone Company* (Hubbard was trustee). Marconi's early work was supported by his father, who seed-funded his son, and this was followed by the angel funding of the Jameson clan and their merchant contacts when, in 1897, the *Wireless Telegraph and Signal Co. Ltd.* was organized. In all of this, his 20-years-older nephew Henry Jameson Davis was his fierce supporter.

So, all three men had quite a lot in common on the personal level of their activities and the people in their close environment. In addition, there were also other factors that created a comparable context for their work:

Antagonistically scientific environment: Although in the Morse case, the technically educated people recognized the value of his invention, that was not the case with Bell and Marconi. Only Morse was encountering a stimulating environment, where the American Congress—true, it took some years—even funded his first experiment with the Baltimore line. But Bell and Marconi were confronted with an uninterested, sometimes even hostile scientific community. In capitalistic America, obsessed by

---

[451] Financing developments before a company is organized requires informal seed funding. Angel funding is the present-day expression for the early start-up capital of a company.

the enterprises creating railroad empires, Bell was faced with the telephone conspiracy and other scientific inventors like Elisha Gray claiming priority. Marconi was facing the hostile British scientific community with Oliver Lodge—also claiming priority—as its major exponent. And the engineers from the British Post Office were—to state it mildly—not that enthusiastic at all when Marconi created his company.

Creating broad awareness: Morse was received by scientists, politicians, and those in political power. He gave demonstrations before his fellow people at the university where he worked, and to (sceptical) members of Congress and the president. In addition, Morse travelled and gave many demonstrations to the (enthusiastic) public. In contrast, Bell and Marconi were opposed by people in political power who—not too unusually—were not really interested, did not share the inventors' visions, and had to be convinced by example. Bell undertook a massive campaign of lectures to show his invention 'to the people'. Marconi did the same by involving the 'monarchy and aristocracy' in the enormous publicity he created with his public experiments.

Hostile business environment: All three inventors became entrepreneurs when they organized a company. For Morse and Marconi, the alternative had been selling their patented invention to the government (ie the Post Offices). For Bell and his partners in the association, that was never an option. The alternative of selling it to a telegraph company like Western Union failed due lack of interest. Bell and Marconi were facing a hostile business environment: the hostility of the vested interests of the monopolies in telegraphy. Bell had to fight a David-Goliath battle with the telegraph monopoly of the almighty Western Union. Marconi had to face the state monopoly of the British Post Office and the opportunistic British War Office, Royal Army, and Royal Navy, as well as the British telegraph service industry. And in America, the fierce business opposition came from the telegraph establishment.

System developers: Despite the lack of interest (Morse) and opposition (Bell, Marconi) all three inventors, with the help of early angel investors financing their ambitious projects, created communication systems that succeeded in business. Morse created a system that was widely adapted, but did not create a range of companies based on his patent rights. Nevertheless, the *Morse system* was widely copied. Bell created the *Bell system* used by many companies (the Bell companies) licensed for his patents and using his equipment for telephonic communication. Marconi created the *Marconi system* also used by many companies licensed to his patents and using his equipment for wireless communication.

Morse, Bell, and Marconi were children of their times: Morse, in the pioneering days of electricity, and Bell and Marconi, both active in the time that electricity had found its way into communication and was maturing rapidly. For Bell, it was the time when telegraph engineers were eagerly looking for ways to improve upon the efficiency of telegraphy: the hunt was for multiplex-telegraphy in which several messages could be transmitted at the same time (eg in the United States by Western Union). And for Marconi, it was the time of the search for alternative ways of transmission without cables (eg in Britain by the Post Office). Again, their consequential technical and entrepreneurial activities also show quite a similarity.

Patent protection and defence: For all three men, protecting their invention by a patent was important. In 1840 Morse was granted US Patent № 1,647, which he applied for in 1837, but he failed to get European patents. Bell was granted US Patent № 174,465, dated March 7, 1876, and obtained GB Patent № 4,765 in 1876 in Britain. Marconi was granted the British Patent № 12.039 on June 2, 1897, and got a range of similar patents all over the world. Securing the patent rights of their invention was just a beginning that brought all of them on a path of massive litigation: the ensuing patent wars, a fight that was about the commercialization of the patents they obtained: the *Morse monopoly*, the *Bell monopoly*, and the *Marconi monopoly*. Morse was seriously challenged by individuals like Jackson and O'Reilly and had to go to court a dozen times. Bell fought for his patent protection in over 300 court cases. Marconi was involved in more than 30 court cases.

Inventor-entrepreneurs: All inventors were faced with technical, organizational, and marketing issues of some magnitude. This resulted in a wide range of activities that illustrates the intellectual broadness of the inventor-entrepreneurs. All were curious, ingenious, creative, and determined people able to combine the art of invention as well as the art of business. They had a vision, they had the guts and endurance, and they created and exploited their inventions that changed the world. They were all involved in the early entrepreneurial exploration of their inventions, both by selling patent licenses as well as their equipment. Morse did not consider himself a businessman; he licensed the manufacturing to others. He did not reap the fruits when his Morse system became widely adopted in Europe, where he was without patent protection. By 1881, half a decade after his invention, both Thomas Watson and Alexander Bell, being financially independent by that time, had left the company and gone their own ways. Only Marconi stayed strongly involved in his company for a longer time—until the Great War influenced his life and the future of his company greatly.

From these observations one can conclude that although each of the inventions took place in its own timeframe and geographic location, within its specific context, apart from their personal individuality, the three inventors had quite some similarities in their work, a similarity that seems to be enhanced when we look more in detail at how their inventions came to be in existence during their *Act of Invention*.

The Acts of Invention of Morse, Bell and Marconi

In the case of communication engines, we paid attention to the development of the communication technology in three phases. Looking at who did it and what he did, one can observe the following activities grosso modo (Figure 275).[452]

Conceptualization: Stimulated by specific events, developments, meetings, etc. (in short 'external inputs'), there is this moment that an 'idea' is created. The inventor in the making recognizes a technical possibility, and the entrepreneur in the making recognizes an opportunity to fulfil a need, something like: 'If we could combine … and … we could make …

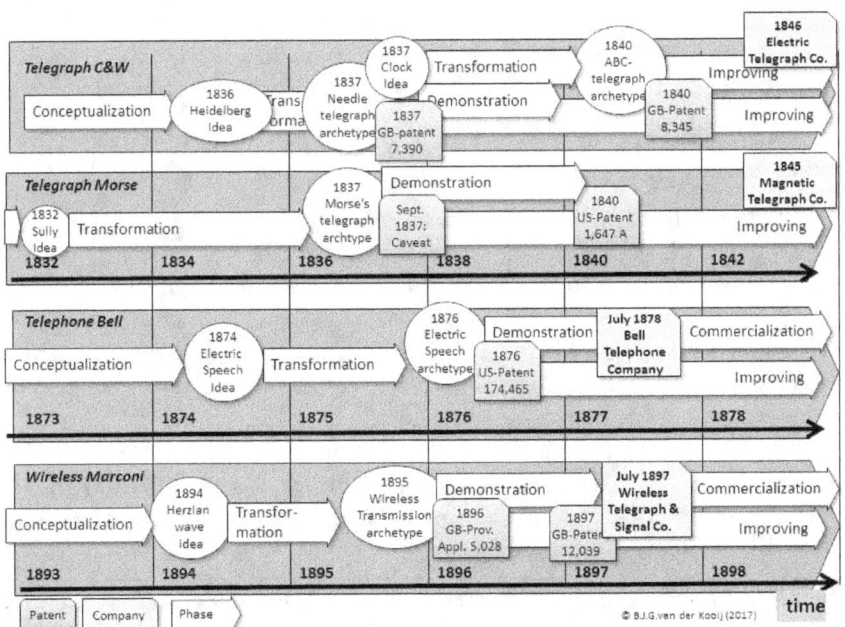

Figure 275: Overview of the acts of invention for telegraphy, telephony and wireless.

Figure created by author.

---

[452] The reader is referred to the other case studies in the *Invention Series* for an extensive analysis of the facts and events used here as an example.

and sell it as....' Following the early idea, the mental picture matures into a concept of specific functionality. In the case of telegraphy, Morse had his 'Sully idea' (1832), Cooke his 'Heidelberg idea' (1836), and Wheatstone his 'clock idea' (1837). In the case of telephony, Bell had the 'electric speech idea' (1874), and Marconi had his 'Hertzian wave idea' in 1894.

Transformation: Following the first conceptualization comes transformation of the rough outline of the first 'idea' into a working model, often a lengthy and random process of many prototypes that ultimately results in functional and working archetypes. That experimental and evolutionary process of trial and error becomes a patentable object that the inventor considers to have novelty. He files as extensive a patent specification as possible and makes a broad claim. When others in the Patent Office agree with him on the novelty, he is granted a patent protection for his claims. That patent already can be the cause of many disputes and controversy. Morse became embroiled in the Great Telegraph Patent Case after he was granted his '1,647' pioneering patent in 1840. Bell was faced with the Telephone Patent Conspiracy and later with massive litigation after his '465' patent in 1876. Marconi was opposed by the British scientific community (Lodge et al), who claimed priority, and later had to fight his '039' patent rights extensively.

At a given moment in time, depending on the context, this chain of activities is complemented by further activities of a different nature:

Demonstration: When the invention has obtained patent protection, not only the inner circle of the technical community, but also the general public and specific interest groups (being future clients) can be informed. Depending on the novelty, this can take some time. Cooke and Wheatstone demonstrated their invention to the railroad companies they saw as potential users. Morse demonstrated his invention to Congress to obtain funding for a first telegraph line. It took him some five years before he succeeded. Bell showed his invention at the Centennial Exhibition of 1876, and gave a range of lectures. Marconi demonstrated his invention to the Post Office and British military, soon creating massive publicity in the general press.

Improving: As the first patented archetype is still an immature and unproved product, a lot of attention has to be paid to the technical improvement. This constant process of experimenting leads often to new patents that protect those improvements. Sometimes the new patent (aka basic patent) creates a cornerstone for the commercial activities that follow. Wheatstone got a patent for the ABC pointer telegraph in 1840, which soon replaced the needle technology. Bell got his second telephone '787' patent in 1877, which created the Bell monopoly. Marconi was granted his '7,777' patent for his syntonic telegraph, which created the Marconi monopoly.

Commercialization: The archetype is protected by a patent, broad claims are made, experiments improve the product and system, and demonstrations show its feasibility. That unavoidably arouses commercial interest in the business community. Moreover, angel funding is depleted by now, and additional experimenting and demonstration have to be funded, so commercialization of the patent rights becomes an issue. The choice is twofold: either one sells the patent rights to another party, or one decides to implement a commercial organization on one's own. Remarkably, the inventor-entrepreneurs of the Communication Era chose for the latter. For Cooke and Wheatstone, it took till 1845 before they created the *Electric Telegraph Company*. Morse started the *Magnetic Telegraph Company* in 1846. Bell, stimulated by his father-in-law, created the *Bell Telephone Company* in 1878, and Marconi created the *Wireless Telegraph and Signal Co.* in 1897.

This outlines the total process from idea to concept and archetype we call the *Act of Invention* process that starts with the birth of an idea, subsequently turns that idea into a concept and archetype, and ends with the blessing of the Patent Office.

The Dominance of Infrastructural Networks

Our inventors all invented a *system* for communication. Along with the equipment to transmit and receive the communications, they created an infrastructure enabling the communication over long distances. An infrastructure that was the focus of attention of the early 'service providers': the telegraph services, the telephone services, the wireless services, up to the internet service providers of our days. An infrastructure that started simple, then grew into local networks that were later interconnected on a regional and national basis[453]. And those networks were using the communication systems of their developers.

Samuel Morse, after getting his experimental line funded by Congress in 1843, started the *Magnetic Telegraph Company*. This company supported initiatives from business associations to creating telegraph services on a commercial basis. He supplied the equipment and the license to use it in exchange for part ownership in the companies. This strategy created an

---

[453] This development can also be recognized by the electricity distribution networks. See: B.J.G. van der Kooij, *The invention of the Electro-motive Engine* (2015) pp. 200-218.

abundance of local telegraph networks stretching over a thousand miles by 1847.

Alexander Graham Bell, after the earlier Patent Association, started the Bell Telephone Company (GH Trustee) in 1878. From that company emerged two companies: the *National Bell Telephone Company* (1879) and the *International Bell Telephone Company* (1879). These companies leased the equipment and the license to use it to independent service providers for specific geographic areas. By 1879 some 185 agreements had been granted covering the major cities of the United States. These Bell companies, in cooperation with local investors, were also created in Europe (such as Sweden).

Guglielmo Marconi, after the creation of the *Marconi Wireless Telegraph Co. Ltd* and the *Marconi International Marine Communication Company* in 1899, started a range of coastal station in Britain under the Lloyd's contract of 1901. By 1903, in several countries in Europe (including Italy), later followed by Spain and Portugal (1910), licenses were granted to local Marconi companies, which built the numerous wireless stations. In 1912 there were 13 overseas Maroni companies: alongside the United States and Canada, there were companies in France, Argentina, Russia, Italy, and Spain. He had headquarters in Brussel, Paris, Rome, Madrid, and Buenos Aires. A network of dozens of stations covered all continents, from North and South America to Europe, Asia, Africa, and Australia.

Thus, for all the systems of communication, the *network* was a dominant part of it. Next to the technical difficulties creating that network, local issues also complicated its establishment. And as those companies creating and participating in the network payed license fees, they created revenues for the Marconi Wireless Telegraph Company and its subsidiary, the Marconi International Marine Communication Company. But the national interest in wireless communications proved too strong, and the networks became part of geopolitical conflicts forcing Marconi out.

The man that was considered to be a foreigner in Britain, 'British' in Italy, and 'not American' in the United States, had to wait until after the Great War before his shortwave system would conquer the world again.

> *The sensational success of the beam system elevated Marconi to an unprecedented height in the public eye. It no longer mattered who had done what, how, or when in the messy history of wireless. Marconi was now ensconced in the pantheon of technical innovation with a handful of peers such as Bell and Edison, and he would remain there for the rest of his life. The giant corporations build on his inventions, and their rivals, the governments who sought to control the powerful forces they unleashed, would continue to wrestle in the commercial and political,*

national and transnational arenas of the 1920s and '30s, but Marconi's stature—if nothing else—was assured. He had provided the technical, and oversaw the political and commercial basis, of the worldwide networked system of wireless communication he had first imagined twenty-five years earlier. The development of the shortwave was his last defining moment.

The arrival of the beam system [1924] also disrupted the sleepy coexistence that had settled in between the cable companies and wireless during the previous two decades. Thanks to shortwave, wireless was now incontestably quicker, cheaper, and just as efficient as cable telegraphy: it was also the basis for the international extension of broadcast radio, which thanks to short-wave could now be transmitted worldwide. With the arrival of beam communication—the precursor of what we now call 'telecommunication'—, every country in the world also had to decide between private and public enterprise, and choose a company to undertake or partner in its construction and operation. This was invariably one of the four giant companies that had already divided the world in spheres of interest. … It consisted of the Radio Corporation of America, Marconi's Wireless Telegraph Company, France's Compagnie Generale de Telegraphie sans fil, and Germany's Telefunken. (Raboy, 2016, p. 504)

The Era of Radio that would start and boom after the Great War was to be the fourth phase of the Communication Revolution. A phase that would see the emergence of radio and television broadcasting. But that is another story…

The Invention of Wireless Communication

The invention of wireless communication was the logical extension of the invention of earlier communication engines like the wired telegraph and telephone. Marconi's contribution as a basic invention had been discussed in the preceding part, so we will wrap up our analysis by zooming on the question of 'who invented the wireless telegraph'.

Who Invented Wireless Telegraphy?

The value of Marconi's contribution (Figure 276) to the Communication Revolution (Figure 271) is undisputed. With many others (Figure 284) he contributed to the development of the communications engines, but trying to answer who was/were the inventor(s) of the wireless telegraph is a complex affair. This priority question about the wireless system has been a topic of discussion among historians since Guglielmo Marconi obtained his British Patent № 12.039 in 1897. We will not attempt to mingle in this discussion, as it is not relevant to our explorations, but we will just try to present the different points of view within the context of that period of time.

Facing the Big Question

When trying to address the question of who is the inventor of wireless communication, one is faced with some problems of interpretation.

One is in the sphere of *semantics*: what exactly is an invention? What is the difference with a discovery? Is invention different from innovation? This is a problem that seems to be solvable by defining the words (see Preface). However, one soon realizes that there is quite some semantic confusion as one looks up the different definitions scholars have been using at different moments in time. The same goes for the subject: what is wireless communication exactly? Is it the early rudimentary spark-based signalling, or is it the later use of tuned electric continuous waves? Or is it 'radio': the transmission of single-shortwave frequency-based information (eg speech and music)?

Another problem is in the *moment* of establishing priority: the question of 'who is the first' when he did what he did: his idea, his artefact, his prototype. Sometimes this is a question that is related to the moment when someone—observing the phenomenon—made the *discovery* and published a paper or presented a demonstration about it. Sometimes this question is related to legal priority (related to the patent situation) of the person who created the *invention* and received a patent for it, a problem solvable by establishing the priority date of the patent.

In addition, the *consequences* of the priority play a role. Sometimes it is about honouring somebody who made a range of important contributions. Then it is about the honour to be 'the father of'. But the consequences can also be more serious, as commercial interests play a role. Then the priority discussion can result in litigation about infringements of the patent rights.

Take the following example, where Tesla was posthumously recognized as discovering radio:

> *In a lecture-demonstration given in St. Louis in the same year—two years before Marconi's first experiments—Tesla also predicted wireless communication; the apparatus that he employed contained all the elements of spark and continuous wave that were incorporated into radio.* (Harkins, 2008, p. 755)

A final problem is related to the *perspective* one uses to address the question. Is it a contemporary point of view heavily influenced by personal interests, ambitions, and prestige? Or is it a later point of view claiming a national interest, as can be observed by nearly all countries that want to honour their own national heroes? Or is it a present-day historiographic point of view that reflects on history, trying to clarify from a historical point of view?

Clearly, there are many different—sometimes opposing and contradicting—interpretations about who is the inventor of wireless communication. They may differ, but nevertheless, all those interpretations have the process of 'change and novelty' in common. But again, even in that process, there are differences. Take the 'invention', the result of the process of invention as carried out by the specific person—the inventor—that creates change and novelty. Next, there is the 'innovation', the result of the process of innovation creating a working artefact that—as the result of entrepreneurial activities—can be brought to the market by subsequent organizational activities. Both those processes, invention and innovation, make up the *Act of Invention*. However, then again, confusion arises. Is the interpretation concerned with the original 'idea' (as in the mental picture), is it the first model or prototype (as in the archetype), or is it the product/system that is successfully commercial implemented (ie brought to market)?

Definition Problem

When discussing the priority aspects of an invention, it is important to define the word 'invention' with care. Many different interpretations exist, but the most common understanding—at that moment—in time was colourfully described by John Ambrose Fleming (1849–1945) in 1919:

It is necessary, however, to keep clearly in mind the true meaning of "invention". Invention does not consist in displaying a few brilliant and original ideas. Neither does it consist in outlining a certain set of requirements and broadly defining the means by which certain ends may be attained. Invention consists in overcoming the practical difficulties of the new advance, not merely talking or writing about the new thing, but in doing it, and doing it so that those who come after have had the real obstacles cleared out of their way, and have a process or appliance at their disposal which was not there before the inventor entered the field. In most cases, however, the removal of the obstacles which block the way is not entirely the work of one person. The fort is captured only after a series of attacks, each conducted under a different leader. In these cases the inventor who breaks down the last obstruction or leads the final assault is more particularly associated in the public mind with the victory than are his predecessors, though his intrinsic contribution may not be actually of greater importance. (Fleming, 1919, p. 514)

Invention by that time was obviously more than just having 'the idea'; there was also the implementation aspect. There, the accumulated knowledge was applied in the next step of 'capturing the fort', often in the form of experiments leading to an, often cumbersome but successfully working, archetype of artefact. An opinion already vented on many occasion in court cases: Such as by Judge Shepley:

No more difficult task is imposed upon the court in patent causes then then of determining what constitutes invention, and drawing the line of distinction between the work of the inventor and the constructor. [454]

The preceding illustrates that the construct of invention/innovation behind the words gives already rise to confusion. That being said, we will try to address the question about the inventor from different perspectives.

Contemporary Points of View

Looking back in time, seen from our present day's perspective, Marconi's invention was surprisingly simple. Both the transmitter and the receiver were not too complex, with quite simple electromechanical components. The transmitter was made of a battery, induction coil, spark balls with antennae, and a sending key. The receiver was an antenna, coherer, relay, and Morse printing instrument (see Figure 136). That was also the case with Morse's invention of the telegraph[455] and Bell's invention of the telephone.[456] Each of the components was known, invented, or developed by someone else: the battery was Alessandro Volta's invention,

---

[454] Source : Case Pearl v. Ocean Mills (1876). United States Supreme Court Records and Briefs. p.436.
[455] See: B.J.G. van der Kooij, *The Invention of the Communication Engine 'Telegraph'* (2015).
[456] See: B.J.G. van der Kooij, *The Invention of the Communication Engine 'Telephone'* (2015).

the induction coil came from Michel Faraday, the coherer from Branly, etc. It was a system built to use 'electric waves' discovered by Heinrich Herz. If a system was so simple, then what was the novelty?

Seen in the perspective of those days, Marconi's archetype certainly had novelty. The discovery of the phenomenon of 'electric waves' itself may belong to others, among whom were Heinrich Hertz and many participating in the scientific wireless mania. The creation of components for creating and detecting 'electric waves' also belonged to others. But Marconi went further as his invention was not so much related to the phenomenon and the separated components, but it was a practical system based on—at that time—known components that realized a new principle: a complete system wireless signalling using the newly discovered phenomenon of electric waves.

In those days many other scientists were active studying the so-called Hertzian waves, such as Edouard Branly, whose coherer stimulated Marconi to build his 'sensitive tube'. Surely Marconi's work, expanding on the earlier work of other experimenters such as Augusto Righi, had been an important contribution. As was acknowledged by Branly:

> *Monsieur Branly, in the course of a communication to the "Societe Frangaise de Physique" in 1896, made the following statement: "Although the experiment which I have always prospected as the main experiment of my study of radio conductors (battery element, iron filing tube, and galvanometer, making a circuit where the current passes after an electric spark has been flashed at a distance) is the image of wireless telegraphy, I have no pretence to have made this discovery, as I never thought of transmitting signals." (Bulletin de la Societe Francaise de Physique, Resume des Communications, seance du 16 decembre, 1896, p. 78 du volume de 1898.)* (Dunlap, 1941, p. 18)

On the other hand, there were people who claimed priority after Marconi had been granted his patent, such as Oliver Lodge, who was also known for his detector of electric waves (aka coherer), based on Branly's work. On August 14, 1894, he had given his Oxford memorial lecture on the work of Heinrich Hertz. There, he had demonstrated the quasi-optical nature of the 'Hertzian waves' and demonstrated their similarity to light and vision, including reflection and transmission at distances up to 50 meters. Sometime later Lodge would, after the widely published lecture Preece gave on Marconi's invention at the Royal Institution on June 3, 1897, write to the editor of the *Times* and claim to be the inventor of wireless signalling:

> *It appears that many persons suppose that the method of signalling across space by means of Hertzian waves received by a Branly tube of fillings is a new discovery, who has recently been engaged in improving some of the details. It is*

well known to physicists, and perhaps the public may be willing to share the information, that I myself showed what was essentially the same plan of signaling in 1894. My apparatus acted very vigorously across the college quadrangle, a distance of 60 yards up to the limit of half a mile. ... the only important discovery about the matter was made in 1888 by Hertz; and on that is based the emitter of waves; the receiver depends on cohesion under electrical influence and has been re-observed in other forms by other experimenters, including the writer in 1890. (Jolly, 1972, pp. 45-46)

There were other views on the topic of wireless signaling such as given by Preece from the British Post Office on several occasions. However, the Preece lecture heralding Marconi as the true inventor of a new system of telegraphy has to be seen in context with the controversy that raged between Preece and the Maxwellians, the controversy between the men of theory and the man of practice.

On June 4, 1897, William Preece held a lecture to a large audience at the Royal Institution in London on 'Signalling through Space without Wires". He devoted considerable time to the work of Marconi, acknowledging its novelty:

Mr. Marconi utilizes electric or Hertzian waves of very high frequency, and they depend upon the rise and fall of electric force in a sphere or spheres. He has invented a new relay which, for sensitiveness and delicacy, exceeds all known electrical apparatus. The peculiarity of Mr. Marconi's system is that, apart from the ordinary connecting of wires of the apparatus, conductors of very moderate length only are needed, and even these can be dispensed with if reflectors are used.
...

It has been said that Mr. Marconi has been doing nothing new. He has not discovered any new rays: his transmitter is comparatively old; his receiver is based on Branly's coherer. Columbus did not invent the egg, but he showed how to make it stand on its end, and Marconi's has produced from known means a new electric eye more delicate than any known electrical instrument, and a new system of telegraphy that will reach places hitherto inaccessible. (Preece, 1897, pp. 892, 896)

Another scientist who questioned Marconi's priority was the Russian scientist Alexander Stephanovich Popov, who claimed in 1897 in a letter to the British magazine *The Electrician* that Marconi's receiver was a reproduction of his own (Figure 153, Figure 154). The dispute remained friendly, both parties respecting each other's work, and Marconi claiming he had not been aware of Popov's work (that was published in the Russian language).

In July 1902, when Marconi was experimenting on board of the Italian cruiser Carlo Alberto at anchor in the Russian Imperial Navy base at Kronstadt, a Russian visitor came to see him and said: "I am Alexander Popov, I want to pay my respects to the father of wireless." (D. Marconi, 2001, p. 144)

He was not the only one to pay his respects, as also the German professor Adolphus Slaby of the Technical High School at Charlottenburg, Berlin, declared in an article in an American magazine, *The Century Magazine*, in April 1898:

In the English professional journals an attempt has been made to deny novelty to the method of Marconi. It was urged that the production of Hertz rays, their radiation through space, the construction of his electrical eye--all this was known before. True; all this had been known to me also, and yet I never was able to exceed one hundred meters. In the first place, Marconi has worked out clever arrangement for the apparatus which by the use of the simplest means produces a sure technical result. Then he has shown that such telegraphy (writing from afar) was to be made possible only through, on the one hand, earth connection between the apparatus and, on the other, the use of long extended upright wires. By this simple but extraordinarily effective method he raised the power of radiation in the electric forces a hundredfold. The upright extended wires work like the pierced tube of a watering-cart; the rays of electric force spurt, as it were, in every direction upright to the wire; they cause a great part of space to be drawn into sympathy. Now, since these wires are the essence of Marconi's discovery, the term « telegraphy without wires » is really erroneous; more correctly should it be called telegraphy by sparks, in opposition to the term used hitherto, « telegraphy by circuits » (Stromtelegraphie). (Slaby, 1898)

That was the opinion of the scientific and engineering community involved in the new phenomenon of electric waves and wireless signalling. But there were also other opinions by non-technical people who became involved in the priority question: the officers of the court.

In 1905, as a result of the lawsuit *Marconi vs DeForest Wireless Telegraph Company* that had started in 1901, the judgment given by Judge Townsend[457] also gave an indication of how Marconi's invention was seen:

"It would seem, therefore, to be a sufficient answer to the attempts to belittle Marconi's great invention that, with the whole scientific world awakened by the disclosures of Hertz in 1887 to the new and undeveloped possibilities of electric waves, nine years have elapsed without a single practical or commercially successful result, and that Marconi was the first to describe and the first to achieve the transmission of definite intelligible signals by means of the Hertzian waves." (Raboy, 2016, p. 258)

---

[457] Source : United States Supreme Court Records and Briefs. p.469.

Later Points of View

The most famous radio engineer in the period from the late 1910s through the mid-1950s was Edwin Howard Armstrong, best known for the development of the superheterodyne receiver, among many other achievements. And he had no doubt about Marconi's priority in the initial development of radiotelegraphy, as he declared:

> *Had Marconi been more of a scientist and less of a discoverer, he might have concluded that his critics were right, and stopped where he was. But like all the discoverers who have pushed forward the frontiers of human knowledge, he refused to be bound by other men's reasoning. He went on with his experiments; and he discovered how, by attaching his transmitted waves to the surface of the earth, he could prevent them from traveling in straight lines, and make them slide over the horizon so effectively that in time they joined the continents of the world. Several years were to pass before agreement was reached on the nature of Marconi's great discovery, though Marconi himself understood very well how to apply it and to employ it usefully; and it proved to be the foundation upon which the practical art of wireless signaling was built. Marconi's claim to the invention of wireless telegraphy is beyond challenge.* (A. Douglas, 1990, p. 289)

Some hundred years after the 'Atlantic Crossing', Francesco Parsec Marconi wrote about the invention of his grandfather:

> *What was essential is that he had brought each previously known component of his system to the best level of performance possible with the technology of the day, a system that was very precisely suited to his one and very simple and almost obsessive objective: sending and receiving intelligible signals over the greatest possible distances.*

> *Also later in time, when radio broadcasting had succeeded the early wireless transmission of telegraphic signals, in the discussion of who was the inventor of the 'radio'—radio being the word used for wireless transmission—that took place in America, the Supreme Court judged in 1943 on case that had been initiated in 1916 when the American Marconi had claimed governmental infringement on his tuning patent:*

> *The Supreme Court acknowledged "Marconi's reputation as the man who first achieved successful radio transmission." In an about face, the Court brushed off any evidentiary weight to this reputation: That reputation, however well-deserved, does not entitle him to a patent for every later improvement which he claims in the radio field. Patent cases, like others, must be decided not by weighing the reputations of the litigations, but by careful study of the merits of their respective contentions and proofs.* (Harkins, 2008, p. 759)

Subject of the discussion was the early work of Nikola Tesla, the eccentric inventor of Serbian origin who immigrated to America in 1884 and was active on many fields of electricity. Tesla was already famous when he heard about young Marconi, and soon the battle between Tesla and Marconi was part of considerable press coverage. Tesla, occupied with many other fields of electricity (eg the induction motor) was also active in space telegraphy, the domain of high-frequency electricity, duplicating Hertz's experiments. His primary interest was the distribution of power, and his experiments in the 1890s resulted in the Wardenclyffe Tower project. But he also developed a radio-controlled boat, which he demonstrated in 1898. He *predicted* wireless communication, but did not create a wireless communication system.

> *In a lecture-demonstration given in St. Louis in the same year - two years before Marconi's first experiments - Tesla also predicted wireless communication; the apparatus that he employed contained all the elements of spark and continuous wave that were incorporated into radio...*

> *Tesla filed and received the earliest radio patent issued by the United States Patent and Trademark Office. Tesla's United States Patent №. 645,576, entitled "System of Transmission of Electrical Energy," issued on March 20, 1900 from an application filed September 2, 1897. In that patent, Tesla declared that "the apparatus which I have shown will obviously have many other valuable uses - as, for instance, when it is desired to transmit intelligible messages to great distances.* (Harkins, 2008, p. 755)

So, Tesla had already had the 'idea' and showed his apparatus to transmit 'wireless energy' before Marconi got his British patent for 'wireless telegraphy'. Tesla's lecture, demonstration, and later published patent anticipated several features of the Marconi patent. It would become a major element in the court case *Tesla vs Marconi*, which started in 1916 but did not reach a Court conclusion until 1943.

> *Thus, notwithstanding Marconi's success in commercializing the first radio, the Court concluded that Marconi was not the first inventor. The first radio patent belonged to Tesla.* (Harkins, 2008, p. 759)

But there was more to the Supreme Court's 1943 decision than just Tesla's supposed contribution to wireless communication.

> *The decision also elided the fact that in 1943 the United States was at war with Italy, and the US government was in no mood to pay a settlement to a British company over a forty-years old dispute involving the inventions of a dead "Italian scientist" (as Marconi was called in a lower court ruling). It highlighted the continuing political resonance as well as the mystique associated with Marconi.* (Raboy, 2016, p. 670)

Technical Points of View

As noted before, many of the components of the wireless system that Marconi created were the result of work done by others scientists of those days. Not surprisingly, Marconi—reflecting on the work he had done—saw his invention as an improvement on the work of others:

> *My discovery was not the result of long hours and logical work, but of experiments with machines invented by other men to which I applied certain improvements. I used a Hertzian radiator and the Branly coherer. ... The improvements which I made were to connect both the receiver and sender with first the earth and second the vertical wire insulated from the earth. The later was by all means the more important of the two innovations.* (Dunlap, 1941, p. 53)

He was not the only one who took this perspective, being so heavily involved in the technicalities of the invention. Indeed, he had made improvements on the component level, but there was more to be said when one looked at the totality of his invention: the working *system* of wireless telegraphy over long distances. And Marconi had found a new *method*—also called technology, or 'knowing how to make the system'—of creating such a system. Just as he had claimed in his provisional application in 1896 (Figure 170), later to become his patent № 12,039 of 1897:

> *I declare that what I claim is—*
> *1. The method of transmitting signals by means of electrical impulses to a receiver having a sensitive tube or other sensitive form of imperfect contact capable of being restored with certainty and regularity to its normal condition substantially as described.* (text of patent application)

This view was later supported by others, among whom was Francesco Parsec Marconi, the grandson of Guglielmo Marconi, who wrote in his *Personal Reflections of 'An Italian Adventurer'*:

> *The core issue in Marconi's case was that he had not invented anything really new or at least easily recognizable as such but what was new was the use to which he put the old concepts and techniques in order to exploit them for a very practical purpose. His, then, would be a very difficult device to patent where priority of discovery of each component is emphasized as in the US and Italy, for example, especially since his predecessor's results (mainly Oliver Lodge) were of common knowledge in the world at the time. In England, on the other hand, what mattered most at the time was the possibility to claim the method itself as property and that was exactly what Marconi did with his crucial patents. These were used by him and his company to protect them, at least for the crucial first years, from potentially fatal attacks from powerful competitors.* (F. P. Marconi, nd)

Fellow Human Points of View

During his lifetime, Marconi was certainly lauded by many for his work in wireless telegraphy. For example, at the annual dinner of the American Institute of Electrical Engineers on January 13, 1902, some 300 members— among whom was Alexander Graham Bell— heralded the grateful guest of honour Marconi. There, he had to mention Britain's stance on his endeavours in relation to the welcome he had received in Canada.

> *Mr. President, Ladies and Gentlemen: I can hardly find words to express my gratitude and thanks for the reception I have had here to-night. I thank you very much for the appreciation of the work which I have been fortunate enough to carry out. I feel myself highly honored to be entertained by such a great body as the American Institute of Electrical Engineers. I think it is well-known all over the world that Americans stand first in applied electrical engineering. I feel myself greatly honored to be among so many eminent men, whose names are household words in the whole civilized world. ...*
>
> *It is rather strange, I may mention, that in England the British Government has the monopoly of the telegraph, and I am glad to say that it has encouraged my experiments rather more than prevented them. Certainly, when it comes to commercial work between stations in England--not between stations in England and ships, but between stations in England--we have come to a suggested arrangement; and I do not think the British Government would have thought of interfering with experiments, the object of which was the advancement of science. ...*
>
> *I wish to state here before you all that I am very greatly indebted to the governments of Newfoundland and Canada for the encouragement, sympathy and assistance they have given me in my work, and I trust that the future results will bear out our hopes and expectations.[458]*

In later years, Marconi received many accolades for his work, including honorary doctorates from several universities and the Nobel Prize for Physics, which he shared with Karl Ferdinand Braun in 1909. That was a great honour from the scientific community, but he also received honours from many other nations that recognized the value of his contributions.

For example, the Italians heralded him on many occasions and made him a senator in the Italian Senate. In Britain, he was appointed Honorary Knight Grand Cross of the Royal Victorian Order and received the Albert Medal of the Royal Society of Arts, the John Fritz Medal, and the Kelvin Medal. He was decorated by the tsar of Russia with the Order of St. Anne, and the king of Italy

---

[458] Source http://www.fgm.it/documenti/articoli/didattica/insegnanti/9.pdf (pp. 22–26).

created him Commander of the Order of St. Maurice and St. Lazarus, and awarded him the Grand Cross of the Order of the Crown of Italy in 1902. Marconi also received the Freedom of the City of Rome (1903), and was created Chevalier of the Civil Order of Savoy in 1905, and he received the hereditary title of Marchese in 1929.

In the end, Aitken best sums up the significance of the first wireless pioneer being an inventor-entrepreneur combining the Act of Invention with the Act of Business:

> *What differentiated Marconi from his contemporary rivals was not his scientific knowledge, nor, initially, the distinctive excellence of his technology. It was his sense of the market, of where a demand for is new technology existed or could be created. A creative genius in electronic engineering Marconi could have been; but he was also a commercial entrepreneur. And his entrepreneurship was a vital element in the creation of the radio communications industry precisely because the nature of the technology itself did not unambiguously indicate the economic used to which it could be put.* (Aitken, 2014b, p. 306)

Today's Points of View

More than a century after Marconi's experimental work in the Villa Griffone—where his basic idea of using Hertzian waves to transmit over distance was implemented by experiments of transmission over increasing distances, the overcoming of natural obstacles, and the transmission of a decipherable signal—Marconi is considered to be the inventor of the wireless engine.

> *…what made Marconi the real forerunner of radiotelegraphy were not the important technical improvements he introduced to his system in 1895 (first of all the aerial ground system), but the early development of his idea (through the progressive increase in distance distances, the overcoming of natural obstacles and the transmission of a decipherable signal). His idea was already clear and perfectly defined at the time of his first and famous experiment at the Villa Griffone, when a signal was sent over Celestini hill. …*
>
> *Therefore the frequent controversy on the priority and merits of one invention over the other loses a large part of its value when compared to Marconi's complete idea, a project that not only was achievable and achieved, but also endless improvable. Neither Tesla, nor Lodge or Popov, nor Branly or others were able to conceive it before Marconi. For some time they did not even understand its profound implications.* (Falciasecca, 2010, p. 12)

This view is also expressed by others in more recent analyses of Marconi's contributions to wireless telegraphy, as described in the epic work of Raboy:

Marconi was not the first to experiment with Hertzian waves, nor the first to try to send telegraphic signals without wires, but by combining the two he became the first person to effectively use what we now call the radio spectrum for communication. (Raboy, 2016, p. 28)

Simply put, Marconi discovered and developed a practical use for the electromagnetic spectrum. It was at least another decade before other experimentalists learned to transmit voice by wireless—and also discovered that they had to deal with Marconi's ironclad patents in order to commercialize it. (Raboy, 2016, pp. 48-49)

What set Marconi apart from the rest was that he saw wireless communication, in his mind's eye, quite literally as telegraphy without wires. Thus in his experiments he set out to reproduce as closely as possible the physical conditions condition of telegraphy, and use them to the same ends. ... It was Marconi who made the leap from Hertz's lab experiments to practical wireless telegraphy using electromagnetic waves as the medium of communication. This was his original contribution, and what caused him to be labelled a genius in the heroic version of his life story. Marconi's entire career was built on this one idea. ... Marconi not only invented mobile communication, it shaped and defined his world. (Raboy, 2016, p. 52)

Considering these points of view, both Marconi's contemporary and that of the present day, one can conclude that Marconi's contribution is similar to those of Samuel Morse and Alexander Graham Bell. Each had a different contribution to the Communication Revolution, but what they have in common was that their contribution (aka their 'invention', in our terminology, their 'basic innovation') was part of a cluster of innovations. Their work was the merger of different trajectories of developments: the insights created by the more abstract scientific contributions and the applications from the engineering contributions, building on earlier practical experiences. It was the new combination, the result of the act of insight and the act of skills that created the novel thing: in the case of Marconi, a system of wireless telegraphy.

The Cluster of Innovations of Wireless Telegraphy

Marconi's contribution, although fundamental in its nature, was just one of a range of contributions by many thinkers and tinkerers (Figure 145). These contributions sprouted from the phenomenon of 'lightning' observed by the electro-physicists (Figure 149) who were focused on the new phenomenon of 'electric waves' (Figure 156):

Experimental and theoretical contributions to electric waves: The attention to the side-effects of the electric sparks—and the great thunderous sparks of lightning—creating light, resulted in the discovery of electric waves. It was Heinrich Hertz's contribution in 1888 that started the broad interest in the Hertzian waves travelling with the speed of light through the 'ether'.

Components for creating and detecting electric waves: Other experimenters had developed components related to the creation of electric waves. These devices—called spark oscillators—created in the invisible frequency spectrum a burst of un-tuned electric waves that could be detected over increasing distances. These waves were detected by other devices related to the conductivity of electricity, devices later called 'coherers'.

System contributions: By that time the communication system using electricity as carrier of information—the cabled telegraph and telephone—had grown into maturity. Facing technical problems of their own, it was obvious that communication over distance was important to society. The contributions of many people to those systems improved their performance, but there was still one barrier to pass: the use of cables.

At that moment in time, the contribution of Guglielmo Marconi, deceivingly simple, created an alternative in the form of wireless telegraphy. After Marconi had invented his system for wireless communication a range of contributions improved on the communication system of tuned spark telegraphy. Once again, these were improvements to the components and the overall system:

Component improvements: Over time the devices to create and detect electric waves were improved upon by many contributors. Among them was Marconi's reliable detector, nicknamed 'Maggie'. This was later followed by completely new devices like Fleming's thermionic valve.

System improvements: The early system for wireless communication, due to the use of sparks to create the electric waves, had many inherited problems. The transmission was improved upon when syntonic systems were developed by Lodge, Braun, Slaby/Arco, and Marconi, and the early continuous-wave spark systems emerged.

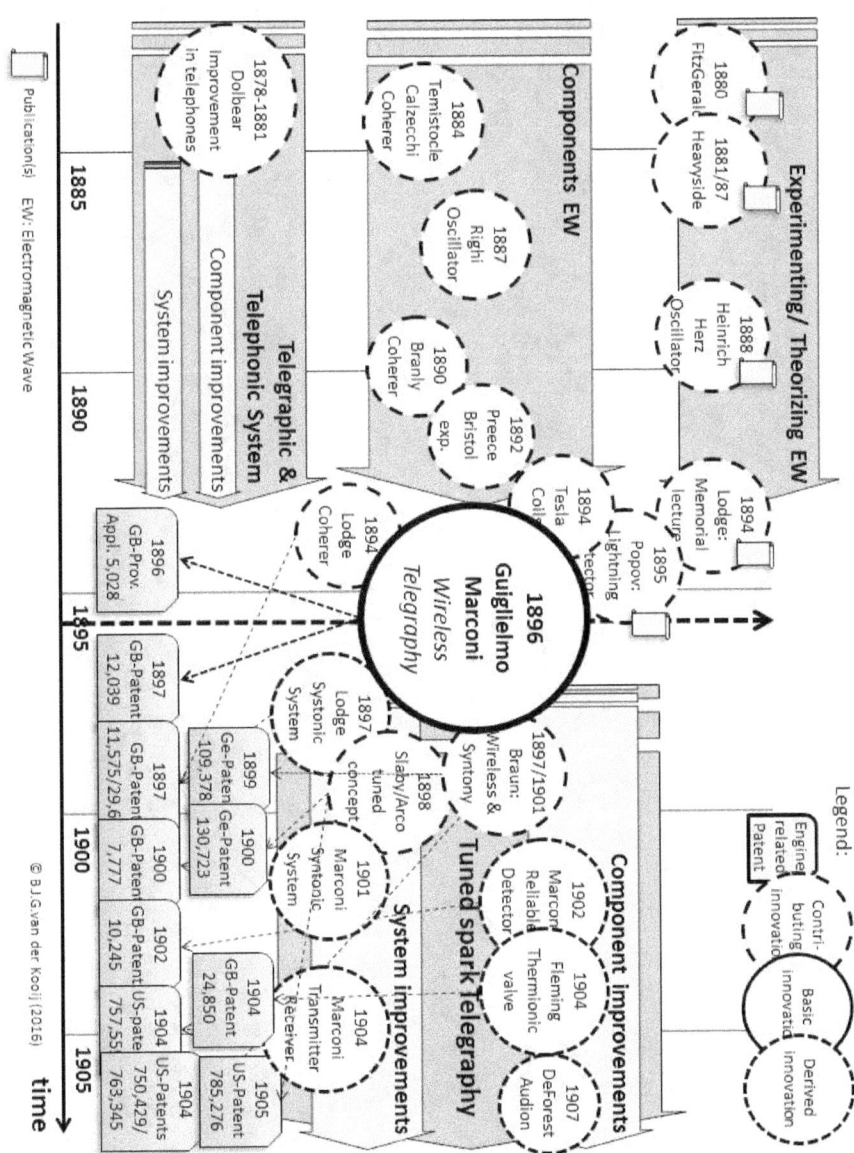

Figure 276: The cluster of innovations around Marconi's invention of a system for wireless telegraphy.

Figure created by author.

It was a cluster of innovations, with Marconi's contribution at its core, that created wireless telegraphy (Figure 276). This cluster of innovations was similar those involved in the invention of the telegraph by Morse and the invention of the telephone by Alexander Graham Bell. Together, they created the foundations of the Communication Revolution.

Conclusion (Part 3)[459]

For someone living in the pre-electric era, it would be hard to image how one would verbally communicate over longer distance the way we do today. For somebody living today—2017—in a world where people are always, one way or the other, 'online',[460] it is hard to image what the world looked like without those powerful communications engines we have available today. The fruits of developments in information, computation, and tele-communication (ICT) brought together in one 'smart' device called smartphone. The smartphone is with us when we work and recreate, when we relax, walk and travel, or when we ride a bike in Amsterdam (Figure 277). In terms of communication, we

Figure 277: Girl with smartphone on bike in Amsterdam (2014).

Source: http://bicycle-amsterdam tumblr. com/page/2

have come a long, long way from the Indian Smoke Message System (SMS).

---

[459] This conclusion is preceded by Part I and Part II, which related to the development of the communication engines 'telegraph' and 'telephone'. See: B.J.G. van der Kooij, *The Invention of the Communication Engine 'Telegraph'* (2015), p.442; B.J.G. van der Kooij, *The Invention of the Communication Engine 'Telephone'* (2016), p.317.
[460] The expression 'online' in our Computer Age indicates a state of being connected by means of a communication engine (aka the smartphone) to a communication infrastructure (aka the Internet).

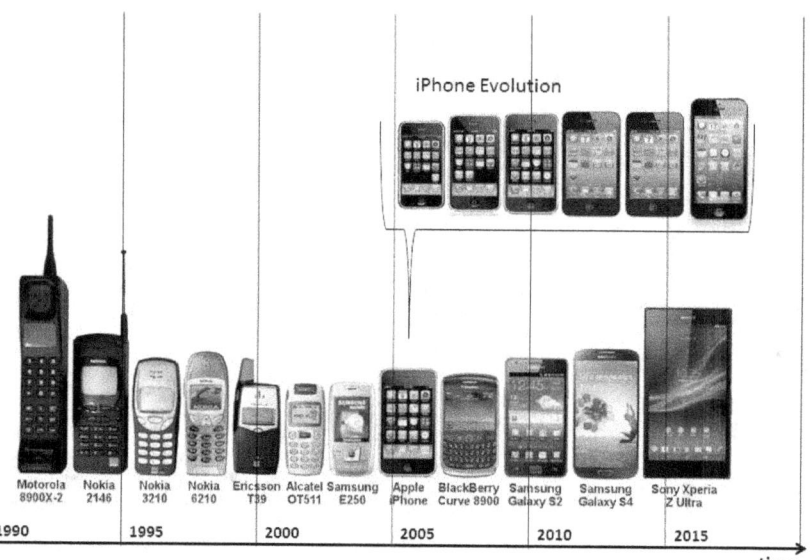

Figure 278: Evolution from cell phone to smartphone

From left to right, the mobile phone evolution is shown in different models, from the Motorola brick phone (1994), the Nokia 6210 (2000) and the Apple iPhone (2007) to the Samsung Galaxy S2 smartphone (2010) and Sony Xperia (2013). Insert shows evolution of Apple's iPhone.

Figure created by author based on material found on Internet.

Maybe even harder to imagine is what it would be like to live in a world without electricity, which powers the Information Age that created the smartphone. What would the world be without electricity to power today's mobile telephone; the device that brought *distant writing* (electric writing, formerly known as telegraphy) and *distant speaking* (electric speech presently known as telephony) without cables (aka *wireless* communication) together?

The road to today's smartphone started in the 1970s with the bulky 'transportable' cell phones and an infrastructure of small, dedicated wireless telephone networks. Soon the mobile telephones developed into the smaller handheld 'brick' mobile phones of the 1990s (Figure 278). Then emerged the early feature phones in the 2000s: simple mobile telephones to which had been added the functions of a PDA,[461] along with texting facilities

---

[461] A PDA (personal digital assistant) is a device able to connect with a computer network that emerged in the mid-1980s. It was a small handheld, dedicated personal computer used for the management of personal information, originating from a specific application software program called personal information management (PIM). It realized the former paper-based

(SMS). Soon, along with speech and text, the mobile phone integrated a photographic camera and video recorder. Then the combination of *communication functions* and *computing functions* began the development of 'smart' phones. They were interconnected by rapidly expanding telephone networks that with every 'generation' (eg G2, G3, G4 and G5 networks) offered more functionality, more capacity and higher speeds.

Figure 279: Icons to access mobile apps on a smartphone (2017).

Source: Shutterstock (fair use is claimed).

Developing in parallel with computer networks (aka the Internet) which enabled a range of information functions, the devices added more and more functionality. These functions were realized by the mobile 'apps'[462]—short for mobile application programs (Figure 279)—for specific applications: from info on public transportation to weather forecasting and to financial banking. Mobile apps that were available in online stores and could be installed at the choice of the smartphone owner.

The roots for this 'online' phenomenon and 'smartphone' development, which came into existence recently in the twenty-first century, are to be found quite a while ago. It was the nineteenth century that saw the dawn of the Age of Communication when the telegraph was invented. That invention of electric telegraphy occurred in the aftermath of periods of considerable turmoil, including the American Revolution (1765-1783) and the French Revolution (1789-1799). In addition, it was the time of early industrialization called the First Industrial Revolution (1760-1850).

Some decades later, the by-then-matured telegraphy would facilitate the next phase with the arrival of electric speech: telephony was born. Technical developments created the technical devices (the telephones) as well as the technical infrastructure (the telephone network). Fast growth resulted in a business boom with monopolistic business practises (America), a state monopoly (Britain), and state-dominated companies (Europe). It was

---

functions of a calendar, schedule, and address book, and added new functions like e-mail communication.

462 Mobile apps basically are similar to application software for desktop computers (eg text processing). Specific groups of apps, such as the one for information retrieval (news, weather, stock market data), communication (text, speech and video processing) and data processing (financial and banking apps) contributed highly to their functionality

in the early days of the Second Industrial Revolution (1870-1914) that the GPT-Electricity contributed so much. Electricity that, along with being able to transport *power* en masse along its cables, was becoming increasingly better at the transmission of *information* along cables.

Again some decades later, the next phase started with the new-born wireless telegraph: cables were replaced by electric waves that went with lightning speed through the ether. Understandably, the existing communication industry (ie the service providers and equipment manufacturers of the telegraph and telephone) with their considerable infrastructure that had required enormous investments, resisted the arrival of this new disruptive technology. A technology that would mature in two decades, proving its value in a range of application areas, from maritime communications to military communications. This was a period of fierce business competition and patent wars. However, it would take another type of war—the Great War of 1914–1918—to eliminate the barriers for further progress.

The first century of the Communication Revolution (Figure 271) may have started with the Era of Telegraphy, followed by the Era of Telephony, but it was not to take off until the Era of Wireless Communication matured in the 1920s. Then the technological progress would really pick up speed in the form of developments like radio broadcasting, television broadcasting, and ultimately the wireless telecommunication of our days. It was a chain of technological inventions that were the basis for this remarkable revolution in the Era of Wireless Communication.

It's understandable that one might wonder what made all that possible. What were the mechanisms of change and novelty, and what were the requirements that made invention and innovation possible? Or even broader: what makes change and novelty happen? In this last part of our analysis we will try and conclude on some basic patterns found in our observations. Patterns that govern change and novelty.

First, let's step back in history and look at the work of some economic philosophers who were concerned with the topic of economic growth: Adam Smith, Robert Solow, Paul Romer, and Gerhard Mensch.

The Invisible Hand of Innovation

As we have seen, a consequence of the emerging wealth of technological inventions was the creation of new businesses by (inventor-)entrepreneurs. Not only did the new communication networks create the communication infrastructures that facilitated the national economies, they also created employment where the new 'telecommunications' industries grew and prospered. However, when a new industry emerged as the result of a disruptive new technology, there was also another effect noticeable: the old industry came under pressure as companies failed and economies contracted and unemployment rose. This process of creative destruction was encapsulated in the specific context of its time, both with its Spirit of the Time and Madness of the Time. It resulted in the business cycles with their booms (economic growth) and busts (economic depression) (Schumpeter, 1934, 1939).

The booms and busts appeared at different moments in time, in different forms, and with different intensities. Some were more locally contained; some had a worldwide effect (eg the Bank Crisis/Panic of 1873 that caused the Long Depression). Some coincided with railway manias accompanying the industrialization of that time. In each of the major economies, they created their own effect: from the Great Victorian Boom in Britain to the Gilded Age in the US, La Belle Époque in France, and the Gründerzeit in Germany/Austria (Figure 115).

Remarkably enough, these developments were accompanied by the new phenomenon of the world expositions, where people were shown—among others topics—the wonders of technical progress, such as the London Great Exhibition of 1851, the Paris International Exhibition of Electricity of 1881, and the London Inventions Exhibition of 1885.

All that industrialization obviously created jobs. But there was more than just the employment. As we have seen, the industrial revolution fueled by industrial capitalism had created massive financial wealth. This private wealth sought its way back into the economy, for example, in investments in railroad projects and early telegraph projects. This process had to cope with its own problems, such as the Panics/Crises of 1847, 1857, 1866, 1873, 1882, and 1893. All in all, both the technological wealth and financial wealth created economic cycles with times of prosperity and times of depression. Governments became more and more concerned and looked for ways to control these effects with (financial) policies.

559

It was the pioneer of political economy Adam Smith (1723–1790),[463] a philosopher conribtuting to the Scottisch Enlightenment and professor at the University of Glasgow, who—after exploring the history of astronomy, ancient physics and the ancient logis and methaphysics (Smith, 1880)— wrote *An Inquiry into the Nature and Causes of the Wealth of Nations* (1776). He was interested in what made 'wealth' happen in a time where the wealthy land-owning aristocrats exploiting 'land and labour' were rivaled by the trade-based wealth and the colonial wealth of the 'nouveau riche'. Subsequently, Adam Smith developped a different economic view; free trade not hindered by financial barriers. He stated:

> *It is the maxim of every prudent master of a family, never to attempt to make at home what it will cost him more to make than to buy.... If a foreign country can supply us with a commodity cheaper than we ourselves can make it, better buy it of them with some part of the produce of our own industry, employed in a way in which we have some advantage.* (Smith, 1817, pp. 264-265)

In contrast to the protectionist policies of mercantilism,[464] he stated that free trade created competition, cheaper goods, and higher profits, and he argued that governmental policies should stimulate that free trade— one reason being that this would be to Britain's advantage, as the fruits of its industrialization would be exported to other countries following that same approach of free trade.[465] Smith's views on less-protectionist economic policies emerged in the times of early industrialization where the division of labour in the factory system had created industrial capitalism.[466] In terms of economic theories, it was the view of Smithian economic growth, where the economic 'production function'[467] included, next to labour and land, also

---

[463] Adam Smith created 'political economics' by studying the creation of wealth in states; it became the science of the laws governing the production and exchange of the material means of subsistence in human society.

[464] Mercantilism is a range of governmental policies aimed at protecting the home market and charging foreign imports of finished goods with high tariffs. The import of raw materials was not or was minimally charged, as was the case with the exportation of finished goods. Mercantilist policies went hand in hand with colonialism; colonies provided the mother country with access to resources and raw materials and, in return, would act as a market for industrial products made in the mother country.

[465] Britain and France were rivals in the Scramble for Africa as well as the colonisation of Asia. However, by that time Britain was negotiating a commercial treaty with France (aka Eden Treaty of 1786). The treaty ended a period of economic war and created a system of reduced tariffs on goods from either country.

[466] It was also the time where the American Revolution (1765–1783) would end the ties between the 13 colonies and Britain. France had just lost its Canadian and Indian colonies, and was in the run-up to the French Revolution.

[467] A concept used by economists to understand how economies function. It relates the output (ie growth of an economy) of a black box system to a range of inputs (eg labor, capital, etc.).

capital. Together land, labour and capital were considered to create the wealth of a nation.

In fact, he offered more than just an economic theory, as he observed the role of the individual in his 'Invisible Hand of Economy':

As every individual, therefore, endeavors as much as he can both to employ his capital in the support of domestic industry, and so to direct that industry that its produce may be of the greatest value; every individual necessarily labors to render the annual revenue of the society as great as he can. He generally, indeed, neither intends to promote the public interest, nor knows how much he is promoting it. By preferring the support of domestic to that of foreign industry, he intends only his own security; and by directing that industry in such a manner as its produce may be of the greatest value, he intends only his own gain, and he is in this, as in many other cases, led by an invisible hand to promote an end which was no part of his intention. (Smith, 1817, pp. Book IV, Chapter II. 9).

Smith's 'Invisible Hand' was about individual actions that created unintended social benefits for the society as a whole. It was part of his view that resulted in 'laissez faire' economic policies and that contrasted with the protectionism favouring privileged society at that time. With these privileges—aka monopolies—the few had obtained riches at the expense of the masses. Smith's ideas were the foundations of classical free-market economic theory, and they marked the dawn of the development of liberal economic theories resulting from the Scotish Enlightenment movement, which tried to encapsule the rules governing economic progress after the First Industrial Revolution.

However, the economic growth could not be explained by land, labour, and capitalism alone.[468] There was more that contributed and was left unexplained. By the 1950s economists—such as Robert Solow and Paul Romer—were trying to explain the so-called 'residual factor' of the production function: the part that could not be explained by capital accumulation and increased productivity. Robert Solow's growth theory recognized that Ricardian[469] factors—more workers, more capital, or more capital per worker—could not wholly explain economic growth. Innovation mattered, so Solow's economic theory of growth included, along with land, labour, and capital, technical progress. In terms of the economic theories, Solovian growth added technological development to the production function. But in what way did that technical progress influence economic

---

[468] Later in time other economists—such as Nicolai Kondratieff, Simon Kuznets, Clement Juglar, Joseph Kitchin, and others— created theories about those economic cycles, but they failed to get a grip on its origins, as much of economic growth was left unexplained.

[469] David Ricardo developed the theory of comparative advantages, the law of diminishing returns, and the theory on economic rent.

cycles? It was Gerhard Mensch (1937), in his book *Stalemate in Technology: Innovations Overcome the Depression* (1975/1979), who related the emergence of basis innovations to upswings of the economic cycle:

> *The changing tides, the ebb and flow of the stream of basic innovations explain economic change, that is, the difference in growth and stagnation periods. ... The lack of basic innovations seems to have caused the slide into past technological stalemates, and innovative surges finally brought the stalemate to an end.* (Mensch, 1979, p. 135).

So, economic growth was—partly—a result of technical progress. And technical progress was related to basic innovations. Mensch was not the only scholar to make the same judgement: "The basic notion in all these theories is the same. Innovations are important drivers of economic growth and of economic cycles." (de Groot & Franses, 2005, p. 6).

Thus, the technology-driven *basic innovations*—within their cluster of innovations—were linked to economic growth through the cluster of businesses (Figure 280). And as innovations are the result of human curiosity, ingenuity, and creativity, it is that specific human behaviour that is

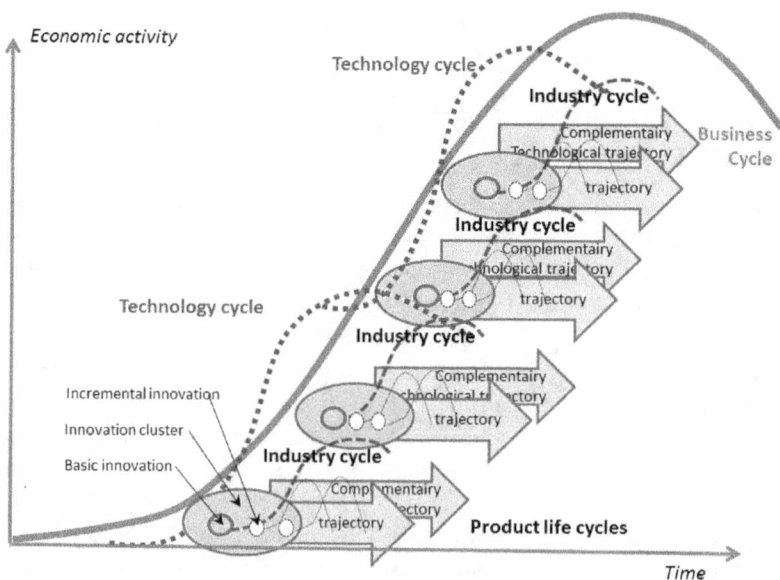

Figure 280: Technology as driver of Business Cycles.

Figure created by author

the creator of individual and collective wealth.[470] Human behaviour we observed so clearly in the *Act of Invention* (Figure 232, Figure 275) and the *Act of Business* (Figure 233) of our inventors.

Having established the dominant mechanism of economic growth, it is now the time we looked more in detail at how that human behaviour[471] was related to technological innovation.

Origin of Innovative Human Behaviour

Given that human behaviour is the motor of innovation, one could wonder what the fuel for that motor is. What motivated our inventors? What were the more specific personal characteristic's in the individual behaviour of our inventors that were triggered during their Acts? Could common elements be found in what caused their inventive and entrepreneurial behaviour? Let's try and take a—careful—effort to analyse the 'innovative behaviour' of the inventors of the Communication Revolution.

A common denominator of all the inventors was their *personality* (shaped by their nature and their nurture[472]). Reflecting on the 'nature' of our inventors is outside the scope of our study, as it is part of the interpretation by their biographers,[473] but in general terms their personalities can be identified. Marconi's personality, for example, is described as an irresistible combination of magnetic charm, singular vision, and ruthless grand ambition;

> *Marconi could be tough and single-minded … Marconi was also obsessive about being kept informed of every aspect of the company's business.… And he could be merciless … had little interest in anything outside wireless … He desired female*

---

[470] Behaviour is a broad term that describes the interaction with a system and its environment: human behaviour, social behaviour, etc.

[471] In Psychology Human Behaviour refers to the full range of physical and emotional behaviours that humans engage in; biologically, socially, intellectually, etc. and are influenced by culture, attitudes, emotions, values, ethics, authority, rapport, persuasion, coercion and/or genetics. This type of behaviour is scientifically viewed as being without specific meaning, unlike social behaviour that is influenced by the expectations, norms and values of others. 'Good social behaviour' reflects that the human interaction conforms to norms set by society.

[472] This refers to the Nature versus Nurture debate about whether human behaviour is determined by the human genes or by the human upbringing. The former refers to behavioural genetics that covers heritable individual behaviour such as cognitive abilities and personality. The latter refers to the influence of one's upbringing on one's behaviour and was described by John Locke, who stated that every human being started with a tabula rasa ('blank slate').

[473] Such as Marc Raboy's biography of Marconi, Grosvenor and Wesson's biography of Alexander Bell, and Silverman's and Mabee's biographies of Samuel Morse.

attention and companionship, and was constantly infatuated, often with very young woman. He wanted more than anything to be loved. (Raboy, 2016, pp. 669, 284, 675).

Alexander Graham Bell's 'was one of driving energy, insatiable scientific curiosity, independence of thought and individuality of action' (Osborne, 1943, p. 16). Morse is reported to have had the gift of friendship, leadership and…a personality prone to extreme paranoia, bitterness, and a strongly held sense of cultural elitism. Whatever the case, it was their personalities that drove their actions in the context of their time.

Personality dictates Innovative Behaviour

So, a personality that had a vision of the future possibilities that their invention encompassed, was a common denominator of all the inventors' *individual behaviours*. Given the socio-economic context of their lives—that created the different conditions for their inventive activity—each basic innovation was the result of the inventor's individual actions, a creative, explorative behaviour that originated from fascination with an idea based on a perceived concept. Maybe, when combined with some dissatisfaction with their ongoing life, this new thing certainly challenged them when they were mentally travelling through time into the future, seeing the contours of the opportunities their idea of a future development offered. Opportunities that further fuelled their embryonic foresight of a future to come. It created their specific visions, which were maybe sometimes based on unrealistic dreams, but also on practical and realizable conceptions.

Next to that element of vision, more was needed to succeed. Their individual behaviour was also characterized by some moral courage and perseverance to implement their idea and vision. They were not dreamers, they were doers that had the actual courage 'to do it'. They were action-oriented, driven to invest their time and energy, to fanatically pursue their vision, to fight for their (patented) rights, and to battle their numerous (business/scientific) opponents. This all done in the pursuit of their original vision.

Notwithstanding all the barriers that society created, their individual Acts of Innovation and Acts of Business ultimately benefitted society.[474]

Today we would consider someone like Marconi part of the scientific-technological elite, but what really set him apart in his day was the vision he articulated for the benefits of global communication. Decades before McLuhan coined the phrase,

---

[474] The Enlightenment philosopher Thomas Hobbes, in his work *Man and Citizen* describes courage and prudence as strength of mind as opposed to goodness of manners. These virtues are always meant to act in the interests of the individual while the positive and/or negative effects of society are merely a by-product.

Marconi was living in a global village of his own making, personifying a utopian idea of the liberating power of communication technology. He was the first to believe in, understand and express the power of communication. (Raboy, 2016, p. 676)

Undoubtedly, along with creativity and being action-oriented, many more attributes to one personality can be found, but we leave that to scholars more versed in this topic.

Environment dictates Innovative Behaviour

Considering the influence of 'nurture', Morse, Cooke, Bell, and Marconi were children of their time growing up in their specific dynamic social-political environments that were described before. That environment formed the traveller (Morse/Cooke in Europe), the new immigrant (Bell in Canada), and the foreigner (Marconi in England). That environment also offered the opportunities for their Acts of Innovation and Acts of Business. And one part of that total environment was the family in which they were brought up.

One can assume that their family backgrounds may also have provided stimulus for their later actions. Morse was the son of a notable father, tutor at Yale, pastor in Charlestown, and publisher. Bell's father was a well-known Scottish phonetician, professor, and creator of a system to teach the deaf. Marconi came from a mixed Anglo-Italian background, his father a landowning business man and his mother related to Irish businessmen. Their family backgrounds also influenced their educational upbringing; none of them came from an illiterate background. Although quite diverse, they all pursued an education: Morse as painter, Bell as elocutionist, and Marconi with private tutoring. However, at the moment of their contributions, none of these men were educated as 'electricians'. They had no specific expertise in the new technology of electricity, either as theorizing scientists nor as practical engineers, but they were fascinated by that new phenomenon of electricity.

But there is more than the direct environment of family and relatives that influenced their behaviour. They lived in the societies of their days. Societies that created an infrastructure for their activities. A cognitive infrastructure with accumulated knowledge and knowhow (eg in the field of electricity). A financial infrastructure where accumulated wealth was available to finance their expanding activities (eg the capital market). And an institutional environment that made their corporations and their financing possible (eg the institutions like banks and stock markets), that gave them their patents and the patent protection that goes with it (eg the institution of the Patent Office).

Innovative Behaviour: The Invisible Hand of Innovation

Initiated by dissatisfaction and opportunity, sparked by a mental picture, enforced by the conceptualization of an idea based on an evolving vision, having the courage and endurance to implement and fight, with the conditions created by their contexts of upbringing, their Acts of Innovation and Acts of Business constituted the *Invisible Hand of Innovation* that contributed so much to the general purpose technology of electricity. Acts that were determined by the overall context of their times; the days of the Industrial Revolutions.

The contributions of our inventors where the result of their individual actions. They lived in a social environment of their family and their business compatriots; financers willing to take financial risks who believed in them and other people who complemented their capabilities. And they were living in a specific period of time called the Industrial Revolution, with both its specific Spirit of Time and Madness of Time.

Looking from a satellite perspective, this Invisible Hand of Innovation seems to be governed by a range of mechanisms and regularities which seem to contain certain ground rules (aka laws).

Such as the *First Law of Innovation*[475], which states that innovation—on the individual level and to be defined as stepwise change in a function—is only going to take place when a certain set of requirements are met. There is the requirement of *discomfort* with the existing situation (either due to a perceived opportunity or due to an actual dissatisfaction). There also has to be an individual *vision* of what the new situation could be, such as a challenging perspective. Next, there is the requirement of individual *courage* to undertake the change, as the path of innovation to be followed is not going to be an easy one. And finally, the individual *circumstances* have to be right, as they create the environment in which the change is going to take place.

In addition, there seems to be the *Second Law of Innovation* that states that innovation can take place on different levels: on the level of individual persons, the level of collectives/groups and the level of societies/ nations. Take for example applying the first Law of Innovation on the level of a group (such as a company). Then, there also is the requirement of a collective *discomfort* with the existing situation (either due to economic/technical performance, or stimulated by the perspective of a

---

[475] In my book *Innovatie: van Onbehagen tot Durf* (Innovation, from Discomfort to Courage, only available in Dutch) which investigated changes in the Dutch society of the late twentieth century, I formulated in 1989 these three Laws of Innovation for the first time (p.286).

perceived commercial opportunity). There also has to be a collective *vision* of what the new situation could be, for example in the form of an—implicit or explicit—corporate strategy. Next, there is the requirement of collective *courage* (eg allocating funds to a risky activity instead of using for existing business). And finally, the collective *circumstance*s have to be right, as they create the environment in which the change in the group/collective is going to take place.

And finally, there seems to be a *Third Law of Innovation* that states that the appearance of innovation is related to its context: the social/economic/political/technical/scientific environment in which it takes place. A context that has a material, cognitive, and institutional component creating the total infrastructure geared for change and novelty. The material infrastructure is characterized by the availability of resources (eg financial). The cognitive infrastructure is the availability of knowledge and knowhow. And the institutional infrastructure can be found in the norms and values, and laws and rules of a collective/society.

These three laws seem to rule human behavior in relation to change and novelty. So, as the next and last step in our concluding analysis, let's look at that human element in more detail.

The Human Element in Innovation

In this analysis, we looked at the *spawning* of the general purpose technology of electricity. Along with electric light, electrical telegraphy, and electrical telephony, wireless telegraphy was one of the many inventions that originated from the general purpose technology of electricity. A technology that proved to be very pervasive in spawning new application areas. This spawning attribute was to become one of the many factors that contributed to the Second Industrial Revolution.

We observed how the development of wireless telegraphy expanded the Communication Revolution, a period in time where communication over distance was transformed from the classical means and methods—such as the classic postal mail and the optical semaphore—into lightning-fast communication with the new electric telegraphic instruments. The wireless telegraph, again, was a development that was realized through the contributions of many people, from the tinkering and thinking engineers and scientists to the entrepreneurs creating business activities large and small.

While reflecting on the massive social changes that originated from the contributions of so many people willing to devote their creative and entrepreneurial efforts to changing the world, we will look at the human related properties that resulted in the development of the communication engines.

Human Curiosity, Ingenuity, and Creativity

One observation stands out among the many that can be found: the obvious but easy-to-miss observation that innovation is about human activity. The creative, engineering and entrepreneurial behaviour of people resulted in all these individual contributions. It were their Acts of Invention and Acts of Business that created the 'clusters of innovations' and the 'clusters of businesses'.

The driving force behind that individual behaviour is curiosity, one of the dominant characteristics of human nature.[476] It is the curious nature of man that has led him to wonder, ponder, and then learn. Curiosity is the building block of our common knowledge structure, the key that opens new vistas of thought. It is curiosity in man's nature that drives him to wonder and ponder and understand different phenomena in life.

---

[476] Curiosity is a quality related to inquisitive thinking such as exploration, investigation, and learning. Curiosity as a behaviour and emotion is attributed over millennia as the driving force behind not only human development, but developments in science, language, and industry.

However, being curious is not enough. More is needed to realize invention and innovation. After obtaining knowledge and insight, there is the creative act, where ingenuity[477] creates the new combination: the moment the invention is born or the innovation is conceptualized. It is like a mental image in the creative mind of the new thing to be created. That creative act, sometimes described as getting the 'idea', is only the beginning. The 'idea" may be conceived, but before the concept is converted into an innovation that is ready to be introduced to the market, quite a bit of additional efforts have to be made. The analogy with the creation of new human life comes to mind. After the moment of conception, the pregnancy period is needed to create the new human being. After its birth, however, it needs quite some time of development before it 'can stand on its own'. Similarly, when innovation is the case, after the prototype is developed and before it is to be put out as a commercial product on the market— especially if the market does not exist at that moment—quite a bit of additional creativity and entrepreneurial efforts are needed.

In addition, one has to realize that all this curiosity, ingenuity, and creativity happened in the context of its time. That overall context dominated the developments to come, as a world in turmoil (as in the American Revolution and the French Revolution, but also the Madness of the Time) has a different influence on human behaviour than a world at peace and undergoing economic progress with its Spirit of Times. This case clearly shows how this human behaviour resulted in the contributions of so many people towards the development of wireless telegraphy in the second half of the nineteenth century.

Curiosity into the Nature of Lightning

Just as the telephone developed from early investigations into the nature of sound, the trajectory that led to the invention of the wireless telegraphy started with scholarly curiosity of a totally different phenomenon: the Nature of Lightning. This was not too surprising, as the atmospheric lightning that accompanied storms and thunders had people frightened from early times, but it also made them wonder. This was also the case with so many other forces of nature: such as investigations into the nature of fire and the power of heat had led to the steam engine.

In older times people attributed the cause of lighting and thunder to the angry gods, such as Thor throwing his hammer (Figure 281). Later, when the reasoning of the Scientific Revolution had complimented religious

---

[477] Ingenuity is the ability to invent things or solve problems in clever new ways. Ingenuity involves the most complex human thought processes, bringing together our thinking and acting both individually and collectively to take advantage of opportunities and/or overcome problems.

beliefs, atmospheric lightning became an object of the curiosity of the natural philosophers who investigated its nature and the related 'electrical fluid' as a phenomenon in its own right (Figure 19).[478] Then, by the eighteenth century, lightning became the focus of scholars who explored its specific properties. Some delved into the visual phenomena of the lighting and ultimately discovered the arc light.[479] Others touched on the phenomena that were part of what would later be called 'electric waves' (Figure 284, left).

This curiosity, part of human behaviour originating in the art of survival, created insight into the observed phenomenon. It was this early understanding that started a collective learning process, one of

Figure 281: The Germanic god Thor wielding his hammer, creating fire and thunder in the skies.

Source: Wikimedia Commons. Artist: Mårten Eskil Winge.

experimentation that could sometimes be quite dangerous, as Georg Richmann (1711–1753) found out on August 6, 1753, when he was electrocuted while experimenting with atmospheric electricity (Figure 282).

But it was also a process where the phenomenon at hand was replicated on a smaller scale: the electric spark that could be reproduced by discharging Leyden jars over a Rhumkorff coil, which created 'sparks' of bright white light quite uncomfortable to the eye. From this experimentation developed the trajectories of the discovery of electric light as well as the trajectory of electric waves.

Figure 282: Georg Richmann killed by lightning experiments (1753).

Source: Wikimedia Commons.

---

[478] See: B.J.G. van der Kooij, *The Invention of the Electro-motive Engine* (2015), pp. 32–62.
[479] See: B.J.G. van der Kooij, *The Invention of the Electric Light* (2015), pp. 37–41.

It was Felix Savery (1797–1841), professor at the Polytechnique in Paris, who concluded in 1827 from his experiments that a charged Leyden jar would discharge in a damped oscillatory manner. He concluded this from the magnetization of steel needles placed aside a two-meter-long wire loop used to discharge the jar. That magnetizing of a steel needle at this distance was early evidence of high-frequency transmission and detection. One could say that he made the electric waves visible.

Ingenuity in Science and Engineering

Where curiosity caused people to study the phenomena and resulted in their discoveries, it was another property that led people to wonder what they could do with their discoveries. They started to experiment, investigate, and find ingenious solutions to the problems their curiosity had identified. The results came in different shapes and forms: from the failures of dead-end technologies to the more successful inventions that changed the world. Whatever the case, their ingenuity to create solutions resulted in a considerable body of knowledge and knowhow that contributed to the invention of the wireless engine.

Theorizing about electric waves: The theoretical scientists that created insight into the Nature of Lghtning with their experimenting also created early artefacts that could produce sparks. Their work had already resulted in artefacts like Hertz's experimental transmitter/receiver (Figure 284, top right) and artefacts to detect those electric waves. The time was ready for the combination of 'telegraphy' and 'electric waves'.

Practicing electric waves as carrier of information: With the devices developed for creating and detecting Hertzian waves, a new means to carry information very fast over a long distance had emerged. Previously it had been coded information and speech transmitted by the cabled infrastructure; now it was the transmission by electric waves as experimented on by the engineering scientists (Figure 284, bottom). With their hands-on experience, they developed the basic knowhow of the wireless communication technologies.

Experience from earlier communication infrastructures: Both the work of the scientists and the engineers had taken place in the context of the fast-growing telegraph and telephone infrastructures in the second half of the nineteenth century. With the creation of this communications network, a lot of technical experience was built up in the telegraph industry, which could be used when other communication engines were developed. In addition, it was the telegraph industry, where the need to optimize transmission capacity had initiated a fury of activity, that

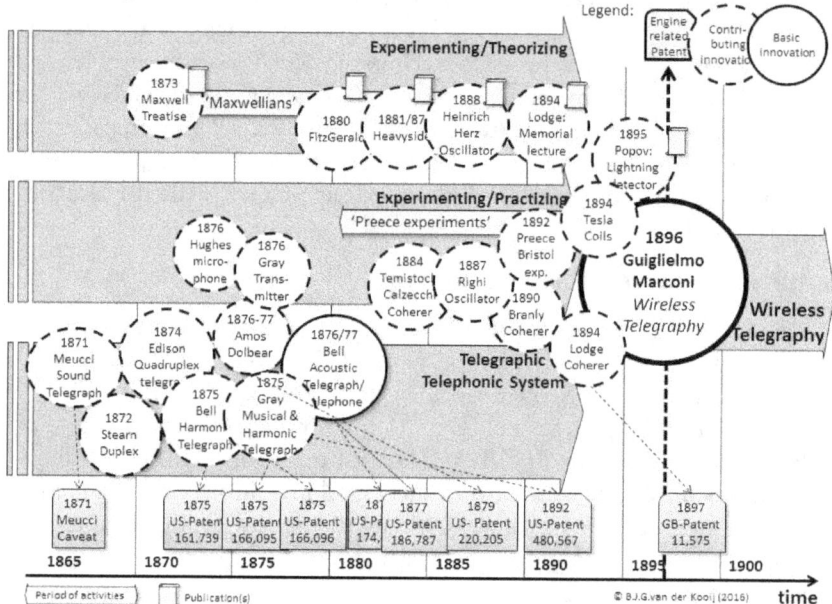

Figure 283: Trajectories contributing to Marconi's act of invention.
Figure created by author.

initiated the search for the high-capacity harmonic telegraph. The result was the telephone. The same was the case which the cable-problem, that created the need for an alternative solution for communication over distance. The result was wireless telegraphy.

To put it simply it was science, engineering, and previous infrastructures, that had created the fertile soil for the development of wireless telegraphy from the roots of the Hertzian waves (Figure 283).

Creativity in Science, Engineering and Business

The Hertzian waves became the focus of the wireless mania among experimenting engineers and theoretic scientists (Figure 151, Figure 230). They all, in one way or the other, added to the increasing understanding of the phenomena at hand.

Scientists: There were those natural philosophers—later called scientists—who were primarily intellectually interested in the phenomena. They were the thinkers experimenting in their laboratories, and by publishing and lecturing about their results and observations, they spread the knowledge, in the process enhancing their academic (eg Lodge) or professional stature (eg Fleming). And from time to time they stumbled on something important: a discovery that gave fundamental insight into the observed phenomenon.

Depending on the context in which their work took place, we see a different additional aspect that followed those discoveries. In the New World of capitalistic America, an entrepreneurial climate—not restricted by historic constraints—flourished. It was the time of early industrialization. The first tycoons (aka robber barons) of the railroads had emerged, trying to realize their *American dream.*[480] In that context, many of the wireless scientists became involved in commercial activities, either in an effort to exploit their patent or trying to combine forces with industrialist entrepreneurs with the purpose of 'earning a buck' from their work. In the Old World a different climate was noticeable. There, industrial entrepreneurial activity, restricted by the historic corset of the societal elite (eg France) or encapsulated in the national public interest (Britain), was

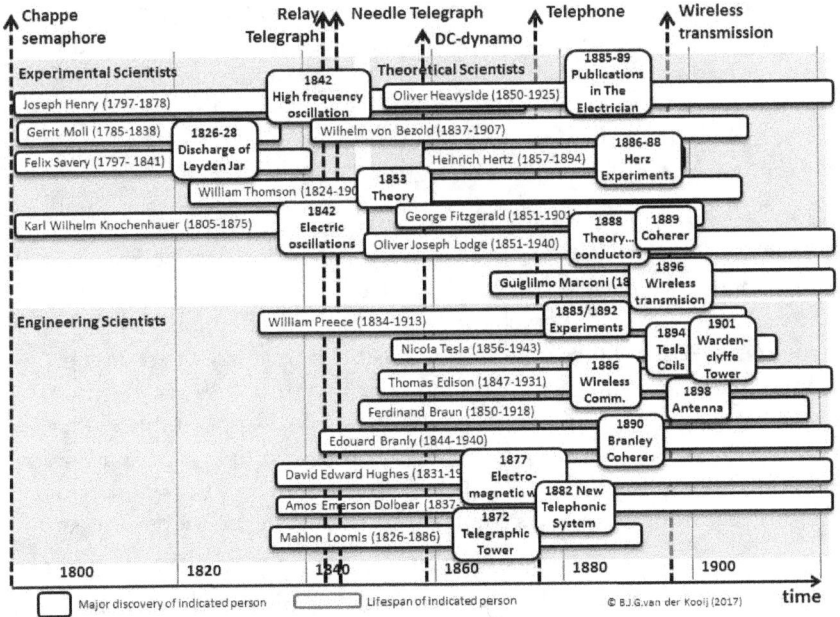

Figure 284: Overview of contributors to wireless technology.

Figure created by author.

---

[480] The American dream was the national ethos of the United States, the set of ideals (democracy, rights, liberty, opportunity, and equality) in which freedom includes the opportunity for prosperity and success, and an upward social mobility for the family and children, achieved through hard work in a society with few barriers.

battling societal resistance to change, so many scientists left it to others to wage the unavoidable battles of business (eg Fleming advising Marconi).

Engineers: There were those practical philosophers—later called engineers—who were primarily interested in applying the discoveries made about the phenomena. They also had their 'idea', but that mental image was more practical. They were the tinkerers, and their work was at the core of engineering, the learned analysis (convergent thinking) and synthesis (divergent thinking) that became combined with their individual 'practical' creativity (Cropley, 2016).

Morse, the painter used to conceptual thinking, saw after being confronted with electricity, a mental picture of 'the messages traveling with the speed of light over electric cables'. Nevertheless, it took him years to create a crude working artefact, built on an easel. Bell associated the human ear with the sound receiver. In his creative mind, there was a cross-fertilization of his knowledge of speech and the opportunities electricity was offering: sound and electricity were the foundations of his new combination. Marconi was believed to have had the mental image of wireless communication in Sanctuario di Oropa: 'While in the Alps I got the idea that not merely signals but actual words and voices could be transmitted from one place to another without wires.' (Raboy, 2016, p. 31).

Inventor-entrepreneurs: There were those people—later called inventor-entrepreneurs—who, next to the technical affection with their experiments, were eager to exploit the fruits of their work by themselves. They had—by chance or by origin—affection with the activities we described as the Act of Business: organizing, arranging, commercializing, and demonstrating their invention. As the electric technologies proved so rich in spawning business opportunities, they created the organizational structures that financed their work and exploited the (patented) fruits of their work.[481] In most cases, they did not do this alone but were stimulated by people in their environment.

Morse was financially and politically supported by Francis O.J. Smith, a member of Congress with some capital, and by the manager Amos Kendall, former postmaster general. Bell was stimulated and supported in his business by Hubbard and Sanders. Marconi was encapsulated in an entrepreneurial network, his parents and their

---

[481] Thomas Edison (1847–1931) is an example of this category. Holding some 1,093 US patents in electric light, telegraphy, and telephony, he created numerous companies (the Edison Companies such as the Edison Electric Light Company, the Edison Machine Works, etc).

Jameson connections, during his pioneering years in Britain. These inventors were the ones who responded to the external stimulus and became the core of the evolving business activities.

Change and Novelty

It was in this timeframe that Guglielmo Marconi entered the scene and baffled both the scientists and the practitioners. But that entrance was not greeted with too much enthusiasm, not by those who had vested interests in the preceding technology, not by those who explored the same subject, nor by those who wanted control over its application. One could wonder why there was such reluctance and resistance, and to try and find an answer, we have to back to the basic need for communication. This brings us to another observation that encompasses the development of the GPT-Electricity: the techno-economic-scientific context in which the development of the communication technologies took place.

Economic Change: Competition Everywhere

Communication has been a basic need for humans from historic times on. People, belonging to families, clans, and societies, have always needed to communicate to survive. Communication is not only a practical tool to execute individual behaviour and express individual feelings, but also a means to work together and to secure a place in the hierarchy of a group. If these collectives are small, limited to the family living in one shelter or the clan in one hamlet, communication limited to distances as far as the voice carries: the *spoken word*. When collectives expand into regional societies, another form of communication is needed: the *written word* that can be transported by courier (running or on horseback). So the need for communication over distance increases when societies expand and interact over larger regions. Over time other systems were developed, such the semaphore system in Napoleonic times that supported the French geographic expansion and that gave rise to the *coded word* visually transmittable over larger distances. Then the new technologies of electricity facilitated first *distant writing* and then ultimately made *distant speech* possible. Every technology developed with its own limitations and drawbacks, up to invention of the wireless telegraphy: the *transmitted word* transported throughout the spaces of the universe.

Seen in this perspective, communication at a distance is a basic human need, and the traditional forms of communication are complimented by the spoken word, the written word, and the transmitted word. Communication is a tool for societal interaction: a tool to learn, a tool to share, a tool to create influence, and a tool to exercise power. And as societies compete among each other for space and resources, communication becomes a tool for military activity. Societies also interact more peacefully, trading their surpluses and acquiring the things they need, and communication then becomes a tool for business activity.

Now, on to the new technology of wireless communication: its fast penetration in society was the consequence of the fulfilment of that basic need, expanding on the earlier developments already in place (the massive use of telegraphy and telephony). Its first successful deployment in maritime communication might have been in the vacuum left by the cabled infrastructures to solve the incommunicado problem. Its next deployment was as a military tool fulfilling the need for communications in military operations. And it was to be used to govern a large empire: the 'tool of the empire' for the British imperialism. And after all those needs were fulfilled, it became a business tool.

The control over and the possession of the wireless tool was important as became clear when nations conflicted and wars loomed on the horizon. Wireless became a national interest, hence the rise of competition between nations to have control over wireless technology. But only the technically more advanced nations such as Britain, Germany, and the United States were able to execute that control by creating their own national wireless industries, which were soon competing with each other. The consequence was business competition in the application of wireless technology for maritime and governmental use. Due to the disruptive character of the new wireless technology—it threatened seriously the massive investments in cabled telegraphy (eg. the Atlantic cables)—the new technology met large resistance with the vested interests (eg the telegraph industry). Hence the fierce competition between the old and the new technology in the application of wireless technology for business use.

Scientific Change: Increasing Knowledge

Another result from all that human curiosity and ingenuity was the increasing understanding of the mystic phenomenon of electricity, basically undetectable by human sensory organs like the eye.[482] From the early times of the experimenting with the different forms of electricity (eg the frictional electricity, the biological electricity, the electrochemical electricity) people tried to understand what they were dealing with. Was electricity something of a 'fluid', such as the two-fluid theory created by the Frenchmen Cisternay du Fay in the 1730s, or was it a single fluid, as Benjamin Franklin in the 1740s proposed? Others like Henry Cavendish (seventh duke of Devonshire) also performed electric experiments. The resulting

---

[482] Only when one touched a source of electricity, such as the electrically charged torpedo fish (aka electric ray, crampfish, numb fish), was it notable.

comprehensive theory of electricity was mathematical in form and was based on precise quantitative experiments. In 1771 Cavendish published an early version of his theory, based on an expansive electrical fluid that exerted pressure (in analogy with the well-observable fluid of water). Thus, he conceptualized the concept of electric potential (which he called the 'degree of electrification').[483]

Over time, the experimentation created increasing insight, and the theories of electricity started to develop.[484] These grew out of the mechanical thinking that had emerged in the Age of Mechanization, a thinking that was dominated by movement, contacts, and forces.

Contact theories: One of the first electric theories was the theory of 'contact electrification', which explained why electricity was created by mechanical friction in the electrostatic generators of that time (eg Hauksbee). Next, in the 1780s emerged the electrolytic-metallic contact theories. These theories surfaced when electricity was created as the result of a contact between a metal and a chemical substance. This gave rise to Galvani's theory of bio-electricity based on the animal electric fluid (as created by the electrical organs of the torpedo fish). However, his theory was contradicted by Volta's contact theory, which explained the creation of electricity as a chemical process. Although quite popular in its time, these theories were soon to be replaced as the understanding of the phenomenon of electricity improved.

Action at a distance: In line with the (mechanical) theory of attraction (ie Newton's law of gravity), in the 1780s Charles-Augustin de Coulomb explained the laws of attraction and repulsion between electric charges and magnetic poles. These laws later initiated the 'action at a distance' theories for electromagnetism. By the 1820s the electro-dynamic theories started to surface. People like Hans Christian Oersted (1777–1851), the Frenchman Andre-Marie Ampère (1775–1836), and the Englishmen Michael Faraday (1791–1867) experimented with the phenomenon of electricity and discovered its properties in the realm of electromagnetism. Oersted did his compass needle experiment and formulated Oersted's Law, which stimulated so many other scientists. Ampère explained the mechanism of electricity in his general theory connecting electric currents with magnetic forces. He created a mathematical and physical theory and postulated his force law (Ampere's Law, 1823) about the magnetic forces between two parallel electric wires: 'The mutual action of two lengths of current-carrying wire

---

[483] Maxwell named the physics laboratory at the University of Cambridge after Cavendish.
[484] A (scientific) theory is an explanation of how nature works based on the scientific method, creating a body of knowledge.

$e + \dfrac{df}{dx} + \dfrac{dg}{dy} + \dfrac{dh}{dz} = 0$	(1)	Gauss' Law
$\mu\alpha = \dfrac{dH}{dy} - \dfrac{dG}{dz}$ $\mu\beta = \dfrac{dF}{dz} - \dfrac{dH}{dx}$ $\mu\gamma = \dfrac{dG}{dx} - \dfrac{dF}{dy}$	(2)	Equivalent to Gauss' Law for magnetism
$P = \mu\left(\gamma\dfrac{dy}{dt} - \beta\dfrac{dz}{dt}\right) - \dfrac{dF}{dt} - \dfrac{d\Psi}{dx}$ $Q = \mu\left(\alpha\dfrac{dz}{dt} - \gamma\dfrac{dx}{dt}\right) - \dfrac{dG}{dt} - \dfrac{d\Psi}{dy}$ $R = \mu\left(\beta\dfrac{dx}{dt} - \alpha\dfrac{dy}{dt}\right) - \dfrac{dH}{dt} - \dfrac{d\Psi}{dz}$	(3)	Faraday's Law (with the Lorentz Force and Poisson's Law)
$\dfrac{d\gamma}{dy} - \dfrac{d\beta}{dz} = 4\pi p'$ $\quad p' = p + \dfrac{df}{dt}$ $\dfrac{d\alpha}{dz} - \dfrac{d\gamma}{dx} = 4\pi q'$ $\quad q' = q + \dfrac{dg}{dt}$ $\dfrac{d\beta}{dx} - \dfrac{d\alpha}{dy} = 4\pi r'$ $\quad r' = r + \dfrac{dh}{dt}$	(4)	Ampère-Maxwell Law
$P = -\mathcal{E}p \quad Q = -\mathcal{E}q \quad R = -\mathcal{E}r$		Ohm's Law
$P = kf \quad Q = kg \quad R = kh$		The electric elasticity equation ($E = D/\varepsilon$)
$\dfrac{de}{dt} + \dfrac{dp}{dx} + \dfrac{dq}{dy} + \dfrac{dr}{dz} = 0$		Continuity of charge

Figure 285: Maxwell's equations in his original notation in 'A Dynamical Theory of the Electromagnetic Field' (1865).

Source: http://www.ieeeghn.org/wiki/index.php/
STARS: Maxwell's_Equations

is proportional to their lengths and to the intensities of their currents.' Ampère also applied this same principle to magnetism, showing the harmony between his law and Coulomb's Law of magnetic action. Faraday's experimenting in the 1830s with electromagnetic induction—with the Faraday coil and creating his 'lines of force'—ultimately managed to realize electromagnetic rotation based on induction.

Then James Clerk Maxwell (1831–1879) expanded in the 1860s on the work of Faraday and summarized it in Faraday's Law, part of a set of equations which is accepted as the basis of all modern theories of electromagnetic phenomena (Figure 285, Figure 286). And in 1888 Heinrich Hertz, after his work in contact mechanics and verifying Maxwell's predictions, developed his theory on electric waves and contributed to the electric 'action at a distance' theories. Hertz's proof of the existence of airborne electromagnetic waves led to an explosion of experimentation with this new form of electromagnetic radiation, the so-called 'Hertzian waves'.

This overview, in short rough brushstrokes, paints a colourful picture of the development of scientific insight—aka knowledge—in electricity leading up to the work of Marconi. Following the line back in time, we see that Hertz's, Maxwell's, Faraday's, and Ampere's views on the nature of electricity are intertwined. It shows the dominant role Maxwell fulfilled when he applied the new language of mathematics—stimulated by the preceding work of Henry Cavendish (1831–1810)—to electric phenomena.

Maxwell first met Faraday in 1860, shortly after he assumed his place as professor at King's College. His contact with Faraday at the Royal Institution, where Maxwell lectured in 1861, made him an admirer not only of Faraday the man, but also of Faraday the experimenter. Maxwell was engaged, in particular,

1. $\nabla \cdot \mathbf{D} = \rho_v$

2. $\nabla \cdot \mathbf{B} = 0$

3. $\nabla \times \mathbf{E} = -\dfrac{\partial \mathbf{B}}{\partial t}$

4. $\nabla \times \mathbf{H} = \dfrac{\partial \mathbf{D}}{\partial t} + \mathbf{J}$

Figure 286: The four Maxwell's equations in a simplified version.

Source: www.maxwells-equations.com/

by Faraday's concept of the nature of the space or field existing around a magnetized or electrified body... In applying his analytic mind and mathematical command to electromagnetic problems, Maxwell did not follow the French school (Coulomb, Laplace, Poisson, and Ampère), which regarded electrical and magnetic phenomena as instances of action at a distance. Faraday's experiments in giving reality and form to magnetic and electrostatic fields with their lines of force emanating from a magnetic pole or a charged electric point prompted Maxwell to examine the physical properties of the surrounding space (Dibner, 1964).

It was the mental image of those 'lines of force' that created the concept of the electromagnetic field. In 1873 Maxwell published his magnum opus: *Treatise on Electricity and Magnetism*, and in the preface, he wrote:

The fact that certain bodies, after being rubbed, appear to attract other bodies, was known to the ancients. In modern times, a great variety of other phenomena have been observed, and have been found to be related to these phenomena of attraction. They have been classed under the name of Electric phenomena...Other bodies, particularly the loadstone, and pieces of iron and steel which have been subjected to certain processes, have also been long known to exhibit phenomena of action at a distance. These phenomena, with others related to them, were found to differ from the electric phenomena, and have been classed under the name of Magnetic phenomena...

In the following Treatise I propose to describe the most important of these phenomena, to show how they may be subjected to measurement, and to trace the mathematical connexions of the quantities measured. Having thus obtained the data for a mathematical theory of electromagnetism, and having shewn how this theory may be applied to the calculation of phenomena, I shall endeavour to place in as clear a light as I can the relations between the mathematical form of this theory and that of the fundamental science of Dynamics, in order that we may be in some degree prepared to determine the kind of dynamical phenomena among which we are to look for illustrations or explanations of the electromagnetic phenomena...It appears to me, therefore, that the study of electromagnetism in all its extent has now become of the first importance as a means of promoting the progress of science.

He did what he'd intended to do, understand the relation between electricity and magnetism, and he explained the nature of electromagnetism in mathematical terms: Maxwell's equations (Figure 286). He did that in a language hardly comprehended by normal human beings: mathematics.

Technical Change: Increasing Knowhow

It was not only the development of electric theories that created insight into the characteristics and properties of electricity. The more practical experimenting by the practitioners added to the 'knowing how to make things' based on electricity. After understanding the basic characteristics, they created the *electric circuits* that were the basic structure of all the apparatus that were to be developed. These circuits were composed of passive and active components, and that resulted in a specific electric behaviour of the system.

Passive Components: In the wave of the development of the electromagnets (eg Henry, Sturgeon) some fundamental components such as the 'coil' and transformer (based on Faraday's Coil) appeared. Now, based on the previously developed concept of 'resistance' (Ohm's law as created by Georg Simon Ohm), the inductor (L) was applied in circuits. The same goes for the Leyden jar, which evolved into the capacitor (C). Together with their derivatives—the variable resistor (rheostat) and variable capacitor—the resistor, inductor, and capacitor became the passive building blocks of electric circuits (Figure 287). In addition, there emerged the passive components based on a specific effect, such as the quartz crystals that oscillated due to the piezoelectric effect.[485] In different combinations, these passive components were used in electric circuits. First, they were used in telegraphic circuits, soon also in telephonic circuits, and later in wireless circuits (Figure 140, Figure 142).

Active Components: After Fleming invented his thermionic valve (a diode), introducing the vacuum bulb technology to electric systems, it was Lee de Forest who created the first active element: the Audion (a triode) that could amplify an electric signal. That component would initialize a whole new development of wireless communication, as sparks were no longer needed to create high-frequency oscillations.

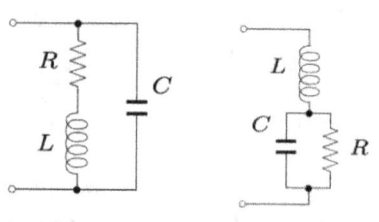

Figure 287: LCR circuits composed out of passive components.

The vacuum tube made it possible to control the behaviour of the electrons in a vacuum by creating a unidirectional flow (diode) or an amplified flow (triode) of electrons. It gave rise to a new type of electronic circuits (Figure

[485] The piezoelectric effect is the interaction between a mechanical and electric state of a crystalline material. In daily life it is applied in the ignition source for cigarette lighters.

288). Over time the circuits became more complex and created a much more complex electric functionality (eg. Armstrong's super-heterodyne receiver).

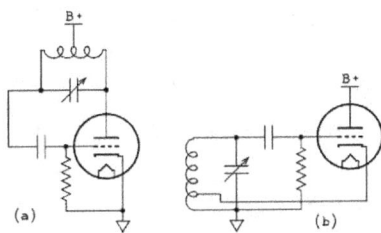

The result of the development of the passive and active elements gave rise to the development of increasingly complex electric RLC-circuits. This process started with the

Figure 288: LCR oscillator circuits composed of passive components and an active element (triode).

circuits used in wireless telegraphy to 'tune' the system and the circuits to enhance the sensitivity of the receiver. These circuits were followed by those that would resonate using a quartz crystal and could be used to receive electric waves of a fixed frequency. It was the dawn of the new electronic technologies that controlled the actions of electrical currents and the characteristics of the electric waves. And with those electronic circuits came the new calculating techniques, such as the technique to calculate the behaviour (eg frequency, damping) of a resonant circuit. These techniques provided the tools to the electric engineers to build their systems. All this technological innovation resulted in massive novelty.

Social Change: Increasing Novelty meeting Resistance

When the early telegraph services were offered, times were different. There was little resistance against this new form of communication—except maybe from the organizers of the Pony Express (America) and the Royal Mail (Britain). But they did not have an inkling of what was going to happen, most probably, when the first telegraph poles were erected. At first, communication along the new telegraph infrastructure—also called the Victorian Internet—made little direct impact on most people's lives. But soon it had a profound indirect influence on the way society functioned. The non-private telegraphy soon facilitated business communication, news communication, and governmental communication. National and international markets became more transparent as information on prices became widely available (eg stock markets, commodity markets). There was a news explosion as information about events was transmitted within hours. That faster flow of news from around the world created a new market for daily newspapers. Feeding these newspapers—often with the word 'telegraph' in their name—with much of their news was a new breed of businesses for the telegraph age: the wire agencies. And last but not least, the telegraph became a tool for the military and governments. They used it on the battlefields (eg the Crimean War 1853–1856), in diplomacy, and to govern their empires.

> *The most important fact about the telegraph is at once the most obvious and innocent: It permitted for the first time the effective separation of communication from transportation. This fact was immediately recognized, but its significance has been rarely investigated. The telegraph not only allowed messages to be separated from the physical movement of objects; it also allowed communication to control physical processes actively. The early use of the telegraph in railroad signaling is an example: telegraph messages could control the physical switching of rolling stock, thereby multiplying the purposes and effectiveness of communication. ... In commerce this meant the decontextualization of markets so that prices no longer depended on local factors of supply and demand but responded to national and international forces. The spread of the price system was part of the attempt to colonize space.* (Carey, 1983, pp. 3, 16)

There was more. The telegraph also added a fundamental change to society, as it facilitated the standardization of time. Public timekeeping had been a local affair; every town had its own time displayed on the bell tower. As long as life was predominantly local, that was not much of a problem. However, already in an early stage, railroad transportation needed synchronization for their timetables: this resulted in the standard railway time (1840s) that was, from 1852 on, transmitted by the Cooke and Wheatstone telegraph system. Time became standardized, although due to local opposition, it took till 1880 before the central Greenwich Mean Time (GMT) was implemented across Britain. The rest of the world followed suit.

The introduction of the telephone was complementary to the telegraph services. However, now the interests of the telegraph service industry were challenged, and they reacted either by entering the telephone market themselves (Western Union) or by claiming a monopoly on it (British Post Office). The societal impact of the telephone was more profound, as it changed daily life in more than one way. No part of society was left out, and every sector was influenced by the new communication medium, as already observed in 1910 by Casson:

> *What we might call the telephonization of city life, for lack of a simpler word, has remarkably altered our manner of living from what it was in the days of Abraham Lincoln. It has enabled us to be more social and cooperative. It has literally abolished the isolation of separate families, and has made us members of one great family. It has become so truly an organ of the social body that by telephone we now enter into contracts, give evidence, try lawsuits, make speeches,*

propose marriage, confer degrees, appeal to voters, and do almost everything else that is a matter of speech. ...

Public officials, even in the United States, have been slow to change from the old-fashioned and more dignified use of written documents and uniformed messengers; but in the last ten years there has been a sweeping revolution in is respect. Government by telephone! This is a new idea that has already arrived in the more efficient departments of the Federal service. And as for the present Congress, that body has gone so far as to plan for a special system of its own, in both Houses, so that all official announcements may be heard by wire. ...

In news-gathering, too, much more an in railroading, the day of the telephone has arrived. The Boston Globe was the first paper to receive news by telephone. Later came the Washington Star, which had a wire strung to the Capitol, and thereby gained an hour over its competitors. To-day the evening papers receive most of their news over the wire a la Bell instead of a la Morse. This has resulted in a specialization of reporters --one man runs for the news and another man writes it. Some of the runners never come to the office. They receive their assignments by telephone, and their salaries by mail. There are even a few who are allowed to telephone their news directly to a swift linotype operator, who clicks it into type on his machine, without the scratch of a pencil. This, of course, is the ideal method of news-gathering, which is rarely possible. (H. N. Casson, 1910, p. 199)

True, many inventions had already influenced life before. Take the examples of the electric light in the same timeframe. It freed mankind of the darkness of the night. The electric motor freed mankind—and many participants of the animal kingdom—from its role as supplier of mechanical force. But telephony was considered to be, already in the early twentieth century, much more fundamental.

This effort to conquer Time and Space is above all else the instinct of material progress. To shrivel up the miles and to stretch out the minutes—it has been one of the master passions of the human race. And us the larger truth about the telephone is that it is vastly more than a mere convenience. It is not to be classed with safety razors and piano players and fountain pens. It is nothing less than the high-speed tool of civilization, gearing up the whole mechanism to more effective social service. It is the symbol of national efficiency and cooperation. (H. N. Casson, 1910, p. 237)

Then came wireless telegraphy, and it was confronted with massive opposition that came from vested interests such as the telegraph industry and the state monopoly. Even worse, wireless telegraphy became a national interest when governments realized its potential. The initial impact of wireless telegraphy was similar to that of the earlier cabled telegraphy. It also became a tool for the military and the state. The same was the case with the long-distance governmental use of wireless telegraphy as tool of the empire. However, due to its nature, wireless telegraphy added to the reach of the cabled telegraphy into areas that were not served before: the seas that were to be served by the maritime wireless telegraphy. Not only did it solve the incommunicado problem oceangoing ships had been facing up to then, it also added to navigation the Greenwich Mean Time, which could be communication directly to ships with radio time signals. And finally, it improved the safety at sea considerably, the more so because international regulation stipulated after 1912 (the year the *Titanic* sank after a collision with an iceberg) that all merchant ships carrying passengers were obliged to use wireless transmission equipment.

However, the real impact of wireless communication would not show up until after the Great War. Then emerged the radio broadcasting of speech and music that could be received by the public with a simple 'radio'.

> *Commercial radio broadcasting took off in 1922; the years 1922 through 1925 were the boom years of early radio. In 1921, one in five hundred households owned a radio; in 1926 that increased to one in six. Stations came and went. Early fare on radio stations consisted mainly of music, variety shows, vaudeville routines, drama, and some news and political programming. Commercial broadcasting stole audience share from other entertainment industries, such as phonograph sales and live entertainment. Radio advertising also became widespread during the economic strains of the Great Depression.*[486]

After that development, it would take some decades more before the next phase of the Communication Revolution wireless telephony (aka the mobile phone as we know it) would emerge and change the way people interact on a daily basis.

In the preceding analysis, we observed the specific techno-economic-scientific-social context for Marconi's contribution. The socio-economic part of that context was based on the fundamental human need for communication and its rapid acceptance. The other part was the increasing Body of Knowledge and the Body of Knowhow that created the techno-scientific context for the contributions of so many further applications of electricity. It was the massive source of novelty that sparked the creativity of so many. By the beginning of the twentieth century, after the Great War, the world was ready for the massive invasion by 'electronics'. These technologies used passive and active components in increasingly complex electr(on)ic circuits to create 'engines' never seen before. It would be the continuation of the Times of Invention.

---

[486] 'Social impacts of Wireless Communication'. Source: https://www.enotes.com/research-starters/social-impacts-wireless-communication

Times of Invention

Looking back from the middle of the nineteenth century, one can conclude that the Communication Revolution is about the several clusters of innovations that created the telegraph, telephone, and wireless engines. It started in 1837 with the Act of Invention of Cooke and Wheatstone in Britain and Samuel Morse in America. These were followed by Alexander Graham Bell's Act of Invention in 1877 and Guglielmo Marconi's Act of Invention in 1895. Each resulting invention had its own significant impact on the consecutive development of society in the form of the clusters of businesses that emerged. They also impacted how society functioned, both in private and professional life. It was technical innovation that fuelled the Communication Revolution, a process that was not without problems of its own.

We analysed extensively what happened with the invention of wireless telegraphy: both Marconi's Act of Invention and his Act of Business. We placed that invention within its techno-economic-scientific context with the contributions of many others and the resulting wireless mania and business creation. We showed the technical improvements that followed Marconi's invention up to the Great War, and we peeked ahead at the broadcasting times to come after the Great War. We noted that much of all those developments had to be seen in the overall context of their times. One of those contexts was the country in which Marconi conceptualized his invention: Italy. The other was Britain, where he commercialized his invention. And the third context was the (Western) world in which he exploited his invention.

The historic context for the development of wireless telegraphy was founded in the First Industrial Revolution. After the dramatic social changes during the *Age of Revolutions* with the American Revolution (1765–1783) and the French Revolution (1789–1799) the overall context for invention and innovation had changed considerably. In the early nineteenth century—when Volta presented his invention of the chemical battery to the world—there was a new societal context in which industrial entrepreneurship could develop. The contributions of all those thinking and tinkering people, free from feudal constraints and religious dogmas, had created an increasing understanding—both in terms of knowledge and knowhow—of that new, fascinating technology of electricity. This culminated in the 1830s, when early inventors succeeded in using electricity for rotative power (the DC motor) and for electric communication: the telegraph (Figure 289).

The Age of Capital: The Second Industrial Revolution

After the Age of Revolution, in the midst of the nineteenth century, the world started to change again. The last Europe-wide revolutions, the Revolutions of 1848, had created the social-political context for the changes to come. Now the change to come was going to be fueled by the *Age of Capital* (1848–1875) (Hobsbawm, 2010a), which would create the foundation for the next Industrial Revolution (1850–1920). This was, again, a period that experienced economic and technical progress of a magnitude not seen before (Figure 289). It was based on an increasing understanding of the Nature of Matter that had developed in the atmosphere created by the Enlightenment. Now the accumulated wealth (eg from mercantilist and colonial trade) was not so much used for warfare, but for entrepreneurial and industrial activities.

National Developments

The industrialization period after the European Revolutions of 1848 is called the Second Industrial Revolution. It developed alongside social, political, and industrial revolutions. In the different countries, the Second

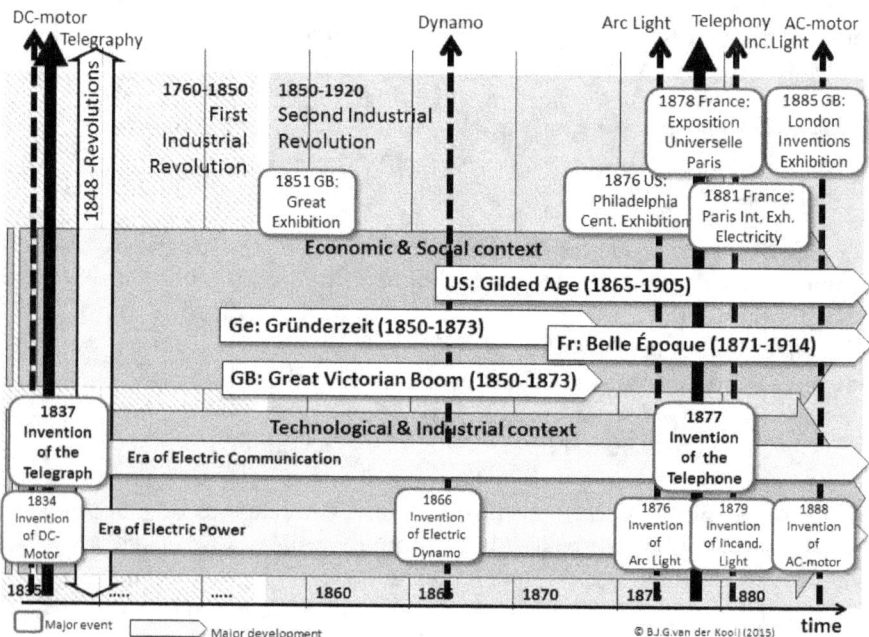

Figure 289: The context related to the invention of wireless telegraphy (Second Industrial Revolution).

Figure created by author.

Industrial Revolution had its own characteristics, but they all had one thing in common: rapid economic growth after the watershed of the European Revolutions of 1848 and the technical change that fueled it. Take, for example, the situation in the following countries:

Britain: Although Britain had no equivalent to the Revolutions of 1848 on the continent, the *Great Victorian Boom* (1850–1873) saw the pace of British economic growth accelerate significantly. The wild fluctuations in the economy that prevailed before 1850—with the Panic of 1847 (the Victorian equivalent of the 1929 Wall Street crash) at the end of the Railway Mania—were replaced by relatively smooth, but considerable, growth (and some minor Panics in 1866 and 1873). The railway boom had created massive employment, the growing transportation infrastructure was facilitating transportation, and steam-powered transportation was everywhere. Export of cotton goods had doubled. Private enterprise, fuelled by industrial capitalism, was expanding as result of the great wealth transferred over decades from colonies. Cheap and willing capital was widely available. Former mercantilist practises—like the protectionist Corn Laws—were replaced by free-trade policies. It was the early phase of the transformation from the First into the Second Industrial Revolution. The 'Great Exhibition of the Works of Industry of all Nations' in London in 1851 may have flabbergasted its many visitors with—among the many other exhibitions—the new wonders of electricity: telegraphy and DC motors. But it was just a prelude of times to come.

America: In America, after the Civil War (1861–1865) was fought, it was the *Gilded Age* (1870–1900), with its rapid economic and industrial growth that picked up in the late 1860s. Labor was scarce, and jobseeking immigrants came by the millions (eg from Italy) to find jobs in farming and mining. Railroad projects were a major business, its lines opening new areas to farming. Mining, oil, and steel production were booming, and industrial monoplies were being created. It was the time of the business tycoons, who dominated the oil industry (eg John D Rockefeller, Standard Oil), the steel industry (eg Andrew Carnegie), the railroad industry (eg Jay Gould), and financial service industries (eg John Piermont Morgan). The electric dynamo—created in the 1860s—supplied an abundance of electricity distributed by local and regional electricitity networks. The telegraph had penetrated society, and electric light increasingly illuminated households, public places, and factories. The Centennial Exhibition of 1876 in Philadelphia showed ten million people—among the many other exhibits—the miracles of electricity: Edison's giant dynamo and Bell's first telephone. America also saw the rise of labor unions and the first labor strikes (like the Great Railroad Strike of 1877).

France: On the continent, France saw the period of the *Bell Époque* (1871–1914). After the French lost the Franco-Russian War at the Battle of Sedan (1870), a period of peace, politcal stability, and economic prosperity began. Again, it had been the Paris Commune that had revolted in 1871, and the Second Empire was succeeded by the Third Republic. Soon railroads were constructed, radiating from the heart and soul of France, Paris. The city of light (ie 'gaslight') was renewed by the urban architect Hausmann, and department stores became an 'en vogue' stimulation of consumerism. And the exhibitts on the Expositon Universell of 1867 in Paris flabergasted the millions of vistors with the wonders of industry, among which was the magical device called telegraph.

Germany: The 1848 Revolutions that took place in Germany and Austria gave rise to the *Gründerzeit* (1850–1873), which ended abruptly with the Gründerkrach in 1873. It was a time with massive industrialisation: steel and machine industries (steam engines) fuelled by the rich coal deposits in the area around the river Ruhr (aka das Ruhrgebied). The resulting urbanization created a boom in housing, and a new civil class (das Bürgertum) evolved, creating many new enterprises.

The Great Victorian Boom and the German Gründerzeit emerged from the 1850s, both the Gilded Age and the Bell Epoque started in the late 1860s, and they all encompassed a remarkable period that saw massive technological inventions. In many fields—such as metallurgy, chemistry, transportation, machine tooling, etc—but certainly in the application of electricity, technological inventions had great societal impact. These include the invention of the arc light and the incandescent lamp, the AC induction motor and the AC distribution network, the telegraph and the telephone each with the networks of cables. It was an upbeat beginning to the next technological revolution: the Second Industrial Revolution (aka the 'Technological Revolution'). This was a period of massive *technical change*, in part caused by the general purpose technology of electricity (Figure 290).

The GPT-Electricity and the Second Industrial Revolution

Certainly, Marconi's contribution to the total development of communication was important, but—seen in the larger scheme of the second half of the nineteenth century—it was just one of many others that occurred in a dynamic period, one that saw a wealth of other techno-economic developments. These developments were initiated by the Scientific Revolution creating a different way of looking at the Nature of

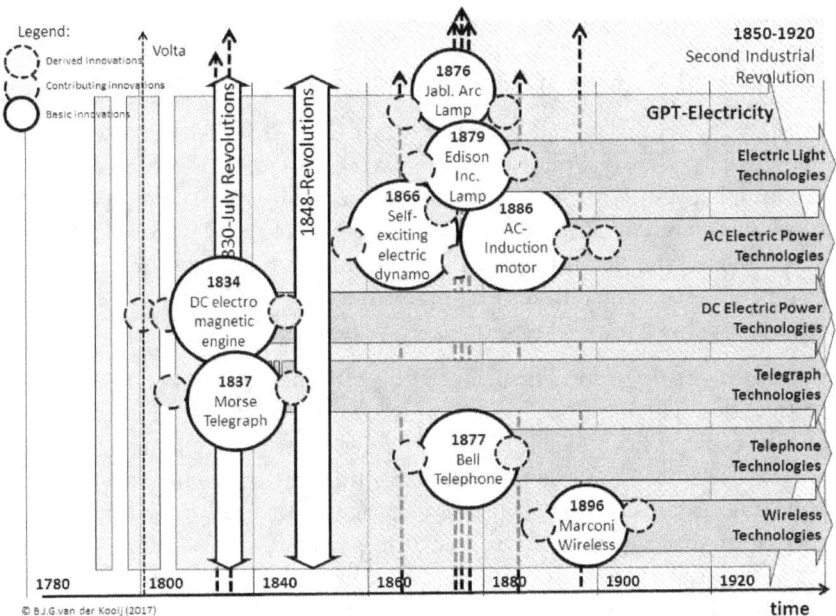

Figure 290: The spawning of the general purpose technology of electricity into other related technologies.

Figure created by author.

Matter, as well as the tangible results of the Enlightenment movement, which created the freedoms of body, soul, and spirit.[487] After all the changes that characterized the first Industrial Revolution (1760–1850), it was now the time of the Second Industrial Revolution, to which the spawning general purpose technology of electricity contributed so much, resulting in:

Power for the people: The GPT-Electricity's main trajectory was about the creation of rotary motion, both the DC motor and the AC motor, freeing humankind even more from the need for animal and natural powers. In addition, rotary motion contained the creation of electricity in the reciprocal version of the electric motor: the electric dynamo (both DC and AC). These rotative engines were all fundamental, as soon the power applications of electric technologies complemented those of the steam technologies; the electric locomotive and electrification of railroads were the result. In manufacturing, electricity now started to

---

[487] See: B.J.G. van der Kooij, *Context for Innovation: British (R)evolutions in Perspective* (2016), pp. 561–570.

power machine tools, revolutionizing the manufacturing process as it replaced the cumbersome steam-powered belt-and-shaft system. In private life it powered the numerous home appliances that freed women of tedious labor and helped give rise to female emancipation.

Light for the people: Additionally, the impact of the GPT-Electricity was enhanced by the spawning electric light technologies. Electric light, after the initial lighting by often quite-dangerous gaslight, made mankind even more independent of the pattern of daylight. Not only were social and cultural life extended into the evenings, but the manufacturing days were also prolonged. The darkness of night that had rule human life in times before was beaten by technological innovations.

Communication for the people: The GPT-Electricity spawned in even more trajectories. The hardly existing long-distance communication exploded as the result of new telegraph technologies, telephone technologies, and wireless technologies. News spread around the world with the speed of light, the process of doing business changed when telegraph and telephone made instant communication possible, ships sailing the oceans were no longer incommunicado, and solving geopolitical conflicts was supported by new ways of communication.

These examples illustrate the impact of the GPT-Electricity on mankind: on individual lives as well as on societies, economies, and conflicts in the period of the Second Industrial Revolution. Sure, the developments in the GPT-Electricity were not the only contributions to that techno-economic revolution. Other technologies had developed similarly important artifacts, such as the internal combustion engine that would soon power automobiles, ships, and airplanes. And in the same period of time there was also a range of other inventions (eg the progress made in the fields of chemistry and medicine) with enormous social consequences. They all created the Times of Invention.

The GPT-Electricity contributed, with many other technologies, to the rise of the Second Industrial Revolution (1850–1914) that was characterized by so many inventions that we could call it the *Times of Invention*. This Second Industrial Revolution, accompanied by the Great Victorian Boom, Gründerzeit, Gilded Age, and Belle Époque, spreading out over Europe and America as described before, was itself the result of many underlying *revolutionary* developments in the preceding times (Figure 291). For example, there was the massive industrialisation of the first Industrial Revolution (1760–1850) and major social revolutions such as the American Revolution (1765–1783), the French Revolution (1789–1792), and the European Revolutions of 1848.

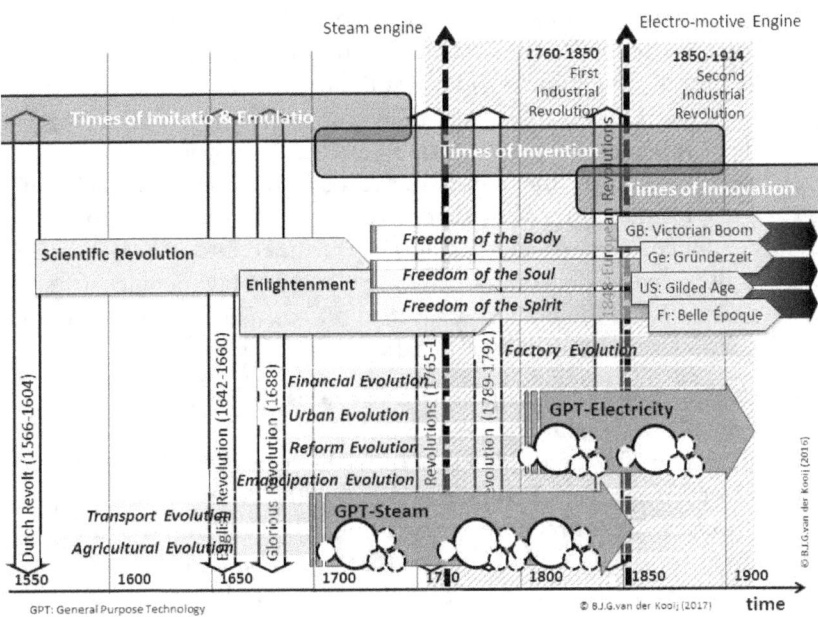

Figure 291: Overview of the contributing evolutions to the Times of Invention.

Figure created by author.

That industrialization spreading out over Europe and America was in itself the result of a range of evolutionary changes. One evolutionary change was from 'cottage industry' to 'manufacturing industry' powered by the steam engines, a development that had brought people from the countryside to the growing cities (ie the urban evolution). Manufacturing of goods had become organized when mechanization of the production process was introduced (ie the factory evolution). Factories became financed through a banking system distributing the wealth collected by merchant trades and from colonies (ie financial evolution). Complementing those evolutions were the rise of parliamentary sovereignty and diminishing absolute monarchy (ie the emancipation evolution) and the declining influence of the Roman Catholic Church (ie the reform evolution). Less visible, but of similar importance, had been the changes in agricultural practises that fed the increasing populations (ie the agricultural evolution) and in transportation by the networks of roads, canals, and railroads (ie the transportation evolution).

After the 'Times of Imitatio and Emulatio' that so characterized the Middle Ages, the Scientific Revolution and the Enlightenment had initiated the new 'Times of Invention'. Looking at the natural world they were living in, people —freed from the suffocating dogma of religion and the physical constraints of ruling powers—increasingly better understood the Nature of Matter. Such as there was the Nature of Heat and the Nature of Lightning. This gave rise to the 'Times of Invention', when the properties of basic natural phenomena were discovered and transformed into use. As was the case when the power of heat became controlled by 'steam engines' (the GPT-Steam) or when the power of 'electric fire' caused by lightning became controlled by 'electro-motive engines' (the GPT-Electricity).

This investigation of the natural world gave rise to an equivalent investigation of the social world people were living in. Now the power of people emerged, as they were liberated from the physical and mental restrictions of feudal times. People could move, travel, and work freely when they obtained the Freedom of the Body. The Freedom of the Soul left them to choose their own religious affairs. And being able to speak their minds, share their opinions with others, and stand for their beliefs, gave them the Freedom of the Spirit.

Future to Come

This concludes of our extensive content analysis of the invention of the wireless communication engine and the context in which that invention took place. We looked in quite some detail at the events around Marconi's Act of Invention with its cluster of innovations. The direct contextual picture is painted in rather rough brushstrokes, specifically looking at the developments that led to the invention of the wireless telegraph. The totality is illustrated though the overall context for change that existed around the turn of the twentieth century. This was a context created by scientific, social, and political revolutions, and the technical change resulting in technological and industrial revolutions.

Although we have focused on technological innovations related to electricity, one has to realize that it was not electricity alone that ceated the Second Industrial Revolution.

> *The most obvious drama of this period was economic and technological: the iron pouring in millions of tons over the world, snaking the ribbons of railways across the continents, the submarine cables crossing the Atlantic, the construction of the Suez Canal, the great cities like Chicago stamped out of the virgin soil of the American Midwest, the huge streams of migrants. It was the drama of European and North American power, with the world at its feet.* (Hobsbawm, 2010a, p. 16)

At first sight, technical change and social change seem to be entirely dependent on one another; that is, each is both the cause and the effect of the other. But there is more, as they also seem to have a sequential pattern where social change prepares for technical change and technical change, in turn, causes social change. And at the core of those technical changes are all those curious and ingenious minds living in the social-political context of their times.

The wireless communication engine in all its forms and appearances was the work of many. But among his equals, one person stands out: a young Italian called Guglielmo Marconi, driven to bring his invention to maturity. A man captured in the Spirit of his Time and frustrated by the Madness of his Time. A man with a vision that hardly encompassed the future that would evolve from his contribution: the invention of the wireless communication engine. He certainly could not have foreseen that the later developments in the next phase of the *Communication Revolution* would be so dominant. A phase were the further development of wireless or 'mobile' telephony would result in the 'smartphone' of the early twenty-first century, the communication device that dominates the social and business lives of our day, when everybody is permanently 'online'. But that is another story...

References

Aitken, H. G. (2014a). The continuous wave: Technology and American radio, 1900-1932: Princeton University Press.

Aitken, H. G. (2014b). Syntony and spark: The origins of radio: Princeton University Press.

Aliprantis, C. (n.d.). A Creeping Army of Denunciators: The Secret Police in Neoabsolutist Austria. Retrieved from Academia website: https://www.academia.edu/16703859/_A_Creeping_Army_of_Denunc iators_The_Secret_Police_in_Neoabsolutist_Austria_1849-1859

Arapostathis, S., & Gooday, G. (2013). Patently contestable: Electrical technologies and inventor identities on trial in Britain: MIT Press.

Baker, D. (2009). Relations between East and West in the Middle ages: papers: Transaction Publishers.

Baker, R. S. (1902). Marconi's achievement: Telegraphing across the ocean without wires. McClure's Magazine, 18(4), 4-12.

Baker, W. J. (1970). A History of the Marconi Company 1874-1965: Routledge.

Baron de Montesquieu, C. d. S. (1965). Considerations on the Causes of the Greatness of the Romans and Their Decline. Translated, with Notes and an Introduction, by David Lowenthal: Collier-Macmillan: London.

Beauchamp, C. (2001). History of telegraphy: IET.

Bergek, A., Jacobsson, S., Carlsson, B., Lindmark, S., & Rickne, A. (2008). Analyzing the functional dynamics of technological innovation systems: A scheme of analysis. Research Policy, 37(3), 407-429. doi: http://dx.doi.org/10.1016/j.respol.2007.12.003

Bernardi, W. (2000a). The controversy on animal electricity in eighteenth-century Italy: Galvani, Volta and others. Nuova Voltianas, Bevilacqua and Fregonese, eds.(Milan: Hoepli), 101-114.

Bernardi, W. (2000b). The controversy on animal electricity in eighteenth-century Italy: Galvani, Volta and others. Nova Voltiana: Studies on Volta and His Times, 1, 113.

Bevir, M. (2012). Governance: a very short introduction (Vol. 333): Oxford University Press.

Bigazzi, M. (1995). Young Marconi's Experiments. Univeristas, Newsletter of the International Centre for the History of Universities and Science, Vol. 7 (January 1995).

Blair, V. (2002). Venice: Napoleon's Italian Thorn. Research Subjects: Governments & Politics. Retrieved from The Napoleon Series website: http://www.napoleon-series.org/research/government/diplomatic/c_venice.html

Blom, P. (2016). After the Crash and before the War: Culture and Society in Europe in the 1930s. In H. Williams (Ed.), A World Transformed: Studies in the History of Capitalism (Vol. 2, pp. 50-53): Legatum Institute. Retrieved from https://lif.blob.core.windows.net/lif/docs/default-source/publications/a-world-transformed-studies-in-the-history-of-capitalism--pdf.pdf?sfvrsn=2.

Bond, J. (1926). In the Pillory: The Tale of the Borgia Pope. Rome Retrieved from https://en.wikisource.org/wiki/In_the_Pillory:_The_Tale_of_the_Borgia_Pope.

Bowman, T. (2005, April 2015). A discussion of the origins of the electric telegraph. Retrieved June 2015, from www.academia.edu/7415642/A_discussion_of_the_origins_of_the_electric_telegraph

Brown, B. (2003). The History of the Corporation BF Communications Inc.

Bruton, E. M. (2012). Beyond Marconi: the roles of the Admiralty, the Post Office, and the Institution of Electrical Engineers in the invention and development of wireless communication up to 1908. University of Leeds.

Burns, R. W. (2004). Communications: an international history of the formative years (Vol. 32): IET.

Carey, J. W. (1983). Technology and ideology: The case of the telegraph. Prospects, 8, 303-325.

Casson, H. N. (1910). The history of the telephone: Books for Libraries Press.

Casson, H. N. (1910). The history of the telephone. Chicago: A.C.McClurg & Co.

Chittolini, G. (1989). Cities, "City-States," and Regional States in North-Central Italy. Theory and Society, 18(5), 689-706.

Clerk Maxwell, J. (1864). On Faraday's lines of force. Transactions of the Cambridge Philosophical Society, 10, 27.

Crookes, W. (1892). Some possibilities of electricity. Scientific American, 33, 13504-13505.

Cropley, D. H. (2016). Creativity in engineering Multidisciplinary Contributions to the Science of Creative Thinking (pp. 155-173): Springer.

Crowther, J. G. (1974). Guglielmo Marconi. New Scientist, 166-168.

de Groot, B., & Franses, P. H. (2005). Cycles in basic innovations (Vol. Report 2005-35): Econometric Institute Erasmus University Rotterdam.

Della Rocca, M. (2008). Spinoza: Routledge.

Devezas, T. C. (2005). Evolutionary theory of technological change: State-of-the-art and new approaches. Technological Forecasting and Social Change, 72(9), 1137-1152. doi: http://dx.doi.org/10.1016/j.techfore.2004.10.006

Dibner, B. (1964). James Clerk Maxwell. Spectrum, IEEE, 1(12), 50-56. doi: 10.1109/MSPEC.1964.6501276

Dolbear, A. E. (1893). The Future of Electricity: Telegraphing without wires. Donahoe's Magazine.

Dosi, G. (1982). Technological paradigms and technological trajectories: A suggested interpretation of the determinants and directions of technical change. Research Policy, 11(3), 147-162. doi: http://dx.doi.org/10.1016/0048-7333(82)90016-6

Douglas, A. (1990). The Legacies of Edwin Howard Armstrong. Proceedings of the Radio Club of America, 64(3).

Douglas, S. J. (1989). Inventing American Broadcasting, 1899-1922: Johns Hopkins University Press.

Dubilier, W. (1908). The Collins Wireless Telephone. Modern Electronics(August).

Dubilier, W. (1909). Continental Wireless Telephone & Telegraph Company (brochure).

Duggan, C. (2014). A concise history of Italy: Cambridge University Press.

Duijn, J. J. v. (1983). The long wave in economic life: Allen & Unwin London.

Dunlap, O. E. (1941). Marconi, the Man and his Wireless: The Macmillan company.

Dutton, B. D. (nd). Benedict De Spinoza (1632—1677). Retrieved from Internet Encyclopedia of Philosophy website: http://www.iep.utm.edu/spinoza/

Dylla, H., & Corneliussen, S. T. (2005). John Ambrose Fleming and the beginning of electronics. Journal of Vacuum Science & Technology A: Vacuum, Surfaces, and Films, 23(4), 1244-1251.

Fahie, J. J. (1899). A history of wireless telegraphy 1838-1899. . Edingburgh and London: William Blackwood and Sons.

Falciasecca, G. (2010). Pragmatics of an Invention. In M. Giorgi, M. Righi & B. Valotti (Eds.), Guglielmo Marconi, Wireless Laureate (pp. 107). Bologna: Bononia University Press.

Faraday, M. (1832). Experimental researches in electricity. Philosophical transactions of the Royal Society of London, 122, 125-162.

Fawcett, W. (1902). The Latest Advance in Wireless Telephony. Scientific American, 86(21), 363-363.

Fayant, F. (1907). Fools and their Money. Success Magazine, 9-11, 49-52.

Fessenden, R. A. (1908). Wireless telephony. American Institute of Electrical Engineers, Transactions of the, 27(1), 553-629.

Fitzgerald, G. F. (1880). On the electromagnetic theory of the reflection and refraction of light. Philosophical transactions of the Royal Society of London, 171, 691-711.

Fleming, J. A. (1919). The principles of electric wave telegraphy and telephony: Longmans, Green.

Friedewald, M. (2000). The beginnings of radio communication in Germany, 1897–1918. Journal of Radio Studies, 7(2), 441-463.

Garratt, G. R. (1994). The early history of radio: from Faraday to Marconi (Vol. 20): Iet.

Garratt, G. R. M. (1994). TheEarly History of Radio: From Faraday to Marconi: IET.

Geddes, L. A., & Hoff, H. E. (1971). The discovery of bioelectricity and current electricity The Galvani-Volta controversy. Spectrum, IEEE, 8(12), 38-46.

Goldthwaite, R. A. (2009). The economy of renaissance Florence: JHU Press.

Granata, M. (2015). Smart Milan: Springer.

Grandin, K., Mazzinghi, P., Olander, N., & Pelosi, G. (2012). A Wireless World. Royal Swedish Academy of Sciences and Firenze University Press, Italy.

Guagnini, A. (2002). Patent Agents, Legal Advisers and Guglielmo Marconi's Breakthrough in Wireless Telegraphy. History of Technology, 24, 171-202.

Guagnini, A. (2009). John Fletcher Moulton and Guglielmo Marconi: bridging science, law and industry. Notes and Records of the Royal Society, 63(4), 355-363.

Harkins, C. A. (2008). Tesla, Marconi, and the Great Radio Controversy: Awarding Patent Damages Without Chilling a Defendant's Incentive to Innovate. Mo. L. Rev., 73, 745.

Headrick, D. R. (1991). The invisible weapon: Telecommunications and international politics, 1851-1945: Oxford University Press.

Heilbron, J. L. (1979). Electricity in the 17th and 18th Century: A Study of Early Modern Physics: Univ of California Press.

Henderson, L. J. (1914). The Fitness of the Environment. The American Journal of the Medical Sciences, 148(3), 433.

Hertz, H. (1894). Untersuchungen über die Ausbreitung der elektrischen Kraft (Vol. 2): JA Barth.

Hill, J. R. (2005). The Revolutions of 1848 in Germany, Italy, and France. (Senior Honors Theses), Eastern Michigan University. Retrieved from http://commons.emich.edu/cgi/viewcontent.cgi?article=1044&context=honors

Hills, J. (2002). The struggle for control of global communication: The formative century: University of Illinois Press.

Hobsbawm, E. (2010a). Age of Capital: 1848-1875: Hachette UK.

Hobsbawm, E. (2010b). Age of Empire: 1875-1914: Hachette UK.

Hobsbawm, E. (2010c). Age of Revolution: 1789-1848: Hachette UK.

Hong, S. (2001). Wireless: From Marconi's black-box to the audion: MIT Press.

Howeth, L. S. (1963). History of Communications Electronics in the United States Navy: For sale by the Superintendent of Documents, US Govt. Print. Off.

Hunt, B. J. (1983). " Practice vs. Theory": The British Electrical Debate, 1888-1891. Isis, 341-355.

Iball, E. (n.d.). The Role of the Barbarian Invasions in the Fall of the Roman Empire. Retrieved from Academia website: http://www.academia.edu/7030739/The_Role_of_the_Barbarian_Inva sions_in_the_Fall_of_the_Roman_Empire

Jacot, B. L., de Boinod, B. L. J., & Collier, D. M. B. (1935). Marconi-- master of Space: An Authorized Biography of the Marchese Marconi: Hutchinson & Company.

Jolly, W. P. (1972). Marconi. London: Constable.

Jolly, W. P. (1974). The making of the man. Electronics & Power, 20(8), 312-314.

Jones, P. M. (2008). Industrial Enlightenment: Science, technology and culture in Birmingham and the West Midlands, 1760-1820: Manchester University Press.

Kiernan, V. G. (1957). Foreign mercenaries and absolute monarchy. Past & present(11), 66-86.

King, W. J. (1962). The Development of Electrical Technology in the 19th Century. Washington Governement Pr.: Smithsonian Institution.

Kingsbury, J. E. (1915). The telephone and telephone exchanges: their invention and development: Longmans, Green, and Co.

Kipnis, N. (1987). Luigi Galvani and the debate on animal electricity, 1791– 1800. Annals of Science, 44(2), 107-142. doi: 10.1080/00033798700200151

Lee, A. (2015). The Ugly Renaissance: Sex, Greed, Violence and Depravity in an Age of Beauty: Anchor Books.

Lee, B. (2012). Wireless, its Evolution from Mysterious Wonder to Weapon of War, 1902 to 1905. AWA-Review, Antique Wireless Association, 25, 37.

Lipsey, R. G., Carlaw, K. I., & Bekar, C. T. (2005). Economic transformations: General purpose technologies and long-term economic growth. Oxford: Oxford University Press.

Lopez, G. (2012). Milan, A Short History. Milano: Musia Editore S.p.A.

Machiavelli, N. (1908). The Prince (T. b. W. K. Marriott Ed.): London; EP Dutton & Company: New York.

Marcoccio, A. M. G., A. (2003). Da Cruto a Philips. Alpignano.

Marconi, D. (2001). My Father, Marconi (Vol. 16): Guernica Editions.

Marconi, F. P. (nd). Personal Reflections of "An Italian Adventurer". Marconi Family Biographies. Retrieved from Marconi Society website: http://marconisociety.org/about/family-biographies/#toggle-id-3

Marconi, G. (1901). Syntonic wireless telegraphy. Scientific American, 51, 21304-21305.

Marconi, G. (1902). The progress of electric space telegraphy. Scientific American, 54, 22208-22211.

Marconi, G. (1909). Wireless telegraphic communication. Nobel Lecture, December, 11.

Marconi, G., Gill, F. N. G., & Barcroft, H. (1901). Journal of the Society for Arts, Vol. 49, no. 2530. The Journal of the Society of Arts, 49(2530), 505-520.

Marconi, M. C. (2002). Marconi my beloved: Branden Books.

Marks, P. (2011). Dot-dash-diss: The gentleman hacker's 1903 lulz. New Scientist(2844).

Marshall, H. E. (1920). The Story of Europe: Ostare Publications.

Massie, W. W., & Underhill, C. R. (1908). Wireless Telegraphy and Telephony popularly explained: D. Van Nostrand Co.

Mensch, G. (1979). Stalemate in technology: innovations overcome the depression: Ballinger Cambridge, Mass.

Moffett, C. (1899). Marconi's wireless telegraph. McClure's Magazine, 13(2), 4-17.

Mokyr, J. (2011). The enlightened economy: Britain and the industrial revolution, 1700-1850. London: Penguin UK.

Monroe, A. E. (1919). The French indemnity of 1871 and its effects. The Review of Economic Statistics, 269-281.

Moravia, S., & Breidenbach, F. (1969). An Outline of the Italian Enlightenment. Comparative Literature Studies, 6(4), 380-409.

Munro, D. C. (1916). The Popes and the Crusades. Proceedings of the American Philosophical Society, 55(5), 348-356.

Munro, J. H. (2011). The Rise, Expansion, and Decline of the Italian Wool-Based Textile Industries, 1100-1730: a study in international competition, transaction costs, and comparative advantage Studies in Medieval and Renaissance History (Vol. 3rd Series, Vol. 9): University of Toronto.

Nairn, A. (2002). Engines That Move Markets: Technology investing from railroads to the internet and beyond: John Wiley & Sons.

O'Hara, J. G. (2007). A 'Horrible Conflict with Theory'in Heinrich Hertz's Experiments on Electromagnetic Waves. European Review, 15(04), 545-559.

Osborne, H. S. (1943). Biographical Memoir of Alexander Graham Bell. Retrieved from National Academy of Sciences website: http://www.nasonline.org/publications/biographical-memoirs/memoir-pdfs/bell-alexander-graham.pdf

OX-1. (2017, March 2017). Marconi Collection: Company. [Marconi Company]. Marconi Archive, (Box 168, Box 243, Box 381). Bodelian Libraries, University of Oxford.

OX-2. (2017, March 2017). Marconi Collection: Patents. Marconi Archive, (Box 411, Box 416). Bodelian Libraries, University of Oxford.

OX-3. (2017, March 2017). Marconi Collection: Annual Reports. Marconi Archive, (Box 401). Bodelian Libraries, University of Oxford.

Paoletti, C. (2008). A military history of Italy: Greenwood Publishing Group.

Parker, G. (2013). Global crisis: war, climate change and catastrophe in the seventeenth century: Yale University Press.

Pera, M., & Mandelbaum, J. (1991). The Ambiguous Frog: The Galvani-Volta Controversy on Animal Electricity: Princeton University Press.

Piccolino, M. (1998). Animal electricity and the birth of electrophysiology: the legacy of Luigi Galvani. Brain research bulletin, 46(5), 381-407.

Preece, W. H. (1897). Signalling Through Space Without Wires. Science, 6(155), 889-896.

Raboy, M. (2016). Marconi: The Man Who Networked the World: Oxford University Press.

Robertson, J. H. (1947). The story of the telephone: A history of the telecommunications industry of Britain: J. Pitman.

Rosenthal, L., & Rodic, V. (2015). The New Nationalism and the First World War: Springer.

Rowlands, P. (1990). Oliver Lodge and the Liverpool Physical Society: Liverpool University Press.

Russer, P. (2009). Ferdinand Braun—A pioneer in wireless technology and electronics. Paper presented at the Microwave Conference, 2009. EuMC 2009. European.

Russer, P. (2012). Ferdinand Braun: a pioneer in wireless technology and electronics. A Wireless World, 228-247.

Rutherford, E. (1897). A Magnetic Detector of Electrical Waves and Some of Its Applications. Philosophical Transactions of the Royal Society of London. Series A, Containing Papers of a Mathematical or Physical Character, 189, 1-24. doi: 10.1098/rsta.1897.0001

Rybak, J. P. (nd). Oliver Lodge: Almost the Father of Radio. Retrieved from Antique Wireless Association website: http://www.antiquewireless.org/uploads/1/6/1/2/16129770/48-oliver_lodge.pdf

Sarkar, T. K., & all, e. (2006). History of wireless (Vol. 177): John Wiley & Sons.

Satia, P. (2010). War, wireless, and empire: Marconi and the British warfare State, 1896–1903. Technology and Culture, 51(4), 829-853.

Schumpeter, J. A. (1934). The theory of economic development: An inquiry into profits, capital, credit, interest, and the business cycle (Vol. 55): Transaction publishers.

Schumpeter, J. A. (1939). Business cycles; a theoretical, historical, and statistical analysis of the capitalist process (Fels) (1st ed.). New York, London,: McGraw-Hill Book Company, inc.

Schumpeter, J. A., & Opie, R. (1934). The theory of economic development; an inquiry into profits, capital, credit, interest, and the business cycle. Cambridge, Mass.,: Harvard University Press.

Scott Baker, N. (2011). Medicean metamorphoses: carnival in Florence, 1513. Renaissance Studies, 25(4), 491-510.

Scudder, M. (1926-1934). Manual of Extinct or Obsolete Companies. 1926-Vol. I, 1928-Vol. II, 1930-Vol. III, 1934-Vol. IV. Retrieved May 10, 2013, from http://earlyradiohistory.us/extinct.htm

Sharlin, H. I. (nd). William Thomson, Baron Kelvin. Encyclopedia Britannica, (May 2017). Retrieved from Encyclopedia Britannica website: https://www.britannica.com/biography/William-Thomson-Baron-Kelvin

Simon, H. A. (1996). The sciences of the artificial: MIT press.

Simons, R. W. (1996). Guglielmo Marconi and early systems of wireless communication. Gec Review, 11(1), 37-55.

Slaby, A. (1898). The New Telegraphy. Recent Experiments in Telegraphy with Sparks. The Century Magazine, 867-874.

Smith, A. (1817). An Inquiry into the Nature and Causes of the Wealth of Nations (Vol. 2): Рипол Классик.

Smith, A. (1880). The principles which lead and direct philosophical enquiries: illustrated by the history of astronomy: Basel Mission Press.

Snobelen, S. (2012). The myth of the clockwork universe: Newton, newtonianism, and the enlightenment. The persistence of the sacred in modern though. University of Notre Dame Press, Notre Dame, IN, 149-184.

Soeiro, D. (2013). On Artificial and Animal Electricity: Alessandro Volta vs. Luigi Galvani. Journal of Philosophy of Life, 3(3), 212-237.

Sterling, C. H., & Kittross, J. M. (2001). Stay tuned: A history of American broadcasting: Routledge.

Susskind, C. (1962). Popov and the Beginnings of Radiotelegraphy. Proceedings of the IRE, 50(10), 2036-2047.

Susskind, C. (1964). Observations of Electromagnetic-Wave Radiation before Hertz. Isis, 55(1), 32-42. doi: 10.2307/227753

Thompson, E. (1892). Future Electrical Development The New England Magazine.

Thompson, S. P. (1890). The Electro-Magnet. Science, 16(October 10, 1890), 107-202.

Thomson, E. (1920). The epoch-making discoveries of the years 1819 and 1820. American Institute of Electrical Engineers, Journal of the, 39(12), 1021-1027.

Trainer, M. (2007). The role of patents in establishing global telecommunications. World Patent Information, 29(4), 352-362. doi: http://dx.doi.org/10.1016/j.wpi.2007.03.005

Tushman, M. L., Anderson, P. C., & O'Reilly, C. (1997). Technology cycles, innovation streams, and ambidextrous organizations: organization renewal through innovation streams and strategic change. Managing strategic innovation and change, 34(3), 3-23.

Usher, A. P. (1929). A history of mechanical inventions. New York: McGraw-Hill Book Company.

Valotti, B. (2014). Beyond the Myth of the Self-taught Inventor: The learning Process and Formative years of Young Guglielmo Marconi. In I. Inkster (Ed.), History of Technology (Vol. 32): Bloomsbury Publishing.

Volta, A. (1800). On the Electricity Excited by the Mere Contact of Conducting Substances of Different Kinds. In a Letter from Mr. Alexander Volta, F. R. S. Professor of Natural Philosophy in the University of Pavia, to the Rt. Hon. Sir Joseph Banks, Bart. K. B. P. R. S. Philosophical Transactions of the Royal Society of London, 90, 403-431. doi: 10.2307/107060

Wenaas, E. P. (2007). Radiola: The Golden Age of RCA, 1919-1929: Sonoran Publishing.

Whewell, W. (1858). History of inductive sciences. London: John W. Parker.

White, T. (1902). Our Wonderful Progress: The World's Triumphant Knowledge and Works : a Vast Treasury and Compendium of the Achievements of Man and the Works of Nature.

White, T. H. (1996). Early Radio Industry Development (1897-1914). 2013(12 May). Retrieved from United States Early Radio Industry website: http://earlyradiohistory.us/sec006.htm

White, T. H. (2015). Fakes, Frauds, and Cranks (1866-1922). Retrieved June 2015 from http://earlyradiohistory.us/sec021.htm

Wills, I. (2009). Edison and science: a curious result. Studies in History and Philosophy of Science Part A, 40(2), 157-166. doi: http://dx.doi.org/10.1016/j.shpsa.2009.03.006

WP-staff. (1903). Industrial Affairs: INTERNATIONAL WIRELESS CO. Various Sub-Companies Merged in One Corporation. Wall Street Journal(April 22, 1903,), 5.

Abbreviations

OX 1–3: Marconi Archives in the Bodleian Library in Oxford, England.

MM: Museo Marconi, Fondatzione Guglielmo Marconi, Pontecchio Marconi, Italy

GH : Guildhall Library, London

About the author

Drs.Ir.Ing. B. J. G. van der Kooij (b. 1947) obtained his MBA in 1975 (thesis: *Innovation in SMEs*) from the Interfaculteit Bedrijfkunde (nowadays part of the Rotterdam Erasmus University). In 1977 he obtained his MSEE (thesis: *Technology forecasting of Microelectronics*) from the Delft University of Technology.

He started his career as assistant to the board of directors of Holec NV, a manufacturer of electrical power systems employing about 8,000 people at that time. His responsibilities were in the field of corporate strategy and innovation of Holec's electronic activities. Travelling extensively to Japan and California, he became well known as a Dutch guru on the topic of innovation and microelectronics.

From 1982–1986 he was a member of the Dutch Parliament (Tweede Kamer der Staten Generaal) and spokesman on the fields of economic, industrial, science, innovation, and aviation policy. He became the first member to introduce the personal computer in Parliament, but his work on legislation like the *TNO Act*, *Patent Act*, *Chips Act*, and others went largely unnoticed.

After the 1986 elections and the massive loss for his party (VVD) he was dismissed from politics and became a part-time professor (Buitengewoon Hoogleraar) at the Eindhoven University of Technology. His endowed chair was the Management of Innovation. Additionally, in 1986 he started his own company, Ashmore Software BV, developer of software for professional tax applications on personal computers. After closing these activities in 2003, he became a real estate project developer, and in 2009 a real estate consultant till his retirement in 2013.

Innovation being the focus of attention all his corporate, entrepreneurial, political, and scientific life, he wrote three books on the subject and published several articles. In his first book, he explored the technological dimension of innovation (the pervasive role of microelectronics). His second book focused on the management of innovation and the human role in the innovation process. And in his third book, he formulated 'Laws of Innovation' based on the Dutch societal environment in the 1980s.

In 2012 he started studying the topic of innovation again. His focus is on the theory of innovation, and his aim is to develop a multidimensional model explaining innovation. For this, he is creating extensive and detailed case studies observing the inventions of the steam engine, the electromotive engines, the communication engines, and the computing engines, studying their characteristics from a multidisciplinary perspective (economic, technical, and social). In 2013 he was accepted to the TU-Delft by Prof. Cees van Beers as a PhD candidate. They agreed on his research proposal to investigate the Nature of Innovation by looking at real-world case studies, which are presented in the *Invention Series*. However, in 2016 he was left without a promoter, as Van Beers—for reasons of his own—lost faith in his work. Luckily, others at the university still believe in and are supportive of his work.

INNOVARE NECESSE EST

Acknowledgments

This scholarly work has been created by myself alone. Any errors made are mine. It is a work based on secondary research into 'technical/economic/scientific/political/social' phenomena investigated and described by others from their point of view (as family member, journalist, biographer, or as scholar), and published in the traditional way in books and articles in professional journals. According to the fair use clause in the Copyright Acts, we took great care to acknowledge the sources, but are bound to fail due to their multitude. Next to all the substantive contributions of the scholars acknowledged and mentioned extensively in the quotes and references, we were able—through the modern digital infrastructure called the Internet—to get access to many primary sources provided on numerous websites and blogs owned and maintained by scholars, writers, museums large and small, and other institutions. They gave me the highly appreciated 'open access'[488] to their work, and were often too numerous to quote, but if the occasion arose, I acknowledged them in a footnote for reference. Realizing they observed our mutual subject of interest from their own perspectives, that diversity gave me the opportunity to look at the case from different angles. And to finalize this enumeration of sources, I used for background orientation much of the information available on the digital encyclopaedia called Wikipedia. There, we found the great treasury of media known as Wikimedia Commons. We are in debt to all those unknown people who contributed material that gave me the opportunity to illustrate my cases. If the original creator was known, we included their names. To the others, we apologize.

This work is about content and context of the invention of the wireless engine: in short, the Marconi case. Thus, we found our geographical domain of interest in both Italy and Britain. In Italy, we visited both the Marconi Museum in Pontecchio, near Bologna, and the Science Museum in Milan. In Britain, we visited the Museum of History of Science. In the footnotes, the reader will find the people who were 'graciously'—as they were mostly female—willing to share with me their knowledge and insights. In addition, we visited Oxford, where we were assisted by the kind people of the Bodleian Library that houses the special collection called the 'Marconi Archives'.

---

[488] Open access refers to online research outputs that are free of all restrictions on access (e.g. access tolls) and free of many restrictions on use (eg certain copyright and license restrictions). The aim of our University of Technology in Delft is that TU Delft researchers publish each of their publications in open access.